Investigating Groundwater

INTERNATIONAL CONTRIBUTIONS TO HYDROGEOLOGY

29

Series Editor: Dr. Nick S. Robins
Editor-in-Chief IAH Book Series, British Geological Survey, Wallingford, UK

INTERNATIONAL CONTRIBUTIONS TO HYDROGEOLOGY

Investigating Groundwater

Ian Acworth

University of New South Wales, Australia

CRC Press
Taylor & Francis Group
Boca Raton London New York

CRC Press is an imprint of the
Taylor & Francis Group, an **informa** business

A BALKEMA BOOK

CRC Press
Taylor & Francis Group
6000 Broken Sound Parkway NW, Suite 300
Boca Raton, FL 33487-2742

First issued in paperback 2020

© 2019 by Taylor & Francis Group, LLC
CRC Press is an imprint of Taylor & Francis Group, an Informa business

No claim to original U.S. Government works

ISBN: 978-1-138-54249-5 (hbk)
ISBN: 978-0-367-73120-5 (pbk)

Library of Congress Cataloging-in-Publication Data
Applied for

Visit the Taylor & Francis Web site at
http://www.taylorandfrancis.com

and the CRC Press Web site at
http://www.crcpress.com

Contents

Foreword

Hydrogeology covers many different aspects of science and engineering and has been taught internationally in both Geology Schools and Engineering Schools, each perhaps with an appropriate bias. Soil scientists, biologists, microbiologists, geophysicists, chemists, organic chemists, botanists and water engineers are all concerned to some degree with the movement of water through semi-pervious membranes, and a wide and diverse literature has developed which is all, to some degree, relevant to the study of hydrogeology. Fully understanding hydrogeology requires that the reader is aware of all the different branches of science and engineering and sufficiently skilled that they can understand the way in which some of the important concepts have been defined in parallel using completely different concepts and definitions. Comprehension is further complicated by the widely different educations that a reader brings to the text. For this reason a large number of images and figures are included in this book in the firm belief that a picture really is worth a thousand words!

The subject area is large and no one book can hope to cover all aspects of this field. Some rationale behind the choice for what to include and what has been left out is therefore appropriate. The material has been developed in Australia with the use of many Australian examples. Although the methods and applications have a general applicability, some appreciation of the Australian environment is useful, and this is presented in the first chapter. It is now appreciated that many groundwater resources are of considerable age; however, there is little understanding of the prevailing climate that occurred in Australia at the time groundwater was recharged. Australia is unlike European and North American environments where soil and water were very significantly changed and reset during the last ice age. A brief review of climate change is therefore also presented in the first chapter. There can be little appreciation of the different aquifer types in which groundwater occurs, so the remainder of the chapter presents a review of the occurrence of groundwater in different geological environments.

Chapter 2 is concerned with surface water processes. It is included as the hydrogeologist, particularly from a geological background, is frequently unaware of the important processes and measuring techniques required to estimate rainfall, evaporation, evapotranspiration and runoff. These are the major fluxes of water that occur at the surface of the earth and groundwater recharge is difficult (or impossible) to predict without the ability to quantify these fluxes. Chapter 3 reviews the interconnectivity of surface water and groundwater which leads to a discussion on groundwater recharge processes and calculations.

Chapters 4 to 7 cover basic hydrogeological concepts, while Chapters 8 to 11 provide detail of appropriate geophysical techniques. The narrative then moves to subsurface investigation with an account of drilling and sampling methods (Chapter 12) and geophysical well logging (Chapter 13).

In the interests of brevity and recognizing that there are excellent texts already available, hydrochemistry and isotopes are not included in great detail but are summarized in Chapter 14. This is not in any way a suggestion that the material is not of great importance, and examples of detailed changes in chemistry with depth are described.

Chapter 15 gives an account of well hydraulics and the interpretation of pumping test data. A description of the finite difference method as applied to 1-D radial flow modelling is included in the discussion of well hydraulics in Chapter 15 to illustrate the potential power of numerical techniques.

A high proportion of engineers that become involved in the field of groundwater resources make very extensive use of groundwater modeling. However, as this is an application of hydrogeological theory hopefully assisted with data made available by application of techniques for investigating groundwater described in this book, a detailed discussion is not included. Excellent texts are available that cover groundwater modeling in detail. The same can be said for the design and construction of abstraction wells. As this process follows on from a successful application of *Investigating Groundwater*, this subject is also excluded.

Acknowledgements

A large number of photographs and illustrations have been used in the preparation of this book. I would like to particularly acknowledge Martin Andersen, Ross Beazley, Tony Bernardi, Gary Dowling, Will Glamour, Landon Hallorhan, Colin Hazel, Bryce Kelly, Catherine Kerr, Heather McDowell, Des McGarry, Gabriel Rau, Wendy Timms and Moya Tomlinson for permission to include photographs they have taken. Unless stated otherwise, all figures have been specially prepared for this book by the author. The graphics capabilities of GRAPHER and SURFER have been widely deployed.

Aspects of the research work which appear in many of the figures reporting field studies would have not been possible without the support of colleagues at the Water Research Laboratory (WRL) in Manly Vale, Sydney. The WRL is a part of the School of Civil and Environmental Engineering at the University of New South Wales, Australia. In particular I would like to acknowledge and thank Mark Groskops for endless support in the field in earlier years and Ron Cox for financial support as Manager and Director of the WRL for a significant period.

More recently, colleagues within the Connected Waters Initiative (CWI) at UNSW received major funding from the Cooperative Research Centre for Cotton Catchment Communities (CRC CCC) that allowed us to spearhead work into the Maules Creek site where connected waters processes have been investigated. Major funding was also received to establish the UNSW portion of the ARC/NWC National Centre for Groundwater Research and Training (NCGRT) and the National Collaborative Research Infrastructure Strategy (NCRIS) – Groundwater Program. Both of these programs contributed major financial input and to achieving an atmosphere where significant research was possible. The support from the NCRIS Groundwater Infrastructure funds was particularly significant in providing the financial framework for the collection of major data sets at Fowlers Gap, Maules Creek, Wellington and Baldry.

This book has partly evolved from notes prepared for various undergraduate and postgraduate courses in groundwater taught at the University of New South Wales, Sydney, Australia. Students too numerous to recognize individually are thanked for their perceptive comments. Parts of Chapters 2 to 7 have been used in the Australian Groundwater School since 1990.

Chapter 1

Groundwater environments

1.1 EARLY HUMAN MOVEMENTS BASED UPON GROUNDWATER AVAILABILITY

The move out of Africa by *Homo sapiens* must have been strongly influenced by the climate prevailing at the time and also by the depth of sea water covering land bridges between continents. Cuthbert *et al.* (2017) note that reliable groundwater from perennial springs draining large-scale aquifers will also have played a significant part providing refugia along the migration pathways.

Based upon the evidence of global mean temperatures and atmospheric CO_2 data derived from ice cores (Lisiecki and Raymo, 2005), the penultimate last ice age ended approximately 140 ka when an interglacial warm spell developed for the next 17 ka. The warm spell would have meant ready availability of water, abundant vegetation and game. Groundwater recharge would also have occurred, providing regional recharge to aquifers.

Sea levels would have been close to or slightly higher than the present day with all land bridges covered by water. This would have made human migrations more difficult and maybe removed any necessity to migrate, as food was plentiful.

Significant cooling occurred beginning approximately 123 ka with drying and a decline in sea levels. Warmer spells existed between 110 ka and 100 ka, and again between 90 ka and 71 ka with drier spells between 110 ka to 123 ka and again between 90 ka to 100 ka.

By 106 ka, humans had migrated as far as the south-west coast of Oman where chert tools have been found in Dhofar (Rose *et al.*, 2011). Movement out of Africa would have only been possible while there was sufficient water and food available. However, reduced sea levels will have also promoted migration by making land bridges accessible. During these times it is thought that migration routes would have opened up through north-eastern Africa, southern Yemen and Oman and into the Indian subcontinent. It is by this route that the Aborigines are believed to have made the long journey from Africa to Australia, arriving in Australia approximately 65 ka.

There was a significant cooler period between 57 ka and 71 ka that would have been characterized by dry and cold conditions similar to the last ice age between 29 ka and 14 ka. Homo Sapiens arrived in Australia (65 ka) during this cooler and drier phase and after managing the deep sea crossing from Asia onto the Australian Continental Plate, presumably at the narrowest gap somewhere close to Timor E'ste. By 50 ka they had spread across Australia.

While the Aborigines arrived in Australia during fairly mild conditions that permitted the continuation of a hunter gatherer society, it was not long before the climate slowly became colder and drier as the next ice age developed. Knowledge of reliable springs and water holes would have been critical to their survival during the 15 ka of the last ice age (29 ka and 14 ka). Springs are found where groundwater overflows to the surface, so it could be argued that the Aborigines were one of the first, if not the first, groups to be aware of groundwater and the reliable discharge from large aquifer systems that continues for thousands of years after recharge. There was no written script at this time and the groundwater knowledge would have been passed from generation to generation as part of oral history. It is not surprising, considering the lack of appreciation of groundwater systems and processes in the 21st century, that much of this oral history was interlinked with mythology and incorporated into Dreamtime stories and maps. Whatever worked to help people remember the paths to critical water sources would have been essential.

The Holocene climatic optimum extended from approximately 10 ka to 6 ka when temperatures were a little warmer and rainfall a little higher than present. The climate then dried and cooled a little, approximately 5 ka.

Agriculture is considered to have developed independently in three areas approximately 9 ka:

- The fertile crescent at the north end of the Persian Gulf where wheat and barley were domesticated,
- In China where rice and millet were farmed,
- and less well known, in the Papua New Guinea highlands, where bananas, yams and taro were cultivated in fields drained of excess water.

The field terraces in southern Yemen date back to approximately 5 ka, when winter rain falling on the tops of the Red Sea mountains was collected for farming. The extent of terracing is remarkable (Figure 1.1). As the climate continued to dry, dams at the base of the mountains were also developed. The site of the great dam of Marib is upstream (south-west) of the ancient city of Marib, once the capital of the ancient kingdom of Saba. The sluice gates of Marib Dam can still be seen as shown in Figure 1.2.

The kingdom of Saba was a prosperous trading nation, with control of the frankincense and spice routes in Arabia and Abyssinia. The Sabateans built and rebuilt Marib Dam several times, to capture the periodic monsoon rains which fall on the mountains and so irrigate the land around the city. Recent archaeological findings suggest that the first simple earth dams and a canal network were constructed as far back as 4 ka. The main dam was constructed in the 8th century BC and was about 550 m long and pyramidal in cross-section. It was built using fine stone-and-masonry construction, with sluice gates to control the flow of water. It irrigated more than 1,600 ha and supported a densely settled agricultural region, dependent on careful water conservation. The area became a key stopping point for the camel trains that plied their trade from Salala in Dhofar, bringing frankincense and myrrh for sale in the Greek and later Roman parts of the Mediterranean.

The Queen of Shabwa (Sheba) visited King Solomon in Israel in approximately 950 BC Issar (1990). The goods traded by the Sabateans probably also included precious stones carried from Galle in southern Sri Lanka where an ancient port was located on the trading routes to the Far East.

It is also probable that the first water wells would have been developed as water levels declined in a spring and the spring was deepened and the sides supported. The first recorded device for the abstraction of groundwater is a 5.6 ka well constructed from wood in Zhejiang Province in China (Jiao, 2007).

Figure 1.1 Terracing at the top of the mountain to use all available rainfall from the winter rains – Red Sea Mountains, Yemen (Photo: Heather McDowell)

Figure 1.2 Sluice gates of the great dam at Marib, Yemen (Photo: Heather McDowell)

Qanats are groundwater development systems that were developed in ancient Iran by the Persian people approximately 3000 BC. The technology spread eastwards to Afghanistan and westwards to Egypt and Spain (Issar, 1990). In Oman and Yemen qanats are referred to as *falaj* (*afalaj* pl).

The afalaj systems were being constructed at the same time as the first dams. This suggests that Persia, Oman and Yemen may have been at the forefront of water harvesting

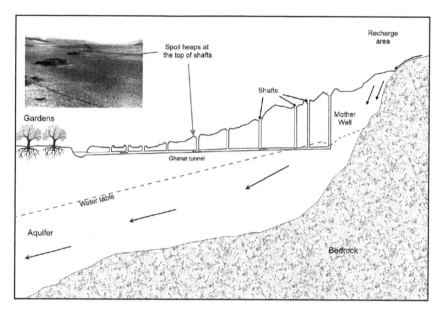

Figure 1.3 Schematic of a qanat (Persian name) or a falaj (so called in Oman and Yemen) (adapted from Amin and Salimi Manshadi (1996))

developments in the world. Considerable understanding of groundwater occurrence is required in the construction of an falaj! Figure 1.3 shows a longitudinal schematic of a qanat (Amin and Salimi Manshadi, 1996). The tunnel systems were constructed into the alluvial fans spreading from the mouths of mountain wadis or from permanent sources of water, such as the lake formed in evaporites shown in Figure 1.4 where the start of the falaj can be seen in the far rock wall. At present there are approximately 300,000 km of qanats in Iran with some as deep as 400 m (Amin and Salimi Manshadi, 1996). The annual flow has been calculated as 7.5 Gm^3.

Figure 1.5 shows the falaj tunnel and the water flowing from the tunnel before flowing over an aqueduct.

Although human observation and utilization of groundwater pre-dates biblical times, a clear appreciation of the occurrence and movement of subsurface water is far more recent and remains to be accepted by significant groups in many societies (Alley and Alley, 2017). The overwhelming increase in our knowledge of the discipline has only been achieved in the 20th century.

Mather (2004) reviewed the early development of the science of hydrogeology and suggested that the first significant work was undertaken in the UK by William Smith (1769 – 1839), who prepared the first geological map for anywhere in the world. Smith reported a good understanding of the stratigraphic control of spring lines. Prior to this work was a general understanding of the importance of water supplies from beneath the ground, even if there was little understanding as to how they got there. The well in a castle was as important as the castle wall in the development of fortified settlements in Europe. Price (2004) records the early work of Dr. John Snow on groundwater contamination, where leakage from a sewer contaminated a water supply well and led to the last major outbreak of

Figure 1.4 Source of a falaj in a collapse lake formed in evaporites. The vertical cut in the far wall allows water to enter the falaj (Photo: Ian Acworth)

Figure 1.5 (a) Falaj tunnel; (b) falaj flowing from the tunnel onto (c) an ancient aqueduct to carry the water over a dry wadi bed and onto a date farm for irrigation in eastern Yemen (Photos: Ian Acworth)

cholera in London in 1854. It has long been understood that from a military perspective, an army requires water, and Rose (2004) records the development of emergency groundwater supplies by the British army throughout the major campaigns of the past 100 years.

It has been estimated that some 18% of domestic water use in Australia comes from groundwater and that approximately 30% of all water used in Australia is groundwater. However, Perth is the only major capital city which relies on groundwater for approximately 60% of its municipal supply and many (>100,000) shallow garden supply bores. Alice Springs relies 100% on groundwater, with abstraction from the Mereenie Sandstone accounting for 70% with the remainder of the supply from two smaller aquifer developments. Many smaller centres rely on groundwater for domestic supplies, with an estimated 600 communities using groundwater as their principal drinking water source, particularly in rural areas.

In addition to the use of groundwater as a drinking water supply, groundwater is widely abstracted to use as irrigation water to support farming communities. During drought times when surface water is not available, groundwater provides a reliable permanent supply that can be drawn down as long as it is allowed to be recharged and to recover during wetter periods when surface water is again used. This conjunctive use of surface water and groundwater is becoming increasingly common, along with managed aquifer recharge, where excess surface water is recharged into groundwater systems for later abstraction during dry times.

The time scales of groundwater flow are extremely long. Water in the Great Artesian Basin is older than 1 million years, but ages of 30,000 years are more common. The often great age of groundwater means that it can frequently only be considered as a renewable resource at time scales of thousands of years. Groundwater which becomes contaminated will naturally be regenerated, if we can wait around long enough!

Occasionally, geological events trap groundwater underground, cutting it off from both a source of supply and from its outlets. Climatic change may also deprive aquifers of any means of recharge, as has happened under a number of regions which are now desert but which were formerly much wetter. Although aquifer systems exist in all continents, not all of them are fed on a regular basis by rainfall. Those in North Africa and the Arabian peninsula were formed more than 10,000 years ago when the climate was more humid and are no longer replenished. Groundwater reserves which are no longer replenished are called fossil water.

1.2 GEOLOGICAL TIME SPAN AND TYPES OF ROCK

The earth is thought to have an age of approximately 4.53 Ga, but most groundwater resources are associated with younger rocks as they still maintain significant primary porosity. Rocks of the Cenozoic (65.5 Ma) age and younger contain most groundwater, although many large sedimentary basins of Phanerozoic age containing Paleozoic and Mesozoic formations also exist.

Geology is the science and study of the physical matter that constitutes the earth. The field of geology encompasses the study of the composition, structure, properties, and history of the planet's physical material, and the processes by which it is formed, moved and changed. Some familiarity with these aspects is required and the reader is encouraged to read appropriately; however, a basic introduction is provided here.

Figure 1.6 A granite with large crystals of quartz (free silica), orthoclase feldspar (pink coloured in Figure 1.6) or dark ferromagnesian minerals such as biotite (Photo: Ian Acworth)

There are three basic types of rock recognized by geologists:

- igneous
- metamorphic
- sedimentary

1.2.1 Igneous rocks

The size of crystals in an igneous rock depends upon whether the magma cooled slowly at depth or was erupted from a volcano. In the first case, the crystals have time to grow and either a granite or a gabbro (or an intermediate rock type) is formed, depending upon the availability of silica (SiO_2). If there is free silica (quartz), then a granite is formed (see Figure 1.6), otherwise a gabbro (see Figure 1.7) is formed. These types of rock are called intrusive igneous rocks.

Where the magma escapes to the surface, an extrusive igneous rock is formed. There are chemically equivalent extrusives to the gabbro and the granite. The equivalent to the gabbro intrusive is a basalt. The lack of silica means that a basalt has low viscosity and can flow a long way from the vent without forming a volcanic cone. Figure 1.8 demonstrates how a very low viscosity flow has moved.

1.2.2 Metamorphic rocks

Metamorphic rocks are rocks that have been heated but not melted to form a magma. There has been some realignment of crystals and the rock commonly has some foliation, but the original structures can often still be seen. Figure 1.9 shows a high-grade metamorphic

Figure 1.7 A gabbro with no free silica. The mineral assemblage is completely different from a granite and weathers more readily. This example comes from a deep-seated magma chamber beneath a Tertiary volcano on Ardnamurchan, northern Scotland (Photo: Ian Acworth)

Figure 1.8 A very low viscosity basalt flow in Iceland that has cooled to solidify the flow marks on the surface. Note the more solid-looking basalt in the background (Photo: Ian Acworth)

rock that has been subject to several deformations. Figure 1.10 indicates less compressive stress (folding) but a greater temperature, as much of the early melting quartz has been mobilized into cross-cutting veins. All primary porosity will have been removed by the process of metamorphism.

Figure 1.9 High-grade metamorphic rock from Connemarra, Ireland. Note the white quartz vein running across the rock in the top left and the quartz recrystallization below. The original sedimentary bands have been compressed by folding (Photo: Ian Acworth)

Figure 1.10 Less folding but greater mobilization of quartz indicating a higher temperature during metamorphism. The original sedimentary layering is preserved (Photo: Ian Acworth)

1.2.3 Sedimentary rocks

Sedimentary rocks are comprised of the products of weathering derived from igneous and metamorphic rocks. These rocks can be formed from sand and gravel or from muds and clays or deposited from sea water as in an evaporite or limestone. Sedimentary rocks may either be unconsolidated, as found at the beach or in flood-plain deposits, or consolidated, as found in sandstones or mudstones. Unconsolidated deposits have the highest porosity and generally contain the most groundwater. Consolidated deposits have had all of the initial water that filled the pore space during deposition squeezed out. During this process, there will have been realignment of the sedimentary particles and the deposition of additional mineral phases derived from percolating groundwater that will further reduce the porosity of the unit. Simply stated, an unconsolidated sand with high porosity is turned into a sandstone with low porosity that can be cut with a rock saw.

1.3   GROUNDWATER IN AUSTRALIA

The occurrence of groundwater in Australia can be subdivided as shown in Figure 1.11 produced by Geoscience Australia (GA) (http://www.ga.gov.au/ Creative Commons 4.0 International license). GA have subdivided the continent into five aquifer types:

1   porous, extensive highly productive aquifers
2   porous, extensive aquifers of low to moderate productivity
3   fractured or fissured, extensive highly productive aquifers
4   fractured or fissured, extensive aquifers of low to moderate productivity
5   local aquifers of generally low productivity

The classifications developed by GA are based on how groundwater is found in the various rock types. There is some overlap here between the hydrogeological and geological classifications. For example, both a sandstone and a basalt can be predominantly fractured with groundwater only recovered from the fractures. However, they are completely different rock types and the geologist would never confuse the two. It may also be easier to group together the groundwaters held in a particular sedimentary basin rather than describe each aquifer type in that basin separately. There is no simple answer to this classification problem.

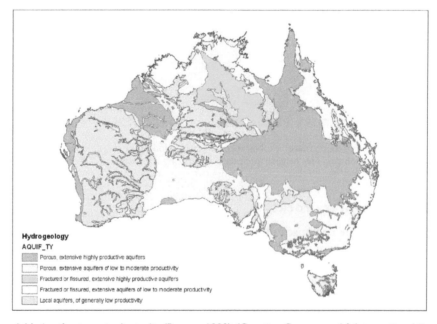

Hydrogeology
AQUIF_TY

- Porous, extensive highly productive aquifers
- Porous, extensive aquifers of low to moderate productivity
- Fractured or fissured, extensive highly productive aquifers
- Fractured or fissured, extensive aquifers of low to moderate productivity
- Local aquifers, of generally low productivity

Figure 1.11 Aquifer types in Australia (Brown, 1989) (Creative Commons 4.0 International license)

Groundwater occurs in the primary and secondary (fractures) porosity of rocks deposited in large sedimentary basins. The dark blue areas in Figure 1.11 represent these basins, where the Great Artesian Basin (GAB) is clearly the largest. A brief description of the three important Australian sedimentary basins is given below. More general comments concerning groundwater in fractured rocks, basalts, paleochannels, unconsolidated surficial deposits and limestone systems follow with examples from both Australia and internationally.

1.3.1   The Paleozoic and Mesozoic sediments of the Great Artesian Basin

The GAB is one of the largest groundwater basins in the world. It spans across three states and the Northern Territory over an area of 1.7 million km^2 (20% of the Australian landmass) and is formed from middle Triassic to Cretaceous aquifers and associated confining beds. Recharge to the aquifers takes place on the eastern margins of the basin in outcrop areas, and environmental isotope studies generally support the contention that this still takes place today. Discharge from the basin takes place as concentrated outflow from springs, from vertical leakage towards the regional groundwater table, inter-basin leakage, and from artificial discharge from artesian and pumped wells (Habermehl, 1982). The water is predominantly fresh and in most areas is under sufficient pressure that flowing artesian bores can be constructed. The first artesian bore was constructed at Bourke in 1878.

Banjo Paterson, an Australian bush poet, composed a poem that neatly encapsulates the mysteries and emotions of the early driller working in the artesian basin. The poem was written in 1896 (Patterson, 2010).

The song of the Artesian Water

Now the stock have started dying, for the Lord has sent a drought,
But we're sick of prayers and Providence – we're going to do without,
With the derricks up above us and the solid earth below,
We are waiting at the lever for the word to let her go.

Sinking down, deeper down,
Oh, we'll sink it deeper down:

As the drill is plugging downward at a thousand feet of level,
If the Lord won't send us water, oh, we'll get it from the devil;
Yes, we'll get it from the devil deeper down.

Now, our engine's built in Glasgow by a very canny Scot,
And he marked it twenty horse-power, but he didn't know what is what.
When Canadian Bill is firing with the sun-dried gidgee logs,
She can equal thirty horses and a score or so of dogs.

Sinking down, deeper down
Oh, we're going deeper down:

If we fail to get the water, then it's ruin to the squatter,
For the drought is on the station and the weather's growing hotter,
But we're bound to get the water deeper down.
But the shaft has started caving and the sinking's very slow,

And the yellow rods are bending in the water down below,
And the tubes are always jamming, and they can't be made to shift
Till we nearly burst the engine with a forty horse-power lift,

Sinking down, deeper down,
Oh, we're going deeper down:

Though the shaft is always caving, and the tubes are always jamming,
Yet we'll fight our way to water while the stubborn drill is ramming
While the stubborn drill is ramming deeper down.

But there's no artesian water, though we're passed three thousand feet,
And the contract price is growing, and the boss is nearly beat.
But it must be down beneath us, and it's down we've got to go.
Though she's bumping on the solid rock four thousand feet below,

Sinking down, deeper down,
Oh, we're going deeper down:

And it's time they heard us knocking on the roof of Satan's dwellin',
But we'll get artesian water if we cave the roof of hell in
Oh we'll get artesian water deeper down.

But it's hark! the whistle's blowing with a wild, exultant blast,
And the boys are madly cheering, for they've struck the flow at last:
And it's rushing up the tubing from four thousand feet below,
Till it spouts above the casing in a million-gallon flow.

And it's down, deeper down-
Oh, it comes from deeper down:

It is flowing, ever flowing, in a free, unstinted measure
From the silent hidden places where the old earth hides her treasure
Where the old earth hides her treasures deeper down.

And it's clear away the timber and it's let the water run,
How it glimmers in the shadow, how it flashes in the sun!
By the silent belts of timber, by the miles of blazing plain
It is bringing hope and comfort to the thirsty land again.

Flowing down, further down:
It is flowing further down

To the tortured thirsty cattle, bringing gladness in its going;
Through the droughty days of summer it is flowing, ever flowing
It is flowing, ever flowing, further down.

Banjo Paterson (1864 to 1941)

The use of bore drains to distribute water and to provide water for stock, while being effective, is very inefficient. It has been estimated that only 5%–10% of the water distributed in this way is actually used; the rest is lost by evaporation and seepage. Approximately 4,770 flowing artesian bores have been constructed to average depths of 500 m up to depths of 2000 m into the lower Jurassic aquifer of the Great Artesian Basin.

The bores were often left to flow into bore drains at the surface and water was used to feed cattle (as noted by Paterson!). Unfortunately, the vast majority of this water evaporates and the overall pressure in the aquifer has declined with many bores that were once artesian, now ceasing to flow. Today there are about 3000 free flowing-bores, 35,000 sub-artesian bores and more than 600 spring complexes around the margin of the basin. An ongoing program aimed at capping and rehabilitating flowing bores was commenced in 1988 and will reduce abstractions by an order of magnitude and allow more efficient usage of the resource. This program is funded jointly by the federal and state governments and by land-owner contribution. The program was considered to be more than 50% complete in 2010.

There are also 20,000 non-flowing bores, mainly in the higher aquifers, extending only up to several hundred metres in depth. Most shallow bores are found in the central parts of the basin, where the artesian systems are too deep for economic abstraction. While there are some very deep bores in the central part, deeper bores generally occur on the margins of the basin.

Springs discharging around the margin of the basin (Figure 1.12) have supported Aboriginal occupation for thousands of years prior to European settlement.

The GAB is predominantly a confined and closed groundwater basin comprising a complex multilayered system separated by largely impermeable units. The aquifers are largely continuous and extend across the basin as shown schematically in Figure 1.13. The aquifers occur at depths up to 3000 m. The discharge point of groundwater flowing in the GAB is frequently a mound spring, as seen in Figure 1.14. Where the flowing bores have been capped and the flow controlled, stock and humans have much better access to the water, as seen in Figure 1.15.

1.3.2 The Tertiary Murray Basin

The Murray Basin occupies an area of about 280,000 km^2, and is composed of Tertiary age sediments which are flat-lying but increase in thickness towards the centre of the basin. The maximum thickness is around 640 m. The basin underlies parts on New South Wales, Victoria and South Australia. A generalized map and cross section of the basin is shown in Figure 1.16. The basin contains a number of aquifers which hold groundwater with quality from good to highly saline and with varying bore yields. An important limestone aquifer underlies an area of around 30,000 km^2, which straddles the SA/Victoria border region. This unit, 50–150 m thick at depths of up to 100 m below surface, provides good quality water for town, industrial, irrigation and stock supplies. In the Mallee district, silts and clays are found along with highly saline water.

In the Murray Basin, groundwater generally is under-utilized in all three states. Presumably, this is partly due to the availability of cheaper surface water resources for irrigation in this important agricultural area. The costs of surface water (in 2009) are reportedly $10–$14/ML, and for groundwater $15–$30/ML. Additionally, there are capital costs for drilling of up to $25,000 for shallow wells and up to $75,000 for deeper production wells. However, it was still economic to use groundwater costing $50/ML, although this is highly dependent on the type of crop. Also, given recent policy changes relating to water pricing and capping surface water allocations, it seems likely that costs of surface water will rise and combined use of surface water and groundwater will become increasingly common, given the availability of the groundwater.

A significant problem has arisen in the Murray Basin from rising groundwater levels, particularly in the southern part of the basin. Extensive land clearing and the development of

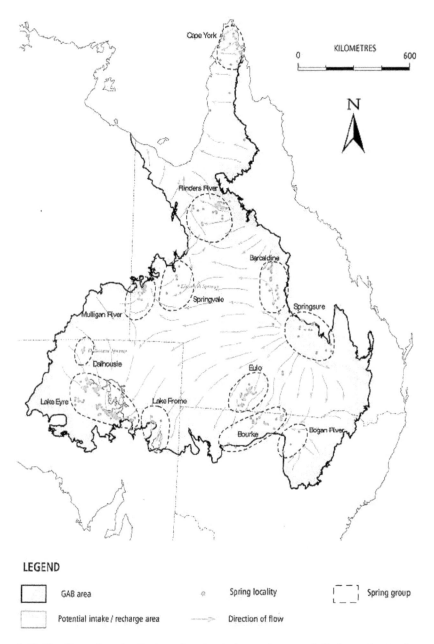

Figure 1.12 The Great Artesian Basin – showing major recharge and discharge (springs) areas and probable flow lines (Brown, 1989) (Creative Commons 4.0 International license)

irrigated agriculture has disturbed the hydrological system. Groundwater recharge has increased with the groundwater level now close to the surface in some areas causing water-logging. Older irrigation areas are reported to have stable "groundwater mounds" while newer areas have continually rising levels. Recent reports suggest that dryland as well as

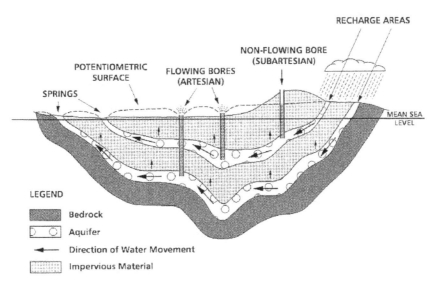

Figure 1.13 Conceptual model of the multi-layered aquifers in the GAB (Brown, 1989) (Creative Commons 4.0 International license)

Figure 1.14 Artesian discharge at Burketown in the Gulf Country of Queensland. The mound deposit is formed from precipitates laid down as the water loses gas (Photo: Catherine Kerr)

irrigated agriculture has given rise to significant salinization within the basin. An estimated 75% of salt loads in streams within the basin are reported to have come from dryland catchments. There is considerable debate as to the most appropriate way to manage salinization within the basin, but it is clear that this is highly complex and there is no one panacea. The complexity and extent of the salinity problem was thought to require a response at government level during the 1990s; however, falling water levels in fractured rock aquifers in the first decade of the 21st century has put this analysis in some doubt. On the Riverine Plain in

Figure 1.15 Capped and piped artesian bore at Rockwell station, SE Queensland. The property name reflects a large well excavated in rock by Aboriginals (though not artesian). Warm enough for a hot bath (Photo: Wendy Timms)

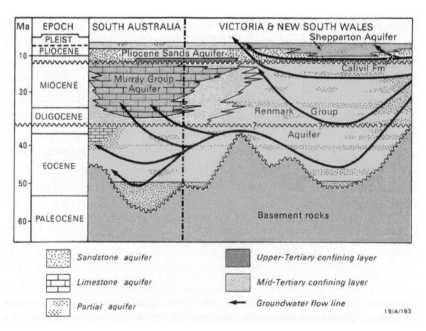

Figure 1.16 West to east cross-section through the Murray Basin (Brown, 1989) (Creative Commons 4.0 International license)

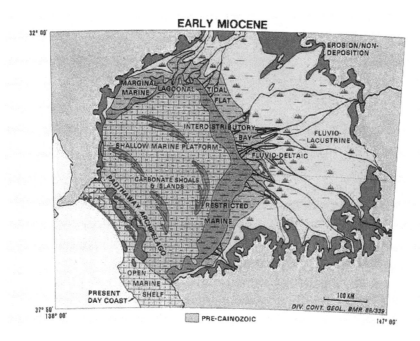

Figure 1.17 Early Miocene geomorphology (Brown, 1989) (Creative Commons 4.0 International license)

Victoria and some alluvial valley systems in New South Wales (Namoi), there is a double problem with rising water levels in shallow aquifers and rapidly declining levels in deeper aquifers.

The geomorphology of the Murray Basin during the Early Miocene is shown in Figure 1.17. The evolution of the geomorphology in Quaternary time is shown in Figure 1.18.

1.3.3   The Late Paleozoic to Mesozoic Perth Basin

The onshore Perth Basin extends from Augusta to just north-east of Geraldton, a distance of 1000 km, covering an area of 51,500 km^2. The basin consists of sedimentary rocks of Silurian to early Tertiary age, overlaid by a veneer of Quaternary surficial sediments. Most sediments are Permian-Cretaceous age, and consist mainly of inter-bedded sandstone and shale, sandstone and coal measures. The major aquifers in the basin are Cretaceous and Jurassic sandstones.

The Perth Basin has been extensively investigated to identify water resources and also for oil and gas exploration. Fifteen groundwater sources have been identified within the basin, from a variety of aquifers, including the surficial formation, the Cretaceous Leederville Formation and the Jurassic Yarragadee and Cockleshell Gully Formations. The mostly confined non-surficial systems are recharged either in outcrop areas where they occur, or by downward leakage from surficial aquifers. It has been estimated that the volume of groundwater in storage in these systems is very large compared with that renewed each year through recharge, and that recharge has varied considerably in the last few thousand years. Given

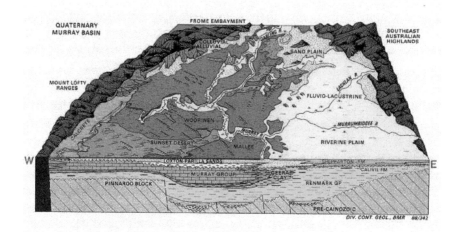

Figure 1.18 Quaternary geomorphology of the Murray Basin (Brown, 1989) (Creative Commons 4.0 International license)

the strategic position of these reserves close to Perth and the numerous other cities and townships along the Western Australian coast, this has generated considerable debate as to whether such large and valuable resources should be utilized through controlled and reversible depletion (as in parts of the Murray Basin) to allow further development of the area (Allen *et al.*, 1992).

1.3.4   Groundwater in Tertiary paleochannels

The Australian land surface was steadily eroded away during the 65 million years of the Tertiary period. The rivers associated with this period of erosion formed their own deep channel systems and should be thought of as paleo river systems. Later in the Tertiary and into the more recent past, some of these river systems became infilled with alluvial deposits, or in some cases, by basalt lava flows. During the Pleistocene (last 1.5 million years), many of these systems were finally covered with a sheet of silt and clay, effectively hiding the old (paleo) river channel.

Very significant groundwater resources are recovered from these paleo valleys when boreholes can be drilled into the maximum depth of the channel. Water for irrigation is currently pumped from many of the paleo drainage channels along the Murray and Murrumbidgee catchments and also from channels close to the some of the tributaries of the Darling, such as the Namoi and Gwyder systems. At Leeton in the Murrumbidgee Irrigation Area (MIA), some bores drilled only to 150 m produce as much as 350 L/s of water with excellent quality (fluid EC [electrical conductivity] of less than 200 μS/cm). Figure 1.19 gives an example of one such bore where the water is being used to irrigate rice.

The management of groundwater in Tertiary paleochannels is a major concern, as they represent the primary water resource for a significant proportion of the irrigated agriculture in Australia. The old river channel is frequently covered by a layer of silt and clay which

Figure 1.19 Irrigation bore for rice growing in the Murrumbidgee Irrigation Area. The bore (not shown) is pumping at 350 L/s (Photo: Ian Acworth)

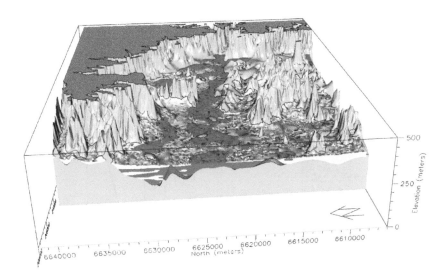

Figure 1.20 Geological model of a paleochannel on the Namoi River in NSW (Geological model constructed by Bryce Kelly, CWI, UNSW)

makes the location of the channel difficult. The silt and clay cover makes the channel difficult to locate, often requiring geophysical methods, and difficult to receive modern recharge. Figure 1.20 shows the results of collecting all available geological data into a geological model of a Tertiary channel on the Namoi River in northern NSW; however, this level of detail is mostly not available.

1.4   GROUNDWATER IN FRACTURED ROCKS

1.4.1   Basement rocks

Crystalline basement rocks crop out widely throughout Africa, Australia, India and parts of South America (Acworth, 1981). The weathered zone of basement rocks provides significant small-scale water supplies for much of the world's population.

The weathering profile developed on crystalline basement rocks in a warm humid and sub-humid climate can be subdivided into several different zones, as shown in Figure 1.21. In regions that are tectonically stable, these zones develop to depths greater than 50 m, often with a layer of laterite developed on the surface, and the lateral distribution of the weathering zones is frequently fairly uniform. By contrast, in regions that are undergoing active erosion due to tectonic uplift, the lateral distribution of the zones is often discontinuous with inselbergs of bare rock occurring surrounded by zones of deeper weathering indicated by the presence of Zone 'a' and 'b' material. The weathering reactions are dominated by hydrolysis with formation of kaolinite (Chamley, 1989).

The various weathering zones have contrasting hydrogeological properties (Figure 1.21 and Table 1.1), with generally decreasing porosity with depth and hydraulic conductivity that increases to a maximum in Zone 'c' and 'd' material (Ollier and Pain, 1995). Weathering reactions produce an increased porosity and, when linked by a fracture network in the underlying Zone 'd' material, form a useful aquifer. The importance of hydrogeomorphological controls over the development of weathering zones has been described for Central

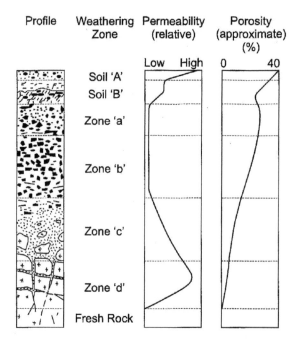

Figure 1.21 Weathering zones in weathered crystalline basement rock and an indication of the changes in permeability and porosity through the weathered profile (adapted from Acworth (1981))

Table 1.1 Range of resistivities for the various weathering grades in weathered crystalline basement rocks. The grades of weathering are shown in Figures 1.21 and 1.22 (adapted from Acworth (1981))

Layer	Description of typical lithology	Resistivity (Ωm)
Soil 'A'	Generally less than 0.5 m thick. Generally a red sandy soil, high porosity, well drained and leached. Laterite seldom present in areas of active erosion.	160–200 when wet; 2000–4000 when dry
Zone 'a'	Few metres thick sandy clay or clay sand, often concretionary.	100–200
Zone 'b'	Massive accumulation of secondary minerals (clay) in which some stable primary minerals remain. Low permeability and high porosity. Usually damp but yields little water.	10–90
Zone 'c'	1 m to 30 m thick. Rock which is progressively altered upward to a granular friable layer of disintegrated crystal aggregates and rock fragments. Intermediate porosity and permeability. This zone frequently contains sub-artesian water, confined by the upper clay-rich material.	60–300
Zone 'd'	1 m to 20 m thick. Fractured and fissured rock. Low porosity but moderate to high permeability in fissures.	600–3000
Fresh Rock	Unweathered migmatite and granite	2000–6000

Sudan (Burke, 1995), Malawi (McFarlane, 1992; McFarlane *et al.*, 1992), Uganda (Taylor and Howard, 2000) and Australia (Ollier and Pain, 1995). Jones (1985), Acworth (1981, 1987) and Wright (1992) describe the development of aquifers in these environments and link the rate of groundwater flow to the formation of clay materials.

The fractures that exist beneath the upper clayey weathered horizons in basement rocks (Hazell *et al.*, 1988; White *et al.*, 1988; Olayinka and Barker, 1990) are not obvious from the ground surface. Examination of fractures in outcrop on inselbergs frequently reveals older fracture trends which, when they occur at depth, are predominantly clay bound and do not form useful aquifers.

The result of weathering that normally follows fracturing in a granite or gneiss (high-grade metamorphic rock) are shown in Figure 1.22. The different grades of weathering (Acworth, 1987, 2001a) give rise to different porosities.

Examples of the rock outcrop are shown in Figure 1.23 and Figure 1.24. Figure 1.23 shows a basement area with active erosion occurring and downcutting that exposes solid rock. Figure 1.25 shows the development of a flat weathering surface and the underlying inselbergs being uncovered by human activity.

Resource evaluation

The location of a significant zone of weathering in a topographically low-lying area in weathered basement rock normally guarantees a reliable yield of high-quality water. A bore drilled into a fracture zone would typically give a yield of 2.5 L/s, sufficient to be developed using a mechanically operated submersible pump. By contrast, a large diameter well constructed into the overlying lower hydraulic conductivity clayey material would only

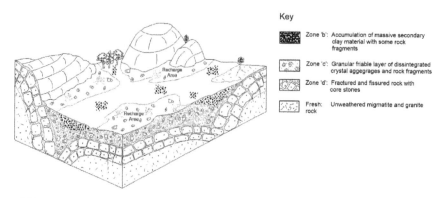

Key

Zone 'b':	Accumulation of massive secondary clay material with some rock fragments
Zone 'c':	Granular friable layer of disintegrated crystal aggregates and rock fragments
Zone 'd':	Fractured and fissured rock with core stones
Fresh: rock	Unweathered migmatite and granite

Figure 1.22 Isometric diagram showing disposition of weathering grades developed in migmatite and granite (adapted from Acworth (1981))

Figure 1.23 Typical crystalline basement rock area in Northern Nigeria (Acworth, 1981) (Photo: Ian Acworth)

produce a yield of approximately 0.1 L/s. The location of zones of weathering is clearly desirable to optimize rural water supply development.

The weathering forms significant clay deposits which are also often close to saturation. The combined effect of surface conduction in the clays and the increased salts contained in the fluid create a significant electrical resistivity contrast between the unweathered rock and the weathered zone. Acworth (1987) gave a range of resistivity values for the differing zones of weathering (Table 1.1) that were derived from extensive forward modelling of apparent resistivity image lines constrained by borehole data. The contrast in bulk resistivities between zone 'b' material (10–90Ωm) and the fresh rock (2000–6000Ωm) makes electrical investigation techniques worthwhile for locating the zones of fracturing and weathering.

Useful groundwater resources are only located in zones 'c' and 'd' weathering and not in the zone 'b' weathered material. The values of resistivity for these zones are intermediate

Figure 1.24 Inselberg showing surface water hole (Acworth, 1981) (Photo: Ian Acworth)

Figure 1.25 Development of a flat erosion surface with underlying inselberg. The clay formed by weathering of the granite is being dug to form bricks for house building (Acworth, 1981) (Photo: Ian Acworth)

between the two end members represented by the fresh rock and the clay-dominated zone 'b'. It will therefore be difficult to resolve accurate thicknesses and resistivities for the two intermediate zones, as there will be problems of both equivalence and suppression in the interpretation of electrical measurements. Notwithstanding this, electrical methods have often been advocated to improve the success of borehole drilling programs (Barker *et al.*, 1992, Carruthers and Smith, 1992).

Records of unsuccessful bores (dry wells or low-yield wells) are not readily available, making reliable statistics of success rates in different environments impossible to determine. However, in the Archean Shield of Western Australia, approximately 2600 bores were drilled in 13 districts as part of a government drought relief program during 1969–70.

The percentage of successful bores (where yield was sufficiently high for the intended use) ranged from 23.5% in one district to 1% in another.

1.4.2 Fractured sandstones

Fractured rock aquifers present unique problems in their investigation, evaluation and management, largely because of their heterogeneous nature, and the dependence of aquifer properties on fracture distribution, connectivity and extent.

Fractured rock aquifers are characterized by high spatial variability in hydraulic conductivity, making traditional hydraulic methods for estimating groundwater flow difficult to apply. The possible range of fractures and fracture sets in rock are described by Cook (2003). The orientation and degree of interconnection of fractures is critical in the development of effective pathways through which groundwater can move.

In Australia, examples of important areas where groundwater extraction from fractured rocks for irrigation has the potential to stress the groundwater resource include:

- The Clare Valley, South Australia, where expansion of the viticulture industry is increasing demand at a rapid rate.
- groundwater extraction from the Howard River Basin for Darwin's water supply may be stressing phreatophytic vegetation in the area, including rain forests and tourist springs.
- in New South Wales, approximately 30 towns in the New England region and central and southern tablelands are partially dependent on water supplies from fractured rock aquifers.
- The Mereenie Sandstone in Central Australia provides water from fractures. The well-known water holes in the McDonald Ranges north of Alice Springs, and the water supply for Alice Springs, come from this aquifer.

On a smaller scale, fracture patterns are less predictable. In the Clare Valley, groundwater availability appears to be controlled by closely spaced (<20 m), small aperture fractures which have no obvious surface expression. On a larger scale, it may be possible to improve the success rate of water well drilling, but bore siting remains problematic in areas where there are no obvious geological structures.

In Central Victoria, there is fracture flow of cold CO_2-bearing mineral water, principally in the Daylesford-Hepborn area in Lower Paleozoic sediments of the Great Dividing Range. These waters effervesce naturally and are sought for drinking and bathing (Weaver et al., 2006) as shown in Figure 1.26. Similar conditions occur at several locations in NSW with a bore at Ballymore shown in Figure 1.27.

Resource evaluation

The evaluation of groundwater resources in fractured rocks is problematical. Well interference is a commonly reported problem where there is a concentration of groundwater users within a small region. The low and variable porosity and hydraulic conductivity of fractured rocks means that drawdown can be very large, even for a low extraction rate. There are numerous reports of extraction from deep irrigation wells drying up shallower domestic supplies (e.g., Dandenong Ranges, Victoria; Clare Valley and Mount Lofty Ranges,

Figure 1.26 Daylesford Springs capped to allow the discharge to be controlled (Photo: Ian Acworth)

Figure 1.27 Carbon dioxide – rich spring water at Ballymore in NSW (Photo: Ian Acworth)

South Australia). Because of the low storage of many fractured rock aquifers, seasonal variations in the water table level tend to be very large, even though recharge may be small. Long-term monitoring may be required to distinguish the effects of well interferences from natural variations of the water table. Where there is a strong, preferred orientation of fractures, the drawdown cone can be highly anisotropic. Well field design in fractured rock aquifers will thus need to consider fracture orientations.

Groundwater quality issues are also different in fractured rock aquifers from those in permeable, intergranular flow systems. High water velocities through fractures mean that contaminants can potentially travel large distances very quickly. The direction of contaminant migration may be difficult to predict from hydraulic head data, particularly if there is a strong anisotropy in hydraulic conductivity.

1.5   GROUNDWATER IN BASALT TERRAINS

Volcanic terrains frequently yield small but reliable supplies of groundwater. Water flows from many discrete seepages and frequently from springs. The water quality is generally good, and spring flow is maintained throughout droughts. Combine the availability of water with the fact that volcanic rocks tend to break down fairly rapidly to good silts and soils and it is easy to see why these areas are favoured for settlement, even if the volcano is still active!

Soils formed from the breakdown products of basalt lava flows form the 'black' soils that are found in various parts of eastern Australia. The agricultural productivity of these soils is very high. The distribution of the soil is dependent upon the distribution of volcanoes – and that requires some understanding of plate tectonics.

1.5.1   Lava types

There are many areas around the world with extensive basalt lava flow complexes. The Columbia River Basalts in the northwest of the USA and the Decaan Basalt Plateau in India are perhaps the best known. Individual flows in the Columbia River sequence can reach 200 km in length and are between 50 and 150 m thick. The flows are gently sloping. The sequence of flows reaches as much as 3000 m in total depth. Flows around Melbourne from more than 300 volcanoes in the past 5 million years are less spectacular but still form a major aquifer unit.

Figures 1.28 and 1.29 show great accumulations of individual basalt flows in Iceland.

The depth of flow and the distance traveled from the source cone is a function of the basalt geochemistry, itself a function of the location of the volcano on the plate.

Figure 1.28 Stacked flows of basalt in Iceland (Photo: Ian Acworth)

Figure 1.29 Basalt columns over basalt flows in Iceland (Photo: Ian Acworth)

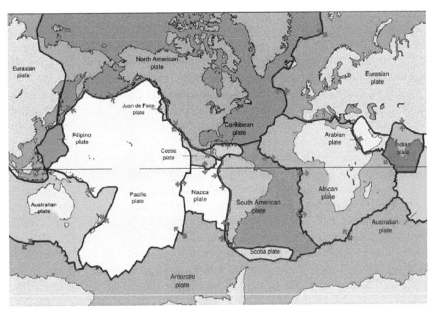

Figure 1.30 Figure showing margins of main plates on the planet with arrows showing convergent or divergent movement (Wikipedia)

Volcanoes are frequently active at plate margins (Figure 1.30) where the type of volcanicity is determined by the type of plate margin (Figure 1.31 from Wikipedia – Creative Commons).

Convergent plate margins indicate that the oceanic plate is diving down into the mantle, dragging some sea bed sediment with it. As the plate descends it heats and some elements of the melt move upwards, mixing with the more silica-rich material in the overlying plate to

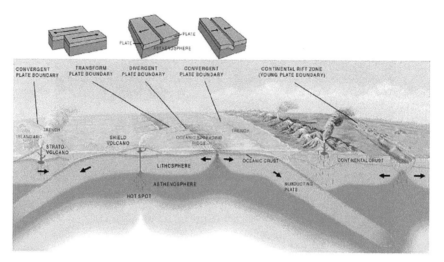

Figure 1.31 The 3D organization of plate boundaries (Wikipedia – Creative Commons)

produce andesitic basalt activity. Andesites are more viscous than oceanic basalt and the lava does not move so far from the vent, thus forming a crater. Note that the Pacific Ocean is rimmed by convergent plate boundaries or zones where the plates are slipping past each other. This means that there is a ring of volcanoes around the Pacific – leading to the term 'ring of fire'. The eruptions of these volcanoes are frequently spectacular and may be devastating. The AD 186 eruption of the Taupo volcano on North Island in New Zealand discharged 9 km^3 of pumice in a stupendous upward blast, at a rate of 100,000 m^3/s, reaching a height of 50 km and leaving the crater lake now occupied by Lake Taupo. It covered the North Island with barren white ash over 1500 km^2 (Sutherland, 1995). A geothermal field now operates on the volcano's north side.

Divergent (spreading) plate margins (Figure 1.31) are typically found in the ocean. Repeated intrusion of molten dike material rises to the surface. The relative motion of the various plates on the earth's surface is shown in Figure 1.32 – data from NASA JPL based upon GPS measurements. Looking to the north of Australia, complex plate motions can be seen that give rise to all the volcanic systems in the area. The GPS data is derived from NASA (http://sideshow.jpl.nasa.gov/mbh/series.html).

A group of volcanoes also occurs that is not associated with plate boundaries. These are the shield volcanoes that seem to form above 'hot spots' in the asthenosphere (Figure 1.33). It is this type of volcano that has left its mark over eastern Australia (Sutherland, 1995).

The most well-established hot spot volcano is considered to be Hawaii. There has been a lot published about this system and the trail (Figure 1.33) it has left. The volcano must make use of a pre-existing weakness to establish a conduit to the surface through the plate. This conduit must be a little elastic, as it remains active for a few million years before finally being snapped and volcanic activity begins again at another location. The initial volcanic activity is in the form of an underwater vent on the sea bed with basalts that form what are known as pillow lava. These build up as an underwater cone until finally the vent breaks the surface and the volcano continues to build above sea level. Pyroclastic material

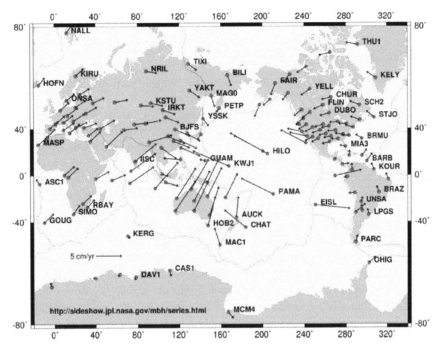

Figure 1.32 Relative plate velocities (NASA JPL)

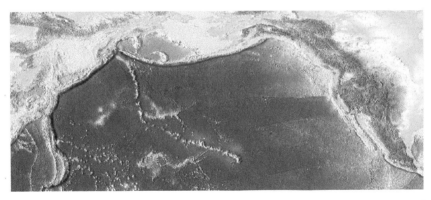

Figure 1.33 Trail of volcanoes left on the surface after the plate moves over a mantle hot spot (NASA JPL)

is not washed away and the volcano grows more rapidly. A full shield-type volcano can result with the final vent more than 3000 m above sea level.

The presence of the magma chamber at shallow depth heats all the surrounding rock and the crust expands and rises as a result. When the plate hot spot finally shifts so far away that the vent can no longer tap the magma reservoir, the volcanic activity wanes and the crust begins to cool and sink. Weathering of the basalt pile is now uninterrupted by new additions of lava and the whole system gets lowered back down to sea level and finally below sea level. Coral reefs will grow on the exposed basalt when the water is warm enough and form a coral key.

1.5.2   Groundwater occurrence

Although there are frequently good supplies associated with volcanic deposits, it is extremely difficult to predict where, or if, the water will be encountered in a borehole. Volcanoes erupt sporadically with hundreds of years between eruptions. During the intervening period, weathering of the top of a lava flow commences and plants and trees become established. The next flow will move by gravity into the lowest part of the landscape. As it does so, it will cover the vegetation that has become established on top of the previous flow. The lava is clearly very hot and any moisture will flash to steam and penetrate into the base of the oncoming flow. These hollows (vesicles) are often poorly connected and while the porosity of this part of the flow can be high, the hydraulic conductivity is low. Pumice is an example of the rapid degassing of a basalt as the pressure is reduced during the eruption. Pumice will float and can often be seen on the beaches of NSW after eruptions in the Indonesian archipelago.

More importantly, significant vegetation (such as trees) will prevent the new lava flow from coming into immediate contact with the unweathered rock in the older flow. The vegetation burns but will often prevent close contact with the underlying flow. Soil will bake but will also prevent close contact. The interflow zone often has a high porosity and the pore space is well connected so that there is good hydraulic conductivity. The interflow zone becomes an aquifer that takes the shape of the contact zone between the flows. The 3D shape of this zone is frequently highly irregular.

In addition to the interflow zone, some basalts will crack as they cool. The cracks become flow zones linking the interflow aquifer zones. Figure 1.34 shows the typical hexagonal cracking associated with basalt cooling.

The varying chemistry of lava gives rise to two different types:

- *Pahoehoe Lava* – developed from silica pore lava. This material runs like treacle. It is the typical lava type of the mid-ocean ridges or the Icelandic-type volcanics. Some hot

Figure 1.34 Hexagonal cracking due to slow cooling of a basalt flow beneath Mount Kaputar in northern NSW (Photo: Ian Acworth)

Figure 1.35 Hexagonal cracking seen in a block of lava flow fallen from Sawn Rocks, NSW (Photo: Ian Acworth)

spot lava is also of this type. The outside of the flow is cooled by contact with the atmosphere or with underlying cooler rocks. The outside begins to congeal and solidify while the inner part of the flow remains fluid. Under these conditions, it is possible for the inner part of the flow to drain away leaving a hollow lava tube. These tubes can become major sources of water, and one of the largest springs in the world actually flows from a lava tube in Hawaii.

- *Aa Lava* tends to have a higher silica and gas content with the result that the flow is blocky and does not flow as readily. The result of the higher viscosity is that volcanic cones are more readily assembled by this lava type. However, the lava may still move considerable distances.

Volcanic eruptions frequently contain ash and tephra deposits that can accumulate to significant thicknesses between the lava flows. These can produce some reasonable yields to bores. The exact locations where groundwater will accumulate depend upon an understanding of the 3D geometry of the flows. Clearly, a new flow will be gravity driven – but without a knowledge of the elevation of the ground surface before the flow begins, little can be determined as to where the next flow will go.

The fine-grained texture of the basalt makes it a reasonably resistant rock to weathering. For this reason, basalt flows frequently cap older sediments and form a plateau. The Alstonville Plateau in northern New South Wales or the Liverpool Ranges are both good examples. On the Liverpool Ranges, the Coolah Tops are almost flat with the edge of the plateau marked by sheer cliffs down to the Breeza Plains. Waterfalls mark the edge of the plateau. There is sufficient discharge from the basalts that the streams draining the area are frequently perennial and flow through most droughts.

Resource evaluation

The hydraulic conductivity distribution within the lava is highly anisotropic and heterogeneous. This makes modelling of the system impossible at a small scale, and only possible at

Figure 1.36 Seepage from the base of a basalt flow on Mount Vincent (Photo: Ian Acworth)

a large scale if major averaging is assumed. The corollary to this is that it is not possible to predict the location of a water-bearing zone in the basalt pile. Clearly, water will move downhill throughout the pile and it would be expected that streams draining the lower parts of the basalt pile will have perennial flow. The large number of individual seeps and springs makes these areas of great ecohydrological importance.

The Mount Vincent Plateau between Mudgee and Lithgow, NSW, is another example. Figure 1.36 shows a typical small seepage that supports stock water and domestic supplies. The springs issue at the base of the basalt. As the basalt is a residual capping with deep erosion into the underlying sandstones and coal measures, the basalt springs feed into streams which then fall over the edge of the plateau's water falls. Another example of lavas giving rise to major water falls is Ebor Falls in northern NSW (Figure 1.37).

In general, it can be seen that the prediction of exactly where in the volcanic pile you should drill for water is very difficult. This difficulty is enhanced by the fact that there is no good geophysical technique available to assist with bore location. The basalt itself has a high resistivity, but if the fractured zone of the basalt also contains good-quality low-salinity water, then there is no electrical contrast. The shape of the resistivity target is also highly variable, so that any interpretation is greatly hindered by equivalence problems. These difficulties have not stopped surveys being carried out – it is just that the results are frequently highly equivocal.

Seismic techniques are of little use, as the unfractured and fresh basalt is a very high-velocity layer that is also frequently not horizontal. The depth to basalt below weathered material can readily be found – but not the depth to the base of the flow where there is an expectation that significant groundwater reserves could occur.

Groundwater investigation in these areas is not straightforward! This has led many farmers to employ 'the local water diviner'. The most appropriate approach is to try and build up a conceptual model of how the basalt flows built up. This will help to determine the most likely location of the lowest flow and the pre-volcanic valley that is probably a good target.

Figure 1.37 River falling over the edge of a basalt flow marking the edge of the New England Plateau (Ebor Falls). Note the columnar jointing in the lava flow to the left of the falls (Photo: Ian Acworth)

Lava flows can frequently block older drainage channels and cause disruptions to drainage. This might simply be the formation of a lake behind a lava dam, or they might cause a complete relocation of the drainage. Eastern New South Wales has many examples of disrupted drainage caused by the accumulation of lavas throughout the Tertiary. In some instances, the lava will have flowed across the top of the old channel – isolating the underlying sediments. If the isolation is complete, then the aquifer now contains fossil water. If the isolation is only partial and the aquifer can still receive recharge laterally, then an important reserve of groundwater can be created.

Quantitative evaluation of the basalt aquifer resource is particularly difficult, primarily because the shape and volume of the aquifer is so poorly known. Pumping tests can be conducted in abstraction bores, but almost invariably encounter boundary (edge) effects within hours of the pump test commencing. Evaluation of this data is possible – but the results are fairly meaningless.

A more pragmatic approach, that also works with complexly fractured basement rocks, is to carry out an inverted step test. Commence pumping at what you feel is a maximum rate based upon the air-lift yield noted during drilling. Continue with this rate until the aquifer is locally exhausted and drawdown is too great. Then allow a recovery to take place and continue at a lower abstraction rate. In this way, you will establish what the maintainable yield is for the resource.

Water quality in these systems is usually very high. There may be enhanced calcium and magnesium – but in general, the salt content will be low.

1.5.3 Australian hot spot chain

Australia has drifted across a hot spot as it moves northward after the split with Antarctica. The pattern of volcanoes is nowhere as simple as the trail left behind on the Pacific Plate, and there is some debate as to whether the very simple hot spot model can explain the Australian pattern. The general concept is that when Australia first began crossing the hot spot in the east of Queensland, the spot moved beneath New South Wales and Victoria and is now in the Tasman Sea. There are 30 central volcanoes that have been created, with the oldest between 33 and 42 million years around Hillsborough in Queensland to 6 to 8 million years in Central Victoria (Sutherland, 1995). The presence of significant volcanic activity in Australia is frequently overlooked, perhaps because there is no mineral value associated with the basalts! However, the basalts are rich in calcium and magnesium and weather to produce very productive soils with a high percentage of smectite clay. The clay absorbs water and holds some of this water at suctions suitable for extraction by plants, further adding to the productive capacity of the soils. The highly productive soils in northern NSW and southern Queensland are developed from the weathering of basalt material. The smectite clay mineral is quite stable and can be moved around by floods forming extensive smectite-dominated soils suitable for farming over much of the River Darling catchment.

1.6 GROUNDWATER IN UNCONSOLIDATED SURFICIAL DEPOSITS

1.6.1   Introduction

Surficial aquifers occur in alluvium, colluvium, dune sands, calcrete and aeolianite, and are mostly of Quaternary age. These exist mostly within 100 m of the surface.

The surficial aquifers are highly productive, with about 60% of abstractions in Australia coming from these, providing both irrigation and urban supplies (Lau *et al.*, 1987). However, some of these aquifers are coming under stress through overuse. Water is stored in intergranular spaces in sands, alluvium, colluvium and gravels. The water levels show marked seasonal fluctuations in response to recharge, as most aquifers are unconfined and recharge takes place over a significant proportion of the aquifers. Hence, unlike some of the deeper groundwater systems, annual recharge can be significant when compared to the volume of water in storage.

Groundwater abstractions are often concentrated close to point of use (a significant advantage of groundwater resources). Locally, abstraction and natural discharge can greatly exceed recharge, giving rise to resource depletion, saline intrusion and knock-on effects on surface water ecosystems which are dependent on groundwater (lakes, wetlands, swamps, etc.) or for base flow (rivers, streams).

These impacts can be alleviated to some extent by artificial recharge, but it has become increasingly obvious that careful management is required for allocation of groundwater supplies from these aquifers.

A schematic of the Botany Aquifer in Sydney (Figure 1.38) shows how some of these impacts can occur.

1.6.2   Swan Coastal Plain

Sand aquifers on the Swan Coastal plain contribute over 70% of water supplies and up to 35% of drinking water supplies for the city of Perth and associated coastal areas. The aeolian sands and coastal limestone (aeolianite) provide excellent aquifers, which are

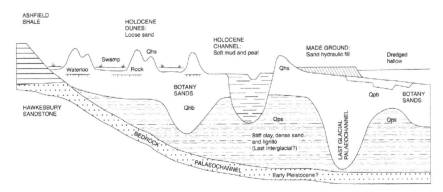

Figure 1.38 Section through the Botany Aquifer in Sydney showing relationship between the aquifer and the (groundwater) lakes

readily recharged by winter rains. The groundwater systems also support extensive wetlands and swamplands throughout the plain and groundwater discharges to both ephemeral streams along the eastern edge of the plain and to the Swan River and Moore River on the central and northern part of the plain as well as to the Indian Ocean in the west.

The pattern of recharge and surface discharge produces groundwater mounds where the phreatic surface is at its highest elevation in the central parts of the plain, and decreases in elevation towards the discharge points at the east and west margins and into westward flowing rivers and streams which traverse the plain. The undulating surface of the sand dunes intercepts the phreatic (water table) surface in dune swales, forming wetlands, damp lands and swamps, which are entirely dependent on groundwater for their existence. Urban and agricultural development on the coastal plain have impacted wetlands to such an extent that most of these in the urban and associated horticultural areas are eutrophic, and although there has been considerable effort put into management of water levels through artificial recharge, conjunctive use of groundwater and surface water, groundwater model- ling and monitoring, water levels have shown a steady downward trend since developments began in the 1950s. Some wetlands have disappeared and others have been inundated from land drainage so that fringing vegetation has died. Phreatophytic vegetation (deep-rooted trees and shrubs which obtain water from the water table) has also been severely affected in some areas close to water supply well fields.

The situation on the Swan Coastal Plain is by no means uncommon, where exploitation of shallow fresh groundwater supplies in unconfined aquifers beneath the expanding city not only draws down the groundwater table, but also gives rise to contamination of water sup- plies from a wide range of activities. Perth is relatively unique in that land-use restrictions have been applied to the developing city suburban areas, to help minimize impacts on groundwater quality. Perhaps the most pressing problem in areas like the Swan Coastal Plain is our lack of understanding of the dependence of natural ecosystems on groundwater, which limits our ability to manage the resource whilst maintaining environmental sustain- ability. It is clear that groundwater on the Swan Coastal Plain maintains a wide range of wetlands, supports phreatophytic vegetation and also now has been identified as a signif- icant driving force expelling nutrients from anoxic sediments into the Swan River, causing nuisance algal blooms.

Figure 1.39 Perched groundwater lake on North Stradboke Island, Queensland (Photo: Ian Acworth)

1.6.3 Sand dune aquifers of the east Australian coast

There are major aquifer systems in wind-blown sands along the east coast of Australia. These include the Botany Aquifer and the Tomago Sands in New South Wales and Moreton Island, Stradbroke Island and Fraser Island in Queensland. Fraser Island is the largest sand dune island in the world.

Each of these aquifers systems contains very large quantities of fresh groundwater and supports complex ecosystems associated with the groundwater discharge. Figure 1.39 shows a major perched groundwater lake (Brown Lake) on North Stradbroke Island.

1.7 GROUNDWATER IN LIMESTONE AND CHALK SYSTEMS

1.7.1 Introduction

Sedimentary carbonate rocks are composed of calcite, aragonite or dolomite. They have great economic significance as reservoir rocks for gas, oil and water. They also have use in agricultural lime, cement, building stone and concrete aggregate. Chalk has been quarried from prehistory, providing building material and marl for fields. Modern coral reefs form major tourist attractions.

Limestone sedimentation is only possible in the absence of detrital silica grains and also in warm shallow water. This means that the environments have to be away from estuaries and in the tropics. The Bahama Banks in the Caribbean have provided much of the environmental data for the deposition of limestone.

Limestone is a sedimentary rock composed largely of the mineral calcite (calcium carbonate: $CaCO_3$). Limestone often contains variable amounts of silica in the form of chert or flint, as well as varying amounts of clay, silt and sand as disseminations, nodules or

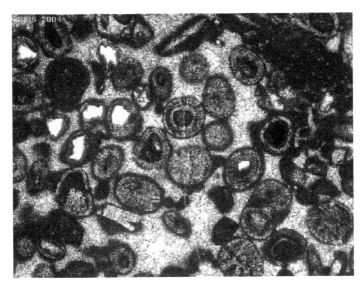

Figure 1.40 Thin section photo-micrograph of oolitic limestone

layers within the rock. Limestone can be recognized in the field by its reaction with hydrochloric acid. A few drops of acid will cause the limestone to fizz with the release of CO_2. Micrite or microcrystalline carbonate is by far the most common constituent in carbonate rocks. The individual crystals in ancient rocks are < 5 μm in diameter and are now commonly calcite. In modern carbonates, the mud is composed of individual crystals of aragonite and are needle-like in shape.

The primary source of the calcite in limestone is marine organisms. These organisms secrete shells that settle out of the water column and are deposited on ocean floors as pelagic ooze or alternatively as a coral reef. Pure limestone is almost white. Because of impurities, such as clay, sand, organic remains, iron oxide and other materials, many limestones exhibit different colors, especially on weathered surfaces. Limestone may be crystalline, clastic, granular or massive, depending on the method of formation. Crystals of calcite, quartz, dolomite or barite may line small cavities in the rock. Another form taken by calcite is that of oolites (oolitic limestone, Figure 1.40) which can be recognized by its granular appearance. Limestone makes up about 10% of the total volume of all sedimentary rocks.

Limestone deposited in coral reef complexes forms a major component of many aquifer systems. Figure 1.41 shows the physical extent of the Great Barrier Reef off the northeast coast of Australia. The growth of these reef systems can only be accommodated if the land surface is gently subsiding to compensate for the growth of the reef. The genesis of the reef material is the intertidal zone. Accumulation of many metres of limestone reef material is then only possible if the land surface is falling or sea levels are rising. The accumulation of coral normally occurs at the edge of the reef with the area inside the coral formed from carbonate muds. As coral dies, it breaks down to give a source of detrital material that can be moved around by the tides. On large carbonate platforms, the tides redistribute material efficiently and often form tidal deltas at the edge of the platform (Blatt *et al.*, 1980).

Secondary calcite may also be deposited by supersaturated meteoric waters (groundwater that precipitates the material in caves). This produces speleothems such as stalagmites and

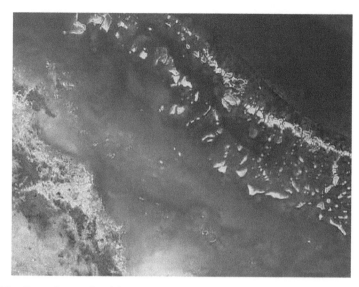

Figure 1.41 The Great Barrier Reef from earth orbit (NASA JPL)

Figure 1.42 Stalagmite form in Cathedral Cave, Wellington, NSW. Red stain due to a leaking metal tank on the surface (Photo: Ian Acworth)

stalactites commonly found in cave systems. Figure 1.42 shows a typical stalagmite from Wellington Caves in New South Wales. These caves are developed in Ordovician limestone, part of the Lachlan fold belt.

For much of the past 100,000 years, the sea level has been below the current level. During the glacial maxima, sea level was between 120 m and 130 m lower than present. That has meant that the shallow limestone reefs and carbonate muds have been exposed to aerial erosion. In particular, high winds and desiccation have caused the formation of carbonate pellets of sand size that have been blown across the exposed shelf and accumulated

Figure 1.43 High-angle cross-bedding in a limestone – an example of redistributed calcite sand-sized grains (Photo: Ian Acworth)

in carbonate sand dunes (calcarenites). These dunes have then been subject to solution by rain. The final result has created cross-bedded calcarenites that look more like sand dunes than limestone (Figure 1.43).

Travertine is a banded, compact variety of limestone formed along streams, particularly where there are waterfalls and around hot or cold springs. Calcium carbonate is deposited where evaporation of the water leaves a solution that is supersaturated with chemical constituents of calcite. Tufa, a porous or cellular variety of travertine, is found near waterfalls. The groundwater draining from a limestone aquifer is saturated with $CaCO_3$ within the system. As it emerges and the partial pressure of CO_2 drops, calcite will precipitate. This gives rise to terraces formed from travertine. The limestone mountains in southern Mexico provide spectacular examples (Figures 1.44 and 1.45).

1.7.2 Chalk systems

Chalk is a soft, white, porous form of limestone composed of the mineral calcite. It is also a sedimentary rock. It is relatively resistant to erosion and slumping compared to the clays with which it is usually associated, so it forms tall steep cliffs where chalk ridges meet the sea. Chalk hills, known as chalk downland, usually form where bands of chalk reach the surface at an angle, so forming a scarp slope. Chalk is formed in shallow waters by the gradual accumulation of the calcite mineral remains of microorganisms (coccolithophores) over millions of years. Embedded flint nodules are commonly found in chalk beds. The source of the silica in the chert is not clear. The chalk makes up the famous cliffs at Dover in the UK. Figure 1.46 shows the vertically dipping chalk at a location known as The Needles in the west end of the Isle of Wight in the UK. Figure 1.47 shows the chalk overlying older Jurassic deposits on the Isle of Wight.

Figure 1.44 Pools formed from travertine in a limestone fed stream at Palenque, Mexico (Photo: Ian Acworth)

Figure 1.45 A travertine shawl – the green material is travertine with an algal growth – deposited from spray falling from a spring in the limestone higher up this canyon wall, Mexico (Photo: Ian Acworth)

Because chalk is porous, made almost exclusively from the calcite remains of coccoliths (foraminifera), the chalk downlands of UK, France and Denmark usually hold a large body of groundwater, providing a natural reservoir that releases water slowly through dry seasons. The rivers Somme and Thames are examples of baseflow released from the chalk aquifer. The chalk aquifers supply some 40 million people with water in southern England and northern France. Soft limestone also occurs around the Mediterranean and in the Middle East.

By contrast, limestone forms a solid slab of material that can be used for building or decoration. The primary porosity of the limestone is practically zero. Secondary porosity is

Figure 1.46 The Needles, Isle of Wight, UK (Photo: Ian Acworth)

Figure 1.47 Cretaceous chalk overlying Jurassic sands, clays and marls, Isle of Wight, UK (Photo: Ian Acworth)

produced by solution of the calcite by acid rainfall. The hydrogeology of these two varieties of limestone are therefore quite different.

The stratigraphy of the chalk differs with locality, and exact correlations are difficult to make. Particularly so because the chalk is fairly featureless in outcrop. There may be fossils throughout – but finding these is time consuming. The chalk has been subdivided into stages based upon lithostratigraphical characteristics (Table 1.2). These stage names are widely used in France, but in England, the older lithological names upper, middle and lower are still widely used by hydrogeologists. The chalk contains several syndepositional hardground horizons which have been caused by the emergence of the chalk above sea level for a time. The chalk was diagenetically altered by rain (fresh) water during this emergence. These

Table 1.2 Stratigraphic subdivisions of the chalk

STAGE		Lithology
Maastrichtian	Maastrichtian	
Campanian		
Santonian	Senonian	Upper Chalk
Coniacian		
Turonian	Turonian	Middle Chalk
Cenomanian	Cenomanian	Lower Chalk

hard grounds form important aquifers. The chalk rock, between the upper and middle chalk and the Melbourne rock between the middle and lower chalk, form significant aquifers.

In the absence of readily recognizable lithological variation, borehole geophysics can be used to determine the location within the chalk sequence (Woods, 2006). Layers of marl are also found within the chalk and represent layers where the bulk resistivity of the chalk is lowered. The marl is thought to be the result of volcanic ash fall into the shallow sea water. The marls are not competent when drilled and frequently squeeze into the borehole, providing problems with borehole construction. The marls are very extensive and can be recognized over hundreds of kilometres in the UK. The Plenus Marl is a particularly well developed marker horizon. The lower and middle chalks have more marls than the Upper chalk which is better characterized by bands of flint (Schurch and Buckley, 2002). The chalk rock is locally a 3.5 m thick hard layer, composed of intensely consolidated and mineralized hard grounds. They consist of several solution-ridden layers, representing a gap in sedimentation and a possible unconformity. Phosphatized and glaconitized chalk pebbles are frequent and give rise to a marked gamma ray anomaly (Schurch and Buckley, 2002).

The chalk has a pronounced dual porosity, in which the matrix provides the storage and the fissures provide preferential pathways for flow. Edmunds et al. (1992) provide a summary of the main flow characteristics of these systems:

1 High permeability is often associated with valleys: transmissivities can be as high as 2000 m^2/d but in interfluves may decrease to around 20 m^2/d.
2 The hydraulic conductivity of the matrix is low (10^{-3} m/d) making a negligible contribution to the transmissivity of the aquifer.
3 Only the upper 50 m or so of the saturated chalk forms an effective aquifer. This is due to the enlargement of fissures by solution processes in the zone of water-table fluctuation, either at the present day or in the past. Pleistocene sea-level changes near coastlines where up to 100 m of effective aquifer may now exist.
4 Preferential flow pathways often exist which have developed in response to lithology, hardness and stratigraphic changes as well as structural controls.
5 Acidic recharge at the Tertiary margins may give or have given rise to karstic features.
6 Paleohydrology has played a part in developing the flow regime. Colder waters in the Pleistocene enhanced rock solubility; Tertiary topography and drainage features influenced the hydraulic regime.
7 Permeability decreases with depth, but enhanced flow may occur at depth from hard-ground horizons.

8   Specific yield is low and ranges from 0.01–0.02 generally (in interfluves) but may be 0.03–0.05 in the upper 50 m of the permeable zone. This leads to large seasonal water level fluctuations (± 20 to 30 m).

The porosity of the chalk represents the arrangement of the coccolith fragments. These fragments have a well-defined size from 1.5 μm to 2.0 μm but their degree of compaction can impact the pore size distribution. Unweathered upper chalk has a porosity of about 0.42, but this varies between 0.26 and 0.45 in the lower and upper chalk respectively as the amount of mineral impurities increase. For the upper chalk, the mean pore radii are about 0.38 μm (Schurch and Buckley, 2002).

The pore space between the coccoliths is very small, with water held by surface tension. There are also fissures developed in the chalk by solution of the calcite. These are connected to the micro-pore space and also through the fissure network, to the atmosphere.

1.7.3   Limestone systems

Unlike the chalk, groundwater movement through limestone is entirely through secondary porosity solution channels, as there is no primary porosity. Carbonate hydrochemistry is widely covered in text books (Appello and Postma, 2005). In essence, rain falling through the atmosphere equilibrates with carbon dioxide in the atmosphere to form a weak solution of carbonic acid. Percolation through the soil brings the water into further contact with biogenically produced carbon dioxide with the resulting solution becoming more acidic. Calcite is soluble and dissolves in the percolating groundwater. The reaction products are carried away by groundwater so that an initial crack in limestone will be continuously widened and deepened by percolating water. In this way, a cave system is developed.

The depth at which solution occurs is determined by the possible discharge of the groundwater. The reaction can only proceed if water can flow through the system and remove the reaction products (calcium and magnesium cations and bicarbonate anion). The implication of this is that groundwater solution does not occur far below sea level, as the sea represents the base level of discharge for the groundwater system.

During the ice ages, the sea level was up to 120 m lower than occurs currently. For this reason, cave systems in limestone close to the coast may be well developed to 120 m below sea level.

In the Mediterranean, much greater depths of solution have been developed, as the Mediterranean Sea dried up completely at the end of the Tertiary. The Messinian Salinity Crisis, also referred to as the Messinian Event, is a period when the Mediterranean Sea evaporated partly or completely dry during the Messinian period of the Miocene epoch, approximately 6 million years ago. Sediment samples from below the deep seafloor of the Mediterranean Sea, which include evaporite minerals, soils, and fossil plants, show that about 5.9 million years ago in the late Miocene period, the precursors of the modern Strait of Gibraltar closed tight and the Mediterranean Sea evaporated into a deep dry basin with a bottom at some places 3.2 to 4.9 km below the world ocean level. Even now, the Mediterranean is saltier than the North Atlantic because of its near isolation by the Straits of Gibraltar and its high rate of evaporation. Local groundwater flow systems around the Mediterranean Basin would have adjusted downward to the new base level where fresh groundwater discharge would have occurred on the plain. In limestone country, this means that there will

Figure 1.48 Limestone pavement on the Burren in Ireland (Photo: Ian Acworth)

now be many deeply submerged cave systems flooded with salt water as sea levels recovered.

The presence of the discharge from these systems can be readily detected by looking at thermal imagery of the sea water. The cooler temperature of the groundwater discharge in summer (or warmer than the sea in winter) is clearly evident. The investigation of interconnection between caves is often carried out using dye.

Limestone dissolves very readily in rainfall with little soil development. A limestone pavement is created (Figure 1.48) that tends to concentrate rainfall into runoff that flows down into the many solution holes (clints and grikes) and into connected cave systems below ground surface. The absence of soil means that recharge is very high in these areas. The groundwater discharge from this system will locally be below sea level (Figure 1.49), and sea water can rapidly move into the cave systems. These conditions are similar to those on the Yucatan Peninsula in Mexico, where extensive limestone formations occur.

1.7.4   Resource evaluation and development

Resource evaluation in limestone rocks must be carried out by pumping test evaluation. There is significant secondary permeability development in these rocks, which means that the representative elementary volume (REV), for which Darcy Law holds, can be many hundreds of metres. The significance of this is that flow at scales smaller than the REV can not be predicted using Darcy Law. You can drill a trial bore and intersect a fracture with copious water, yet sink an abstraction bore close by that is dry. The evaluation of a particular source can only be made after the analysis of long-term pumping-test data. Experience in the UK and France confirms that long-term permanent groundwater supplies can be developed in these aquifers. To overcome the REV problem, shafts were sunk by miners in the UK and adits dug out from just above the base of the shaft. These adits were extended until they encountered sufficient water from fractures. Adits of more than 1000 m have been constructed with a drain at the side of the adit for water to flow to a sump at the base of the

Figure 1.49 Limestone pavement disappearing below the sea. In previous times, the sea level would have been much lower and the caves would now be flooded by sea water (Photo: Ian Acworth)

shaft. Pumping equipment is installed on a ledge above the sump and water is lifted to the surface.

Regional groundwater modelling and resource evaluation works at a larger scale; however, there is frequently a major variation in hydraulic conductivity with depth. In drought conditions, when the water level is reduced, the transmissivity is also significantly reduced.

1.8   GROUNDWATER ON OCEANIC ISLANDS

1.8.1   Introduction

Groundwater is a vital source of water on many tropical islands. The island of O'ahu (Hawai'i) relies on groundwater for 90% of its drinking water and about half of all water used for irrigation. During El Niño episodes, rains often fail in the Pacific and groundwater becomes the only source of water.

The occurrence of groundwater on oceanic islands depends upon the geology of the island. Volcanic islands have groundwater associated with basalt as well as groundwater in calcrete or sand deposits. In the Pacific Ocean, it is frequently the case that limestone has formed on top of volcanic lava producing a complex aquifer system. Tribble (2008) has reported on a number of typical occurrences from the Pacific.

1.8.2   Density differences

The most important factor in the consideration of oceanic island systems is that fresh water has a lower density than sea water and will therefore tend to float on sea water in the aquifer. The interface between the two is susceptible to movement as a result of groundwater

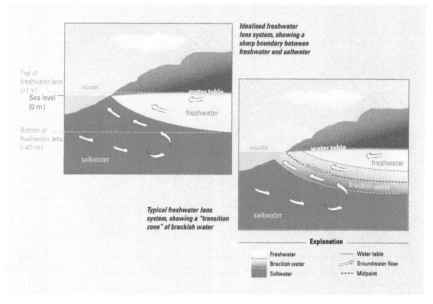

Figure 1.50 Freshwater lens beneath an island (USGS (Tribble, 2008))

abstraction. This makes the management of groundwater lenses a difficult issue that requires careful monitoring. Figure 1.50 shows the relationship between fresh water and groundwater.

1.8.3   Groundwater occurrence

Groundwater either occurs in a basalt environment as a part of an island volcanic chain or in limestone deposits that form a fringing reef to the basalt. These environments have been described above.

The land level rises up at the start of a period of volcanicity and then tends to sink after the volcanicity is completed and the crust cools. This means that a change in sea level occurs. These changes occur in addition to the changes that occur as the result of a global sea-level response to climate change. Separating the effects of sea-level changes caused by plate tectonic processes and the onset of volcanicity from those associated with global climate change can be problematic.

Perched fringing reefs can be found surrounding islands. The reef is formed during a period of stability that is then interrupted by uplift associated with earthquakes before a new period of stability occurs. The result is a perched reef. The sea level has not fallen. In this instance, the land level has risen. Examples of these processes are common on islands such as Vanuatu. In the case of Atturo, north of Dili on Timor'Est, the fringing reefs reach to the top of the island, which is almost 1000 m above modern sea level.

Groundwater may occur at high levels in islands (Figure 1.51), where lithological changes give rise to perched aquifers, or may be more extensive at low levels (Figure 1.52; Tribble, 2008).

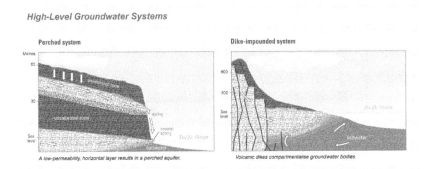

Figure 1.51 Aquifers can be perched well above the regional water level (USGS (Tribble, 2008))

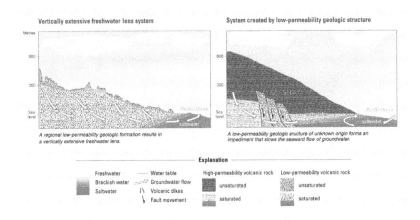

Figure 1.52 Extensive sand dune aquifer systems or systems developed on low permeability layers (USGS (Tribble, 2008))

1.9   GROUNDWATER TERMINOLOGY

Hydrogeology is a young science, and there is still considerable variability in terminology, both nationally and internationally. Even the spelling of groundwater (or ground water) is still debated. As much as English grammar is logical, the term ground is used as an adjective to the noun water. As such, ground water should be two words and should be hyphenated when used as an adjective (e.g., ground-water hydrology or ground-water resources). Common usage has overturned this logic, and the use of groundwater as one word has become widespread. There is, of course, a considerable lack of consistency in usage (e.g., seawater, spring water, bathwater, rainwater or contaminated water).

The definitions in Table 1.3 are adapted from Misstear *et al.* (2006) and modified to reflect Australian usage.

Usage of the term aquifer is not well defined. It means different things to different people, and perhaps different things to the same person at different times. The use is perhaps most

Table 1.3 Well and borehole terminology (adapted from Misstear *et al.* (2006))

Water well	Any hole excavated in the ground that can be used to obtain a water supply.
Drilled well	A water well constructed by drilling. Synonyms are tube-well, production well or production borehole. As drilled wells are the most common, they are often referred to as wells for simplicity.
Hand-dug well	A large-diameter, usually shallow, water well constructed by manual labour. Synonyms are dug well and open well.
Exploratory borehole	A borehole drilled for the specific purpose of obtaining information about the subsurface geology or groundwater. Synonyms are investigation borehole, exploration borehole and pilot borehole.
Observation borehole	A borehole constructed to obtain information on variations in groundwater level or water quality. Also known as an observation well.
Piezometer	A small-diameter borehole or tube constructed for the measurement of hydraulic head at a specific depth in an aquifer. In a piezometer, the section of the borehole (the screened section) in contact with the aquifer is usually very short.
Test well	A borehole drilled to test an aquifer by means of pumping tests.

confusing when referring to a unit above the water table, which could carry water if it were saturated. Is this a potential aquifer? The term must always be viewed in terms of the scale and context of its usage.

Aquifer A saturated permeable geologic unit that can transmit significant quantities of water under ordinary potential gradients.

Aquitard A saturated geologic unit that is capable of transmitting water under ordinary hydraulic gradients but not in sufficient quantities to allow completion of production wells within them.

Confined aquifer A confined aquifer lies between two aquitards. The hydraulic head in a confined aquifer lies above the base of the upper confining layer and is generally referred to as the potentiometric surface or piezometric surface.

In a confined aquifer, the water level (potentiometric surface) in a borehole drilled into the aquifer rises above the top of the aquifer. The well is called an artesian well. In some cases the water level may rise above ground surface, in which case the well is called a flowing artesian well.

Potentiometric surface The concept of a potentiometric surface is only rigorously valid for horizontal flow in horizontal aquifers. Some hydrogeological reports contain potentiometric surface maps based on water-level data from sets of wells that have approximately the same depth but bottom in different aquifers or are not associated with a well-defined aquifer. The potentiometric surface in a confined aquifer may occur at the same elevation as that of the water table in an overlying unconfined aquifer. The two surfaces will respond differently to any disturbance such as recharge or pumping and may be easily differentiated if test data or monitoring information is available.

Unconfined aquifer An unconfined aquifer is an aquifer in which the water table forms the upper boundary. Perched aquifers often occur above an unconfined aquifer, particularly after periods of recharge.

Water table The water table is best defined as the surface on which the fluid pressure P in the pores is exactly atmospheric. If P is measured in gauge pressure, then on the water table $P = 0$.

Saturated zone Occurs below the water table; the soil pores are all filled with water; the moisture content equals the porosity; the fluid pressure is greater than atmospheric pressure; the pressure head can be measured with a piezometer and the hydraulic conductivity is a constant and not a function of the moisture content.

Tension saturated zone Occurs immediately above the water table and is a zone where all the pore space is saturated but the pressure is less than atmospheric pressure because water is held in the pores by surface tension. This zone is very small in coarse-grained media but may be several metres thick in clays.

Unsaturated zone Occurs above the tension saturated zone. The pressure is less than atmospheric and the pore space is only partially saturated. The hydraulic conductivity is a function of the moisture content and decreases rapidly as the moisture content decreases.

Steady-state flow Steady-state flow occurs when, at any point in a flow field, the magnitude and direction of the flux are constant with time. Over long periods of time, the constant flux elements in an aquifer system reach equilibrium. Long-term recharge is balanced by long-term abstraction and discharge. As a result, there are no net changes in aquifer storage. The primary application of steady-state flow techniques is in the analysis of regional groundwater flow problems.

Transient flow Transient flow occurs when, at any point in the flow field, the magnitude or direction of the flux is changing. An appreciation of transient flow is required for the analysis of well hydraulics and many geotechnical and geochemical applications.

Field capacity A soil unit which is at field capacity will allow water added to the top of the soil to displace an equivalent volume of water from the base of the soil unit. Typically, the moisture content is approximately half that of the soil at saturation.

Specific yield This term is used most often by hydrogeologists as a measure of the available water in a unit of saturated aquifer. It represents that portion of the total porosity which is released by the soil before the soil moisture is reduced to field capacity. Soil moisture at values less than field capacity is held in the soil by surface tension effects and can only be extracted by exerting a negative matric pressure (or suction), usually accomplished by plants.

Soil moisture deficit Plants extract moisture from the soil until the root systems are unable to exert sufficient negative pressure to further drain the soil. These suctions can be considerable (see Table 4.6).

Wilting point This is an imprecise measure of soil moisture. However, for recharge studies, the water content represented by the difference between wilting point and field capacity is referred to as the soil moisture deficit (SMD). Water equivalent to the SMD is required to be added to the soil before recharge to the underlying aquifer can take place.

Air dry This moisture content represents a value for the soil in equilibrium with the air, and thus at atmospheric humidity.

Chapter 2

Surface water and the atmosphere

2.1 INTRODUCTION

Surface water and groundwater form major components of the hydrological cycle. Their study must not be separated as water continuously moves between these two domains. If separation does occur then there is the real risk that water will be lost during attempts to model the system. This has long been considered an acceptable risk to the surface-water community as the short-term fluxes of floods greatly outweigh the long-term slow seepage of groundwater. However, the groundwater community is not able to ignore the flux of recharge, determined by water balance methods relying on surface measurements of rainfall, runoff and evaporation, as these underpin sustainable groundwater management.

Groundwater dependent ecosystems (GDE) are established where groundwater rejoins the surface water sphere as base flow to streams and rivers. The recognition of a GDE requires that the flow in the stream or river be understood and quantified at different stages of the river. In some parts of the world, 100% of the river discharge is groundwater. The recent emphasis on surface water and groundwater interconnectivity recognizes these interrelationships and their importance in supporting the associated ecosystems. For these reasons, it is necessary to commence a detailed study of groundwater by recognizing the place of groundwater in the overall hydrological cycle.

2.1.1 The distribution of water on the planet

The oceans occupy 70.8% of the earth's surface, leaving only 29.2% land mass (Figure 2.1). The distribution of land and ocean between the northern and southern hemispheres is strongly biased towards land in the northern hemisphere and ocean in the southern hemisphere. This is shown in Figure 2.2 (USGS, 2013).

In the northern hemisphere, 51.5% of the surface is land and 48.5% ocean. In the southern hemisphere the ratio is only 8.6% land to 91.4% ocean. The data is shown in Table 2.1.

This difference in distribution has a major impact upon climate. In particular, the lack of land between 35° South and 65° South leaves an uninterrupted west wind field around these latitudes. World climate changed markedly during the Tertiary period when Australia finally split from Antarctica about 25 Ma ago, allowing the development of a circumpolar ocean current that effectively isolated the Antarctic continent.

The majority of water on the earth occurs in the oceans, as shown by Table 2.2.

Water is held on the land in rivers and lakes, but not all of this water is fresh. Table 2.3 shows the distribution of surface water on land.

Figure 2.1 The water planet – Earth seen from Apollo 17 (NASA collection)

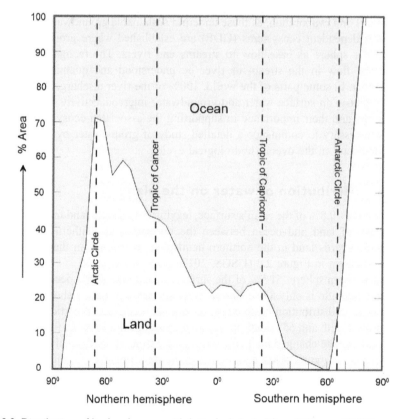

Figure 2.2 Distribution of land and ocean with latitude (adapted from Wiesner (1970))

Table 2.1 Distribution of land and ocean (adapted from Wiesner (1970))

| | Area $10^6 km^2$ | | Percentage % | |
	Ocean	Land	Ocean	Land
Northern hemisphere	154.8	100.3	60.7	39.3
Southern hemisphere	206.5	48.7	80.9	19.1
Earth	361.3	149.0	70.8	29.2
Total	510.3		100.0	

Table 2.2 Distribution of water on the earth (adapted from Wiesner (1970))

	Volume as water $10^6 km^3$	Percentage of total water	Average depth m	Based on area of the
Ocean	1338	97.3	3700	ocean
Ice caps and glaciers	29	2.1	80	ocean
Groundwater	8.4	0.61	56	land
Lakes and rivers	0.23	0.017		
Atmosphere	0.013	0.00094	0.025	planet
Biological water	0.0006	0.000005	0.004	land
Total	1376	100	2700	planet

Table 2.3 Distribution of surface water on land (adapted from Wiesner (1970))

	Volume km^3	Percentage of total water on the Earth
Rivers	0.0017	0.00012
Freshwater lakes	0.125	0.0091
Saline lakes	0.105	0.0076
Total	0.23	0.017

2.1.2   The global water balance

The movement of water between the ocean and the land is in a long-term steady state. The land is not getting either wetter or any drier. USGS (2013) gives a global water balance as shown in Table 2.4. The runoff from the land to the ocean (315 mm) is the excess of the precipitation on land (800 mm) over the evaporation from the land (485 mm). This is balanced by the excess of evaporation from the ocean (1400 mm) over the precipitation over the ocean (1270 mm). Note that land occupies 29% of the planet surface whereas the oceans cover 71%.

2.1.3   Residence times

USGS (2013) has also calculated the mean residence times for water in the various storages on the planet. This data is presented in Table 2.5 and shows that, for example, the atmosphere only contains enough water at any one time to support approximately 8 days of

Table 2.4 Global water balance (adapted from USGS (2013))

	Precipitation		Evaporation		Runoff	
	km^3	mm/yr	km^3	mm/yr	km^3	mm/yr
Ocean	458.9	1270	505.8	1400	−46.9	−130
Land	119.2	800	72.3	485	46.9	315
Planet	578.1	1133	578.1	1133	0	0

Table 2.5 Mean residence time of water on the earth

	Volume km^3	Rate of turnover $km^3 year^1$	Mean residence time
Oceans	1338	0.5058	2600 years
Ice caps and glaciers	29	0.00255	1100 years
Groundwater	8.25	0.0119	700 years
Lakes	0.23	0.00173	13 years
Soil water	0.03	0.0706	155 days
Rivers	0.0017	0.0469	13 days
Atmosphere	0.013	0.5781	8.2 days
Biological water	0.0006	0.0651	3.4 days

rainfall over the planet. Large volumes of water are either saline or frozen, so the importance of the hydrological cycle in moving around the small volume which is left, but which supports life, is critical.

2.1.4   The Australian water balance

The water balance of six continents is shown in Figure 2.3, also from USGS (2013). The vertical scale in this figure is in millimetres of water per year, while the horizontal scale represents the continental surface area in millions of square kilometres. It is clear that Australia is particularly dry. A comparison of global and Australian water balances is given in Table 2.6.

Actual evaporation is less than the global average, not because there is less potential evaporation, but because there is not enough rainfall available to evaporate. This implies that the hydrological balance is particularly delicate in Australia and that small changes can produce major impacts. The significance of salinity in Australia is related to the delicate nature of this balance.

2.1.5   Origin of Water

The present atmosphere of the earth is probably not the first atmosphere on the planet (Ese Encrenaz, 2008). The existence of a hydrosphere in liquid form is seen to be unusual in planetary evolution. It would appear that the release of bound water from the rocks making up the earth's mantle and core is only possible as a result of the very high temperatures and pressures associated with the metallic core of the earth. The release of the water to the surface of the earth occurs as the result of tectonic processes which exist within the mantle. Early atmospheres are not thought to have had the same composition as the present nitrogen, oxygen mixture. It is also interesting to note that a position in the solar system just

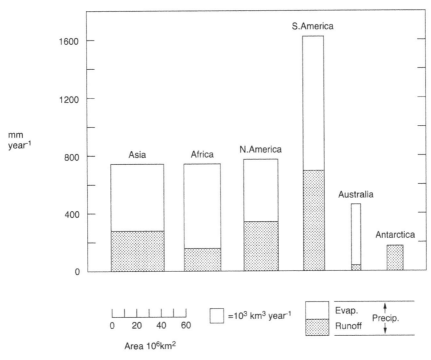

Figure 2.3 Water balance of six continents (adapted from USGS (2013))

Table 2.6 Comparison of Australian and global water balances

	All land surfaces	Australia
Precipitation	800	456
Evaporation	485	419
Runoff	315	37
Potential evaporation	1132	1870
Ratio of runoff to precipitation	0.398	0.08
Ratio of evaporation to potential evaporation	0.43	0.22

5% closer to the sun than present would have resulted in a run-away greenhouse effect and the loss of practically all the atmosphere. Ese Encrenaz (2008) gives a good review of the planetary evolution of water.

2.1.6 Composition of the atmosphere

The permanent constituents of the atmosphere are nitrogen (78%), oxygen (21%) and inert gasses. The variable constituents may be in solid, liquid or gaseous form. Water is the most important and can comprise as much as 4% and be present in all phases. Ozone and carbon dioxide exist as gasses in widely varying amounts. Solid components are particulate dust, smoke, salts and microorganisms. These can reflect or scatter the incoming solar radiation. The detailed components of the atmosphere are shown in Table 2.7.

Table 2.7 Composition of the atmosphere

Gas	Symbol	% by Volume
Permanent gasses		
Nitrogen	N_2	78.08
Oxygen	O_2	20.95
Argon	Ar	0.93
Neon	Ne	0.0018
Helium	He	0.0005
Hydrogen	H_2	0.00005
Xenon	Xe	0.000009
Variable gasses		
Water vapour	H_2O	0–4
Carbon dioxide	CO_2	0.034
Methane	CH_4	0.0001
Ozone	O_3	0.000004
Carbon monoxide	CO_2	0.00002
Sulphur dioxide	SO_2	0.000001
Nitrogen dioxide	NO_2	0.000001
Particulate matter		0.00001

There is no top to the atmosphere. It gradually decreases in density until it is indistinguishable from interstellar space. This occurs approximately at 600 km. There is sufficient gas to obstruct the fall of a meteorite at 300 km and to cause it to glow white hot, but the majority of gas lies below 20 km. The atmosphere is divided into four zones: the troposphere, the stratosphere, the mesosphere and the thermosphere. Most weather phenomena of interest to us occur below the tropopause, in the troposphere. The movement of water as liquid, solid or gas is limited to the troposphere.

2.1.7 Molecular Structure

Despite its common occurrence, water is a highly unusual substance. As it is a hydride of a non-metal, it can be compared with other hydrides. Most of these, however, are not liquid but gaseous (NH_3 for example). Consequently, low temperatures or high pressures are needed to liquefy many hydrides. Yet water is liquid at the relatively high temperature range of 0 °C to 100 °C and freezes readily to form ice. This phenomenon is caused by the structure of the water molecule.

The water molecule consists of two hydrogen atoms joined by single covalent bonds to an oxygen atom. The result of these two covalent bonds is to isolate a "lone pair" of electrons in the outer electron shell of the oxygen atom, which therefore becomes locally electronegative. This localized charge distribution polarizes the whole molecule so that it behaves like a miniature magnet. The hydrogen atoms to the exterior are electropositively charged. Hydrogen bonds are formed between the oxygen atoms of one molecule and the hydrogen atoms of the next forming H_2O in vapour. The angle between covalent bonds in a molecule of pure water is 105°. This imposes a three-dimensional structure on any assemblage of molecules.

In ice, irregular, roughly hexagonal and pentagonal bonding occurs. Liquid water is thought to resemble the structure of ice, except that hydrogen bonding is more ephemeral, being rapidly made and broken between molecules. At any instant in time, the statistical chance of bonds being formed depends upon the temperature. About 15% of the bonds are broken when ice melts at 0 °C, and the total number of bonds broken rises to 30% at 100 °C. Hydrogen bonding can explain not only the liquidity of water at relatively high temperatures but also other important properties such as density, viscosity and surface tension.

2.1.8   Energy Transformations

Energy is consumed in the transformation of ice to water and water to water vapour. Ice may be transformed directly from ice to water vapour. There are, as always, different sets of units used to indicate these transformations. A calorie (cal) is defined as the energy necessary to raise 1 gram of pure water from 14.5 °C to 15.5 °C. A calorie represents 4.1868 joules of energy. At other temperatures it takes approximately 1 cal to increase the temperature of 1 g of water by 1 °C. The evaporation of water requires an input of energy which is called the latent heat of vapourization.

At 20 °C, the latent heat required is approximately 2.454 MJ/kg. The amount of energy required varies with the temperature of the surface. An accurate relationship for the normal range of temperatures is given in Equation 2.1:

$$\lambda = 2.501 - 0.002361 T_s \qquad (2.1)$$

where:
λ is the latent heat measured in MJ/kg, and
T_s is the temperature of the surface from which evaporation is occurring.

The same amount of energy is released by the condensation of water from the vapour phase. In order to melt ice at 0 °C, 333.688 kJ/kg of energy must be supplied to the ice. The phase change occurs without any increase of temperature. This energy is often referred to as the latent heat of fusion.

Water can also pass directly from a solid phase to a vapour phase by the process of sublimation. The energy due to both fusion and vapourization is required. Sublimation therefore requires 2.834 MJ/kg. The opposite process is the formation of a frost, where 2.834 MJ/kg of energy are released during frost formation.

2.1.9   The various types of water

Water can occur as a combination of five different types (Galperin *et al.*, 1993):

1   Water vapour
2   Chemically bound water
3   Physically bound water
4   Ice
5   Liquid

Water can move from one type to each of the other four types.

Water vapour

Water vapour is encountered almost everywhere. Cold vapour is found in the unsaturated zone between the top of the tension saturated zone (above the water table) and the atmosphere. It enters either from the atmosphere or as evaporation from liquid water at the top of the tension saturated zone. The relative humidity in this zone is close to 100% and therefore water vapour moves in response to temperature gradients. There is downward movement in summer and upward movement in winter. There may also be lateral movement away from structures.

A significant proportion of the total water vapour on the planet is associated with steam generated at depth. This may generate geysers as at Yellowstone Park in the US or in Iceland at Geyser (Figures 2.4 and 2.5). However, the main source for water vapour is evaporation from the ocean.

Figure 2.4 Gas pressure building in a geyser at Geyser, Iceland (Photo: Ian Acworth)

Figure 2.5 The same geyser blowing out (Photo: Ian Acworth)

Water vapour is also contained in liquid water. As pressure increases, the proportion of water vapour to liquid water increases until at great depths and temperatures of 600 °C to 700 °C, all the water occurs as water vapour.

Chemically bound water

Chemically bound water is mainly associated with the crystal lattice of minerals. It can exist as single molecules or groups of molecules within the lattice and is generally released by heating up to ≈ 300 °C. Minerals which contain water include $Na_2CO_3 \cdot 10H_2O$ (64% water) and $Na_2SO_4 \cdot 10H_2O$ (55% water).

The OH^- and H^+ ions are also contained within some minerals. On heating, this water is released and forms free water. Aluminum and calcium hydroxides release water on heating to 400 °C. Clay minerals such as kaolinite $Al(OH)_4 \cdot [Si_2O_5]$ and montmorillonite $(AlMg)_2 \cdot (OH)_2[Si_4O_{(10)}] \cdot H_2O$ contain considerable water which is liberated at temperatures of 460 °C to 550 °C. The water contained in these compounds is released during diagenesis of a sedimentary pile and forms part of the process which changes a clay to a slate.

Physically bound water

Physically bound water can be further subdivided into:

* hygroscopic or adsorbed water, and
* loosely bound or pellicular water.

Adsorbed water is characteristic of fine dispersed soils. It is generated by the electrostatic attraction of a layer of water molecules to the surface of minerals. This layer may be 1 to 3 molecules thick and is tightly held. Water in this layer has a density of ≈ 2000 kg m^{-3}, a freezing temperature of -78 °C and a dielectric permeability of between -2 and -2.2 (dielectric permeability of free water is -81) (Galperin et al., 1993). This water has considerable viscosity, elasticity and shear strength. Adsorbed water exists at relative humidities well below 100%. The proportion of adsorbed water in a soil is predominantly a function of the soil mineralogy and particle size. It does not exceed 1% in sand, but it may be 8% in loess (silt) and 18% in clay. This variability is a direct result of the available surface area for water molecules to attach to. The presence of adsorbed water onto the clay platelets significantly increases the bulk electrical conductivity of clays.

The loosely bound layer of water has a thickness of between 10 and 20 molecules of water. This layer is also referred to as the osmotic layer or the solvate layer. Water can be removed from this layer by heating to 100–120 °C. Water can move away from a particle with a thicker film towards a particle with a thinner film of water. This occurs as the result of an ionic gradient in the water films.

The physically bound water characterizes the geotechnical properties of materials. It is a maximum of 1%–7% in sands, 9%–13% in sandy loam and 20%–40% in clays (Galperin et al., 1993).

Experimental data shows that at pressures of 300–500 MPa, almost all physically bound water changes to free water and occurs as solution in pores within the soil/rock. A considerable volume of water occurs in this state with approximately 145 million cubic kilometres (10% of all ocean water) estimated (Galperin et al., 1993).

Ice

Great quantities of ice occur in the polar ice caps and as high-altitude glaciers. The properties of frozen rock are also important in large areas of the tundra in the USSR and Canada and have received considerable attention in these countries. Frozen rock is a multi-component system comprising the mineral skeleton, water, ice and air. Once frozen, the structural properties of the soil are similar to those of rock. The depth of freezing depends upon the thermophysical properties of the soil (heat content and thermal conductivity) and the climatic conditions. The greatest depth of frozen soil occurs in sands and gravels. The least depth occurs in soils with a high humic content such as peat.

Liquid

Liquid water can again be subdivided into:

- Capillary water and
- Free water (gravitational water)

Capillary water fills up pore spaces having diameters of approximately less than 1 mm and cracks opening below 0.25 mm. Capillary water is held in place by surface tension effects. The height of the capillary rise can be calculated from the Laplace formula (Equation 2.2):

$$h_c = \frac{2\gamma}{\rho fgr} \qquad (2.2)$$

where:
h_c is the height of the capillary rise [m],
γ is the liquid-air surface tension (0.0728 N/m at 20 °C),
ρ_f is the fluid density [kg/m^3],
r is the capillary radius [m], and
g is acceleration due to gravity [9.80 m s^{-2}].

The capillary rise can be zero in gravels, 30–40 mm in coarse sands to as much as 6–12 m in clay.

Water free to move under a hydraulic gradient (gravitational water) occurs where the moisture content exceeds the maximum physically bound moisture content and is not restricted by capillarity. Groundwater resource development concerns the movement and abstraction of this free water. Many geotechnical problems are associated with physically bound water. In most cases, there is no free water in clays – it is all physically bound. This has an important impact on the interpretation of piezometer data from clay sites.

2.1.10   The hydrological cycle

The movement of water from the oceans to the atmosphere by evaporation, then over the land as water vapour, through precipitation to the land surface, and back to the ocean as runoff forms the hydrological cycle which supports life on the planet.

The global hydrological cycle is a gigantic distillation machine. The earth forms a vast heat engine and the distillation process is one part of that engine. The efficiency of the engine is related to the physical properties of the water molecule and the energy released as it changes state from one phase to another. Events such as volcanic eruptions, which reduce incident energy by changing the albedo of the atmosphere, can quickly affect the balance, with results which have far-reaching consequences.

2.2   PROPERTIES OF WATER

2.2.1   Water density

The density of a substance is its mass per unit volume:

$$\rho = \frac{mass}{volume} \tag{2.3}$$

The density of water depends both on temperature and the presence of dissolved solids. A knowledge of the density of water is important in calculations of hydraulic head and water pressure. Density differences between two bodies of water may cause flow to occur even where the elevations of the two bodies are identical. Values of density for a range of temperatures are given in Table 2.8.

2.2.2   Water viscosity

The resistance to the movement of one layer of fluid over another adjacent layer is ascribed to the viscosity of the liquid. The dynamic viscosity of a fluid is the property that allows fluids to resist relative motion and shear deformation during flow. The more viscous a fluid, the greater the shear stress at any given velocity gradient. According to Newton's law of viscosity:

$$\tau = \mu \frac{du}{dy} \tag{2.4}$$

Table 2.8 Properties of water at 1 atmosphere

Temperature °C	Density ρ [kg m^{-3}]	Dynamic Viscosity μ [kg m^{-1} s^{-1}]	Compressibility β [m^2 N^{-1}]
0	999.8395	1.787×10^{-3}	5.01×10^{-10}
5	999.9720	1.567×1^{-3}	-
10	999.6996	1.307×1^{-3}	4.78×10^{-10}
15	999.0996	1.139×1^{-3}	-
20	998.2041	1.002×10^{-3}	4.58×10^{-10}
25	997.0449	0.8904×10^{-3}	4.57×10^{-10}
30	995.6573	0.7975×10^{-3}	4.46×10^{-10}
40	992.2158	0.6529×10^{-3}	4.41×10^{-10}
50	988.0363	0.5468×10^{-3}	4.40×10^{-10}
60	983.1989	0.4665×10^{-3}	4.43×10^{-10}
70	977.7696	0.4042×10^{-3}	4.49×10^{-10}
80	971.7978	0.3547×10^{-3}	4.57×10^{-10}
90	965.3201	0.3147×10^{-3}	4.68×10^{-10}
100	958.3637	0.2818×10^{-3}	4.80×10^{-10}

where:

τ is the shear stress

dv/dy is the velocity gradient

μ is the dynamic viscosity

The units of dynamic viscosity μ are Pa.

In many problems involving viscosity, the magnitude of the viscous forces compared to the magnitude of the inertia forces is of interest. Since the viscous forces are proportional to the viscosity μ, and the inertial forces are proportional to the density ρ, the ratio μ/ρ is often involved. This ratio is called the kinematic viscosity, with units of $m^2\ s^{-1}$.

Viscosity is also temperature dependent as it is a function of the number of hydrogen bonds existing. Values of viscosity for a range of temperatures are given in Table 2.8. The value of fluid viscosity influences the rate of fluid movement through a porous medium.

2.2.3 Water pressure

The *fluid pressure* p at any point in a standing body of water is the force per unit area which acts at that point. Under hydrostatic conditions, the fluid pressure at a point reflects the weight of the column of water overlying a unit cross-sectional area around the point:

$$p = \rho g \psi \tag{2.5}$$

where:

ρ is density,

g is the acceleration due to gravity ($9.80\ m\ s^{-2}$ – in Sydney at 34 °S), and

ψ is the height of the water column above the reference plane.

The units of pressure are pascals Pa or $kg\ m^{-1}\ s^{-2}$.

It is possible to express pressure relative to absolute zero pressure, but more commonly it is expressed relative to atmospheric pressure. In the latter case, it is called gauge pressure, as this is the pressure registered by a pressure gauge with atmospheric pressure as the reference.

2.2.4 Water compressibility

The *compressibility* β of a fluid reflects its stress-strain properties. Stress is the internal response of a material to an external pressure. For fluids, stress is imparted through the fluid pressure. Strain is a measure of the linear or volumetric distortion of a stressed material. When a fluid is stressed, the strain occurs as a reduced volume (and increased density) under increasing fluid pressures.

Compressibility is defined as strain/stress $\frac{d\epsilon}{d\sigma}$. It is the inverse of the modulus of elasticity. Units of compressibility are m^2/N.

For a given mass of water:

$$\beta = -\frac{d\rho/\rho}{dp} \tag{2.6}$$

The compressibility of water has a very low value but is still dependent upon temperature. Values for a range of temperatures are given in Table 2.8.

The compressibility of solids is usually designated by α where the units are the same as for fluids. Values of α are several orders of magnitude greater than those for β. This inequality is of importance when aquifer confined storage is calculated.

2.2.5   Water surface tension

In studies of unsaturated flow, the surface tension σ determines the capillary rise in porous media and thus influences water uptake by plants and the size of the tension saturated zone or capillary fringe. The surface tension directly affects the matric potential Φ_m.

The units of surface tension are J m^{-2}. Surface tension is also a function of temperature, decreasing with an increase in temperature.

2.2.6   Summary of Water Properties

A summary of the physical properties of water are given in Table 2.9.

2.3   RADIATION

The hydrological cycle is driven by the input of radiation from the sun to the top of the earth's atmosphere. The distribution of that incoming radiation drives the various weather systems on the planet that, in turn, deliver rainfall to the earth's surface. Radiation is often quoted in a number of different units. Table 2.10 gives a summary of conversion factors.

Table 2.9 Summary of water properties

Property	Value	Units
Liquid Density	998.0	kg m^{-3}
Solid Density	910.0	kg m^{-3}
Vapour Density	1.73×10^{-2}	kg m^{-3}
Heat of Fusion	334	kJ kg^{-1}
Heat of Vapourization	2.45×10^3	kJ kg^{-1}
Specific Heat	1.0	Jkg^{-1}K
Relative Permitivity	80	-
Thermal Conductivity	1.0	W/mK
Kinematic Viscosity	1.0	m^{2}s^{-1}
Surface Tension	72.7×10^{-3}	J m^{-2}

Table 2.10 Radiation – conversion factors

1 cal cm^{-2} day^{-1}	=	0.041868 MJm^{-2} day^{-1}		
1 MJm^{-2} day^{-1}	=	23.884 cal cm^{-2} day^{-1}	=	0.408 mm day^{-1}
1 mm day^{-1}	=	2.45MJm^{-2} day^{-1}	=	58.6 cal cm^{-2} day^{-1}
1 Wm^{-2}	=	0.0864MJm^{-2} day^{-1}	=	2.064 cal cm^{-2} d^{-1}

Table 2.11 Approximate energies of natural phenomena

Event	Joules
Solar energy received by the earth in one day	1.5×10^{22}
Strong earthquake	10^{19}
Average cyclone	10^{18}
Major volcanic eruption	10^{16}
Average squall line	10^{1B}
Average summer thunderstorm	10^{13}
The atomic bomb at Nagasaki	10^{13}
Burning of 7000 tonnes of coal	10^{13}
Average forest fire	10^{12}
Average local shower	10^{11}
Average tornado	10^{10}
Average lighting strike	10^{8}
Average dust devil	10^{6}
Individual gust of wind near the earth's surface	10^{4}

The sun continuously emits short-wave electromagnetic radiation as a direct consequence of the high surface temperature ($\approx$5800 K). The Stefan-Boltzmann Law can be used to determine the energy flux from a body if the surface temperature of that body is known. The relationship is:

$$\Xi = \epsilon \sigma T^4 \tag{2.7}$$

where:
Ξ is the energy flux in Watts per square metre W m^{-2},
ϵ is the emissivity of the surface, which is 1 for a black body and has values in the range $0 \leq \epsilon \leq 1$ for other surfaces,
σ is the Stefan-Boltzman constant, which has the value 5.67 $\times 10^{-8}$ W m^{-2}K^{-4}, and
T is the absolute temperature in degrees Kelvin (K).

The current best estimate of the energy flux at the top of the earth's atmosphere (Ξ_E) from a number of satellite measurements is 1367 W m^{-2}. This value is not a true constant, since it can fluctuate during the course of solar cycles.

It is worth comparing the quantity of solar radiation received by the earth each day with other phenomena (Table 2.11) to get an idea of the immense scale of the sun's output.

A figure for the amount of short-wave radiation reaching the earth's surface in the general case can be calculated. This figure is complicated by three factors:

1   the varying distance of the earth from the sun. The earth follows an elliptical path around the sun, not a circular path,
2   the axis of rotation of the earth is inclined at 23.5° to the plane of the elliptic, and
3   the curvature of the earth's surface.

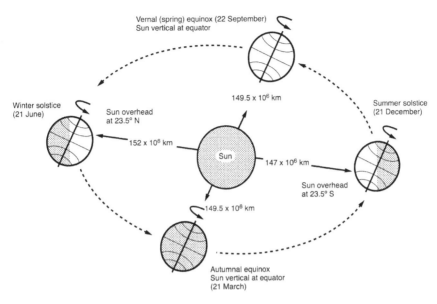

Figure 2.6 Seasonal variation in solar radiation (adapted from Wiesner (1970))

2.3.1   Effect of variations in the earth-sun distance

The earth rotates around the sun in an elliptical orbit. The quantity of incoming radiation is therefore affected by the change in distance between the earth and sun during the year. This is shown in Figure 2.6.

The correction factor for the solar constant (Ξ_E) can be derived in terms of the radius vector. This is the instantaneous earth-sun distance divided by the mean earth-sun distance. The radius vector (d_r) varies from 0.983 in early January, when the earth and sun are closest, to 1.017 in early July, when they are farthest apart (see Figure 2.6). The flux of solar radiation at normal incidence outside of the earth's atmosphere can then be calculated for different seasons if the radius vector is known.

A good approximation for the radius vector is:

$$d_r = 1.0 + 0.017 \times \cos\left(\frac{2\pi}{365}(186 - J)\right) \tag{2.8}$$

where:
d_r is the radius vector of the earth-sun system and
J is the Julian day number – derived from adding the elapsed number of days to the current date starting on January 1.

2.3.2   Effect of earth's rotation

The result of the inclination of the axis of rotation to the plane of elliptic is that the solar zenith angle Z changes with season. The solar zenith angle is defined as the angular distance

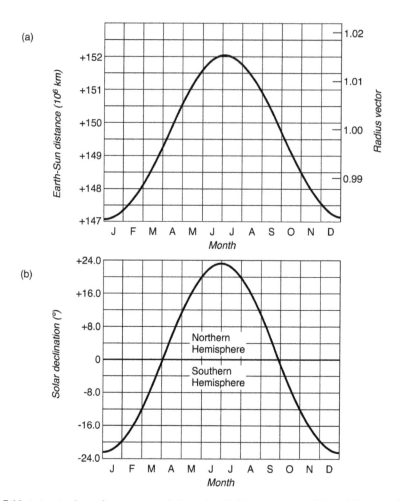

(a)

(b)

Figure 2.7 Variation in the radius vector and the solar declination (adapted from Wiesner (1970))

of the sun from the local vertical. The solar zenith angle varies with the latitude of the location and the time of the solar day. Earth-sun distances and the variation in the solar declination are shown in Figure 2.7.

The solar declination for any given day in the year can be approximated by:

$$\delta = 23.45 \times \cos\left(\frac{2\pi}{365}(173 - J)\right)$$ (2.9)

where:

δ is the declination of the sun in degrees. The declination of the sun is its angular distance north (+) or south (−) of the celestial equator (List (1947).) and

J is the day number.

2.3.3   Effect of latitude and curvature of the earth's surface

The incoming solar radiation at any latitude, day of the year and hour of the solar day (R_a) can be calculated if the latitude and solar time are known. The basic formula for computing the amount of solar radiation (R_a) reaching a unit horizontal area at the top of the atmosphere in time t is:

$$\frac{dR_a}{dt} = \frac{\Xi_E}{d_r^2} \cos z \tag{2.10}$$

where:
d_r is the radius vector,
z is the sun's zenith distance, and
Ξ_E is the solar constant.

The sun's zenith angle z is given by:

$$\cos z = \sin \varphi \sin \delta + \cos \varphi \cos \delta \cos h \tag{2.11}$$

where:
φ is the latitude of the observer and
δ is the sun's declination.

Substituting Equation 2.11 into Equation 2.10 gives:

$$\frac{dR_a}{dt} = \frac{\Xi_E}{d_r^2}(\sin \varphi \sin \delta) + \frac{\Xi_E}{d_r^2}(\cos \varphi \cos \delta \cos h) \tag{2.12}$$

Substituting for t its value in terms of h and integrating with respect to h over the period sunrise to sunset (with the minor approximation that δ and d_r remain constant through the day (List, 1947)), gives:

$$R_a = h\left(\frac{\Xi_E}{d_r^2 \pi} \sin \varphi \sin \delta\right) + \left(\frac{\Xi_E}{d_r^2 \pi} \cos \varphi \cos \delta \sin h\right) \tag{2.13}$$

Rearranging:

$$R_a = \frac{\Xi_E}{d_r^2 \pi}(h \sin \varphi \sin \delta + \cos \varphi \cos \delta \sin h) \tag{2.14}$$

The altitude and azimuth of the sun (List, 1947) are given by:

$$\sin a = \sin \varphi \sin \delta + \cos \varphi \cos \delta \cos h \tag{2.15}$$

where:
a is the altitude of the sun (angular elevation of the sun above the horizon) and
h is the hour angle of the sun (angular distance from the meridian of the observer).

The altitude of the sun is then a function of the latitude of the observer, the time of day (hour angle) and the date (declination). When the sun's altitude is zero, $\sin a = 0$, Equation 2.15 can be solved to give a value of the hour angle:

$$\cos h = \frac{-\sin \varphi \sin \delta}{\cos \varphi \cos \delta} = -\tan \varphi \tan \delta \tag{2.16}$$

or:

$$h = \cos^{-1}(-\tan \varphi \tan \delta) \tag{2.17}$$

or, if the latitude and declination are input in degrees, then Equation 2.18 gives the hour angle (in radians) as required for later computation:

$$h = \cos^{-1}\left(-\tan\left(\frac{\pi\varphi}{180}\right)\tan\left(\frac{\pi\delta}{180}\right)\right) \tag{2.18}$$

The incoming radiation at the top of the atmosphere at the given latitude and time of the year can then be calculated from:

$$R_a = \frac{\Xi_E}{d_r^2 \pi}\left(h \times \sin\left(\frac{\pi\varphi}{180}\right)\sin\left(\frac{\pi\delta}{180}\right) + \cos\left(\frac{\pi\varphi}{180}\right)\cos\left(\frac{\pi\delta}{180}\right)\sin h\right) \tag{2.19}$$

where:
R_a is the extraterrestrial radiation [W m^{-2}],
Ξ_E is the average solar constant [1367 W m^{-2}],
d_r is the radius vector of the earth from the sun,
δ is the solar declination [degrees],
φ is the latitude of the observation location [degrees], and
h is the sun hour angle [radians].

2.3.4   Radiation budget within the atmosphere

The complexities of determining the quantity of radiation actually received at the earth surface and that is then available to evaporate water are shown in Figure 2.8.

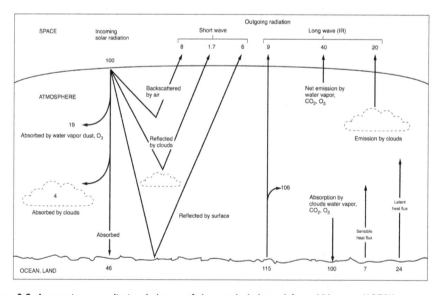

Figure 2.8 Approximate radiation balance of the earth (adapted from Wiesner (1970))

Much of the short-wave radiation that reaches the atmosphere (R_a) is dissipated before it strikes the ground surface. Some is reflected back from clouds, some is absorbed by water vapour, oxygen, ozone and carbon dioxide, some is scattered by particles. The fraction of the solar radiation which reaches the earth is called R_s. It is the sum of the direct and scattered radiation. The radiation flux can be measured by instrumentation or can be estimated from a variety of meteorological observations. A review of these methods is given below. See Smith (1991) for further details.

The coefficients in Equations 2.20 and 2.22 that follow have been derived by regression analysis, comparing measured data at specific sites with radiation estimates (Kowal and Kassam, 1978) for example. A variety of regression constants and empirical approaches to the problem are reported in the literature. This is further complicated by the use of a wide variety of units.

Estimate of incoming short-wave radiation R_c at ground surface

Incoming short-wave radiation can be estimated from measured sunshine hours according to Equation 2.20:

$$R_c = (1 - \alpha)\left(a_s + b_s \frac{n}{N}\right)R_a \qquad (2.20)$$

where:
R_c is the proportion of incoming radiation [R_a] actually received at the ground surface and available for heating the ground, photosynthesis and evaporating water,
a_s is the Angstrom 'a' coefficient – approximated to 0.19,
b_s is the Angstrom 'b' coefficient – approximated to 0.38,
n is measured hours of bright sunshine [hr],
N is total day length [hr],
α is the albedo of the surface, and
R_a is extraterrestrial radiation [W m^{-2}].

The Angstrom 'a' coefficient can be thought of as the percentage of R_a that will strike the earth's surface travelling through the clouds. The 'b' coefficient is the proportion that is impacted by cloud cover. Clearly, the choice of these coefficients will significantly alter R_c.

The total day length (N) can be calculated from Equation 2.21, where h is the sun's hour angle:

$$N = \frac{24\,h}{\pi} + 0.11 \qquad (2.21)$$

The constant 0.11 in Equation 2.21 represents an approximation to the fact that the sun begins to shine, and the earth receives radiant energy, when the top of the sun's disc appears above the horizon at sunrise – and lasts till the whole disc disappears below the horizon at sunset. The spherical geometry is calculated on the position of the centre of the sun's disc.

A proportion of the incoming short-wave radiation is reflected back to the atmosphere as a result of the reflectivity (albedo) of the surface. Values of the albedo (α) for various surfaces are given in Table 2.12 after Jury *et al.* (1991) and Kowal and Kassam (1978).

Table 2.12 Albedo values for natural surfaces

Nature of Surface	Albedo α (%)
Water	4–39
Fresh snow cover	75–95
Old wet snow	45–55
Light sand dunes	30–60
Paddocks	12–30
Forests	5–20
Dry soil during dry season	0.35
Wet bare soil	0.10
Soil Surfaces	
Dry serozem	25–30
Wet serozem	10–12
Dry clay	23
Wet clay	16
Dry chernozem	14
Wet chernozem	8
Crop Surfaces	
Fresh green tropical vegetation	0.20
Corn (New York)	23.5
Sugar cane (Hawaii)	5–18
Pineapple (Hawaii)	5–8
Potatoes (USSR)	15–25
Short green grass (Rothampstead)	25
Pine plantation	14

Estimate of long-wave radiation (R_b)

The earth radiates as a black body and loses radiation to the atmosphere. This is particularly apparent from rocks at night. The amount of black-body radiation is a function of the temperature and is reduced by cloud cover and the humidity of the air. It can be calculated (Wiesner, 1970; Smith, 1991) from:

$$R_b = \sigma T_{av}^4 (0.34 - 0.14\sqrt{e_d})\left(0.1 + 0.9\frac{n}{N}\right) \tag{2.22}$$

where:
R_b is the net long-wave radiation [W m^{-2}],
σ is the Stefan Boltzman constant,
n is the number of hours of bright sunshine,
N is the total possible hours of bright sunshine,
T_{av} is the mean air temperature [K] – the average of the daily maximum and minimum temperatures °C +273.2, and
e_d is the vapour pressure [kPa].

Net radiation

The net radiation available at the earth's surface is thus:

$$R_{Net} = R_c - R_b \tag{2.23}$$

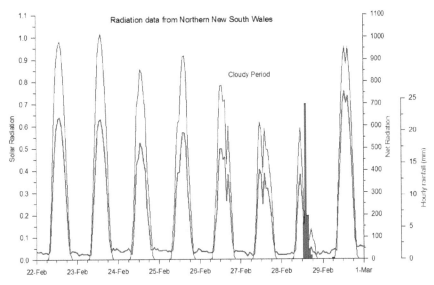

Figure 2.9 Radiation data for a climate station at Maules Creek in northern New South Wales. Data shows hourly measurements of solar radiation, net radiation and rainfall

The radiation received at the surface of the earth or the ocean heats the ground or the sea which in turn heats the overlying air. The movement of this heated air forms the wind. The movement of the heated ocean water drives the ocean currents. Part of the available energy is used by plants to transpire water. Part is also used to evaporate standing water directly to the atmosphere. The net radiation available therefore directly impacts evapotranspiration (evaporation and transpiration) and thus directly impacts the water balance. Quantification of net radiation is therefore extremely important in studies of the water balance.

The calculations shown above provide an estimate for daily net radiation calculations that formed the basis for climate studies. The basic measurements were made around the world at climate stations. In the last 20 years, automatic climate stations have become available that allow measurements to be made more widely.

An example of radiation data over a 7-day period as recorded at a climate station is shown in Figure 2.9. Note in this example that the backward long-wave radiation continues through the night as the earth cools and that the incoming short-wave radiation is limited to daylight hours. The impact of cloudiness is shown by the reduced radiation associated with the rain storm on 28 February 2008.

Ground Flux

Heat is stored in and released from the soil. The amount of the heat flux which is absorbed by the soil will depend upon the soil heat capacity. This will be greater for wet soil than for dry soil. To estimate the soil heat flux for a given period, Equation 2.24 can be used:

$$G = c_s d_s \left(\frac{T_n - T_{n-1}}{\delta t} \right) \tag{2.24}$$

where:

G is soil heat flux [MJ m^{-2}/day],

T_n is temperature [°C] on day n,

T_{n-1} is temperature [°C] in preceding day n−1,

δt is length period [n] in days,

c_s is soil heat capacity [MJm^{-3}/°C], and

d_s is estimated effective soil depth [m].

The magnitude of the soil flux can be significant if the net radiation balance is being calculated daily. It changes very rapidly after rain. Moisture increases the heat capacity substantially over a dry crusted soil.

The net radiation available at the earth's surface is thus:

$$R_{Net} = (1 - a)R_s + R_{sky} - R_{earth} - G \tag{2.25}$$

Substituting the various terms gives:

$$R_{Net} = R_a(1 - \alpha)\left(0.25 + 0.5\frac{n}{N}\right) - \sigma(T_k + 273.2)^4(0.34 - 0.14\sqrt{e_d})\left(0.1 + 0.9\frac{n}{N}\right) - G \tag{2.26}$$

with all terms as previously defined.

2.4 ATMOSPHERIC AND OCEAN CIRCULATIONS OF WATER

The earth receives more energy at low latitudes than it loses. By contrast, it loses more than it receives at high latitudes. The imbalance leads to the formation of the Hadley cell circulation shown diagrammatically in Figure 2.10. Cumulonimbus storm cells at low latitudes produce heavy rain and move the hot air from equatorial latitudes upwards into the top of the troposphere, where it flows northward and southward, cools and descends again to form the mid-latitude high pressure zones. Descending air cools and does not produce rain – leading to the mid-latitude deserts.

The general circulation of winds on the planet is the way in which the atmosphere corrects the large-scale imbalance in the distribution of heat over the planet surface. A daily update to the wind systems of the world is available from http://earthnullschool.net.

2.4.1 General circulation

The intertropical convergence zone (ITCZ) is the zone which defines the meteorological equator. It follows the seasonal shift on the sun and extends around the globe in the equatorial latitudes. Air rises from the surface in the ITCZ, often through towering cumulo-nimbus clouds, and flows towards each pole, before beginning to subside again. The ITCZ does not form a continuous belt of storm clouds. The simple circulation is shown in Figure 2.10.

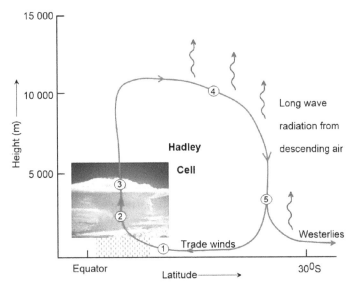

Figure 2.10 Simple 'Hadley cell' circulation (adapted from Wiesner (1970)). See text for explanation of the numbers (Photo credit: NOAA/AOML Hurricane Research Division – NOAA Photo Library fly00890)

Elements in Figure 2.10 are numbered as follows:

1   Moist trade-wind flow being drawn into the ITCZ,
2   air rising into a massive cumulo-nimbus storm cell. These clusters are easily seen on satellite imagery and are responsible for the turbulence often encountered on flights between Sydney and Los Angeles, which cross the ITCZ,
3   latent heat release in the precipitating systems pushes the rising air into the troposphere; the air at high elevations flows towards the poles,
4   cooling causes subsidence of the air at 25 to 30 °S or °N,
5   divergent flow at the base of the subsiding zone, with some flow poleward as south-westerly winds and some flow return to the equator in the trade-wind south-easterly flow.

The upward movement of air, due to heating in equatorial regions, leaves low pressure at ground surface and creates high pressure in the upper part of the atmosphere. This low pressure area is sometimes called the 'Equatorial trough.' Early workers assumed that the trough persisted around the equator as a continuous structure, but satellite imagery has shown that this is not the case.

2.4.2   The Coriolis Force

The simple circulation shown in Figure 2.10 is significantly modified by what is referred to as 'the Coriolis Effect'. In reality, this comes about simply as a result of the differential angular momentum that exists as latitude increases from the equator to the pole. Table 2.13 shows the change in angular rotation.

Table 2.13 Angular velocity as a function of latitude

Latitude	Angular Velocity km/h
Equator	1670
15	1613
30	1446
45	1181
60	835
75	432
Pole	0

The Coriolis Effect causes a deflection of poleward moving air. The air has a higher angular velocity, as it moves north or south from the equator, than the ground over which it is flowing. This has the effect of causing a relative deflection, turning the wind to the left in the southern hemisphere, and to the right in the northern hemisphere. Conversely, air moving towards the equator from the poles has a lower angular velocity than the ground surface over which it finds itself. This has the relative effect of turning the wind towards the left in the southern hemisphere and towards the right in the northern hemisphere.

The Coriolis Effect modifies the air moving towards the equator from the Hadley cell to become the south-easterly trade wind in the southern hemisphere or the north-easterly trade wind in the northern hemisphere. Air moving towards the poles from the subsiding Hadley cell circulation forms the belt of south-westerly winds in the northern hemisphere, responsible for bringing much of the rainfall to Europe, and the north-easterly winds in the southern hemisphere, which blow around the globe as the 'roaring forties.'

Acworth (1981) showed that the development of storms is related to waves in the upper atmosphere easterly wind flow. A planar flow resulted in only the development of small cumulus. Waves resulted in the rapid development of storm cells when flow was towards the equator and clear sky when flow was away from the equator.

2.4.3 Mid-latitude deserts

The subsiding air in the Hadley cell circulation dries and forms the high pressure zones which create deserts around the world at these latitudes. Australia lies in the latitude where this occurs, which is the main reason for aridity in Australia. Similar southern hemisphere deserts occur in South Africa and South America, and in the USA, North Africa and Asia in the northern hemisphere.

2.4.4 Mid-latitude pressure systems

Over much of the planet north or south of 30° latitude, there is no well-organized circulation in the lower atmosphere. The circulation is dominated by large-scale eddies (cyclonic disturbances) that intermittently transport heat and angular momentum between the tropics and the polar regions. The air moving from the Hadley cell circulation towards the poles has to lose the angular momentum acquired at lower latitudes. Table 2.13 shows that air moving from the equator to 60 degrees latitude would have a surface velocity of 800 km/h. Clearly this is not sustainable, and the angular momentum is dissipated by surface friction in the cyclonic disturbances common at these latitudes.

2.4.5   Sub-polar lows

On the poleward side of the sub-tropical highs, surface westerlies flow into regions of low pressure. In the northern hemisphere, there are two fairly stable low pressure zones, one over the Atlantic Ocean (*the Icelandic Low*) and one over the North Pacific Ocean (*the Aleutian Low*). In the southern hemisphere, there is continuous ocean at these latitudes and a continuous belt of low pressure surrounding the Antarctic continent.

Air over the poles is cooled by the surface and generally forms a high pressure zone. Movement of air towards the equator from this high pressure zone is deflected westwards. The winds leaving this circulation are called polar easterlies.

The surface westerlies meet and override the polar easterlies along the polar front. This front is well marked where there is a strong contrast in the properties of the air mass, which often occurs in the northern hemisphere as a result of the disposition of land and ocean. The warmer air overrides the cooler dense polar air to form a 'warm front,' with a clear sequence of clouds and weather. The cyclonic disturbances in the northern hemisphere have closed isobars (lines of equal pressure), and the warm front is followed by a 'cold front' where the polar air again displaces and lifts the tropical air. Closed cyclones are less common in the southern hemisphere where the junction between the two air masses is marked by a continuous succession of waves. The leading edge of the wave is a cold front, and a continuous succession of cold fronts move around the Antarctic continent. This interchange of polar air with tropical air causes the 'Southerly Buster' weather systems which occur along the east coast of Australia.

2.4.6   Typical Australian circulation

The pressure distribution over Australia is affected by the movement of the solar equator. Figure 2.11 shows the pressure distribution for a typical summer period. The high pressure off the coast of eastern Australia is feeding warm moist air into the continent. The low pressure over the north will be supporting storm cell development over the tropics, and a hurricane exists in the low pressure belt to the north-west.

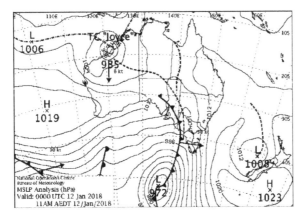

Figure 2.11 Australian summer distribution (Australian Bureau of Meteorology, Creative Commons 3.0)

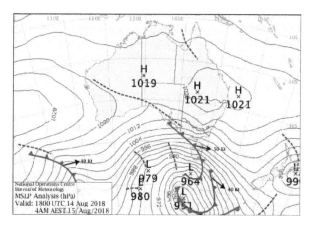

Figure 2.12 Australian winter distribution (Australian Bureau of Meteorology, Creative Commons 3.0)

Figure 2.12 shows the winter circulation. This pressure distribution shows the tropical low pressure belt back to the north of Australia and the development of polar lows and associated cold fronts to the south. The depression over Victoria and NSW will bring a cold outbreak with snow over the Snowy Mountains.

2.4.7   The monsoon

The ITCZ moves north and south with the seasons. The monsoon is caused when the flow of wind between the ocean and the land is reversed due to the movement of the ITCZ. Darwin experiences monsoon rainfall when the ITCZ is located over northern Australia during the period December to March.

A strong anti-cyclonic system builds over Asia in the northern hemisphere winter. Outflow from this system sets up a dry north-easterly flow over India and much of South-East Asia. This is the north-east monsoon. These winds pick up moisture over the oceans and feed storm cell development over northern Australia. During the wet season, there are fluctuations in the storm cell activity, related to the upper atmospheric easterly wind flow. Thunder occurs on nearly half the days from December to March, and even during breaks in the monsoon, thunderstorms are still likely in the afternoon and the night. Monsoonal rains in India and South-East Asia are much heavier in general than those over Australia.

As the ITCZ moves north, dry south-easterly winds replace the monsoon over northern Australia, and the flow over India is reversed causing the monsoon to begin there. Some of the heaviest rainfall in the world is associated with the Indian monsoon. The greatest recorded rainfall in one calendar month is 9300 mm in July 1961 at Cherrapunji in India (Crowder, 1995). A typical monsoon wind circulation for the Indian summer monsoon is shown in Figure 2.13, downloaded from the https://earth.nullschool site on 15 August 2018.

2.4.8   Hurricanes and cyclones

The most violent weather systems occur associated with hurricanes or cyclones. The term hurricane is used in the northern hemisphere and cyclone in the southern hemisphere. The

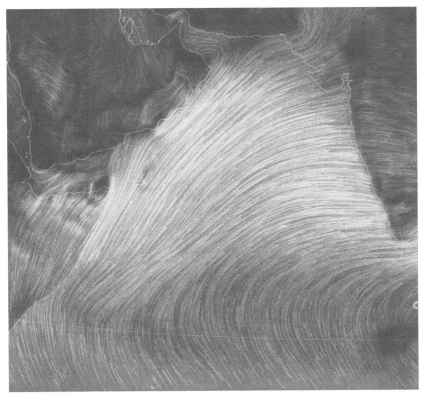

Figure 2.13 The summer monsoon winds blowing strongly over the western side of India on 15 August 2018. Image from the climate modelling site https://earth.nullschool.net. The green vectors show the direction and strength of the winds. They veer from the SE trade winds south of the equator to become SW trade winds blowing onto the west coast of India

circulation of air around these large depressions is anticlockwise in the northern hemisphere and clockwise in the southern hemisphere. This is a direct result of the Coriolis Effect. Cyclones can be responsible for very large amounts of rainfall, not to mention severe damage. Cyclones occur off northern Australia most years.

The wind scale used to compare wind strengths is the Beaufort scale shown in Table 2.14.

On average, six cyclones cross the coast in Australia each year. The highest number in one year was 16, in 1963 (Crowder, 1995). Understanding the processes within cyclones is important in that they can cause major structural damage. From a hydrological perspective, they are important for the generation of storm surges and for the heavy rainfall associated with the cyclone. As the cyclone decays overland, it frequently forms a rain depression which can also cause major flooding. The maximum rainfall associated with cyclones are shown in Table 2.15.

Table 2.14 The Beaufort wind-force scale

Scale No	Conditions at sea, far from land	Prob Wave Ht	Speed km/hr	Grade	Conditions on land
0	Sea like a mirror	0	0	Calm	Calm; smoke rises vertically
1	Scale-like ripples are formed but without foam crests	0.1	1–5	Light air	Wind direction shown by smoke-drift but not by wind-vanes
2	Small wavelets, still short but more pronounced; crests have a glassy appearance and do not break	0.2	6–11	Light breeze	Wind felt on face; leaves rustle; ordinary vanes moved by wind
3	Large wavelets; crests begin to break; foam of glassy appearance; perhaps scattered white horses	0.6	12–19	Gentle breeze	Leaves, small wigs in constant motion; wind extends a light flag
4	Small waves, becoming longer; fairly frequent white horses	1	20–28	Moderate breeze	Raises dust and loose paper; small branches are moved
5	Moderate waves, taking a more pronounced long form; many white horses are formed (chance of some spray)	2	29–38	Fresh breeze	Small trees in leaf begin to sway; crested wavelets form on inland waters
6	Large waves begin to form; the white foam crests are more extensive everywhere (probably some spray)	3	39–49	Strong breeze	Large branches in motion; whistling in telegraph wires; umbrellas hard to use
7	Sea heaps up and white foam from breaking waves begins to be blown in streaks along the direction of the wind	4	50–62	Near gale	Whole trees in motion; inconvenience felt when walking against the wind
8	Moderately high waves of greater length; edges of crests begin to break into spindrift; foam is blown in well-marked streaks along the direction of wind	5.5	62–74	Gale	Breaks twigs off trees; generally impedes progress
9	High waves; dense streaks of foam along the direction of the wind; crests of waves begin to topple, tumble and roll over; spray may affect visibility	7	75–88	Strong gale	Slight structural damage occurs (chimney-pots and slates removed)
10	Very high waves with long overhanging crests; the	9	89–103	Storm	Seldom experienced inland; trees uprooted;

(Continued)

Table 2.14   (Continued)

Scale No	Conditions at sea, far from land	Prob Wave Ht	Speed km/hr	Grade	Conditions on land
	resulting foam, in great patches, is blown in dense white streaks along the direction of the wind; the surface of the sea takes on a white appearance; rolling of the sea becomes heavy; visibility affected				considerable structural damage occurs
11	Exceptionally high waves (small and medium ships might be for a time lost to view behind the waves); the sea is completely covered with long white patches of foam lying along the direction of the wind; everywhere the edges of the wave crests are blown into froth; visibility affected	11.5	104–117	Violent storm	Very rarely experienced; accompanied by widespread damage
12	The air is filled with foam and spray; sea completely white with driving spray; visibility very seriously affected	14	over 117	Hurricane	

Table 2.15 Cyclonic rainfall maxima

Amount (mm)	Duration (hours)	Date	Location
1715	72	Feb 1983	Moolooah, Qld
1260	72	Feb 1958	Finch Hatton, Qld
960	24	Jan 1979	Bellenden Ker, Qld
927	36	Apr 1898	Whim Creek, WA
747	24	Apr 1898	Whim Creek, WA
699	48	Dec 1916	Maltoid, Qld

Coastal groundwater impacts of hurricane storm surge

One of the most significant impacts of a cyclone is on coastal groundwater. The thin lens of fresh water is destroyed by the incursion of sea water onto the land. This occurs wherever there is a major ocean surge caused by the drop in air pressure and the surge of sea water caused by the very strong onshore winds. It takes many years for the coastal processes to re-establish a stable interface between discharging fresh groundwater and sea water. The

Figure 2.14 NOAA GOES-16 satellite image of Hurricane Irma as a Category 4 storm passing the eastern end of Cuba at about 8:00 a.m. EDT on 8 September 2017 (Image Credit: NOAA/CIRA)

before and after images published by the USGS and shown in Figure 2.15 from Hurricane Ike in 2008 dramatically demonstrate the impact of the storm surge. There was a 6.8 m storm surge recorded with widespread coastal flooding.

2.4.9   Ocean circulation

The El Niño-Southern Oscillation has been blamed for many deviations in the weather. The El Niño is only a small part of the overall oceanic circulation which strongly controls weather around the globe. The fact that small changes in the oceanic circulation have such major impact merely reflects how profound the influence is. From a groundwater perspective, the El Niño determines whether a wet or dry period is likely in Australia, with groundwater recharge only probable during prolonged wet weather periods.

It is only recently that knowledge of the deep circulation of the ocean currents has been gathered. The time scale over which water in these currents moves is of the order of thousands of years. Perturbations on this major system of currents can occur. Upwelling of cold ocean water in certain locations supports major fish stock and industries. The presence of cold sea water cools the lower atmosphere, creating stable conditions where there is no possibility of generating rain-producing cumulus clouds. The deserts of Namibia and Peru are generated by this mechanism.

In normal, non-El Niño conditions, the trade winds blow towards the west across the tropical Pacific. These winds pile up surface water in the west Pacific, so that the sea surface is actually 0.5 m higher around Indonesia than adjacent to Ecuador. The sea surface

Figure 2.15 Change caused by the 6.8 m storm surge from Hurrican Ike in September 2008. The images need no further comment (USGS image)

temperature is about 8 °C higher in the west, with cool temperatures off South America due to the cool upwelling. Rainfall occurs associated with the warm air in the western Pacific, with long-term drought in the eastern Pacific.

2.5   METEOROLOGICAL MEASUREMENTS

2.5.1   Temperature

Measurement of temperature requires an instrument that has a known variation with temperature such as thermometers (known expansion rate with temperature) or thermocouples (known change in conductivity with temperature). Thermometers require a good circulation of air but must also be protected from direct sunlight or precipitation. As a result they are generally placed in a screen (Figure 2.21d). Digital temperature sensors based upon thermocouples are also widely used.

Temperature has a diurnal variation with a minimum just after sunrise and a maximum in the early afternoon. Historically, many stations had maximum and minimum thermometers to record this data.

Variation of temperature with height

The rate of change in temperature with elevation is called *the lapse rate* and varies as a function of the moisture content in the air. In the troposphere, the lapse rate is approximately 0.6 °C per 100 m.

Variation in the lapse rate of a particular mass of air will determine if the air is stable or unstable. Unstable air masses are subject to the growth of cumulo-nimbus clouds and storm rainfall.

The variation of many meteorological parameters with height is obtained by a radiosonde balloon. The balloon carries a small radio transmitter aloft with sensors which transmit measurements of pressure, temperature, and humidity as the balloon rises to the top of the troposphere.

The loss/gain of heat from an air mass causes a change in the *environmental lapse rate* for that air mass. Changes occur regularly in response to the diurnal variation in net radiation. Wiesner (1970) shows typical variations of lapse rates with surface heating and cooling during the day (Figure 2.16).

When air rises or falls without any change in moisture content (condensation), the lapse rate is known as the *dry adiabatic lapse rate (DALR)*. The air expands and cools as it rises and contracts and warms as it sinks. The DALR is 1 °C per 100 m.

Typical natural processes which result in adiabatic cooling or heating are turbulence, movement of air up or down a mountain slope and the upward movement of one air mass over another as a result of the movement of a frontal system.

If air becomes saturated with water vapour as a result of adiabatic cooling, condensation will occur accompanied by the release of heat. The release of heat means that the *saturated adiabatic lapse rate (SALR)* will be lower than the DALR. It is not possible to give an exact relationship for the SALR as the degree of cooling will be a function of the moisture content of the air. Figure 2.17 (Wiesner, 1970) shows the variation of the lapse rate with temperature.

The heating/cooling of air is only adiabatic when heat is conserved in the air mass. Once condensation occurs, and the moisture forms drops of rain which fall to the ground, heat is

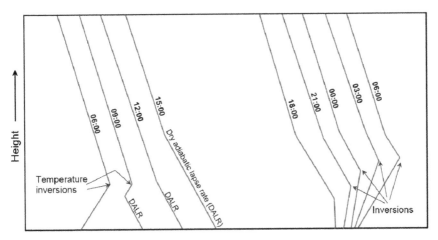

Figure 2.16 Variation of lapse rates with diurnal surface heating and cooling (adapted from Wiesner (1970))

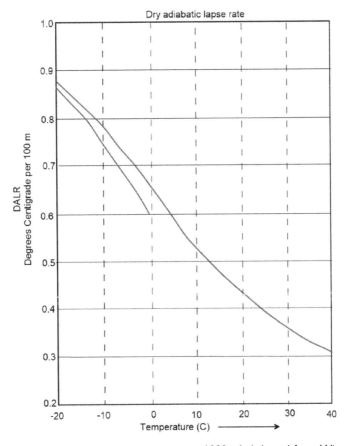

Figure 2.17 Variation of the SALR with temperature at 1000 mb (adapted from Wiesner (1970))

removed from the system. Air which sinks to lower elevations after the release of moisture warms up at the DALR, ending up at the same altitude as it started from, but substantially warmer and drier. This process is common in mountains and can cause a rapid thaw of lying snow. The principles are shown in Figure 2.18. The wind is called a Foehn wind in the European Alps but has many other names around the world.

Stability of unsaturated air

If unsaturated air has a lapse rate which is less than the DALR, it will be stable. If it is caused to rise over a hill, it will cool more than the surrounding air and will therefore sink again as soon as it can. If the air has a lapse rate greater than the DALR, it will be unstable. Wiesner (1970) gives a simple analysis which illustrates this point. Consider the case for two different air masses, each with a surface temperature of 18 °C and environmental lapse rates (ELR), determined by a radiosonde balloons, of 0.8 °C per 100 m and 1.1 °C per 100 m respectively.

If air in either air mass is caused to rise adiabatically, it will cool at the DALR of 1 °C per 100 m. The temperature of this air mass can be calculated at 100 m increments and is shown in Table 2.16 (Case 2). Also shown in Table 2.16 are the temperatures at different elevations in the two air masses, as determined by the radiosonde data (Cases 1 and 3).

Case 1 in Table 2.16 is an example of a stable air mass. The temperature in the rising bubble of air (Case 2) is always cooler than the surrounding air (Case 1) and will tend to sink down again as soon as it can. By contrast, Case 3 is an example of an unstable air mass, because this air, once it has been lifted up, is warmer than the surrounding air, and will continue to rise until moisture condenses.

The stability of saturated air can be calculated in a similar manner, by comparing the SALR with the ELR for the air mass.

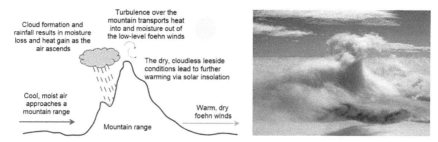

Figure 2.18 Mechanisms that create the Foehn wind and an example from the Arctic (adapted from UK Meteorological Office with photo credit: Andy Elvidge)

Table 2.16 Example lapse rate calculations

Level	Surface	100 m	300 m	400 m	400 m	Condition
Case 1	18	17.2	16.4	15.6	14.8	Stable
Case 2	18	17	16	15	14	DALR
Case 3	18	16.9	15.8	14.7	13.6	Unstable

Cloud Formation

Clouds will be formed as soon as the air mass is cooled below the dew point temperature. This occurs as the result of lifting by some mechanism. Cloud types can be used to indicate the stability, or otherwise, of an air mass. Descriptions of different cloud types can be found in Crowder (1995) and many other texts on meteorology.

The formation of clouds, and cumulo-nimbus storm cells in particular, can be predicted if the ELR of the air mass is known by the use of thermo-dynamic diagrams. Figure 2.19 (Wiesner, 1970) indicates the use of these.

Case 5 in Figure 2.19 shows an unstable ELR dotted while the ascending air particle follows the solid line. The energy available for cloud growth is indicated by the shaded area. In Case 6, the ELR is closer to the SALR and there is less energy for the formation of clouds. This situation would generate *fair weather cumulus* rather than the *cumulo-nimbus* clouds of Case 5.

2.5.2 Wind direction and speed measurement

The wind speed is measured by an anemometer as shown in Figure 2.20 and Figure 2.21b. The anemometer is usually placed at 2 m height. A vane system is used to record the direction of the wind. Both measurements are logged by a data logger as part of a climate station.

Anemometers require regular servicing and may be subject to attack by birds. Plastic ties can be attached to the equipment to deter birds from perching. These can be seen in Figure 2.21 on the beam supporting the radiation measuring sensors.

2.5.3 Net radiation measurement

A net radiometer consists of two pairs of radiation sensors. Incoming short- and long-wave radiation is measured by one pair of sensors, while outgoing long- and short-wave radiation

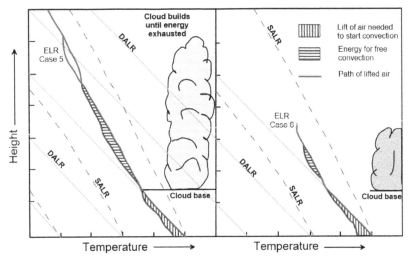

Figure 2.19 Thermo-dynamic diagram showing the formation of cloud when air is lifted – (adapted from Wiesner (1970))

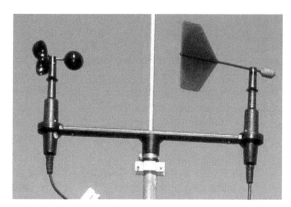

Figure 2.20 Anenometer and wind vane used to record wind speed and direction (Photo: Tony Bernardi)

Figure 2.21 Climate station at an experimental farm: (a) sensors for incoming short-wave radiation, outgoing long-wave radiation and net radiation; (b) anemometer for wind speed and direction; (c) eddy covariance measurement to determine evapotranspiration; (d) air temperature and humidity sensors; (e) short-wave radio communications; (f) on-site logging computer including atmospheric pressure measurement; (g) solar power panels and (h) laser absorption spectrometry transmitter to investigation vapour transport (evapotranspiration)

Figure 2.22 Kipp and Zonen CRN4 radiation sensor installed in a forested catchment at Wellington Caves, NSW: (a) incoming short-wave radiation sensor (pointing towards the sky); (b) outgoing short-wave radiation sensor (pointing towards the ground); (c) outgoing long-wave (far infrared [FIR]); (d) incoming FIR sensor on the upper body of the instrument (not visible in the photo); (e) ground temperature sensor; (f) cable ties attached to the beam to discourage birds from perching (Photo: Tony Bernardi)

are measured by a second pair of sensors. A radiometer installation from Kipp and Zonen is shown in Figure 2.22. The net radiation corrected for temperature is output to the climate station.

By comparison to the net radiometer, in some older climate stations, the number of hours of bright sunshine were recorded by a heliograph. This device focuses the sun's rays onto a card track which burns when the sun is not obscured by clouds. The number of hours of sunshine were then measured manually from the burn trace on the card and recorded in the daily weather log.

2.5.4 Humidity measurement

Humidity is a measure of the amount of water vapour in the air. There are many different ways of expressing humidity:

Absolute Humidity – The mass of water vapour in a unit volume of air. It is a measure of the actual water vapour content of the air.
Specific Humidity – The mass of water vapour per unit mass of air (including the water vapour). It is another measure of the actual water vapour content of the air.
Mixing Ratio – The mass of water vapour per unit mass of dry air (excluding the water vapour)

Vapour Pressure – The partial pressure of the water vapour. (Air pressure is the sum of all the partial pressures of the various gases comprising the atmosphere.)

Saturation – The air is saturated with water vapour when it holds as much water vapour as it can at that temperature.

Saturated Vapour Pressure – The water vapour pressure when the air is saturated.

Dew Point – The temperature to which air must be cooled (at constant pressure and constant water vapour content) for saturation to occur.

Frost Point – When the dew point falls below freezing, it is called the frost point.

Wet Bulb Temperature – The lowest temperature to which air can be cooled by evaporating water into it.

Relative Humidity – The ratio of the actual amount of water vapour in the air to the amount it could hold when saturated; expressed as a percentage.

Humidity is usually measured by using wet and dry bulb thermometers. These are two identical thermometers with one having its bulb covered by a wick saturated with moisture. The greater the humidity is, the less is the cooling capacity of the wet bulb thermometer by evaporation. At a given temperature, there is a maximum amount of water vapour which can be held. The space is then saturated at that temperature, and the partial pressure exerted by the water vapour is the saturation vapour pressure (e) at that temperature. As the temperature rises, more water vapour can be held. If the air is not saturated and cooling occurs, then a temperature is reached when it becomes saturated. This temperature is known as the dew point. If air is cooled below the dew point without condensation, there is a state of supersaturation, or more water vapour is held than is necessary to saturate the space (Wiesner, 1970). A humidity sensor is included with an accurate thermometer in the climate station (Figure 2.21d).

The vapour pressure of the air (e_d), or the saturation vapour pressure at the dew point (T_d), can be measured directly, or deduced from empirical equations which use the dry bulb temperature (T_a), the wet bulb temperature (T_w) and the saturation vapour pressure at the wet bulb temperature:

$$e_d = e_{a(t_{wet})} - \gamma_{asp}(T_{dry} - T_{wet})P \tag{2.27}$$

where:
γ_{asp} is 0.00066 for Assmann aspiration at 5 m/s [°C/s], or
γ_{asp} is 0.0008 for natural ventilation at 1 m/s [°C/s], or
γ_{asp} is 0.0012 for indoor ventilation at 0 m/s [°C/s],
T_{dry} is the dry bulb temperature [°C],
T_{wet} is the wet bulb temperature [°C],
P is atmospheric pressure [kPa], and
$e_{a(t_{wet})}$ is the saturation vapour pressure at the wet bulb temperature [kPa].

The relative humidity can be calculated simply from:

$$R.H. = \frac{e_d}{e_a} \times 100 \tag{2.28}$$

where e_a is calculated from:

$$e_a = 0.611\exp\left[\frac{17.27T}{T + 237.3}\right] \tag{2.29}$$

and where:
e_a is the saturation vapour pressure [kPa] and
T is air temperature [°C].

2.5.5 Pressure measurement

Atmospheric pressure is measured using an accurate transducer with one side of the silicon diaphragm sealed. The atmospheric pressure is commonly recorded inside the climate station (Figure 2.21f). The height of the station above sea level is required.

Variation of pressure with height

Atmospheric pressure falls off with height as there is less air above the ground. Equation 2.30 shows the relationship:

$$\frac{dp}{dz} = -g\rho \tag{2.30}$$

where:
p is atmospheric pressure,
ρ is density,
g is gravitational acceleration, and
z is elevation above sea level.

The variation of pressure with altitude has been specified by the International Commission for Navigation (ICAN). ICAN specified an atmosphere up to 11 km as $p_0 = 1013.2$ mb, $T_0 = 288$ K and a lapse rate of 0.65 °C per 100 m, as:

$$p = 1013.2\left(\frac{288 - 0.0065z}{288}\right)^{5.256} \tag{2.31}$$

The relationship between height, temperature and pressure for the standard ICAN atmosphere is shown in Table 2.17:

Table 2.17 Relationship between height, temperature and pressure for the ICAN atmosphere

Height(m)	Temperature °C	Pressure hPa	Density kg/m³
0	15.0	1013.2	1.225
1000	8.5	898.7	1.112
2000	2.0	794.9	1.007
4000	-11.0	616.3	0.819
8000	-37.0	355.8	0.525
12000	-56.5	192.2	0.309

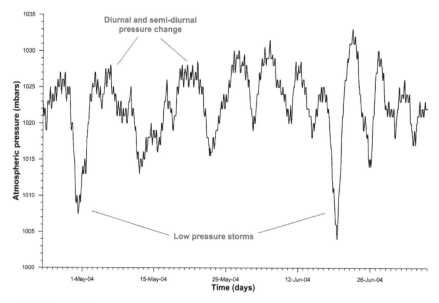

Figure 2.23 Change of atmospheric pressure with time

Variation of pressure with time

The variation of atmospheric pressure with time becomes an important factor in the study of hydraulic head variation. The time variation can best be investigated by transforming the atmospheric pressure record using a Fourier transform. This will be covered in more detail in Chapter 7, but there are two main mechanisms which become important:

- the semi-diurnal and diurnal response to atmospheric tides,
- the response to moving areas of high and low pressure related to atmospheric disturbances.

These characteristics of the variation of atmospheric pressure with time are shown in Figure 2.23.

2.6   RAINFALL

2.6.1   Rain-producing mechanisms

There are a number of rain-producing mechanisms:

Orographic – Orographic lifting occurs when air is forced to rise over a mountain range which acts as a barrier to movement of the air mass. As a result, the windward side of a mountain range is usually a region of high precipitation. Föhn winds occur on the downwind side (Figure 2.18).

Frontal – This form of lifting occurs when a warm air mass rises over a cooler air mass. The boundary between the air masses is referred to as a frontal surface.

Convective – Convective lifting occurs when the air mass is heated from below and rises through the cooler air by convection. The usual result of convective lifting is a thunderstorm, which is the most efficient process of rainfall generation.

The main difference between orographic and frontal systems and convective systems is in the distribution of the rainfall which they produce. Orographic and frontal systems, because they are approximately two-dimensional, produce a fairly well-distributed and uniform rainfall pattern. Convective storms produce rainfall in heavy bursts which have an extremely poor distribution. Rainfall in these events usually varies dramatically with distance.

2.6.2 Rainfall in Australia

There are a few rainfall stations in Australia with more than 100 years of daily rainfall recorded. The rainfall record made at the Barrabra Post Office in north-eastern NSW is one of these, with almost 140 years of data. The mean annual rainfall is 687 mm with a standard deviation of 172 mm. The annual rainfall data is shown in Figure 2.24. Deviation from the mean is also shown that indicates significant variation within the period of the record. A major change in annual rainfall occurred in 1947, preceded by almost 50 years of drier than average conditions. Wetter than average conditions were maintained for the next 50 years, after which average conditions have occurred for the last 20 years. This 50-year cycle was also noted by Ransic *et al.* (2007).

The use of groundwater in Australia is inextricably related to availability of rainfall and degree of aridity of climate. The pattern of rainfall for the last summer wet season is shown in Figure 2.25 and for the last winter season in Figure 2.26.

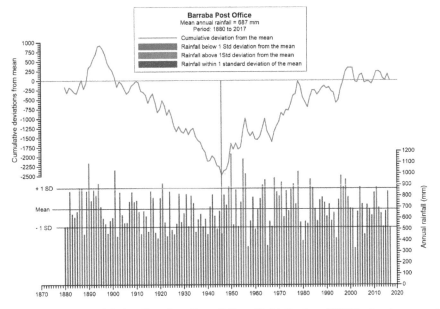

Figure 2.24 Annual rainfall for Barrabra Post Office (BoM Station 54003) showing cumulative departures from the mean rainfall. A switch between drier than average to wetter than average occurs in 1947

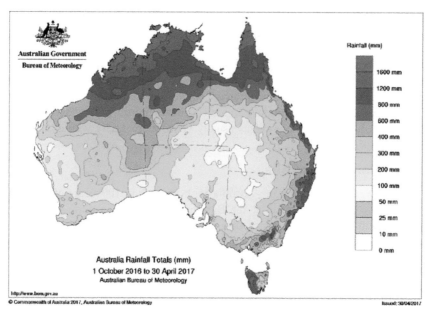

Figure 2.25 Total rainfall for the 2017 Northern wet season − 1 October, 2016 to 30 April 2017
(Creative Commons Attribution Australia license)

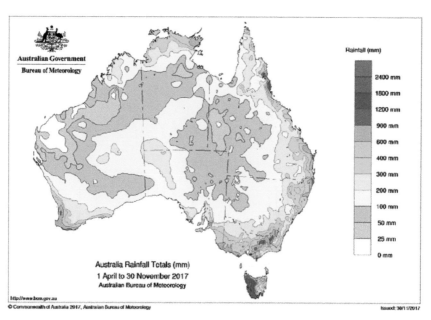

Figure 2.26 Total rainfall for the 2017 southern wet season − 1 April 2016 to 30 September 2017
(Creative Commons Attribution Australia license)

With climate change now accepted as a present threat, it is worth pointing out that the long-term average rainfall taken by the Bureau of Meteorology (BOM) data is for a running average of the last 30 years. The 30-year period from 1970 to 2000 is one of the wettest periods in the last 200 years in eastern Australia and possibly should not be used for long-term planning processes!

The total annual surface runoff from the continent is only about 440,000,000 cubic metres per year (Lau *et al.*, 1987), of which the majority discharges to the Gulf of Carpentaria and into the sea around Tasmania. About one-third of Australia has uncoordinated drainage with no surface runoff.

2.6.3   Rain gauges

The amount of precipitation is usually measured by a rain gauge (or pluviometer). The simplest rain gauge is a daily read gauge (usually read at the same time each day – typically 09:00 in Australia). The gauge consists of an accurately machined funnel which is emptied into a calibrated flask. A manual reading of the water level in the flask gives a measure of rainfall in the previous 24 hours.

An alternative type of gauge is the pluviometer; pluviometers are continuously recording gauges. The tipping bucket rain gauge (or pluviometer) shown in Figure 2.27 records the time at which a bucket of a given size (usually 0.2 or 0.5 mm) has been filled by rainfall. The time and date for each tip are recorded. Daily rainfall is then integrated across the 24-hour period.

2.6.4   Measurement of snow fall

The measurement of snow fall in a standard rain gauge is subject to considerable error due to turbulence around the gauge and the tendency for snow to accumulate in drifts. The snow

Figure 2.27 Tipping bucket rain gauge mechanism – with surrounding bucket removed (Photo: Ian Acworth)

Figure 2.28 Snow gauge in the Snowy Mountains (Photo: Ian Acworth)

that is caught in the gauge is melted and the *water equivalent* reported (Figure 2.28). A water content of 10% of the snow depth provides a rough approximation to the water equivalent, although the density of freshly fallen snow may be much less than this.

In the Snowy Mountains of south-eastern Australia, the accumulation of snow that melts in the spring is the primary source of replenishment for the reservoirs, which supply irrigation water to the agricultural areas in the Murray Basin. The same process is common in many other areas of the world, and an accurate assessment of snow depth before the spring melt is used by hydrologists to assist with management of water allocation for the following irrigation season.

A thick accumulation of snow can also mean a high flood potential, particularly if the snow is caused to melt rapidly by warm rainfall. Melting snow also efficiently recharges soil moisture and underlying aquifers. The slow release of water means that a minimum of water is lost to runoff.

Snow surveys are made periodically through the winter to measure the thickness and density of the accumulated snow. A thin-walled tube with a sharp leading edge is pushed through the snow pack to the ground surface. The tube is then withdrawn and weighed. The weight of the empty tube is subtracted to give the mass of the accumulated snow. A snow survey requires measurements along a traverse which are repeated at regular intervals throughout the winter.

The Bureau of Meteorology reports a regular snow depth at several locations throughout the Snowy Mountains using automated equipment. A gauge at Spencers Creek, out from Charlotts Pass, provides a standard monitoring point for the Snowy River catchment.

The snow pack survey, when combined with an estimate of the extent of snow cover, is used to assess the water content which is held in storage in the mountains. Melting of the snow pack can only begin when the snow temperature has risen to 0 °C. Initial melt water clings to snow granules and is held by surface tension, so that at least 2% to 8% of the snow pack must melt before runoff begins. The slow melt of a thick snow pack can be a very

Table 2.18 WMO recommended minimum rain gauge density

Type of Region	Normal tolerance (area for one station) (k^2m)	Maximum recommended area for 1 station (k^2m)
Flat areas of temperate, Mediterranean and tropical zones	600–900	900–3000
Mountainous regions of temperate, Mediterranean and tropical zones	100–250	200–2000
Small mountainous islands with irregular precipitation	25	
Arid and polar zones	1500–10000	

effective source of groundwater recharge. The rate of melting does not exceed the vertical hydraulic conductivity of the soil so that runoff is not generated.

Lack of access to many mountainous areas means that a measurement of snow depth is not possible. Remote sensing methods are increasingly in use to give an estimate of snow mass. Airborne or spaceborne radar platforms provide a good estimate. This method makes use of the contrast in dielectric properties between bare ground, water and snow.

2.6.5 Minimum density of precipitation gauge network

The World Meteorological Organisation (WMO) have set minimum densities for rain gauge networks if they are to provide a satisfactory basis for hydrology. These are shown in Table 2.18 after Wiesner (1970). Optimum networks are extensive and should allow accurate prediction of average and extreme precipitation in all areas covered by the network. As rainfall is a relatively easy parameter to measure and extensive long-term records exist, much work has been carried out on the statistical analysis of rainfall data and the methods of extrapolation available to cover variation in space and time.

2.6.6 Data presentation

Generally, the analysis of rainfall data requires the consideration of information obtained from rain gauges. These data can be presented in two forms:

Mass Curve – A plot of cumulative rainfall depth as a function of time. The mass curve is the integral of the hyetograph, or the hyetograph is the differential of the mass curve. The instantaneous rainfall intensity is given by the slope of the mass curve. Figure 2.29 shows hyetographs for a number of rainfall stations during a storm event at Fowlers Gap, western NSW.

Hyetograph – The hyetograph is a plot of incremental rainfall depth as a function of time. It is generally shown as a histogram and gives the depth of rainfall over a given time period.

Residual mass curves

In assessing rainfall data, particularly rainfall data over a number of years, it is necessary to check the consistency of the data. This need arises from changes which are rarely, if ever,

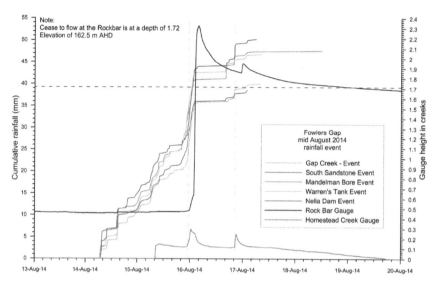

Figure 2.29 Hyetographs for a rainfall event at Fowers Gap, western NSW. Also shown is a stream gauge in the same catchment

published but which will influence and amend the records obtained from the gauge. Typical changes which may occur include:

- changes in the gauge location;
- changes in instrumentation; and
- changes in observational procedure.

The usual test for consistency of data records is the *double mass curve* analysis. This procedure can also be used to test the consistency of other parameters such as evaporation with equal validity. The basis of a double mass analysis is the comparison of a record at one gauge with concurrent records at one or more adjacent gauges. Typically accumulated annual rainfalls are considered, although seasonal or monthly accumulated rainfalls may also be considered. Figure 2.30 shows rainfall stations in the Peel catchment of northern New South Wales being tested for consistency by examining the linear relationship of the curves.

Residual mass curves are used to demonstrate departures from mean annual rainfall. The amplitude of the residual mass curve indicates the extent of the deficiency below, or surplus above, the mean. A small amplitude indicates low variability of rainfall. A residual mass curve for Barabra in northern NSW is shown in Figure 2.24. The data clearly shows a major shift in climate occurring in 1947. This shift is seen in a large number of plots for stations throughout the Tablelands of NSW (Rančić *et al.*, 2009).

Spatial variation

Rainfall over a catchment will vary due to a number of features, such as catchment topography, storm movement, etc. As a result, not all locations within the catchment will receive the same depth of rainfall (or intensity). Analysis of many hydrological phenomena require

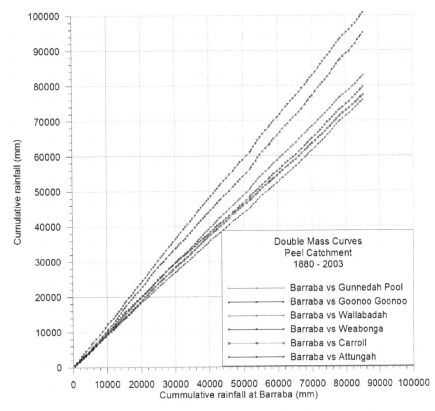

Figure 2.30 Double mass curves for rainfall stations in the Peel catchment, northern NSW

estimates of the average depth over a catchment, or require estimates of the rainfall depth (and pattern) on individual sub-catchments within the catchment area. There are three main methods by which this required information can be obtained (Fetter, 2001):

1   arithmetic mean of all stations;
2   Thiessen polygon method; and
3   isohyetal methods;

Arithmetic mean

In using this method, all records within the catchment area are averaged, stations outside the catchment are discounted and the average rainfall over the area is simply determined as the mean of the gauges within the catchment. This is clearly only a first approximation and should not be used other than for a very rough analysis.

Thiessen polygon method

The polygon method is based on the subdivision of the catchment according to areas closest to a particular gauge, i.e., the area enclosed within each sub-catchment boundary is closer to the gauge within that sub-catchment than to any other gauge within, or close proximity to, the total catchment area. Each sub-catchment area is then assigned a rainfall depth equal to that measured at the enclosed gauge.

The procedure for determining the Thiessen weights (the weighting given to each gauge) is:

1 triangulate between adjacent stations;
2 draw perpendicular bisectors of the lines drawn in the previous step;
3 extend the bisectors until intercepted by another bisector – this forms the polygon; and
4 measure the area within the polygon and divide by the total catchment area.

This method is essentially graphical and was extensively used before the availability of computing resources

Isohyetal method

Isohyets are contours of equal rainfall depth. Hence, determining the average depth of rainfall using the isohyetal method is based on determining contours of equal rainfall depth and undertaking a volumetric determination of the total rainfall over the catchment divided by the catchment area.

The isohyetal method is potentially the most accurate method but is also the most laborious if more than one storm is to be analyzed. This labour requirement results from each storm event having a different isohyetal pattern. In general, subjective interpolation is required between gauges.

Computer packages (e.g., SURFER by Golden Software) can be used to calculate the volume of rain falling over a catchment if sufficient rain gauge data is available.

An illustration of the isoheytal approach is shown in Figure 2.31 using data from the semi-arid research station at Fowlers Gap in western New South Wales.

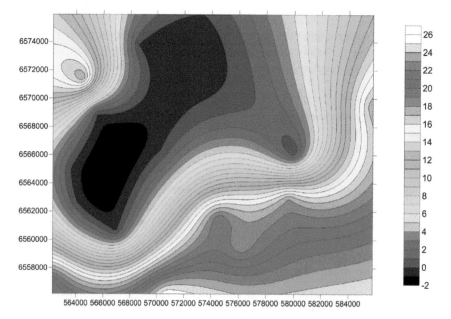

Figure 2.31 Rainfall at Fowlers Gap on 12 August 2003 showing rainfall contours (isohytes)

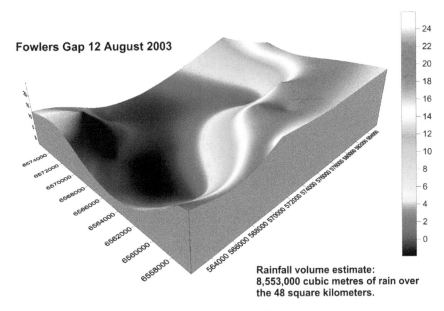

Rainfall volume estimate:
8,553,000 cubic metres of rain over
the 48 square kilometers.

Figure 2.32 Volume of rainfall over the same storm period

2.7   STREAMFLOW MEASUREMENT

Accurate calculation of a water budget must account for water entering and leaving an area of interest. This is a reasonably straightforward procedure for a river that flows inside a well-defined channel and is perennial (runs all year), but it becomes far more complicated for rivers that do not flow continuously (intermittent or ephemeral) and occupy braided channels.

It is extremely difficult to accurately gauge these ephemeral (responding only to rainfall) flow events, yet these are a major factor in the water balance for that area as the river loses water continuously into the ground until, in many cases, the surface flow stops. Ephemeral flow events of this type characterize the drier parts of Australia. Fowlers Gap Creek in western NSW has been equipped with a video camera and stream gauge measuring equipment at a rockbar constriction to the channel. The video record for two successive days is shown in Figures 2.33 and 2.34. The output from the bubbler sensor established in the stream bed upstream of the rockbar is shown in Figure 2.35. The cross-section profile is shown in Figure 2.36.

2.7.1   Stage measurements

Stage is usually used as an index for discharge calculation or to give levels in a reservoir. The measurements are of basically the same parameter but have different applications.

The simplest measurement of river stage in a flowing river is to use a staff gauge that has been surveyed into a local benchmark. The staff gauge is similar to a surveying staff gauge but installed at a fixed location that can be read from some distance away (above the flood).

Figure 2.33 Fowlers Gap Creek gauging site on 9 January 2015

Figure 2.34 Fowlers Gap Creek gauging site on 10 January 2015

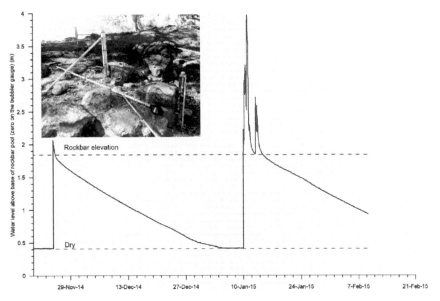

Figure 2.35 Record from the level gauge in Fowlers Gap Creek at the rockbar gauging station. Inset shows the location of the bubbler gauge in the creek bed

Figure 2.36 View of the gauging site at the rockbar (Photo: Ian Acworth)

Figure 2.37 Do not camp or park overnight in a wadi! A flood can occur very rapidly and with almost no warning. This lorry became bogged in dry sand in the river bed the day before. No rain fell at the site overnight! (Photo: Ian Acworth.)

The staff gauge must be constructed from hard and durable material that is also resistant to corrosion, rot and mechanical damage. The essential measurement infrastructure required at a stage measurement site is shown in Figure 2.36.

A river stage board will normally only be read once a day and will not therefore necessarily record the maximum flood that passes by. Various forms of crest gauge are used to overcome this. The crest gauge will record the maximum height of water in the river. The simplest form of crest gauge comprises a vertical tube with a small hole at the base. Inside the tube is a removable measuring stick and some finely granulated cork. Some cork will adhere to the measuring stick and will indicate the maximum river stage achieved. The time of the event is lost.

An alternative method is to use a graduated tape stuck to the staff gauge and impregnated with a soluble dye. The dye will be dissolved by the river water – leaving a reliable tell tale for the top water level reached.

Analysis of major flood events can often be carried out by observing the maximum height at which floating debris is lodged in vegetation. The elevation of these tell tale remnants (Figure 2.38) can be surveyed long after the flood has passed and allow an estimate of the flood peak.

2.7.2 Recording stations

Non-recording stations are manually intensive and not suited to uninhabited areas. They also give very poor resolution of the variation in river stage with time. Typically, some form of automated station is required. The traditional method of recording is to construct a stilling well that is in open contact with the river through a tube that is not so narrow

Figure 2.38 Flotsam caught on top of the borehole structure during the storm (Photo: Gary Dowling)

Figure 2.39 The flood level line inside the instrument box at the same site (Photo: Gary Dowling)

that it damps the measurement of river stage change but not so wide that material gets lodged in it during a flood event. A float is installed in the stilling well, suspended over a pulley and held in position by a counter weight.

As the river stage rises or falls, the pulley rotates. This rotation can be used to drive a pen that writes a continuous line on a chart. The chart is caused to move by a clockwork mechanism that may rotate the chart once a day or once a week.

Figure 2.40 Recording station with a chart recorder or logger installed in the shed that has a head pipe led to it from immediately upstream of the weir (Figure 2.41). (Photo: Ian Acworth.)

Figure 2.41 A compound weir established in a stream bed with a v-notch set in the middle to record low-flow stage height. The structure includes a second much wider weir to collect flood data as well (Photo: Ian Acworth)

Chart recorders have been in widespread use around the world – but suffer from the following:

- pens failing and/or the ink drying out, which is a particular problem in arid areas;
- failure to change the chart at the appropriate time so that the record writes over itself, making accurate reading sometimes difficult;

- the charts have to be read manually. For this reason there remain thousands of unread charts in offices around the world, waiting for staff time to be made available to digitize the chart records.

The simple mechanical float-driven chart recorder was replaced by a bubbler system in many countries. In this system, compressed air is fed to a tube that terminates beneath the lowest river level. A hole in the base of the tube allows the air to bubble slowly out. The pressure in the tube is sensed using a transducer. The pressure is directly related to the weight of water that must be overcome to allow bubbles to escape from the base of the tube.

Chart recorders are being increasingly replaced by data loggers. One simple adaptation of the chart recorder is to replace the pen and chart with an optical/magnetic shaft encoder attached to the pulley. The encoder generates an electrical pulse every time the pulley shaft rotates. Gears can be used to connect a smaller cog to the pulley that will rotate faster and produce a more complete record of the river stage change. Alternatively, the movement of the pulley can be converted to an electrical resistance using a potentiometer. The voltage from the potentiometer can then be recorded.

Data loggers can record the time of the pulse and store the record in memory. Removable memory cards mean that the records can be returned to base for direct reading onto a computer data base.

A disadvantage to digital recording of the data is the requirement for electrical power. This can be sourced from the mains, by replaceable battery packs or, more frequently in a sunny climate, from solar cell recharge to a battery bank.

Pressure transducers are also used to record the height of water above a reference point and do not require the installation of a stilling well. A pressure transducer works by converting the deformation of a thin silicon diaphragm into an electrical signal. The silicon diaphragm is installed in a submersible mounting and fixed at a known depth below minimum water level in the river. As the river stage rises/falls, the diaphragm distorts in response to the varying pressure of the water. Data loggers can then be used to calibrate the pressure change in terms of water level change and record the river stage. Two recording methods are available, either:

- the logger can record the stage at a regular time interval (minutes or hours) or
- the logger can be programmed to continuously check the river stage and to record the time at which the stage changes by a certain amount (typically 10 mm). In this way, the volume of data recorded is much reduced, but slightly more sophisticated processing of the data is required to achieve a record of river stage against time.

2.7.3   Discharge measurement methods

A number of techniques exist for estimating the flow of water past a gauging station. These include:

- volumetric measurement
- dilution gauging
- current meters
- float methods
- hydraulic structures

- water wheels
- acoustic dopler methods

Volumetric measurements

This is the simplest method of making a volumetric measurement but can only be used for low discharges (typically < 50 *L/s*). This can be used to monitor small spring discharges and minor streams. It is necessary to divert the complete flow into a measuring vessel. All that is then required is the time taken to fill the vessel.

Dilution gauging

If a known amount of a conservative tracer is added to a river and turbulent mixing occurs, then the concentration of the tracer can be measured and the discharge calculated. A tracer that is not already in the river must be used, and there must be no chemical reaction between the tracer and the river water. In many countries, sodium chloride is an appropriate tracer and the concentration can be measured by using a fluid electrical conductivity metre. It is also possible to use a coloured dye if this can be safely added to the water.

Subject to the constraints outlined above, two different techniques exist:

1 constant injection and
2 bulk injection.

For the addition of a tracer at a constant rate, the discharge is simply:

$$c = \frac{i}{q+i} \qquad (2.32)$$

where:
q is discharge rate,
i is the injection rate (in the same measurement units as q), and
c is the measured concentration.

Solving for the discharge rate gives:

$$q = \frac{i(1-c)}{c} \qquad (2.33)$$

If a known amount of tracer is suddenly added to the river, then, providing that all the tracer passes the measuring point, the following holds:

$$\int_0^T i(t)dt\, c = I \qquad (2.34)$$

where:
I is the volume of tracer,
T is the time taken for all the tracer to pass the measuring point, and
i(t) is the time dependent tracer discharge at the measuring point.

If $i \ll q$ then:

$$q = \frac{I}{\displaystyle\int_0^T c(t)\, dt} \qquad (2.35)$$

where:
$c(t)$ is the measured concentration.

The bulk injection method is particularly suitable for fast-flowing turbulent rivers where other methods, such as gauging with a current metre, would be difficult.

Current meters

The basic principle in the use of current meters is to measure the flux of water through a river cross-section by determining the velocity of the water at a large number of points in the cross-section. The most commonly used flow meters are the propeller type velocity meters. The current meter can be operated from a cable way, bridge or boat or by wading in the river if it is shallow enough.

The cross-section is defined by a wire with distance markings stretched over the river. The depth to the river bed along the section is determined by probing the bottom, typically at low water when the drag on the cable used to sound the bottom is least.

The discharge is calculated from the mean velocities in vertical cross-sections, usually at 15 to 30 evenly spaced points across the river (Boiten, 2000). A cable way is used to suspend a flow meter for large rivers.

Water flowing in a river is subject to friction as it comes into contact with the bed and sides of the channel. As a result, the fastest water moves through the centre of the channel at the surface. In a standard turbulent velocity profile, the velocity at 0.4 times the water depth above the river bed represents the mean velocity of the river water. A typical velocity profile is shown in Figure 2.42. If only one measurement is taken, then a depth of 0.4 would be chosen. More typically, as many as five velocity measurements are made at the same distance from the bank and at different depths. Close to the bottom, 0.2, 0.4, 0.8 times water depth and close to the surface are appropriate intervals, but more could be selected for deep flowing rivers. If only two measurement depths are chosen, then 0.2 and 0.8 times the water depth would be appropriate as the average of these two is close to the vertical mean (Figure 2.42). It is necessary to wait some time to get an average velocity so that the impact of turbulence is accounted for.

Calculation of flows is then carried out using the following relationships.

The one-point method:

$$v_{av} = v_{0.6} \qquad (2.36)$$

where:
v_{av} is the mean water velocity,
$v_{0.6}$ is the measured velocity at 0.6 d from the water surface, and
d is the stream depth.

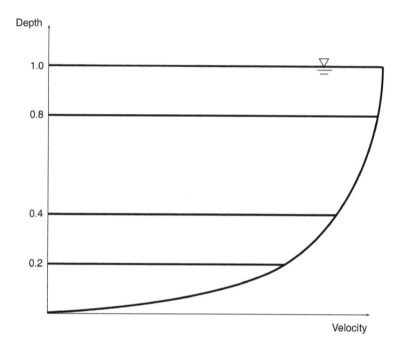

Figure 2.42 Vertical velocity profile in a typical turbulent river (adapted from Wiesner (1970))

The two-point method:

$$v_{av} = 0.5(v_{0.2} + v_{0.8}) \tag{2.37}$$

$v_{0.2}$ is the measured velocity at 0.2 d from the water surface,
$v_{0.8}$ is the measured velocity at 0.8 d from the water surface, and
d is the stream depth.

The five-point method:

$$v_{av} = 0.1(v_s + 3.0v_{0.2} + 2.0v_{0.6} + 3.0v_{0.8} + v_b) \tag{2.38}$$

v_s is the measured velocity at the water surface and
v_b is the measured velocity at the bottom.

Once the mean velocity in each section is obtained, normally by numerical or graphical integration, the total discharge is obtained by weighting each vertical velocity by a representative area. Two methods are in use:

1   the mean section method, or
2   the mid-section area method.

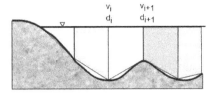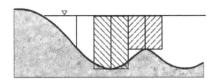

Figure 2.43 Mean section method and mid-section area method (adapted from Wiesner (1970))

THE MEAN SECTION METHOD

In the mean section method, discharge is calculated for the trapezoidal segments between the verticals, using the mean velocity in the confining vertical as a representative velocity. The flow within a trapezoidal segment is then:

$$Q_{panel} = \left(\frac{v_{av}^i + v_{av}^{i+1}}{2}\right) \times \left(\frac{d_i + d_{i+1}}{2}\right) \times b \tag{2.39}$$

The procedure is illustrated in Figure 2.43. For panels that lie against the banks, the representative velocity of the panel can be calculated as $2/3v_{av}$. As the flow towards the bank is parabolic, to use a value of $1/2v_{av}$ would underestimate the velocity.

THE MID-SECTION AREA METHOD

The mid-section area method calculates discharge in a rectangular segment centered on a vertical where the measurements have been carried out. The mean velocity at the vertical times the cross-sectional area is used as shown in Figure 2.43. The flow through a panel is then:

$$Q_{panel} = v_{av}^i d_i \times \frac{1}{2}(b_i + b_{i+1}) \tag{2.40}$$

Float method

The very simplest method of measuring the river velocity is to measure the time taken for a float to move a set distance. Remembering that the mean velocity of the channel will be less than the surface velocity, the speed of the float is multiplied by an empirical correction factor (typically 0.7 to 0.8) to derive the mean river velocity.

Hydraulic structures

Under given conditions, hydraulic structures have known stage/discharge relationships. They can also provide a stable hydraulic control in the channel. There are many different designs for hydraulic structures, and the criteria for choice of design include:

* rating curve relationships,
* capacity/cost relationships,
* available head difference, possibility of raising the head unintentionally upstream and causing flooding,

- the sediment load in the river, and
- fish migration through the structure.

The two main categories installed are:

- raised structures with a full hydraulic jump (sub-critical to critical/supercritical flow) and
- submerged structures with a partial jump (critical/supercritical to sub-critical flow).

The first category includes the standard 90°, 120° V-notch and rectangular weirs used for small flow measurement, while the second is the Crump weir installed in a river bed. Both weir designs involve significant work in the channel bed and may be costly to install.

V-NOTCH WEIRS

An empirical (not subject to dimensional analysis) relationship for 90° weirs is (Fetter, 2001):

$$Q = 1.370H^{5/2} \tag{2.41}$$

where:
Q is the discharge in m^3/s and
H is the height of the backwater above the weir crest in metres.

The weir plate must be accurately machined and maintained such that the edge of the weir is clean and sharp. V-notch weirs are suitable for low-flow measurement but cause too much back up of water if used for large flows.

RECTANGULAR WEIRS

Rectangular weirs can accommodate larger flow rates without causing as much back water build up. The flow rate through a rectangular weir can be determined (Fetter, 2001):

$$Q = 1.84(L - 0.2H)H^{2/3} \tag{2.42}$$

where:
Q is the discharge in m^3/s,
L is the length of the weir crest in metres, and
H is the height of the backwater above the weir crest in metres.

In some areas the range of flow is very great and a combination weir is required with a V notch in the base.

Water wheels

A commonly used device in the irrigation industry is to install a water wheel in the irrigation channel, as shown in Figure 2.44.

Accoustic doppler profile systems

Highly accurate acoustic doppler profiling (ADP) systems are now available (SonTek Riversurveyor M9) specifically designed to measure river discharge, three-dimensional water

Figure 2.44 Water wheel gauge in an irrigation channel (Photo: Ian Acworth)

Figure 2.45 Preparing a SonTek M9 for river gauging (Photo: Will Glamore)

currents, depths, and bathymetry from a moving or stationary vessel. Continuous GPS data allow the exact loaction of the measurements. Real-time interpretation provides the operator with rapid data – in contrast to the somewhat laborious calculations of simple manual systems. The systems are ideal for stable river channels with considerable flow where the instrument can be towed behind a boat (Figure 2.45 and Figure 2.46).

Figure 2.46 Floating riversurveyor (SonTek M9) towed behind a boat (Photo: Will Glamore)

Unfortunately, these systems do not work well if left logging in the semi-arid zone where the channel dries out. Under these conditions, it is necessary to rely on stage measurements and the assembly of a rating curve. This makes the estimate of flood flow where over-bank conditions have developed still very difficult.

Manning equation

In open-channel hydraulics, the average velocity of flow can be found from the Manning equation, *viz.*:

$$V = \frac{1}{n} R^{2/3} S^{1/2} \tag{2.43}$$

where:
V is the average velocity (m/s),
R is the hydraulic radius – the ratio of the cross-sectional area (m^2) divided by the wetted perimeter (m),
S is the energy gradient, which approximates to the slope of the water surface, and
n is the Manning equation roughness coefficient.

The flow velocity depends upon the amount of friction between the water and the stream channel. Smoother channels have less friction and, hence, faster flow. Channel roughness contributes to turbulence, which dissipates energy and thus reduces flow velocity. Values of Manning's coefficient which are shown in Table 2.19 (Fetter, 2001). The flow rate of the river channel is obtained by multiplying the average velocity by the cross-sectional area of the river channel. Note that the hydraulic radius and the gradient are functions of the head

Table 2.19 Roughness coefficients for the Manning equation

Roughness characteristic	Manning Coefficient
Mountain stream with rocky bed	0.04–0.05
Winding natural stream with weeds	0.035
Natural stream with little vegetation	0.025
Straight, unlined earth channel	0.020
Smooth concrete channel	0.012

and that the roughness can change with vegetation growth or erosion/deposition in the river channel.

2.7.4 Rating curves

At each location where a river stage is measured, further detailed flow measurements are carried out to estimate the volume of river discharge at each height of the river. A plot of discharge against stage height is known as a *rating curve* for the river at that location. The rating curve can be used to turn measurements of river stage into river flow and requires measurements of river discharge at a number of different river stages.

It is imperative that the stage discharge relationship be determined over the widest possible range of flow conditions. Measurements at low to medium flow are often well constrained by repeated gauging; however, gauging at high flow is often difficult to achieve due to a lack of access. The rating curve becomes asymtopic to the x axis (flow) at high river stage and therefore inaccuracy in the stage has a major impact on discharge at high flow rates. The river gauging station where the river stage is measured should be selected in a section of the river where the flood is constrained to the river channel. This is typically not a problem in humid regions where rivers occupy permanent well-defined channels – but it becomes a severe problem downstream in flat and/or arid areas where braided rivers exist. A braided river may flow in one of several channels and will typically change the channel with little warning. A major flood will occupy the whole valley, but smaller floods may occupy one of several channels.

2.8 EVAPORATION MEASUREMENT

2.8.1 Evaporation pans

The US Weather Bureau Class A evaporation pans (shown in Figure 2.47) are probably the instruments most widely used the world over to measure evaporation from an open water surface. However, a problem with pan data is that there are several types of pans in use all over the world. For example, until the late 1960s a different type of evaporation pan (the sunken pan – one that was submerged in the ground) was in use in Australia. Although it was changed to the Class A pan, some of the old sunken pans are still the only source of evaporation data in many parts of Australia.

The Class A pan is 122 cm in diameter, 25 cm deep and constructed from galvanized metal. The pan is placed in an open area and fenced to stop animals drinking from it.

Figure 2.47 Evaporation pan on a dusty day in Bauchi, Nigeria (Photo: Ian Acworth)

The level in the pan is maintained constant by adding or subtracting water from the pan daily. The evaporation, less rainfall, is calculated by a simple water balance. The surface of the pan is left open as the loss of water to birds was found to cause less error than that caused by covering the pan by a grill. The pan heats up more rapidly than the surrounding ground and this causes an overestimate of evaporation. On a free water surface such as a lake or reservoir, a correction factor of 0.8 is used, although this will vary depending upon the salinity of the water.

To compensate for the limited coverage of Class A evaporation pans in Australia, there has been considerable research in predicting potential and actual evaporation using remote sensing data and long-term rainfall and runoff data. Maps showing evapotranspiration (ET) have been published by Wang *et al.* (2000) and can be accessed from the web at http://www.bom.gov.au/climate/.

2.8.2   Measurement by Lysimeter

A lysimeter is a device in which a given volume of soil may be planted with vegetation and hydrologically sealed such that water leakage from the controlled system is negligible. It is important that an undisturbed sample of the soil and vegetation is placed in the lysimeter. Change in weight of the soil volume is then monitored and interpreted as a change in water content indicating evapotranspiration loss. Rainfall is recorded close by to estimate recharge – seen as a weight gain (Pruitt and Lourence, 1985).

2.8.3   Regional evaporation estimates

The BOM adapted an areal formulation for aerial potential ET proposed by Morton (1983) that used the Priestly-Taylor equation with modifications to allow for advection. They also

Table 2.20 Mean monthly distribution of rainfall and evaporation for Sydney

Month	Mean Rainfall (mm)	Mean Evaporation (mm)
January	102	213
February	114	179
March	129	138
April	131	90
May	123	62
June	131	41
July	106	47
August	81	65
September	71	91
October	81	127
November	78	153
December	79	192
Total	1226	1398

then used long-term rainfall and runoff data from 77 catchments to obtain by water balance methods, reference estimates of actual ET.

The importance of evaporation can be seen from the data in Table 2.20, which lists monthly average rainfall and evaporation for Sydney.

The water loss from a pan is higher than that which would occur from a surface covered with crops or even from a much larger open water surface. Pan coefficients are used to estimate ET_O from pan data:

$$ET_o = k_{pan} \times E_{pan} \tag{2.44}$$

where:

k_{pan} is the pan coefficient and
E_{pan} is the amount of water lost to evaporation from the pan – measured in the early morning.

2.9 ESTIMATES OF EVAPORATION FROM OPEN WATER

Evaporation occurs from open water surfaces. ET is a large component of the water balance, accounting for a large proportion of the total rain falling on the Australian continent. Penman (1949) developed an equation that allows prediction of open water evaporation.

The atmosphere has a fundamental capacity to evaporate water from the surface. Two processes are involved:

- the radiation input to the ground surface and
- the advection of moisture away from the point of evaporation by the wind.

These processes operate independently from each other. A dry warm wind at night is capable of drying clothing, whereas sunny weather may not achieve drying if the humidity is high. The open water evaporation can be calculated using measurements of standard meteorological variables (Smith, 1991).

2.9.1 Open water evaporation (E)

$$E = \frac{0.408\Delta(R_n - G) + \gamma\frac{900}{T+273}(U_2(e_s - e_d))}{\Delta + \gamma(1 + 0.34U_2)} \tag{2.45}$$

where:

E is open water evaporation [mm/day],
R_n is net radiation at the surface [MJ m^{-2}/day],
G is solar heat flux [MJ m^{-2}/day] – is zero as a first approximation,
T is average temperature [°C],
U_2 is wind speed measured at 2 m height [m/s],
$(e_s - e_d)$ is vapour pressure deficit [kPa],
Δ is slope of the vapour pressure curve [kPa/°C], and
γ is psychrometric constant [kPa/°C].

In this analysis, the net radiation and humidity terms have been considered in detail before. The only additional relationships required are expressions for Δ and γ. These are as follows.
   The slope of the vapour pressure curve (Δ) can be calculated from Equation 2.46:

$$\Delta = \frac{4098e_s}{(T + 237.3)^2} \tag{2.46}$$

where:

Δ is the slope of the vapour pressure curve [kPa/°C] ($\frac{de_s}{deT}$),
T is the air temperature [°C], and
e_s is the saturation vapour pressure at temperature T [kPa].

The psychrometric constant (γ) can be derived from Equation 2.47:

$$\gamma = \frac{c_p P}{\epsilon\lambda} \times 10^{-3} = 0.00163\frac{P}{\lambda} \tag{2.47}$$

where:

γ is the psychrometric constant [kPa/°C],
C_p is the specific heat of moist air = 1.013 [kJ/kg/°C],
P is the atmospheric pressure [kPa],
ϵ is the ratio of the molecular weight of water vapour to that of dry air = 0.622, and
λ is latent heat of water [MJ/kg].

The analysis of the evaporation process is facilitated by grouping the terms in Equation 2.45 as:

$$E = \frac{0.408\Delta(R_n - G)}{\Delta + \gamma(1 + 0.34U_2)} + \frac{\gamma\frac{900}{T+273}(U_2(e_s - e_d))}{\Delta + \gamma(1 + 0.34U_2)} = E_{rad} + E_{aero} \tag{2.48}$$

where:
E is the reference open water evaporation [mm/day].

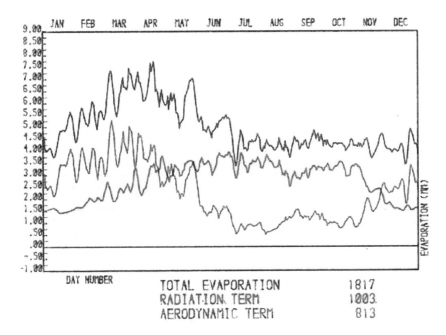

Figure 2.48 Aerodynamic component, radiation component and total evaporation at Bauchi, Northern Nigeria (10 °N). Note the high aerodynamic component during the dry season when the radiation component is lower (Acworth, 1981)

E_{rad} is the contribution from radiation energy input [mm/day] and
E_{aero} is the contribution from aerodynamic energy input [mm/day].

The way in which these two contributions can vary is shown in Figure 2.48 from a sahel-type climate in Nigeria, Figure 2.49 from a mountain climate at 2000 m elevation in Denver, USA, and Figure 2.50 from a tropical island environment on the Seychelles. Note that the aerodynamic term becomes insignificant during the summer months at the height of the rainy season. The humidity is close to 100% for long periods with the air unable to accept any more moisture unless radiation input increases the air temperature and boosts the radiation component. By contrast, the aerodynamic term provides the only energy for evaporation in Denver in the winter when solar radiation is low. Note also the radiative cooling during the winter in Denver.

2.10 EVAPOTRANSPIRATION ESTIMATES

While the equation for the open water evaporation is well established and can be used with appropriate coefficients to estimate evapotranspiration from a crop, Monteith (1965) developed a further model that allows prediction of actual evapotranspiration (ETa). This is set forth below.

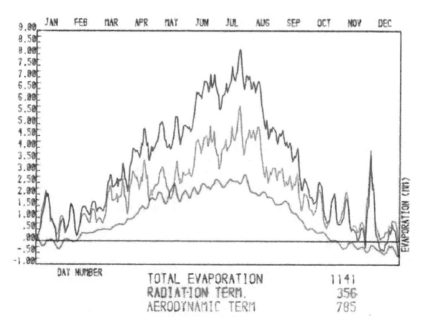

Figure 2.49 Aerodynamic component, radiation component and total evaporation at Denver, USA. Denver is at an altitude of approximately 2000 m and at 40 °N. Note that the aerodynamic and radiation components are each strongly influenced by the season. Note also the negative radiation that occurs in winter as the area cools down (Acworth, 1981)

2.10.1   Radiation balance

Figure 2.51 illustrates the components of the whole energy balance for daytime conditions. With reference to Figure 2.51, these are:

R_n net incoming radiation,
λE outgoing energy as evaporation,
H outgoing sensible heat flux,
G outgoing heat conduction into the soil,
S energy temporarily stored within the soil volume, and often neglected except for forests. It is proportional to temperature changes in the vegetation, air, and shallow soil layer, and to changes in atmospheric humidity,
P energy absorbed by biochemical processes in the plants, typically taken as 2% of net radiation, and
A_d the loss of energy associated with horizontal air movement, significant in an oasis situation.

The various components of the heat budget are often collected as:

$$A = \lambda E + H \tag{2.49}$$

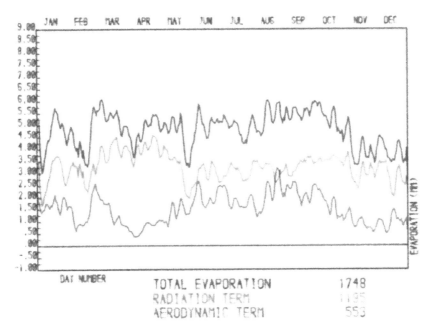

Figure 2.50 Aerodynamic component, radiation component and total evaporation components from a location in the Seychelles (approximately 5 °S). By contrast to the Bauchi and Denver data, the data for the Seychelles shows both the radiation and aeodynamic components as approximately constant throughout the year. This reflects the location close to the equator and the constant high humidity of an island surrounded by ocean (Acworth, 1981)

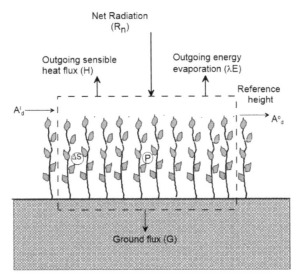

Figure 2.51 Energy fluxes to the plant canopy as shown in Equation 2.50 (adapted from Campbell (1985))

and

$$A = R_n - G - S - P - A_d \tag{2.50}$$

where all terms are in $MJ\ m^{-2}day^{-1}$.

2.10.2 Surface Resistance

Water flows from the soil to roots, through xylem, through mesophyll cells and cell walls, and finally evaporates into substomatal cavities. It then diffuses out of the stomates, through the leaf and canopy layers, and is mixed with the bulk atmosphere. Water potential is highest in the soil and decreases along the transpiration path. This potential gradient provides the driving force for water transport from the soil to the atmosphere. In some texts, the soil, root, xylem, etc. is called the soil-plant-atmosphere continuum, (SPAC). The flow of water through the SPAC can be represented by a series of water potentials and resistances as shown in Figure 2.52 (Campbell, 1985).

With reference to Figure 2.52, the transpiration rate (E) can be written (Campbell, 1985):

$$E = \frac{\psi_{xL} - \psi_L}{R_L} = \frac{\psi_{xr} - \psi_{xL}}{R_x} = \frac{\psi_r - \psi_{xr}}{R_r} = \frac{\psi_s - \psi_r}{R_s} \tag{2.51}$$

where:
E is the transpiration rate [kg/m^2/s],
R is the resistance [m^4 kg/s], and
ψ is the water potential [J/kg].

If the overall resistance is defined as the series combination of all resistance in Figure 2.52, then:

$$E = \frac{\psi_s - \psi_L}{R} \tag{2.52}$$

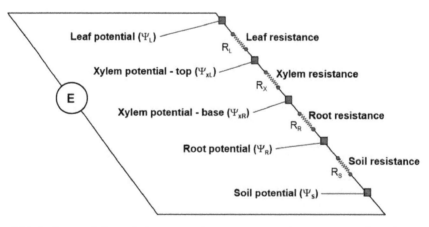

Figure 2.52 Analogue of the soil-plant-atmosphere continuum showing potentials and resistances (adapted from Campbell (1985))

Equation 2.52 can be rearranged to predict leaf water potential:

$$\psi_L = \psi_s - E R \tag{2.53}$$

Equation 2.53 predicts that leaf water potential will be below soil water potential by an amount determined by the transpiration rate and the resistance to transport. When soil moisture is not limiting, E is independent of ψ_L, and R is about constant, so ψ_L varies with E. Until soil water becomes limiting, this response apparently is passive and has little, if any, direct effect on photosynthesis of leaves (Campbell, 1985).

Rough estimates of plant resistance can be made by considering typical potentials and transpiration rates. Campbell (1985) gives the example of a potato crop, after canopy closure. Typical leaf water potentials are around -1200 J/kg when the transpiration rate is 2.4×10^{-4} [kg m^{-2} s^{-1}], so a typical total resistance would be around 5×10^{6} [m^4 s^{-1} kg^{-1}], and the leaf resistance would be 2×10^{6} [m^4 s^{-1} kg^{-1}]. The root resistance is the difference, or 3×10^{6} [m^4 s^{-1} kg^{-1}].

Nomenclature in the various texts describing the resistance terms varies. Smith (1991) refers to a crop canopy resistance term (r_c), which incorporates all the resistances in Figure 2.52. Shuttleworth (1993) refers to the stomatal resistance of the whole canopy (r_s) and incorporates this term in his development of the resistance model for evapotranspiration. The r_s term (Shuttleworth, 1993) is equivalent to the R_L term in Campbell (1985) shown in Figure 2.52.

The stomatal resistance of the whole canopy (r_s) is less when more leaves are present since there are more stomata through which the vapours can be emitted. The movement of water vapour from inside plant leaves to the air is governed by small apertures known as stomata. The air inside the stomatal cavity is nearly saturated (vapour pressure being equal to e_a) while the outside air is usually less saturated. The movement of vapour is controlled by the plant through these cavities; the plant opens and closes these cavities depending on the atmospheric moisture demand ($e_a - e_d$) and the amount of moisture in the soil. Figure 2.53 illustrates the resistance to water movement through the stomata.

Estimation of r_s (or R_L), and the other resistance terms shown in Figure 2.52 is an area of active research (Smith, 1991; Shuttleworth, 1993). The total resistance path is sometimes denoted as r_c and comprises the individual resistances shown in Figure 2.52. For a short green grass crop not short of water, the total resistance (r_c) has been calculated to be approximately 70 s/m. As canopy cover decreases and water availability in the soil decreases, the value of r_c increases to several thousand (Acworth, 1981).

The vapour flux rate (E) of a crop can be expressed as a function of this resistance:

$$E = \frac{k(e_a - e_d)}{r_s} \tag{2.54}$$

where:
k is a constant which accounts for the units used.

The surface resistance of the reference crop of clipped grass 0.12 m high is estimated as:

$$r_c^{rc} = 69 \tag{2.55}$$

where:
r_c^{rc} is the surface resistance [sm^{-1}].

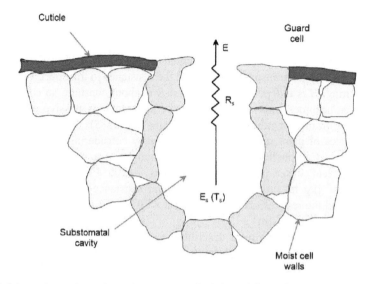

Figure 2.53 Schematic to show the resistance term R_s (adapted from Campbell (1985))

One useful approximation for r_c is to relate it to the leaf area index (LAI) (Allen *et al.*, 1989):

$$r_c = \frac{200}{LAI} \tag{2.56}$$

where:
LAI is the leaf area index.

For clipped grass:

$$LAI = 24 \times h_c \tag{2.57}$$

where:
h_c is the height of the crop in metres.

For alfalfa and other field crops:

$$LAI = 5.5 + 1.5 \ln(h_c) \tag{2.58}$$

For the reference crop of short green grass: $h_c = 0.12$ [m], and the $LAI = 24 \times 0.12 = 2.88$. The canopy resistance becomes:

$$r_c = \frac{200}{2.88} \approx 70 \tag{2.59}$$

For a wet canopy when soil moisture is not limiting, r_c approaches zero.

2.10.3 Aerodynamic Resistance

Another important diffusion process is turbulent diffusion. This refers to turbulence caused by retardation of the horizontal movement of wind over a crop (or water) surface. This is a more important process for exchange of air from close to the ground to higher levels of the atmosphere than molecular diffusion.

The resistance due to this is called the aerodynamic resistance (r_a) and is inversely proportional to wind speed:

$$r_a = \frac{ln[(z_u - d)/z_{om}] \ln[(z_e - d)/z_{ov}]}{(k)^2 U_z} \tag{2.60}$$

where:
r_a is the aerodynamic resistance [s/m],
z_u is the height at which the wind speed is measured,
z_e is the height at which the humidity is measured,
k is the von Karman constant = 0.41,
U_z is the wind speed,
z_{om} is $0.123 \times h_c$,
z_{ov} is $0.0123 \times h_c$, and
d is $0.67 \times h_c$.

Figure 2.54 shows a schematic illustrating the aerodynamic resistance r_a.

The aerodynamic resistance of the reference crop with a crop height of 0.12 m and for all measurements at a standardized height of 2 m, is given as:

$$r_a^{rc} = \frac{208}{U_2} \tag{2.61}$$

where:
U_2 refers to the wind speed measured at 2 m height.

The above concepts of aerodynamic and surface resistances are used to estimate evaporation using resistance networks (quite analogous to electrical circuits), with different

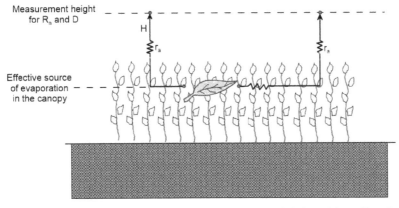

Figure 2.54 Schematic to illustrate the aerodynamic resistance (r_a) (adapted from Campbell (1985))

resistances representing the different plant types and layout patterns (sparse or dense canopy).

2.10.4   The Penman-Monteith equation

This is currently the most advanced resistance model in use. It assumes that:

1   all the energy available for evaporation is accessible to the plant canopy,
2   vapour first diffuses out of the leaves against the surface resistance (r_s), and
3   it next goes into the atmosphere against the aerodynamic resistance (r_a).

The above processes and the energy budget components discussed earlier are used to arrive at the Penman-Monteith equation (Monteith, 1965). The Penman-Monteith equation can be written as follows:

$$\lambda ET_a = \frac{\Delta(R_n - G) + \rho c_p (e_a - e_d)1/r_a}{\Delta + \gamma(1 + r_c/r_a)} \tag{2.62}$$

where:
λET_a is latent heat flux of evapotraspiration [kJ m^{-2} s^{-1}],
R_n is net radiation flux at the surface [kJ m^{-2} s^{-1}],
G is soil heat flux [kJ m^{-2} s^{-1}],
ρ is atmospheric density [kg m^{-3}],
c_p is specific heat of moist air [kJ kg^{-1} °C/s],
($e_a - e_d$) is vapour pressure deficit [kPa],
r_c is crop canopy resistance [s m^{-1}],
r_a is aerodynamic resistance [s m^{-1}],
Δ is slope of the vapour pressure curve [kPa °C/s],
γ is psychrometric constant [kPa °C/s], and
λ is latent heat of vaporization [MJ kg^{-1}].

All terms are as defined above, except the psychrometric constant (γ), which is defined as:

$$\gamma = \frac{c_p P}{\epsilon \lambda} \times 10^{-3} = 0.00163 \frac{P}{\lambda} \tag{2.63}$$

where:
γ is the psychrometric constant [kPa °C/s],
c_p is the specific heat of moist air = 1.013 [kJ kg^{-1} °C/s],
P is atmospheric pressure [kPa],
ϵ is the ratio of the molecular weight of water vapour to that of dry air = 0.622, and
λ is the latent heat [MJ kg^{-1}].

The resistance terms in Equation 2.62 are still subject to research and investigation. If they can be quantified, then this equation can be used to estimate the actual evapotranspiration which occurs. This is the 'one-step' approach recommended by Smith (1991) and Shuttleworth (1993) and will be returned to in Chapter 3.

The one-step approach is not yet accepted and the Food and Agriculture Organization (FAO) have adopted (Smith, 1991) an intermediate procedural step where the ET_a is calculated for an extensive surface of short green grass of uniform height, actively growing, completely shading the ground that is not short of water. This cover is widespread around the world and represents a grass lawn. The reference height of the crop is 12 mm; it has a fixed surface resistance of 70 s/m and an albedo of 0.23. The value of ETa calculated using these parameters and standard meteorological data. This value of ET_a is known as ET_0. The evapotranspiration of a particular crop is arrived at by applying a crop coefficient that will vary as the crop matures. ET_C for a particular crop is therefore:

$$ET_C = K_C ET_O \tag{2.64}$$

where:
K_C is the crop coefficient – often changing with season.

2.10.5 Eddy covariance flux measurements

The Penman-Monteith estimate of actual evapotranspiration relies upon the assumption of laminar flow of moisture up to the evaporating surface. There are a number of uncertainties involved in the calculation. However, the method has served the test of time and remains in use.

It was realized by Monteith that the movement of moisture away from the canopy occurs in the turbulent movement of air above the canopy but that this was difficult to measure. If it could be measured, it would serve as a check on the moisture loss that does actually occur. However, it was not until the turn of the century that equipment became available to quantify those losses. Methods have now been developed that work in the turbulent air above the canopy.

Eddy covariance systems consisting of an ultrasonic anemometer and infrared gas analyzer (IRGA) (Figure 2.55) can be installed on towers above the canopy (Figure 2.56)

Figure 2.55 Eddy covariance sensor at a climate station. 3D anenometers and an infrared gas analyzer are installed. The gas analyzer is focussed on the region where the wind direction is measured (Photo: Ian Acworth)

Figure 2.56 Eddy covariance sensors installed on a temporary tower above a forest canopy (Photo: Ian Acworth)

and can be deployed above a forest cover or across wetlands. These systems can directly measure the amount of CH_4, CO_2 and H_2O that blows in or out of a site during gusts of wind. NASA recognized the significance of these measurements and now fund a global sensor net of eddy covariance stations under the NOAA Observing System Architecture.

The flow of turbulent air above the surface can be conceptualized as a constant stream of eddies, as shown in Figure 2.57 – adapted from Burba (2013).

The infrared gas analyzer works by emitting an infrared beam of light and monitoring the spectrum of the returned beam. The spectrum can be used to determine CH_4, CO_2 and H_2O gas concentrations.

The gas fluxes are monitored continuously at 10 Hz or 20 Hz and summed every 15 minutes to provide a flux over each 15-minute period.

If the characteristics of each eddy can be determined, then the net flux can be resolved. The ultrasonic anenometer is a solid state device that measures the wind speed by

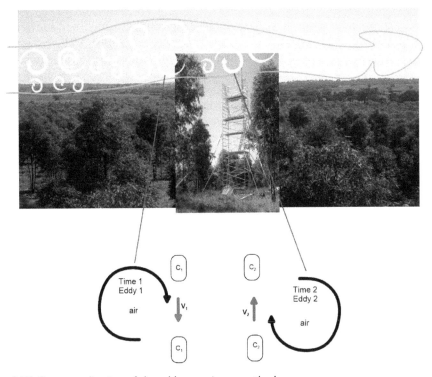

Figure 2.57 Conceptualization of the eddy covariance method

passing ultrasonic sound waves through the moving air. As wind speed changes, the air density changes and this causes a change in the speed of sound. By measuring the delay change, the sensor can determine both the speed and directionof the wind. By making these measurements many times a second (10 or 20 Hz), the characteristics of an eddy can be determined. In Figure 2.57, the first parcel of air moving around Eddy 1 has a direction (downwards) and a velocity of V_1. A second eddy moves the same parcel back upwards with a velocity V_2. If the mass, density and chemical makeup of the parcels are known, then any difference between the two results in either an upward or a downward flux of gas can be averaged over a longer time period (15 mins.) and recorded. The method relies upon the existence of turbulence above the canopy and breaks down under the very calm conditions that can sometimes exist with a stable high pressure zone present or at night when the lack of radiation allows the atmosphere to become calm.

A disadvantage of the technique is that extensive representative expanses of vegetation are required. A second and perhaps more significant disadvantage is that the setup and operation of the equipment requires specialized technical operatives who need to visit the site every 2 weeks or so to clean and download the very large data sets that are recorded. Burba (2013) makes the point a number of times that it is good practice to keep a record of the original 10 or 20 Hz data set so that the data can be checked and re-evaluated as processing techniques improve. This equipment produces very large data sets that are easily disrupted by communication difficulties.

Applications to evapotranspiration determination

Many eddy covariance systems have been installed globally since 2000, and a review of the network in Australia and New Zealand is given by Beringer *et al.* (2016). However, analysis of the data is both complex and time consuming. One of the most highly cited references is Wilson *et al.* (2001), who compare a 5-year record of eddy covariance measurements with sap flow (2 years), water budget (1 year) and catchment water budget (31 years) estimates of evapotranspiration over an uneven and aged mixed deciduous forest in the south-east of the USA. They found good correlation between eddy covariance measurements (average of 571 mm) with catchment water balances (average of 582 mm). Poor correlation between sap flow estimates and eddy covariance was considered to be due to the difficulty of upscaling the sap flow measurements of the wide range of species to the canopy. Soil water budget estimates were positively correlated with eddy covariance and sap flow measurements, but the data was highly variable. In particular, the soil water estimates were poor under very dry conditions or heavy rainfall conditions. Soil water budgeting does not account for interception or rapid bypass of recharge water through root channels.

Sun *et al.* (2008) present results from a variety of forested catchments in northern Wisconsin, USA. Zitouna-Chebbi *et al.* (2018) present results from a Tunisian catchment. Eamus *et al.* (2013) present a review of data from a semi-arid environment and investigate the response to rainfall events.

The ability of the eddy covariance system to measure carbon dioxide makes it an invaluable tool to monitor the carbon budget. Cleverly *et al.* (2016) used eddy covariance to investigate the carbon budget in a semi-arid region as it changed in response to the switch from drought conditions in Australia (2000–2009) to a brief wet period in 2010–2011. During the wet period, vegetation growth was very extensive and served as a significant carbon sink.

Chapter 3

Recharge, discharge and surface water groundwater connectivity

3.1 INTRODUCTION

In the first decade of the 21st century, there has been an increasing emphasis on understanding the connectivity between surface water and groundwater. This is not new understanding, as this connectivity is implicitly assumed whenever the hydrological cycle is described (Acworth, 2009). Rather, it is the increasing realization that we need to more accurately understand all components of the hydrological cycle if we are to manage water efficiently.

A related concern has become the accurate measurement of groundwater recharge – as it has been considered by some jurisdictions that diversion of water for agriculture should be limited to the quantity of naturally occurring recharge. This concept has been linked to that of the *safe yield* of an aquifer. Put simply, it is the belief that if we do divert/abstract less than the recharge of the aquifer, then the system will not be damaged irreparably. This at first sight seems plausible. However, if all the recharge is diverted from the aquifer, then the discharge (baseflow) in the aquifer must also decline, with the result that many groundwater dependent ecosystems will also be negatively impacted. The health of groundwater dependent ecosystems is often protected by law, so the accurate measurement of groundwater discharge is also now required. The specification of recharge, discharge and the interconnected state of surface water and groundwater are all facets of the same problem; the need to better understand the hydrological cycle.

In practise, water resource development in the second half of the 20th century occurred during wetter than average conditions (see for example Chapter 2, Figure 2.24), when there was sufficient water available in surface storages or as groundwater, that inaccuracy in describing the various components of the hydrological cycle did not matter. With the return to more normal climatic conditions, combined with further development of agriculture, it has become necessary to better understand all components of the hydrological cycle so that optimum development can occur.

The disconnect between surface water and groundwater has also developed as a result of the subject area being covered by two different disciplines. Geologists have often described groundwater systems while engineers have worked to understand surface water hydrology. There have been many university courses taught where the two disciplines are presented as completely separate entities!

It is perhaps not surprising that the interconnectedness between surface water and groundwater has been overlooked. As a first approximation, the magnitude of the flux between the two is significantly less than the magnitude of individual components of the flux occurring within either the groundwater or the surface water domains. Water moves relatively quickly through the surface water domain with flood peaks (even in Australia) passing through river systems in days or weeks. Contrast this to the very slow movement of groundwater, where a speed of 150 m a year is fast!

3.2   GROUNDWATER RECHARGE

3.2.1   Definitions

Groundwater recharge is loosely defined as the quantity of water that drains down from the surface and reaches the water table, leading to an increase in the quantity of water stored in the aquifer and therefore to a rise in the water level in the aquifer. Simmers (1988) and Lerner *et al.* (1990) review techniques for the estimation of natural recharge while Simmers (1997) concentrates on arid and semi-arid areas. Scanlon and Cook (2002) provide extensive reviews of groundwater recharge, and Rushton (2003) gives a detailed discussion of recharge calculation with particular emphasis on water balance methods.

Recharge differs from infiltration in that infiltration is the amount of water per unit area that enters the soil profile over a given period. If the infiltration capacity of the soil is greater than the rainfall intensity, then the rate of infiltration will be directly related to rainfall intensity. However, if rainfall intensity is greater than the infiltration capacity of the soil, as frequently occurs after a drought, then excess rainfall is lost as surface runoff.

Once in the soil, water can be lost back to the atmosphere either by direct evaporation or, more importantly, by evapotranspiration. The dynamics of this soil plant atmosphere system were described in Chapter 2. Deep drainage (also known as percolation) is the amount of water per unit area that moves vertically downwards below the base of the root zone. There may be considerable thicknesses of material below the soil root zone that are partially saturated by water and where water is held in tension. The fate of water in this unsaturated zone is often very difficult to determine. If there is a low permeability layer below the root zone, some of the deep drainage may move laterally and enter a stream system without entering the aquifer of interest. In a multiple aquifer system, some of the recharge may occur via leakage from other aquifers.

3.2.2   Recharge to confined aquifers

As a confined aquifer is fully saturated with a pressure surface that is above the top of the aquifer (by definition), it is clearly *impossible* to recharge a confined aquifer by downward drainage of rainfall infiltration. The only way in which a confined aquifer can be recharged is by lateral movement from parts of the aquifer where unconfined conditions exist.

3.2.3   Recharge to semi-confined aquifers

Recharge is possible to a semi-confined aquifer if there is an appropriate downward hydraulic gradient and leakage out of the aquifer at some point to make way for the newly recharged water. However, in general terms, there cannot be recharge to an aquifer if the aquifer is already full.

3.2.4   Time lags

If the groundwater response to rainfall is not quick, then hydrograph techniques become more difficult. An example of this is in the Mallee region of SE Australia, where the increased recharge due to clearance of native vegetation for agriculture may take 50 years to reach the water table. In such cases, soil techniques need to be used. Even in less extreme cases, delays in wetting fronts moving through the vadose zone causes hydrographs to be muted.

Another important time lag is the time required for tracers to move through the ground-water system. The mean residence time for aquifers can range from hundreds to millions of years. Hence, groundwater tracers are often more relevant to estimating paleo recharge, rather than current recharge. This is important in the case of the large sedimentary basins of Australia which received recharge under past wet phases of the Pleistocene.

3.2.5   Spatial variability

The mechanisms by which recharge occurs will vary and hence lead to a number of classifications of recharge as shown in Figure 3.1 (Acworth *et al.*, 2016b). Three classifications are possible:

Direct recharge Only vertical movement of water occurs, and after direct infiltration to the soil, moisture adds to soil moisture storage (SMS). Plants can reduce the SMS by evapotranspiration, and the volume of water held in the SMS is also related to the rooting depth of the plant. It is considered that evapotranspiration can occur at the full potential rate dicated by the atmosphere as long as the SMS contains sufficient water. As the

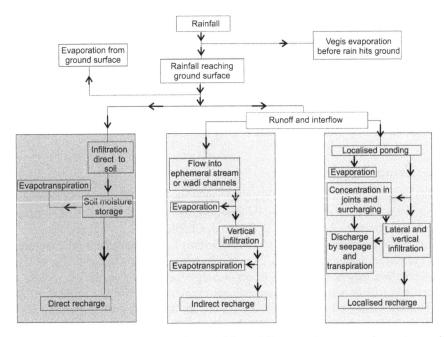

Figure 3.1 Different possible pathways for rainfall to follow on the way to becoming recharge (Acworth *et al.*, 2016b)

quantity of the SMS is reduced, at some point, the plants can no longer transpire at the maximum rate and vigour in the plants is reduced. When there is no longer any moisture left, the plant dies. Related to the SMS is the concept of the soil moisture deficit (SMD). A soil that is freely draining has no SMD. An SMD is created by evapotranspiration. Additional rainfall causing infiltration is required before a soil SMD can be eliminated and free drainage becomes possible. These concepts are fully developed in daily calculations of the soil moisture balance and can be used quite successfully to predict recharge in a reasonably developed soil as long as the soil remains moist and does not dry out. Soil moisture accounting is described by Rushton (2003) and was used by Acworth (1981) to successfully model recharge in a savanna climate.

Indirect recharge If rain falls onto rock or onto a very dry soil, there can initially be little or no downward movement. Most likely, the rainfall becomes surface runoff and moves laterally into ephemeral channels and collects in a wadi-type drainage network. There may be losses from these channels by evaporation, but most likely there will be vertical infiltration into the colluvium beneath the channel. If the colluvium is deep and has an SMD caused by evapotranspiration, moisture will be retained and there will be no final downward movement to the water table. It may also be that clay deposits within the colluvium cause a perched aquifer to occur for a time. In general terms, the quantity of indirect recharge is small compared to the initial storm event with the majority of the rain becoming runoff that leaves the area – possibly as flood flow.

Localized recharge Localized ponding of rainfall may also occur that in turn is subject to evaporation from the pond, or concentration in joints that transmit water downwards quickly. The quantity of this type of recharge is also not considered to be very great as a percentage of the initial rainfall. The majority of the rainfall will become runoff that fills and leaves the local ponding site as overflow.

If the soil is very dry after a long period with no rain, then the infiltration capacity will be much reduced and surface flow can commence after relatively small storm events. The fact that there is a large soil moisture deficit is not significant if the rainfall cannot get into the soil. This was frequently observed to be the case at Fowler's Gap in western New South Wales, where the average annual rainfall was approximately 140 mm. Rainfall of as little as 20 mm was observed to generate wadi flow. In fact, at that station, no recharge was observed, even from a storm of 170 mm falling in 36 hours.

3.2.6   Temporal variability

Recharge is generally not constant in time but varies according to rainfall. In semi-arid and arid areas, recharge is often dominated by infrequent large events (or series of events over a short period of time) or by flooding as shown in Chapter 2.

The recharge contribution from large rainfall events is often called episodic recharge and will be indirect or localized in nature. Episodic recharge contrasts to conditions in less arid areas where there may be continuous direct recharge (say, from a stream) or regular seasonal direct recharge (when precipitation exceeds evapotranspiration for some months each year). For both flood and episodic recharge, there are large year-to-year variations.

Land use is also often used to describe the type of recharge. Recharge in dryland, irrigation and urban areas are respectively called dryland, irrigation and urban recharge. In irrigation areas, recharge can occur through a number of mechanisms including irrigated agriculture, supply and drainage channels, storage and re-use basins. In urban areas, sprinklers, leaking pipes and stormwater runoff may contribute to recharge and again may be much higher than in the surrounding agricultural area.

3.2.7 Factors impacting recharge

Many different factors affect recharge (Rushton, 1988, 2003).

1 At the land surface:

- topography,
- rainfall: magnitude, intensity, duration and spatial distribution,
- runoff and ponding of water, and
- cropping patterns, the ratio of actual to potential evapotranspiration.

2 Irrigation:

- nature of irrigation scheduling,
- losses from channels and water courses, and
- uneven irrigation bays.

3 Rivers:

- hydraulic conductivity of river channel bank and bed material,
- relative level of the river channel and the surrounding groundwater, and
- rivers gaining water from or losing water to the aquifer.

4 Soil zone:

- nature of the soil, depth of soil, and soil hydraulic properties,
- the variability of the soil, both spatially and with depth,
- rooting depth in the soil and seasonal changes in rooting depth, and
- cracking of the soil as it dries out and expansion when it wets.

5 Unsaturated zone between the soil and the aquifer:

- flow mechanisms through the unsaturated zone,
- zones with different hydraulic conductivities, and
- hysteresis effects in soil.

6 The aquifer:

- the ability of the aquifer to accept water (is the aquifer full?) and
- variation of aquifer condition with time (impacts of local abstraction).

3.3 RECHARGE ESTIMATES

There are a range of techniques used for estimating recharge depending upon the accuracy, time and spatial scale required. Broadly speaking, they can be divided into four categories, with a number of techniques and variations possible within each category:

- Groundwater hydrograph methods
- Tracer methods
- Physical methods
- Modelling methods

3.3.1　Groundwater hydrograph methods

If rain falling over a shallow unconfined aquifer reaches the water table within hours, then the rise in water table can be used to estimate the direct recharge that occurred. To achieve this, however, it is necessary to know the aquifer specific yield. More often, the inverse occurs and it is the rise in the water table that is taken to indicate that direct recharge has occurred, and if the quantity of rainfall is known then the specific yield can be calculated. An example of this approach is given in Figure 3.2.

It is important that the groundwater response to rainfall is measured after entrapped air has escaped. The response to recharge in an unconfined sand aquifer should be a simple step. However, in most cases it is a peak in a similar manner to a flood peak. The peak is created as a complex confined response occurs because of a layer of trapped air. An additional complication is the drainage away from the site across a near boundary. This prevents the hydraulic head from fully rising to the height that it would if the boundary was a long way off or was a no-flow boundary. These aspects are shown in Figure 3.3 where the results have been calculated using the RADFLOW program. Two plots are shown, one for each boundary condition where the boundary was 200 m from a pump well. Recharge of 150 mm was applied for the first day and then no recharge for the second day. The specific yield used in the simulation was 0.25.

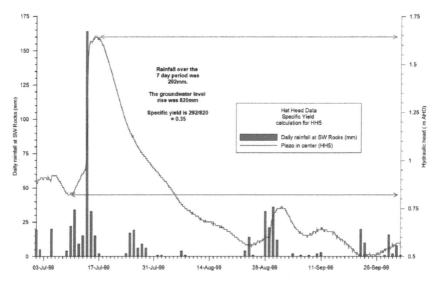

Figure 3.2 The rise in water level in a coastal sand aquifer is used to calculate the specific yield of the aquifer. In reality, the water level may well have been higher over the 7-day period if there had not been a loss of water at the ocean boundary of the system. The specific yield calculated could therefore have been lower. The value of 0.35 is close to the total porosity of the sands and the actual specific yield was more likely to be 0.25

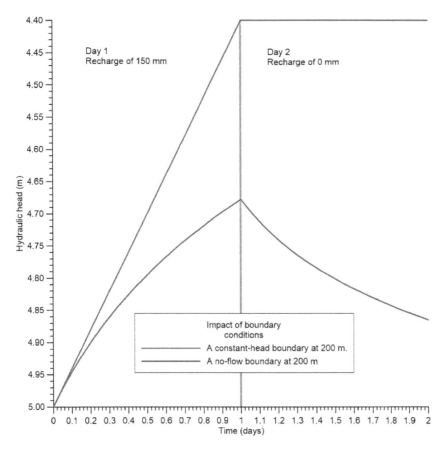

Figure 3.3 Hydraulic head variation due to differences in direct recharge. Recharge of 150 mm on day 1 causes a 0.6 m rise in water levels in an aquifer with a specific yield of 0.25. There is no loss of water at the boundary where a no-flow boundary is set using the RADFLOW program. Recharge appears as a step function. In contrast, the second curve shows the impact of a constant-head boundary at 200 m. The hydraulic head never reaches the same level as for the first condition as the result of loss of water across the boundary

An accurate value of specific yield would only be recovered from the no-flow boundary condition in Figure 3.3. The use of the head change in the second scenario would greatly overestimate specific yield, producing a value of 0.46.

A further area of confusion is created when the aquifer is considered to be unconfined but is actually confined. This can happen frequently in silts, sandy silts and clays that have a high loading efficiency (Chapter 7). Rainfall appears to produce a rapid rise in the water level but this is actually a piezometric surface expressing a higher loading caused by the rainfall but no actual advective flow of water to depth. Timms *et al.* (2001) give an account of this error from the Liverpool Plains. Figure 3.4 demonstrates the occurrence. Clearly, no water is advected downwards to 35 m depth even though the water level in the piezometer responds to the rainfall at the surface.

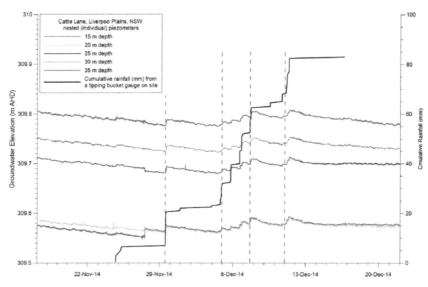

Figure 3.4 Sequence of water levels measured in a nest of piezometers at Cattle Lane (Acworth *et al.*, 2015b) where the additional loading from rainfall at the surface is rapidly transmitted to at least 35 m depth in these high-loading efficiency sediments

3.3.2   Tracer methods

The detonation of atomic bombs during the 1950s produced enhanced levels of tritium ^{3}H in the atmosphere. The tritium is radioactively unstable and decays in a predictable manner. It has a half-life of 12.3 years. The concentration of tritium relative to hydrogen is small and is measured in terms of tritium units (TU). A TU is equivalent to 1 atom of tritium per 10^{18} atoms of hydrogen. A peak of approximately 500 TU occurred during 1963–64. Prior to the start of testing, the natural background TU was 5 to 10 units.

By measuring the amount of tritium in groundwater, it is possible to determine an approximate age of the water relative to the atomic testing programme. Measurements of tritium are made in local rainwater to establish a background level (Stewart and Morgenstern, 2001). Values of tritium greater than this in groundwater indicate rainwater which has fallen since the 1950s as recharge. Low values of tritium in groundwater indicate water which has been recharged before the 1950s.

Tritium decays spontaneously. If there are no further nuclear detonations, then use of this technique based upon bomb-pulse tritium will soon become unusable as the level of tritium in water will not be significantly different from pre-bomb pulse water. However, as the accuracy of detection techniques has improved, there is now the possibility of using cosmo-genically produced tritium in a similar way. In effect, if there is any tritium in the groundwater, then it is evidence of more recent recharge, as natural decay reduces the tritium concentration from the already low levels in rainfall to close to zero.

The presence of chloride in the soil has also been taken as an indicator of recharge conditions. This approach is particularly valid where extensive vegetation cover is cleared for agriculture. It is argued that before clearing, chloride builds up beneath the root zone due to anion exclusion during transpiration. The vegetation cover transpires all the available

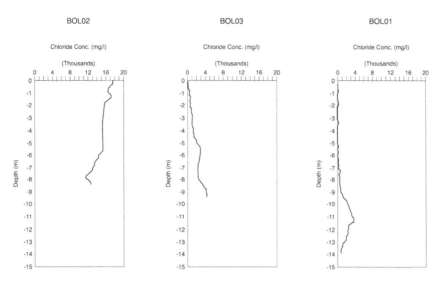

Figure 3.5 Unsaturated zone chloride profiles prior to clearing (a) and after clearing (b and c) (Cook et al., 1992)

moisture with no downward leaching. When the vegetation is cleared, rainfall leaches the chloride downward which can be detected as a pulse in the soil on the way down to the water table. For this method to be used successfully, it is assumed that the soil profile is uniform and that there are no other sources of chloride.

The method has been used successfully on the Australian Mallee where sandy soils are covered by deep-rooted natural vegetation. The climate is semi-arid and the vegetation arguably uses all the available rainfall with no deep recharge. Clearing of this deep-rooted vegetation for cultivation of cereal crops allows rain to infiltrate during wet and fallow periods when the vegetation is no longer present to use the moisture. Deep drainage then occurs with chloride leaching as the recharge pulse moves downward. The measurement of chloride profiles over a period of years following clearing indicates a slow downward movement of the chloride bulge. This is shown in Figure 3.5.

Geophysical techniques have been used to detect changes in the bulk electrical conductivity of the soil profile which have been interpreted to represent changes in the chloride content as recharge occurs. Thus, a low bulk EC represents a recharge zone, whereas a high bulk EC represents stable chloride and no downward leaching. However, this method is not recommended as it is dependent upon all the other factors that impact the bulk electrical conductivity remaining unchanged.

3.3.3 Physical methods

Lysimeters

Lysimeters can be used to quantify the recharge physically by isolating a representative element of the soil and determining the deep drainage (percolation) which occurs. The lysimeter approach is clearly expensive and applies to a small and somewhat disturbed sample. The construction of a lysimeter such that it is truly representative of the local

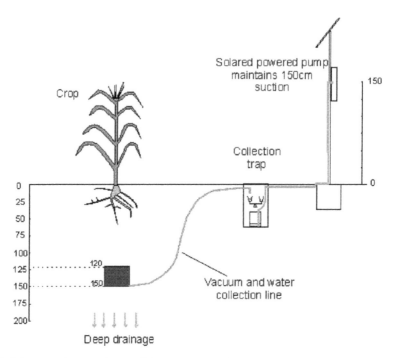

Figure 3.6 Schematic showing design of the mini-suction lysimeter

conditions is also a problem. They are most successful when used to estimate drainage beneath a crop where the soil conditions can be prepared uniformly and the crop planted on the lysimeter to represent the local area around the site. Calculation of recharge beneath natural vegetation is a problem in as much as preparation of a lysimeter disturbs the ground to the extent that the lysimeter installation no longer represents the natural condition.

Mini-lysimeters have been constructed by the Queensland Department of Primary Industries (QDPI) (Des McGarry, pers. comm.) to detect drainage below the root zone of irrigated crops. Figure 3.6 shows a schematic of the mini-lysimeter design. These devices can be used to quantify deep drainage. They are installed beneath the crop rooting zone at between 1.2 and 1.5 m depth. Figure 3.7 shows a soil core with the vacuum equipment fitted at the base so that a constant suction of approximately 150 mm can be applied. The suction is maintained through a nylon tube running back to a base station, where deep drainage water is measured using a tipping bucket gauge and collected for analysis. A constant suction of 1500 mm is applied through the nylon lines to the wicks installed at the base of the soil core to mimic the soil suction that would naturally occur in the absence of the lysimeter.

The data for the 2007/08 growing season for lysimeters draining 1.5 m beneath a crop are shown in Figure 3.8. The lysimeter midway between the head ditch and the tail of the irrigated plot was installed into much lower hydraulic conductivity material. Considerable deep drainage – or potential recharge – is shown beneath the head ditch. This was considered to be the result of high hydraulic conductivity beneath the irrigated field. Some evidence of that is seen in the electrical image (Figure 3.9) carried out over the irrigated field where

Figure 3.7 Soil suction 'trident' installed in silica at the base of the soil ped (Photo: Des McGarry)

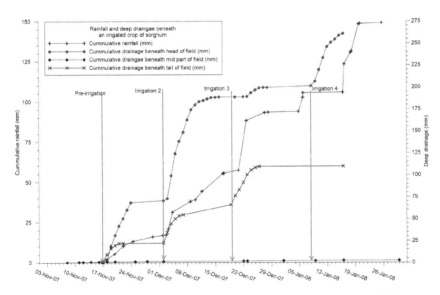

Figure 3.8 Deep drainage beneath a crop of sorghum subject to a pre-irrigation and three irrigations during the growing season. Rainfall at the site is also shown

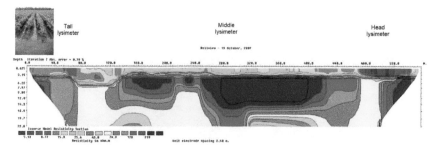

Figure 3.9 Electrical image (Chapter 9) through the irrigated field showing the locations of the lysimeters. The tail of the irrigation plot is shown in the inset

the locations of the lysimeters are shown. The field had been laser graded and some black soil imported to fill old channels on the flood plain close to a river. The old soil surface is clearly seen in the electrical image.

Samaru catchment water balance

There have been many paired catchment experiments where two similar catchments have been instrumented before some change in vegetation (often logging) occurs in one catchment. The first of such experiments were carried out in the highlands of Kenya where tea was cropped. The various components of the water balance change in response to the vegetation change. The Commonwealth Scientific and Industrial Research Organisation carried out detailed work in Western Australia as a part of salinity investigations in the 1980s, and another detailed study was carried out in Northern Nigeria (Kowal and Kassam, 1978).

Kowal and Kassam (1978) instrumented a small catchment on crystalline basement rocks on the campus of Ahmadu Bello University at Samaru, close to Kaduna in Northern Nigeria, in the 1960s. The topography and geology are very similar to that shown in Figure 1.21 or Figures 1.23 and 1.24. The area of the Samaru catchment was 640 ha and rainfall was monitored at 25 gauge sites across the catchment. Evaporation was estimated using the Penman method described in Chapter 2.

Rainfall in the savanna lands occurs as intense summer storms which last from May to September. The variability in rainfall is well known and would have been clear from the gauge density of 1 per 0.25 km^2 that was achieved at Samaru. At the height of the rains there is an excess of water everywhere, but within a couple of months of the end of the rains the soil has dried, crops have been harvested and the climate is dominated by dry winds blowing from the Sahara Desert. Direct groundwater recharge occurs in the wet period, but the amount is highly variable. The reasons for this variability emerge from the detailed water balance at Samaru.

The water budget was calculated from Equation 3.1, modified slightly from Equation 3.2. The catchment had a small water supply dam with a spillway at the downstream end. The dam overflowed during the wet season and the discharge was carefully monitored at the spillway using a curved venturi flume weir installed in the dam crest:

$$P - Q - \Delta W = ET_a \tag{3.1}$$

where:

P is total rainfall measured over the catchment basin of 640 ha,

Q is the total runoff measured at the spillway, comprising surface runoff and seepage draining into the lake,

ΔW is changes in soil moisture which were eliminated by making calculations over a complete water year. This was clearly identified by the rise in the dry season water table which occurs when the soil moisture deficit has been eliminated after a few weeks at the beginning of the rains, and

ET_a was estimated by difference.

The water level in a series of shallow bores was monitored to determine groundwater level changes and is shown in Figure 3.10. Groundwater level measurements were not carried out during the dry season. This data set is an early indication of the value of groundwater monitoring to better understand actual process.

The magnitude of the water budget components over a 6-year period is shown in Table 3.1. Note the great variation between the years. In years of similar rainfall (1967 and 1968 had 990 mm and 998 mm, for example), the groundwater level began rising 2 weeks later in 1968; the cumulative rainfall before the rise was almost 100 mm more in 1968; flow over the spillway commenced a month later in 1968; total water lost over the spillway was 109 mm more in 1967; total runoff was 160 mm more in 1967 and direct recharge was calculated to be 51 mm more in 1968. Very clearly, there is another important factor that influences the catchment response that is not represented by the observations made in Table 3.1, and this factor dramatically impacts the outcome of the rains. Note again that the highest rainfall occurred 1966 (1382 mm) and that produced 200 mm of recharge. Yet, the lowest rainfall year of 1971 (775 mm) produced 238 mm of recharge.

It is very clear that there can be no simple relationship between annual rainfall and annual direct recharge. Although it is common practice in groundwater modelling studies to take recharge as 10% of rainfall, based on the careful analysis by Kowal and Kassam (1978), this is demonstrably a gross oversimplification that should not be entertained – however

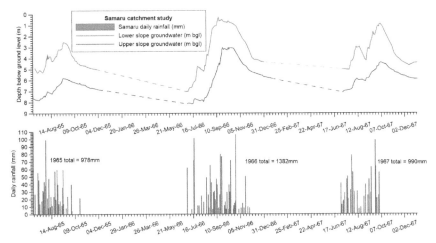

Figure 3.10 The groundwater level response to daily rainfall at Samaru for three wet seasons (1965, 1966 and 1967) (Kowal and Omolokum, 1970)

Table 3.1 Water balance study for the Samaru catchment basin (Kowal and Kassam, 1978)

Water year	1966	1967	1968	1969	1970	1971
Duration (days)	377	350	371	390	330	360
Rainfall (mm)	1382	990	998	1196	884	775
Wet season duration (days)	190	191	169	209	143	129
Rain days	95	83	77	88	77	70
Date of rise in dry season water table	11/7	15/7	27/7	29/6	30/7	31/7
Cumulative rainfall (CP) before rise	506	419	516	381	460	391
Start of flow over spillway	11/8	17/7	15/8	25/7	25/8	26/8
Duration of flow over spillway (days)	124	118	62	113	56	54
Total spillway discharge (mm)	429	239	130	450	275	257
Lake deficit (mm)	30	30	30	30	30	30
Total runoff and seepage (Q)	459	269	160	480	305	287
Runoff (mm)	259	168	8	180	81	49
ET_a (mm)	923	721	838	716	599	488
ET_a CP	417	302	322	335	119	97
Recharge (seepage) (mm)	200	101	152	300	224	238

easy it is to calculate. To make the same point graphically, total rainfall is plotted against direct recharge in Figure 3.11. The correlation coefficient between rainfall and recharge for these 6 years is only 0.01. From a statistical perspective, there is *no* correlation!

While it is easy to criticize this data as being too short, even considering the rainfall over a small catchment, Acworth *et al.* (2016b) have shown that at least 6 years is required to achieve a minimum variance in the rainfall. Yet, the water balance works over a single rainy season producing very different growing season lengths and hydrological variation. It is the change over a single season that will determine the value of the grain crop or the quantity of food produced. For this reason, the longer term, perhaps more statistically valid, data sets are of little value.

The average value of ET_a was 712 mm or about half the annual potential evaporation. The average value of actual evapotranspiration during the wet season was 4.1 mm/day.

3.3.4   Modelling methods

Modelling generally augments other techniques but must not be used by itself to estimate recharge. Simply stated, there are too many degrees of freedom to achieve a valid result. Modelling approaches can be subdivided:

1   Major groundwater models where the climate, rainfall, soil properties and the subsurface are all modelled in 3D.
2   Water balance studies at a field or crop scale.

Full 3D modelling

Very sophisticated modelling studies can be conducted using a 3D model of the complete hydrological system. Such models require a valid rainfall distribution to drive the hydrology; the application of a complex routing model similar to that shown in Figure 3.1 at

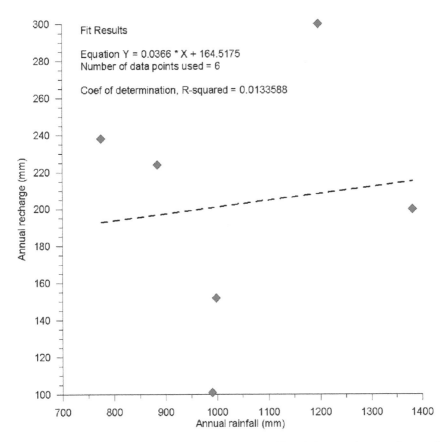

Fit Results

Equation Y = 0.0366 * X + 164.5175
Number of data points used = 6

Coef of determination, R-squared = 0.0133588

Figure 3.11 The lack of any relationship between annual rainfall and direct recharge at Samaru for the years 1966 to 1971

each and every surface node in the discretized grid used; a model to account for runoff; a detailed description of the subsurface distribution of specific storage and hydraulic conductivity and, finally, an accurate description of the boundary conditions around the model.

The Système Hydrologique Européen (SHE) model produced by Danish Hydraulic Institute (DHI) is capable of representing all the variables and at least is capable of conserving mass. The model has been used successfully to investigation processes in a humid climate. Non-linearity between moisture content and hydrogeological variables make the application in a semi-arid environment difficult.

Water balance modelling

The water balance modelling technique attempts to characterize the moisture flux balance within the soil zone. Input to this balance is rainfall, and output can be runoff, evapotranspiration (including both evaporation and transpiration) and downward seepage when the soil moisture is full.

A general equation describing the water balance at the soil surface can be expressed as:

$$P + I - R = ET_a + D + \Delta W \qquad\qquad (3.2)$$

where:
P is precipitation – including snow, frost and dew,
I is applied irrigation water,
R is surface runoff,
ET_a is evapotranspiration,
D is deep drainage or groundwater recharge, and
ΔW is the change in soil moisture stored in the soil profile.

Each of the terms on Equation 3.2 represents flows or changes over a given time period. The drainage term may be either positive or negative depending upon the direction of the hydraulic gradient in the soil beneath the surface. It is extremely important to investigate the balance of the various fluxes over an appropriate time span. A maximum of a daily time step is required, otherwise recharge will be significantly underestimated (Howard and Lloyd, 1979; Rushton, 2003).

The approach of estimating potential evaporation then using moisture availability to correct the value has worked well in many areas for predicting crop water requirements. The FAO have published detailed recommendations for this analysis (Allen *et al.*, 1998), and Rushton (2003) made several suggestions for technique improvement for savanna locations where seasonal rainfall occurs in a distinct wet season separating many months with no rainfall.

Ideally, if the climatic data and the soil characteristics are known, then it should be possible to calculate the actual evapotranspiration directly and to use this in water balance studies. The resistance term r_s in the Penman-Monteith equation (Allen *et al.*, 1998) is a measure of the resistance to moisture flow from the water table to the evaporating surface on the leaf. It is directly related to the unsaturated zone hydraulic conductivity, which, as we have seen, is a function of the soil matric potential and decreases as soil moisture decreases. It should be possible therefore to relate the r_s term to the soil moisture deficit calculated using the water balance. This is the 'one-step' approach favored by the FAO.

If the modelling is successful, then it should be possible to regenerate the observed groundwater hydrograph. An example from Northern Nigeria is given in Figure 3.12. Rainfall was measured at a Nigerian government meteorological station at Bauchi (Lat.: 9.8E

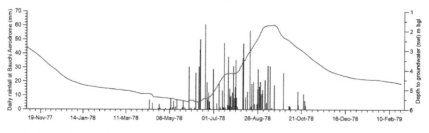

Figure 3.12 Groundwater fluctuation in response to daily rainfall at Bauchi (Acworth, 1981)

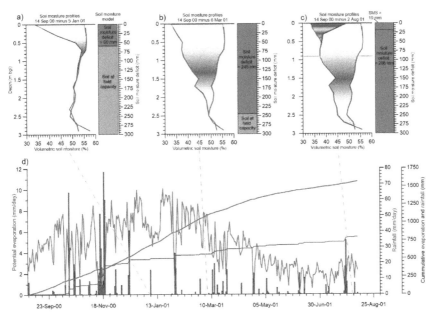

Figure 3.13 The concepts of a soil moisture deficit (SMD) and a soil moisture store (SMS) illustrated using soil moisture and climate data measured over a deep profile of smectite-dominated (98%) clay derived from weathered calc-alkali basalts at the edge of the Liverpool Ranges, NSW. A crop of lucerne was planted at the beginning of September 2000 and left in the ground throughout the experiment. Data from Plot 6 of the Hudson Farm experiment (Acworth et al., 2005)

Long.: 10.2N) and water level measurements were made daily at an observation borehole drilled into the weathered bedrock.

There are a number of significant features to the groundwater level–rainfall relationship, shown in Figure 3.12, that require inclusion in any model of recharge:

1   The groundwater level continues to fall for many weeks after the beginning of the annual rains.
2   A maximum groundwater level is reached towards the end of the wet season when the groundwater level is close to the surface.

Soil moisture deficits (SMD) and soil moisture stores (SMS) The volumetric soil moisture can be measured *in situ* using a neutron soil moisture gauge as shown in Section 13.6.2. If the balance between evapotranspiration and rainfall is kept on a daily basis and if evapotranspiration for that day is greater than rainfall, then the soil beneath the plant must be dried further with a soil moisture deficit created that is equal to the lost moisture. Correspondingly, if on a given day, rainfall occurs that is greater than the evapotranspiration that occurs, then the SMD is decreased by the balance of rainfall after the evapotranspiration for the day has been deducted. By convention (Smith, 1991), the SMD is a positive value.

These concepts can be seen in the field data shown in Figure 3.13. The three plots (a, b and c) represent changes in volumetric moisture between the beginning of September 2000 and (a) 5 January 2001, (b) 3 March 2001 and (c) 2 August 2001. In part (d) of the figure, the daily evaporation, estimated from a climate station on the site, and rainfall are shown along with the cumulative totals for these two parameters. The plot of deep black cracking clay soil (smectite) had been left fallow for several years prior to planting a crop of lucerne at the beginning of September. It is considered that the soil was at field capacity prior to the planting at the end of winter. Experimental data demonstrated that the soil moisture content at saturation was approximately 58%. At the beginning of the growing season in the spring of 2000, the development of the SMD is inhibited by a wet period where the crop would have become well established and started to develop a root network. Daily transpiration is matched by rainfall until the end of November. The dry summer months allowed a significant SMD to develop as the crop rooting network deepened and the high moisture content soils did not inhibit growth. Daily evaporation exceeded 10 mm per day during the summer and little rainfall occurred. The volumetric soil moisture was measured monthly using the neutron soil moisture gauge. Details of the site are given in Timms *et al.* (2002).

At the end of winter (July in the southern hemisphere), the maximum SMD was reached and the reduced evaporative demand saw rainfall at the end of July finally begin to wet the soil. Prior to that time, there was a negative hydraulic gradient through the soil profile with moisture moving upwards. After the rainfall at the end of July, a zero flux plane was established with soil moisture beginning to move down into the profile. Figure 3.13c indicates the establishment of the soil moisture store in effect, on top of the SMD that exists.

No recharge to the base of the soil profile can occur until the SMD is eliminated by the growing SMS. The soil profile at the surface was deeply cracked by the end of winter, and the complete profile had shrunk by up to 200 mm. This was established using soil dilation measurements.

Greve (2009) demonstrated that, despite wide cracks occurring throughout an experimental lysimeter filled with the same smectite clay soil type, the application of rainfall at 100 mm per hour did not produce drainage at the base of the profile. Drainage only occurred when the profile was returned to field capacity (Greve *et al.*, 2010). Figure 3.14 shows the results of crack monitoring as the clay dries out over several weeks. The final image (Figure 3.14e) shows the initial crack network in pink and established after drying with a second network in orange superimposed on the first. It is clear that the subsequent wetting removes any 'memory' in the clay from the location of cracks formed during the first wetting and drying episode.

Figure 3.14 Time-lapse pictures of crack development: (a) an enhanced image of the crack development after drying a first time – cracks are shown in pink; (b) a natural colour image of cracks prior to applying 100 mm water per hour; (c) second crack network shown in orange over the first crack network; (d) natural colour of the second crack network developing and (e) the second crack network fully developed and shown over the first crack network. Data and images from Greve (2009)

It is clear that the smectite-dominated soils, which are well known for their shrink-swell behaviour, can store a large quantity of water. Lucerne is a deep rooting crop that has developed an SMD of 300 mm over a growing season. This is considerably greater than more common crops, such as wheat, that develop perhaps a 250 mm SMD (Rushton, 2003).

The daily derived soil moisture deficit can be taken to vary the basic resistances in the Penman-Monteith equation. In Chapter 2 it was explained that the resistance term r_c represents the difficulty of moving water from the soil through the plant to the atmosphere. It seems reasonable then to increase that resistance term as the SMD increases. This will, in effect, reduce ET_a in a fashion that directly represents the increasing difficulty of moving water through a drying soil. Similarly, the albedo changes between 0.1 for a clear wet soil to 0.35 for a mature vegetation cover (Table 2.12). Also, the aerodynamic resistance is directly proportional to the height of the crop.

Finally, it is never clear exactly what value of SMD to commence the analysis with. In a dry Sahelian environment, it was taken as the beginning of January, where it is assumed that the SMD is at a maximum. The crops have been harvested 3 months earlier and the ground surface is parched dry in the middle of the dry season.

Evapotranspiration occurs at the full rate dictated by Equation 2.62 for as long as there is sufficient soil moisture to support this demand. A level is reached at which the actual evapotranspiration (ET_a) falls below ET_o calculated by Equation 2.62. This occurs at a soil moisture content referred to as the readily available water (RAW) or, in some references, readily evaporable water (REW). Evapotranspiration then proceeds at a reduced rate until the moisture is completely depleted. This soil moisture state is referred to as the total available water (TAW) or total evaporable water (TEW). A linear decrease between the RAW and the TAW is often assumed (Rushton, 2003). Values of RAW are impacted by the physical depth of the soil and the rooting depth of the plants.

A fairly straightforward daily balance can then be struck. Assuming the initial value of the SMD is known and that there is no rainfall that day, the SMD increases by the ET_o for that day. If the value of the SMD lies between the RAW and the TAW, the quantity of ET_o is reduced accordingly. Once the SMD reaches the TAW then no further evapotranspiration can take place. For rain days, the value of rainfall is reduced by the ET_o and the balance is deducted from the SMD. In the event that the SMD is reduced to zero, the balance of rainfall is removed from the system and is assumed to be recharged. The SMD commences the next day with a value of zero. It is noted that the terms RAW and TAW have replaced the earlier terms used by the UK Met Office of 'C' and 'D'.

In climates with a long dry season, the SMD will stay close to the maximum for several months. This will be approximately 300 mm beneath mature vegetation that is adapted to taking water from a deep soil profile. Rainfall at the beginning of the wet season will not immediately remove this deficit, yet crops can be planted in the SMS that builds at the top of the soil profile. This SMS can be depleted on days without followup rainfall and, if no further rain occurs before the crops have transpired the available moisture, the SMD returns to the full value and the early planting fails. This important feature is not recognized in the FAO approach. Acworth (2001a) and later Rushton (2003) proposed a modification to account for this initial rainfall. The modification produces results that better represent observations in the Sahel, and Acworth (1981) demonstrated that in 8 years of daily observations at Bauchi in Northern Nigeria, the results of the model closely matched the observations made at Samaru (Kowal and Kassam, 1978).

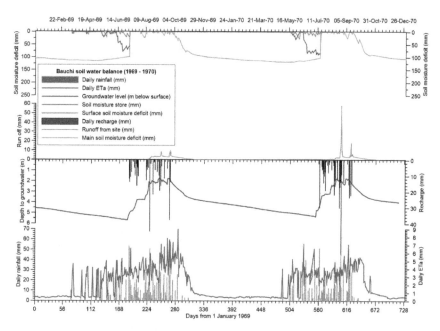

Figure 3.15 Groundwater recharge calculated using the Penman-Monteith equation and modified to include a surface soil moisture store for a Sahel climate (Acworth, 1981). Note that the early season rainfall initiates crops but can be burnt off if no followup rains occur

Results for a Sahel location The results of an approach along these lines is shown in Figure 3.15. Data in Figure 3.15 have been calculated for a soil moisture resource of 50 mm, the equivalent of a short grass crop.

Although values of RAW and TAW are still used, the opportunity is taken to vary some of the physical measurement data to better reflect changes as the SMD approaches and passes the RAW. Details are provided in Acworth (1981).

The results from 8 years of data analyzed daily are presented in Table 3.2. The data in Table 3.2 indicates the variability of components in the water balance equation and the dependence on rainfall distribution. Note particularly the data for 1970 and 1972. The rainfall in 1970 occurred in a number of major storms rather than spread out over several wet periods. The result is that the soil system was rapidly filled, allowing recharge to occur rather than leaving the water in the soil zone to be available for evapotranspiration. The major characteristics of the system are very similar in timing to those observed at Samaru (Table 3.1), indicating that this approach does satisfactorily reproduce observed conditions.

The grouping of rain days is facilitated by disturbances in the upper atmosphere easterly wind flow. If this wind flow is laminar, then conditions in the upper atmosphere do not support the growth of cumulo-nimbus storm cells that bring heavy rain. By contrast, if waves develop in the upper atmosphere easterly winds, then two possible scenarios exist (Acworth, 1981):

Winds blowing away from the equator Air moving away from the equator will be forced to speed up as the result of the Coriolis Effect. This will cause the pressure to increase in

Table 3.2 Variability of recharge derived from water balance studies at Bauchi, 10 degrees north

Year	1969	1970	1971	1972	1973	1974	1978	1979
Rainfall	1067	946	1102	929	739	1154	1182	991
No. of rain days	80	67	60	68	61	93	92	83
Cumulative rainfall before rise	300	282	296	419	316	329	211	319
Day number of rise	188	198	190	202	195	185	16	193
Day number of seepage start	206	208	196	223	251	204	178	200
Day number of seepage end	326	322	326	295	285	319	306	306
Runoff	5	92	149	33	0	132	174	0
Base flow	217	207	276	118	15	247	263	162
ET_a	848	655	682	801	742	767	755	837
Length of growing season	138	120	112	124	124	123	134	131
Day number RAW value (80) reached	297	284	278	265	280	293	304	287
Day number TAW value (100) reached	348	331	318	316	313	327	343	318
Recharge	330	398	524	245	111	481	536	263

the upper atmosphere, and any vertical instability caused by surface heating and rising air will be suppressed. Cumulo-nimbus storm cell development will not occur or at least will be short lived.

Winds blowing towards the equator Air moving towards the equator will effectively slow down as a result of the Coriolis Effect. A zone of lower pressure will be formed in the upper atmosphere into which cumulo-nimbus storm cell development can occur and is encouraged by upper atmosphere conditions. The result is that large storm cells will develop as the result of intense surface heating in the morning with the onset of heavy rain in the afternoon. The storms continue for several hours until the atmosphere stabilizes again at night.

The length of a period of consecutive rain days is then a function of the length of the waves in the upper atmosphere easterlies and the speed at which those waves move around the earth.

The water balance budgeting model has been demonstrated to replicate observed conditions by comparing the Samaru data with the Bauchi data. It has been shown that it is the grouping of rain days that has the most significant impact on the quantity of aquifer recharge. Therefore, the annual recharge can be linked to the degree to which waves in the upper atmosphere easterlies develop. The conditions that facilitate this development require better understanding.

3.4 GROUNDWATER DISCHARGE

3.4.1 Types of discharge

Groundwater discharge is the loss of water from the aquifer over a given period. Groundwater dependent ecosystems (GDE) exist where groundwater discharge occurs. Passive discharge of groundwater occurs where phreatophytic vegetation directly abstracts groundwater through deep-rooted systems. The quantity of water removed may be sufficient to halt active (spring or base flow) discharge.

Groundwater discharge can occur by a number of means:

- point discharge as a spring,
- discharge to the ocean, a stream or another aquifer,
- discharge to vegetation that uses the water to transpire. Such systems are often important groundwater dependent ecosystems,
- discharge to a lake or wetland, or
- through groundwater abstraction by pumping.

Springs

The most readily observed groundwater discharge occurs as spring flow. A spring is a point where groundwater flows out of the ground, usually along a fissure from a fractured system or simply as seepage where the water table reaches ground surface. Depending upon the constancy of the water source (rainfall or snowmelt that infiltrates the earth), a spring may be ephemeral (intermittent) or perennial (continuous).

Springs have played a major role in the settlement of many lands and were probably instrumental in the change from a hunter-gatherer society to a farming based society some 5000 years ago as the climate became drier. Springs can become the subject of religious attention, as shown in Figure 3.16.

Currently, there is much ongoing debate as the bottled water industry labels their water as spring or mineral water when it is in fact groundwater. In practice, many sites where water is taken for bottling are sites in which the chemical quality of groundwater is good. Abstraction boreholes in the aquifer allow the water to be abstracted before it flows to the surface as a spring.

Figure 3.16 Every spring has a spirit to look after it in some parts of the world (Photo: Ian Acworth)

Ocean Discharge

The quantity of discharge to the ocean is extremely difficult to establish. Wherever lime-stones aquifers exist adjacent to the sea, this discharge can be very large. It is possible to detect groundwater discharge to the ocean by the use of thermal imagery.

3.5   CONNECTIVITY OF SURFACE WATER AND GROUNDWATER

3.5.1   Introduction

Winter *et al.* (1998) from the US Geological Survey published a seminal book "Ground Water and Surface Water: A Single Resource".

It can be downloaded as a .pdf from http://pubs.usgs.gov/circ/circ1139/pdf/front.pdf.

Base flow

Streams in catchments overlying aquifers will be continuously fed with base flow from the aquifer, as long as the hydraulic head in the aquifer is above the stream water (Figure 3.17).

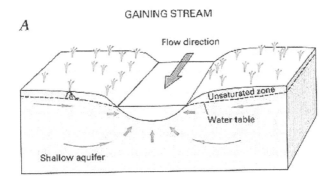

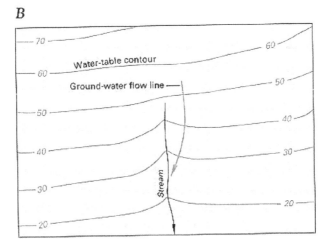

Figure 3.17 Gaining stream (Winter *et al.*, 1998)

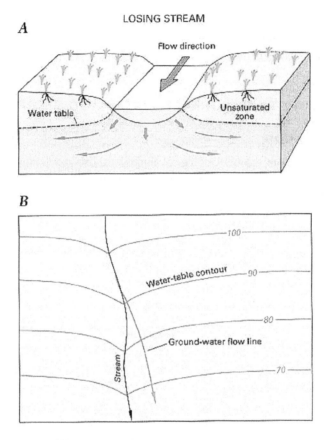

Figure 3.18 Losing stream (Winter *et al.*, 1998)

These streams are known as *gaining streams*. The quantity of base flow will depend upon the hydraulic gradient between the aquifer and the stream (Winter *et al.*, 1998).

In a gaining stream, the water table slopes towards the stream so that there is a hydraulic gradient towards the stream. Conversely, in a losing stream, the water table slopes away from the stream. A stream which is normally a gaining stream may temporarily become a losing stream during a flood. The water that moves into the aquifer is referred to as bank storage by hydrologists. This water returns to the stream after the flood peak has passed and the hydraulic gradient returns to its normal configuration. Figure 3.18 shows a losing stream.

Groundwater abstraction close to a river may reverse the normal hydraulic gradient, creating a recharge boundary for the aquifer. Water then flows continuously from the stream to the aquifer and towards the abstraction bore.

There may be more substantial flow into the beds of rivers from groundwater discharge or the spring flow is so large that it forms a stream straight away. The major springs associated with basalts around Rotorua in New Zealand come into this category, as shown in Figure 3.19.

The type of rock in a catchment will have a major impact upon the base flow response in a stream. If rocks have little porosity and low storage, then the response to rainfall in the stream will be very spikey. By contrast, rain falling on a highly permeable limestone

Figure 3.19 Major spring flow through the base of the stream emanating from volcanic rocks around Rotorua in New Zealand (Photo: Ian Acworth)

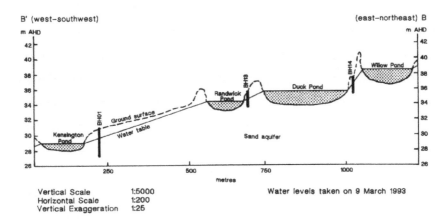

Figure 3.20 Water level in lakes and bores in Centennial Park, Sydney (Dudgeon, 1993)

aquifer rapidly infiltrates, and even major storms will have a very subdued response in the streams which continue to discharge water from storage.

3.5.2   Interaction of lakes or wetlands with groundwater

When lakes and wetlands occur in sands and gravels, they will be inextricably linked to the groundwater system. The Lachlan Swamps or the ponds in Centennial Park provide good examples in the Botany Sands Aquifer (Figure 3.20), where flow through the Botany Aquifer in Centennial Park is shown to be directly related to the presence of ponds. Townley and Trefry (2000) gave a detailed assessment of the 3D head distribution around lakes that can be used to better understand the movement of groundwater into,

Figure 3.21 Multi-channel manometer board in use to measure vertical head differences (Photo: Ian Acworth)

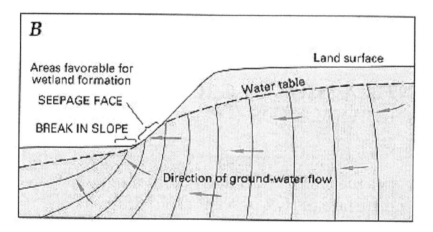

Figure 3.22 Discharge at the break in slope

through or from lakes. These authors carried out extensive work on the Swan Coastal Plain. Acworth and Jorstad (2006) introduced multi-channel manometry techniques for mapping the curvature of flow lines close to lakes. This equipment is shown in Figure 3.21. Acworth (2007) and Winter *et al.* (1998) give further examples (shown in Figures. 3.22 and 3.23).

3.5.3   Hyporheic zone

Some gaining streams have reaches that lose water to the aquifer under normal conditions of stream flow (Winter *et al.*, 1998). This local recycling of water is considered to be very important in cycling nutrients from the ground into the stream system. The zone of

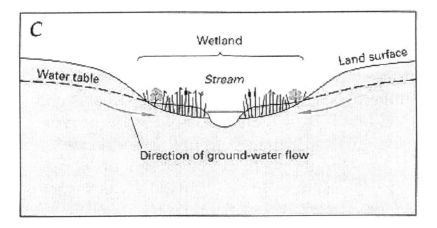

Figure 3.23 Gaining stream

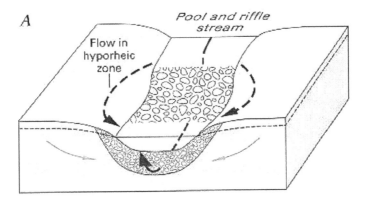

Figure 3.24 Schematic of a pool and ripple system

interchange, which occurs vertically in a pool and riffle system (Figures. 3.24 and 3.25) or horizontally in a meander loop (Figure 3.26), is called the hyporheic zone.

3.5.4 Groundwater discharge to the ocean

Introduction

One of the most complex zones where groundwater mixes with surface water is at the beach. The USGS Circular 1262 by Barlow (2003) is an excellent summary and can be downloaded from the web at http://water.usgs.gov/ogw/gwrp/saltwater/index.html.

Figure 3.27 gives a typical representation of the groundwater/sea water interface zone drawn for the Atlantic margin in the USA, but it could equally represent any beach zone adjacent to a non-destructive plate margin. The east and west coasts of Australia fall into this category.

Historical records of freshwater springs are evident as far back as 60 BC with recorded details of utilization and collection by separate colonies for land and ship uses. The Roman

Figure 3.25 Pool and riffle system in a stream

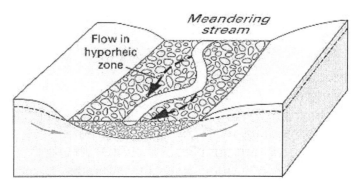

Figure 3.26 Hyporheic flow through the banks of a meander belt

geographer Strabo (63 BC–AD 21) recorded a fresh submarine groundwater spring near the island of Aradus (4 km from Latakia, Syria, located in the Mediterranean); this water was collected by boat, using a lead funnel and leather tube, and transported to the city as a source of fresh water (Taniguchi *et al.*, 2003). Water vendors in Bahrain collected potable water from offshore submarine springs for shipboard and land use (Taniguchi *et al.*, 2002). The open structures of weathered limestone carbonates located along the coastal regions of the Mediterranean, Persian Gulf, Black Sea and Red Sea are the host of submarine flows with upwelling offshore plumes where ancient mariners are believed to have restocked their freshwater supplies. Pliny the Elder (1st century) recorded submarine "springs bubbling fresh water as if from pipes" along the Black Sea, and writings by Pausanius (2nd century) tells of the Etruscans using the coastal springs for 'hot baths' (Taniguchi *et al.*, 2003, 2002). The idea of fresh underground springs is not a new concept, as the historical evidence shows us. However, it is only in the last couple of decades that the concept has earned increased attention as it is becoming understood that that this process may represent a potentially important role in coastal and freshwater management.

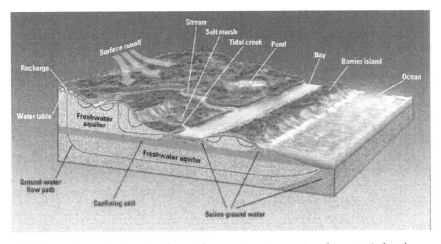

Figure 3.27 Typical groundwater interface with sea water at a passive plate margin beach

Confined or artesian aquifers are common off many coasts and have been reported to outcrop from the ocean floor at any depth and any distance from the shore (recorded as far as 120 km offshore). The Atlantic Margin Coring Project in 1979 revealed "relatively fresh groundwater occurs beneath much of the Atlantic ocean shelf" (Johannes, 1980). Off the coast of Western Australia, an artesian aquifer was struck 6 m below the sea floor in 20 m of water (Johannes, 1980); and offshore from Jacksonville, Florida, a geologic exploratory well 43 km from land struck an artesian aquifer that flowed up the drill pipe and had a hydrostatic head of 9 m above sea level (Taniguchi *et al.*, 2003). The variability in the factors concerning submarine groundwater discharges (SGDs) – confined or unconfined; channelled or permeable sediment; seeping or flowing; tidal oscillations or hydraulic heads; and precipitation or urbanization runoff – are just some of the questions defining the parameters of the SGD that need more detailed research to better define the risk of seawater intrusion.

The occurrence of fresh water beneath the ocean is another indication that sea levels around the world have been on average much lower than they are currently. When the sea level is lower, recharge and discharge processes are also impacted. The rapid rise in global sea level that occurred approximately 13,000 to 11,000 years BP will have swamped coastal groundwater systems generated in the previous 50,000 years. As groundwater moves so slowly, it is not surprising that we still see offshore springs.

The dynamics of the tides and waves at the beach directly impact the discharging groundwater (Turner *et al.*, 1996; Acworth and Dasey, 2003) with groundwater being set up at an elevation close to the mean high tide mark. This observation has a major implication for models of groundwater flow that use the beach as a constant-head boundary.

Many possible configurations of fresh water and groundwater are possible. The controlling factor will be the variation in hydraulic conductivity of the sediments. Figure 3.28 shows relationships in the Florida aquifers (Barlow, 2003).

On a beach scale, there are many important processes that are under investigation (Taniguchi, 2002; Moore, 1999; Hays and Ullman, 2007), with the chemical implications of mixing proving particularly interesting. There is now the indication that shallow groundwater flux to the oceans is a major source of contaminant and nutrients. Figure 3.29 shows the

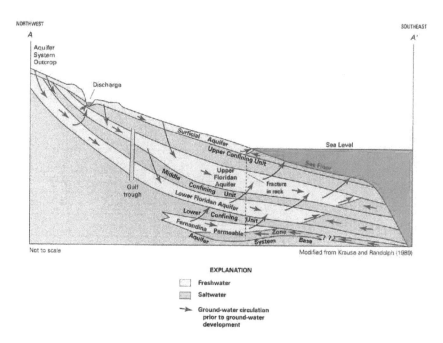

Figure 3.28 Multiple interfaces between fresh groundwater and sea water in Florida

dependence of vegetation on the beach to different salinity water. There are a also major implications for the biota.

3.6   GROUNDWATER DEPENDENT ECOSYSTEMS

The recent references by Hancock *et al.* (2005) and Eamus *et al.* (2006) provide a good summary of groundwater dependent ecosystems (GDEs). The importance of GDEs has only recently been appreciated by the ecological community. From a groundwater resource perspective, the concept and importance of base flows and spring discharge to rivers or the ocean has long been acknowledged (Freeze and Cherry, 1979). As an example, Figure 3.30 shows a permanent groundwater-fed pool in the McDonald Ranges west of Alice Springs in the Northern Territory. Groundwater seeps from the Mereenie Sandstone to balance losses from the pool by direct evaporation from the pool surface and evapotranspiration by the associated vegetation. Figure 3.31 shows a semi-permanent stretch of Maules Creek upstream of its confluence with the Namoi River in northern New South Wales. Shallow groundwater commences discharging to deep pools in the river bed alluvium less than 300 m upstream from the location shown in the figure. Flow continues along the stream channel for 10 kms before it disappears back below the alluvium surface. The stream provides an important source of very clean cool groundwater for many months after a major recharge event and supports various endemic species.

What is a groundwater dependent ecosystem? Eamus *et al.* (2006) provide the following definition:

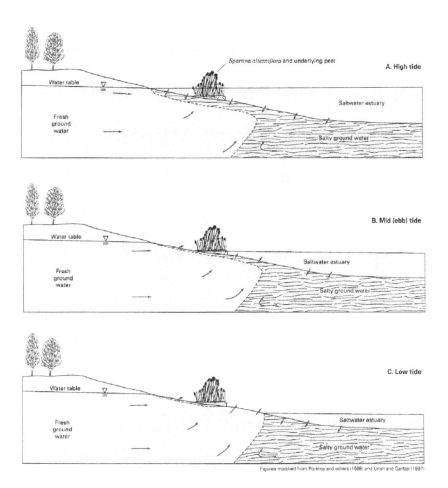

Figure 3.29 Beachface dynamics

An ecosystem can be defined as a set of living organisms that interact with themselves and with their environment at a given site (Eamus *et al.*, 2006). An ecosystem will seldom have any clearly defined boundary.

In semi-arid and arid environments, ecosystems are almost entirely dependent upon the presence of groundwater. This is why our ancestors initially settled around springs. In humid environments, the dependency is somewhat less but may still be significant.

Eamus *et al.* (2006) define three simple classes of GDE. These are:

1   aquifer and cave systems,
2   ecosystems dependent on the surface expression of groundwater, and
3   all vegetation dependent upon the subsurface presence of groundwater.

While noting that classifications such as those suggested by Eamus *et al.* (2006) are necessary to further discussion, it will be accepted that the differences between classes 2 and 3

Figure 3.30 Seepage from the Mereenie Sandstone in the West MacDonald Ranges west of Alice Springs supports this permanent pool of water in an otherwise arid region (Photo: Ian Acworth)

Figure 3.31 Semi-permanent discharge from an alluvial aquifer supports flow in a tributary of the Namoi River in northern New South Wales (Photo: Ian Acworth)

above are frequently unclear. How do we classify ecosystems supported by variable ground-water depths? What do we mean by groundwater? Does this comprise only water below the depth of permanent saturation? What about zones that are periodically saturated with groundwater? Are these aquifers, etc. etc. etc.?

Figure 3.32 Karst development in limestone (Photo: Ian Acworth)

Figure 3.33 Small endemic communities of fish living in the spring (Photo: Ian Acworth)

This uncertainty leads to a further sub-classification:

obligate dependency – In species with obligate dependency, the *absence of groundwater* results in the loss of the species. The fish in Figure 3.33 show obligate dependency. If the spring failed, the fish would be wiped out.

facultative dependency – Species showing facultative dependency will use groundwater or surface water when they are available. As groundwater is more stable that surface water, it is the presence of groundwater that supports vegetation throughout long drought periods.

3.6.1   Aquifer and cave systems

Aquifer and cave systems where stygofauna (organisms living in groundwater) exist within or associated with the groundwater form a major subdivision of GDE. Karstic systems are clearly the easiest to comprehend here as the secondary porosity (the holes shown in Figure 3.32) in the limestone karst are big enough to allow fish to swim in. This is well demonstrated by Figures 3.33 and 3.32, where unique communities of fish exist at the karst spring and are completely limited to this location, as the stream produced by the spring does not flow very far down the river bed. Fractured rocks may also contain stygofauna of a similar size, but sands and alluvium can only contain microscopic stygofauna. Examples of these are shown in Figures 3.34 and 3.35.

Stygofauna that require groundwater to live are referred to (by the biological community) as stygobites (obligate groundwater inhabitants). The hyporheic zone falls into this category because these ecotones support stygobites.

Hypogean life exists in a continuum through different types of karstic, cave, porous and fissured aquifers. Gilbert (1996) asserted that life there is as diversified as in the surface biotic milieu and that aquifers form the most extensive array of freshwater ecosystems on our planet. That these ecosystems exist within the groundwater makes the nature of their dependence absolute, albeit poorly understood and appreciated.

Obligate inhabitants of groundwaters, referred to as stygobionts – collectively stygofauna – are extremely diverse in Australia, especially in the north-west (Humphreys, 1999a; Wilson and Keable, 1999). They are characterized by a series of convergent morphological, behavioural and physiological features termed troglomorphies (cave form) including loss of eyes and pigment and elongation of appendages, thus appearing white, fragile and translucent. Recently a start has been made in determining the composition and regional variation of the stygofauna in Australia (Humphreys, 1999a; Eberhard and Spate, 1995), but knowledge of the basic biology and ecology is largely lacking.

From studies on the biology and ecology of stygofauna in Europe and America (Gibert et al., 1994), several generalizations can be made concerning stygobites. By comparison with their lineage, stygal species characteristically have sparse populations, are long lived, have slow growth rates, mature late and have a small clutch size (often one or two) and are adapted to low food energy levels. In consequence, the population processes may be easily disrupted by anthropogenic impacts to the ecosystem.

Protected below the ground, the stygofauna contains entire major lineages (orders or classes) unknown on the surface globally since the early Mesozoic era. In near coastal groundwaters alone, 10 new families of Crustacea have been described in the last decade. Numerous major taxa (classes, orders, families and genera), previously unknown in Australia, several even unknown in the southern hemisphere, have recently been found inhabiting groundwaters in arid north-western Australia (Humphreys, 2000).

Aquifers important to stygofauna include near coastal groundwater affected by marine tides (anchia-line systems especially those in limestone and basalt Sket (1996), Humphreys (1999b), or Iliffe (2000)). Stygofauna also occur in karstic systems mainly in carbonate rocks and in river gravels. Groundwater calcrete aquifers (Humphreys, 1999a) associated with recent hydrogeochemical cycles within the paleodrainage channels of arid Australia (Morgan, 1993) have also proven important. These stygofauna, especially those in saline waters, have not been reported outside Australia.

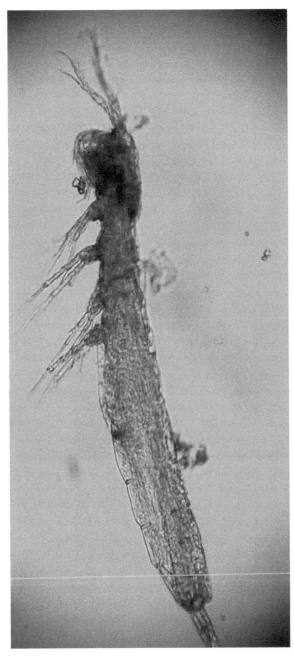

Figure 3.34 Harpacticoid copepod from a cobble bar at the foot of a sandstone cliff, Running Waters close to Alice Springs, NT. Fluid EC in the hole was 5,034 μS/cm (Photo: Moya Tomlinson)

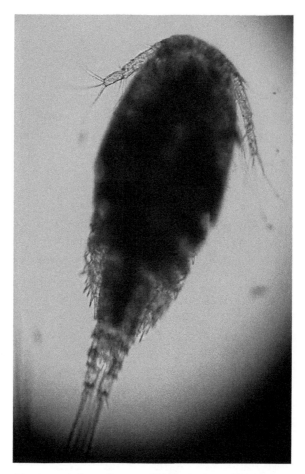

Figure 3.35 Cyclopoid copepod from Ilamurta Well, close to Alice Springs, NT (Photo: Moya Tomlinson)

While most areas that have been sampled yield at least some stygofauna, certain areas are notable. The Nullarbor and the southwest karst areas are unexpectedly impoverished, while the Cape Range/Barrow Island karst, and the groundwater calcretes of the arid Western Shield region, are remarkably rich in stygobionts (Humphreys, 1999a).

3.6.2 Determining groundwater dependence of ecosystems

It is a general property of ecosystems that should a physical resource be available within the biosphere, it will be exploited. This is true for light, for nutrients, and for water. Further, where that resource is at all limiting or in excess, there will develop some functional dependence on the timing and extent of its availability or super abundance.

In the limiting case, most of Australia has either seasonal drought or a semi-arid to arid climate, and if groundwater exists within the reach of organisms, ecosystems will develop which are to some extent reliant upon it. Reducing the availability of groundwater may

Figure 3.36 River flow supported only by groundwater discharge from a limestone aquifer (Photo: Ian Acworth)

result in a gradual, proportional decrease in the health or aerial extent of a given ecosystem; in other cases, a threshold of availability may be reached at which the entire system collapses. Knowledge regarding this response for particular Australian ecosystems is limited.

Thus it is unlikely that there are ecosystems in Australia that exist in the presence of groundwater that have no dependence upon it, and were groundwater to be made more or less available, would not suffer. Nevertheless, intuitively one assumes that the magnitude of the dependence on groundwater is proportional to the fraction of the annual water budget that ecosystem derives from groundwater. This aspect of dependence is the principal criterion used to assess and rank significance.

In Australia, there is a tendency to reflect a vegetation-oriented bias as vegetation health generally underpins the health of an ecosystem. However, it is also true that some systems have faunal components with a more direct dependence on groundwater. For instance, fauna may rely directly on groundwater as a source of drinking water in the dry season via seepage, springs, etc., or as the direct source of habitat in the case of aquatic animals (e.g., crocodiles, turtles, fish, macro-invertebrates). The role of groundwater in maintaining and supporting fauna which are not related to vegetation *per se* is very poorly known and understood.

Groundwater use can be concluded by a positive answer to one or more of the following:

*   A stream or river flowing all year through an arid or semi-arid zone. Figure 3.36 provides an excellent example of such a case. The water is even coloured blue from the calcite dissolved in it – indicating that it comes from discharge of a limestone aquifer.
*   Does the fluid EC in the river decrease downstream? This is a clear indication of dilution by influent fresh groundwater. The reverse, fluid EC increasing downstream, may also indicate addition of salts by influent groundwater of a lower quality, as for the Murray River, but may also indicate evaporative concentration.
*   Does the quantity of river flow increase downstream even where there are no tributaries?

- Do plants flower and seed in the dry season? The mango in Africa is a good example of this. Towards the end of the dry season, the tree produces copious fruit that are produced by tapping groundwater reserves. The presence of a mango is then taken as an indication of groundwater at depth. As actively growing vegetation absorbs infrared radiation, the difference between infrared activity between two pictures, one taken in the wet season when general vegetation growth is apparent and the second taken towards the end of the dry season when only the vegetation that can tap groundwater grows, is a good way to locate groundwater.
- Do plant water use measurements indicate more water use than that available by rainfall? This is difficult and time consuming to establish. There is also a problem in the definition of groundwater implicit in the question (Eamus *et al.*, 2006). The plants could simply be using moisture in the deep soil zone above the permanent zone of saturation that is often taken to represent the water table and the start of groundwater. Work in the Pilbara with stable isotopes has shown that only some individuals in a population of *E. camaldulensis* use groundwater at depths of 25 m while others in the same population use soil moisture. It is dangerous to generalize on water use at a species level.

The use of stable isotopes (oxygen and deuterium) can also be used to indicate the origin of water in the ecosystem. However, this technique relies upon the groundwater having a different isotopic signature to rainfall – which is not always the case.

Chapter 4

Physical properties of soil and the hydraulic head

4.1 PHYSICAL PROPERTIES OF SOIL

Notes in this section are based on soil physics texts (Campbell, 1985; Jury *et al.*, 1991). Soil is a three-phase system made up of solids, liquids and gases. The solid phase contains mineral particles and organic materials. The liquid phase consists of water and dissolved minerals. Liquids immiscible with water are not considered here. The gas phase is composed of about 80% nitrogen, with oxygen and carbon dioxide making up most of the remainder.

4.1.1 Surface area of soil particles

The surface area of particles directly impacts upon the amount of water adsorbed to the particle and therefore to a range of other properties such as the bulk electrical conductivity and the cation exchange capacity. Systems composed of dispersed particles of small size have a very large surface area per unit mass of material.

The specific surface area of a particle is the surface area divided by the mass of the particle. Consider a spherical particle of density ρ and radius R, then the surface area of the particle is $a = 4\pi R^2$. The mass m of the sphere can be expressed as $m = \rho V = \rho 4\pi R^3/3$, where V is the volume. Thus, the specific surface area s is given by:

$$s = \frac{a}{m} = \frac{3}{\rho R} \tag{4.1}$$

The soil organic fraction often has an extremely high specific surface area (often as high as 1000 $m^2\ g^{-1}$). Specific surface areas for generalized particles are given in Table 4.1 (Jury *et al.*, 1991). A representative soil particle density of 2650 kg/m^3 has been used. Specific surface areas and cation exchange capacities for selected clays are given in Table 4.2 (Jury *et al.*, 1991). Note that the data is greater than the indicated data for generalized particles in Table 4.1 because, apart from kaolinite, the clays are multi-layered.

4.1.2 Bulk density, water content and porosity

In a given amount of soil, the total mass M_t is divided between the mass of gasses M_g, the mass of liquid M_l and the mass of solids M_s. Similarly, the total volume V_t is divided between the volume of gasses V_g, the volume of liquid V_l and the volume of solids V_s. Note that the volume of fluid V_f is the sum of V_g and V_l.

Table 4.1 Specific surface areas of generalized particles found in soil. Note the clay data is calculated for a flat disk of negligible thickness (Jury et al., 1991)

Particle	Effective Radius (m)	Mass (kg)	Area (m²)	Specific Surface Area (m² kg⁻¹)
Gravel	1.5×10^{-3}	3.746×10^{-5}	2.827×10^{-5}	0.754
Medium sand	1.5×10^{-4}	3.746×10^{-8}	2.827×10^{-7}	7.457
Medium silt	1.00×10^{-6}	1.110×10^{-11}	1.257×10^{-9}	113.207
Clay	2.5×10^{-6}	1.734×10^{-13}	1.963×10^{-11}	755

Table 4.2 Specific surface areas and cation exchange capacity for selected clay minerals based upon experimental data (Jury et al., 1991)

Clay Mineral	Specific Surface Area (m² kg⁻¹)	CEC (mE kg⁻¹)
Kaolinites	5×10^{3} to 2.0×10^{4}	30–150
Micas (Illites)	1.0×10^{5} to 2.0×10^{5}	100–400
Vermiculites	3.0×10^{5} to 5.0×10^{5}	1000–1500
Smectites	7.0×10^{5} to 8.0×10^{5}	800–1500

From these definitions, the following relationships follow:

$$M_t = M_g + M_l + M_s \tag{4.2}$$

and:

$$V_t = V_g + V_l + V_s \tag{4.3}$$

$$V_f = V_g + V_l \tag{4.4}$$

Particle density [Mg m⁻³]:

$$\rho_s = \frac{M_s}{V_s} \tag{4.5}$$

Dry bulk density [Mg m⁻³]:

$$\rho_b = \frac{M_s}{V_t} \tag{4.6}$$

Total porosity:

$$\phi_f = \frac{V_f}{V_t} = \frac{V_g + V_l}{V_t} \tag{4.7}$$

Gas-filled porosity:

$$\phi_g = \frac{V_g}{V_t} \tag{4.8}$$

Void ratio:

$$e = \frac{V_f}{V_s} = \frac{V_g + V_l}{V_s} = \frac{V_f}{V_t - V_f} \tag{4.9}$$

Mass wetness or mass basis water content:

$$w = \frac{M_l}{M_s} \tag{4.10}$$

Volume wetness or volumetric water content:

$$\theta_v = \phi_l = \frac{V_l}{V_t} \tag{4.11}$$

Gravimetric water content:

$$\theta_g = \frac{M_l}{M_s} \tag{4.12}$$

Degree of saturation:

$$s = \frac{V_l}{V_f} = \frac{V_l}{V_g + V_l} \tag{4.13}$$

4.1.3   Relationships between variables

The following relationships can be derived from Equations 4.5 to 4.13:

$$\phi_f = 1 - \frac{\rho_b}{\rho_s} \tag{4.14}$$

$$\phi_g = \phi_f - \theta_v = 1 - \frac{\rho_b}{\rho_s} - \theta_v \tag{4.15}$$

$$e = \frac{\phi_f}{1 - \phi_f} \tag{4.16}$$

Mass wetness and volume wetness are related by:

$$\theta_v = w\frac{\rho_b}{\rho_l} \tag{4.17}$$

and:

$$w = \frac{\theta_v \rho_l}{\rho_b} \tag{4.18}$$

Table 4.3 Densities of soil (Jury et al., 1991)

Component	Density [Mg m^{-3}]	Component	Density [Mg m^{-3}]
Quartz	2.66	Orthoclase	2.5–2.6
Clay minerals	2.65	Mica	2.8–3.2
Organic matter	1.30	Limonite	3.4–4.0
Water	1.00	FeOH$_3$	3.73
Air (20 °C)	0.0012	Ice	0.92

Porosity can be determined on a field sample of known volume by noting:

$$\phi_l = \theta_v = \frac{V_l}{V_t} = \frac{M_t}{V_t} \times \frac{V_l}{M_t} \qquad (4.19)$$

where V_l is determined by weighing the sample.

Bulk density (ρ_b) can be related to fluid porosity by:

$$\rho_b = \rho_m(1 - \phi) \qquad (4.20)$$

The bulk density can be determined by field measurement. A known volume of soil is taken from the field using a core tube and transferred to the laboratory where it is heated for 24 hours at 105 °C in an oven. The liquid is driven off and the dry sample weighed. Equation 4.6 gives a value for ρ_b.

Bulk density can also be determined on intact clods by coating them with a thin layer of paraffin wax and measuring the volume by displacement.

Impact of adsorption on bulk density

In a review of the Russian literature concerning experiments to determine the bulk density of natural materials, Galperin *et al.* (1993) reported that adsorption of water onto the clay matrix can produce bulk density values of 1400 Mg m^{-3}. They also reported that the physical properties of the water, such as the dielectric permittivity and the freezing point of water adsorbed onto clay, differed markedly from pure water. These results were not universally accepted. However, Zhang and Lu (2018) have published an extensive theoretical review that establishes upper and lower bounds to bulk density. The upper bound is around 1872 Mg m^{-3} and the lower bound is around 0.995 Mg m^{-3}. The impact of this theoretical understanding will take time to work through the literature.

4.1.4   Typical values of physical properties

Table 4.3 gives particle densities for typical soil constituents.

4.2   WATER CONTENT

4.2.1   Introduction

Water content is usually measured on a mass basis by drying a sample at 105 °C and then computing the ratio of mass of water lost to dry mass of the sample (Equation 4.10).

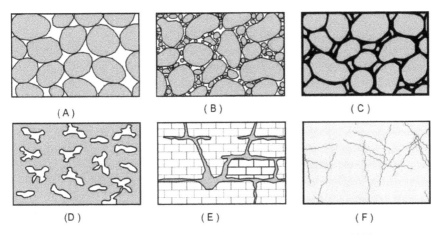

Figure 4.1 Relation between texture and porosity (adapted from Meinzer (1923))

Oven-dry soil is assumed to be at zero water content. Air-dry soil is in equilibrium with atmospheric moisture and therefore has a moisture content greater than zero. The highest water contents attained in the field in a freely draining soil are below the saturation value. Some pores always remain air filled. A typical maximum water content for a soil with a porosity of 0.5 would be between 0.45 and 0.48. Near-saturation conditions are only obtained for short periods following rain or irrigation. The soil quickly approaches a water content termed field capacity. Field capacity water contents vary greatly but often are approximately half of saturation values. The terms saturated and unsaturated require some definition. The volumetric water content (Equation 4.11) is the ratio of the water-saturated pore space to the total volume.

The degree of saturation, described by Equation 4.13, is the ratio of the water-saturated pore space to the total pore space. For a fully saturated medium, the degree of saturation (s) is equal to unity and to the total porosity (ϕ_f). At the contact between the lower part of a dry porous material and a saturated material, the water rises to a certain height above the top of the saturated material. This gives rise to the tension saturated zone or capillary rise zone. The driving force for this rise is the surface tension of the water.

The water table is the point at which the pressure in the formation is exactly atmospheric. The intermediate zone lies above the tension saturated zone and consists of water in the form of thin films adhering to pore space linings. This water is free to drain downwards under the force of gravity. The soil, intermediate and tension saturated zones are collectively referred to as the unsaturated or vadose zone.

4.2.2 Porosity of different rock/soil types

Primary porosity is due to the properties of the soil or rock matrix.
Secondary porosity may be due to secondary solution or fracturing of the rock mass.

Various texture and porosity relationships have been described (Meinzer, 1923). In Figure 4.1, six examples of porosity development are given, *viz.*:

Figure 4.2 Solution channels in a limestone showing major anisotropy in distribution (E) (Photo: Ian Acworth)

Figure 4.3 Environment producing a very uniform sand deposit (A). Note the size of the trees at the base of the dune for a scale (Photo: Ian Acworth)

A Well-sorted sedimentary deposit having high porosity.
B Poorly sorted sedimentary deposit having low porosity.
C Well-sorted sedimentary deposit whose porosity has been diminished by the deposition of mineral matter in the interstices.
D High-porosity deposit – such as a lava flow with vesicles – but the pore space is not connected.
E Rock rendered porous by solution.
F Rock rendered porous by fracturing.

An example of a limestone showing major anisotropy in pore distribution is shown in Figure 4.2.

A range of possible porosity values is shown in Table 4.4 below.

Figure 4.4 Poorly sorted material (conglomerate) (B) (Photo: Ian Acworth)

Figure 4.5 Poorly sorted material (conglomerate cemented with calcrete) (D) (Photo: Ian Acworth)

Table 4.4 Range of porosity values

Material	ϕ (%)
Unconsolidated deposits	
Gravel	25–40
Sand	25–50
Silt	35–50
Clay	40–70
Rocks	
Fractured basalt	5–50
Karst limestone	5–50
Sandstone	5–30
Limestone, dolomite	0–20
Shale	0–10
Fractured crystalline rock	0–10
Dense crystalline rock	0–5

4.3   WATER POTENTIAL

4.3.1   Introduction

Water potential plays a similar role in water flow theory to the role of temperature in heat flow problems or voltage in electrical circuit theory. Water flows in response to gradients in water potential. When water potential is uniform across a boundary, no water will flow, even though water content may be different on the two sides of the boundary.

The water potential is the potential energy per unit mass of water in the system, compared to that of pure, free water at atmospheric pressure. The units of potential are joules/kg and the dimensions of these units reduce to L^2/T^2.

The total water potential is the sum of several component potentials:

$$\Phi = \Phi_m + \Phi_o + \Phi_u + \Phi_p + \Phi_\Omega + \Phi_g \tag{4.21}$$

where:
m is the matric potential,
o is the osmotic potential,
u is the velocity potential,
p is the pressure potential,
Ω is the overburden potential, and
g is the gravitational potential.

In general, only two or three of these potentials require consideration in any particular problem. The required combination will depend upon whether a saturated problem, an unsaturated problem or a combination of the two is being studied.

4.3.2   Osmotic potential

Water molecules have a slightly unequal distribution of electrical charge as the result of the small angle between the molecular bonds in the water molecule (105 °C). The unequal distribution of charge is referred to as a dipole moment. Other ions are attracted by the electric field surrounding the individual water molecules and tend to cluster around them. The result of the clustering is to reduce the energy state of the water. The hydrostatic pressure which balances the ionic attraction of the water molecules at equilibrium in a solution is referred to as the osmotic pressure. The osmotic pressure is a significant component of the total water potential in soil plant systems. The potential may be approximately calculated from the following relationship:

$$\Phi_o = -C\chi\eta RT \tag{4.22}$$

where:
T is the temperature in degrees Kelvin,
R is the universal gas constant (8.3143 J/K mol),
Φ_o is the osmotic potential (J/kg),
C is the concentration (moles/kg]),
η is the number of particles in solution for each mole of solute,
η would equal unity for a non-ionizing solute, and 2 for a salt such as NaCl, and
χ is an osmotic coefficient.

If mixtures of solutes are present, then the total osmotic potential is the sum of the individual contributions from the components. The osmotic potential can also be estimated from the electrical conductivity of the soil saturation extract:

$$\Phi_o = -0.36\sigma_{se}$$
(4.23)

where:
σ_{se} is the soil extract conductivity measured in mS/m.

4.3.3   Matric and pressure potentials

Texts on soil physics (Campbell, 1985; Hanks and Ashcroft, 1980) include reference to matric potential as a separate entity which is always less than or equal to zero, and reference to a pressure potential which is always equal to or greater than zero. Some authors (Freeze and Cherry, 1979) combine these potentials as one, which is less than zero in the unsaturated zone and greater than zero in the saturated zone. There is no consensus regarding this use of terminology, and both approaches appear in the literature.

Matric potential

The matric potential is important as a driving force for flow in unsaturated soil and in the cell walls of root cortex and leaf mesophyll tissue. The matric potential is measured using a tensiometer.

Matric potential is defined as the amount of work, per unit mass of water, required to transport an infinitesimal quantity of soil water from the soil matrix to a reference pool of the same soil water at the same elevation, pressure and temperature.

The matric potential under a curved air-water interface, such as exists in a capillary tube or soil pore, is given by the capillary rise Equation 4.24:

$$\Phi_m = -\frac{2\sigma}{r\rho_w}$$
(4.24)

where:
r is the radius of curvature of the interface,
σ is the surface tension, and
p_w is the water density.

Note that the sign of the matrix potential is always negative. This indicates that water is held in position by surface tension effects and is not free to drain out of the pore space. Equation 4.24 can be used to find the equivalent radius of the largest water-filled pore in a soil at a given matric potential. For example, in a soil at $\Phi_m = -100$ J/kg, you would expect pores with radii larger than $r = -2 \times 7.27 \times \frac{10^{-2}}{10^3} \times (-100) = 1.45\mu m$ to be air filled.

At some water potential, probably less than 10^{-4}, the capillary analogy breaks down as most of the water is adsorbed in layers on particle surfaces rather than being held in pores between particles.

Pressure potential

The fluid pressure potential is calculated from:

$$\Phi_p = \frac{p}{\rho_w} = g\psi \tag{4.25}$$

where:
ψ is the depth of the water column above the reference point,
p is the fluid pressure measured in Pascals,
p_w is the fluid density, and
g is the acceleration due to gravity.
The fluid pressure is always positive as ψ is always greater than zero.

4.3.4 Gravitational potential

The gravitational component of the water potential is fundamentally different to any of the other components since it is the result of "body forces" applied to the water as a consequence of its position in the earth's gravitational field. The gravitational potential is calculated from:

$$\Phi_g = gz \tag{4.26}$$

where:
z is the distance from the point of measurement to a reference level where Φ_g is taken as zero. The reference level is usually taken at mean sea level by hydrogeologists and the soil surface or the water table by soil physicists.

4.3.5 Overburden potential

The overburden potential is related to the pressure potential in that it is the increase in potential of water in a porous system due to weight. Pressure potential is related only to the weight of the water column above a point. In porous media which are capable of deformation, the matrix itself can impart a pressure to the water in the pore space. In other words, the water 'feels' the weight of the overlying water and some part of the weight of the formation containing the water.

When pressure is applied to the matrix, some of the pressure is borne by the solid structure of the matrix itself, but a part may be transferred to the water in the pore space of the matrix. The overburden potential can be present in an unsaturated porous medium. It can become a significant component of the total potential in expanding layer clays (smectites).

In the Murray Basin alone there are approximately 250,000 square kilometres of swelling clay soils. They represent about 10% of the basin and almost 50% of the irrigated soils, and they support irrigated and dryland crops which were worth more than $A 1.5×10^9 annually in 2000 (Smiles, 2000). The field measurement of overburden pressure is complicated by the fact that the aquifer skeleton is not rigid but distorts (Talsma and van der Lelij, 1976; Talsma, 1977; Domenico and Schwartz, 1997; Smiles, 2000). It is also worthy of note that the great majority of groundwater models use a mathematical formulation that assumes the aquifer skeleton is rigid and therefore ignores the overburden pressure.

4.3.6 Velocity potential

The velocity potential arises from the energy per mass which the water has as a result of its motion (kinetic energy). The velocity potential is calculated as:

$$\Phi_u = \frac{u^2}{2} \tag{4.27}$$

where:
u is the velocity of the groundwater.

The velocity potential is normally several orders of magnitude less than the pressure or gravity potentials, as the velocity of groundwater is generally very slow. The term is included here to facilitate comparison with hydraulics (the Bernoulli equation).

4.3.7 Systems of units for the total soil water potential

There are several alternative systems of units in which the total potential and its components can be described, depending upon whether the quantity of pure water is expressed as a mass, a volume or a weight. These alternative systems can readily confuse, and a summary table is provided (Table 4.5).

As noted above, water flows from higher potential energy to lower potential energy. Indicative values of water potential at different moisture contents are given in Table 4.6.

Table 4.5 Systems of units for the water potential

Description	Symbol	Name	Dimensions	SI Unit
Energy per mass	Φ	Water potential	L^2T^{-2}	J/kg
Energy per volume	p	Water pressure	$ML^{-1}T^{-2}$	N/m^2
Energy per weight	h	Hydraulic head	L	m

Table 4.6 Range of values for the pressure (matric) potential

Moisture Content Field State	Matric Potential Φ_m		
	(J/kg)	(Bars)	(m)
Saturation $\theta = \varnothing$	−0.098	-9.8×10^{-4}	−0.01
Field capacity	−9.8	−0.098	−1.0
Wilting point of many plants	−1470	−14.7	−150
Air dry (relative humidity = 0.85)	-2.16×10^4	−216	−2,200

4.4 HYDRAULIC HEAD

4.4.1 Introduction

In many problems involving prediction of flow in fully saturated aquifers, the water potential reduces to the sum of the pressure and gravitational potentials:

$$\Phi = \Phi_p + \Phi_g \tag{4.28}$$

From Equations 4.25 and 4.26, the total potential can therefore be written as:

$$\Phi = \frac{p}{\rho_w} + gz \tag{4.29}$$

If the density is constant, then $\frac{p}{\rho_w}$ may be substituted by $g\psi$, where ψ is the height of the water column above the reference point and z is the distance from the reference point to the datum plane.

The gravitational constant g does not vary significantly with surface location and therefore, as long as density is constant, Equation 4.29 can be divided by g to give:

$$h = \psi + z \tag{4.30}$$

where:

h is the hydraulic head $\frac{\Phi}{g}$,

ψ is the pressure head, and

z is the elevation head.

The dimension of these head quantities is length. This fundamental head relationship is basic to an understanding of saturated groundwater flow. The relationships between hydraulic head, elevation head and pressure head are shown for a piezometer and a flowing well in Figure 4.6.

4.4.2 Gradient of the hydraulic head

Water moves in response to a hydraulic gradient. If there is no hydraulic gradient, the water will remain stationary. The hydraulic gradient can be simply calculated as the change in hydraulic head between two points divided by the distance between the two points.

The hydraulic gradient is often calculated in the x-y plane and referred to as the horizontal hydraulic gradient. In many important situations, the hydraulic gradient varies vertically. That is, if you could measure the hydraulic head at different depths below the same location, the hydraulic head would change.

In a recharge condition, where water is entering the groundwater system, the vertical hydraulic gradient is negative (z taken as positive upwards). The head will decrease as depth increases.

Under discharge conditions, where water is leaving the groundwater system, the vertical hydraulic gradient is positive. The head will increase as depth increases.

Piezometers are usually installed in groups designed to investigate either the horizontal or vertical movement of groundwater. The hydraulic gradient, either in the horizontal or vertical direction, can be established from measurements on a group of piezometers as shown in Figure 4.7.

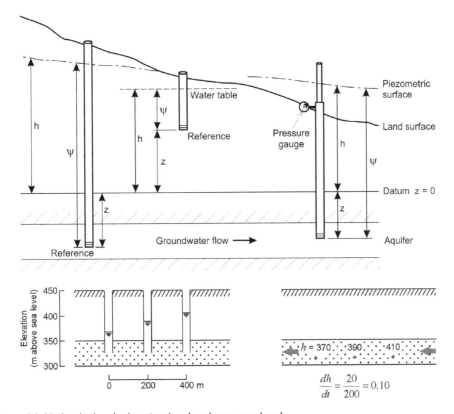

Figure 4.6 Hydraulic head, elevation head and pressure head

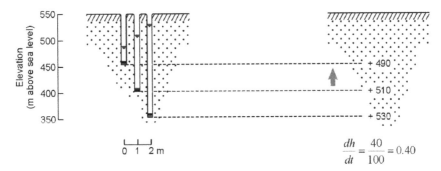

Figure 4.7 Measurement of horizontal and vertical hydraulic gradients (adapted from Freeze and Cherry (1979))

Where a number of piezometers are installed close together in a shallow aquifer, it is possible to connect them all together by a common manifold and apply suction to the manifold that lifts all the levels above ground for easy analysis. This is shown in Figures 4.8 and 4.9. The vertical hydraulic gradient can then be calculated (Acworth, 2007).

Figure 4.8 Multi-piezometer manometer board in use at the side of a creek at Hat Head (Photo: Ian Acworth)

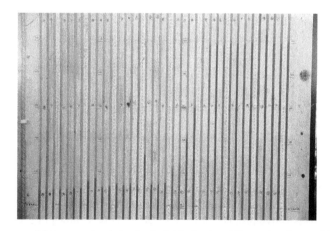

Figure 4.9 Close-up of the water levels in the individual tubes (Photo: Ian Acworth)

Mathematically, the gradient of the hydraulic head can be written as:

$$\nabla h = \frac{dh}{ds} \tag{4.31}$$

where:
s is a distance parallel to the direction of ∇h.

If the value of ∇h is zero, then there will be no movement of water irrespective of the value of hydraulic conductivity. The direction of groundwater flow is a function of the hydraulic conductivity and ∇h. For an isotropic medium, ∇h is perpendicular to the equipotential and groundwater flows in the direction of ∇h. For an aquifer with anisotropic hydraulic conductivity, it is necessary to construct a hydraulic conductivity ellipse (in two dimensions) or ellipsoid (three dimensions) to predict the direction of ∇h, the direction of groundwater flow. An account of the method is given by Fetter (2001). The effect of anisotropy on the direction of groundwater flow can be considerable and should not be ignored.

4.4.3   1D vertical profiles of hydraulic head

Heads can be calculated as the sum of elevation and pressure head as shown in Table 4.7. Movement in the vertical direction can then be determined by the vertical gradient in hydraulic head.

Note that there is no gradient on the hydraulic head $\left(\frac{dh}{dx} = 0\right)$.

The hydraulic head does not stop at the water table, it simply becomes negative! If water is evaporating from the soil surface, moisture must be moving upwards from the water table to the soil surface. There must be a negative hydraulic gradient to achieve this. The matric potential at the top of the tension saturated surface will be close to zero, but slightly negative. Using the same units from Table 4.7, we can estimate values of ψ_m in the unsaturated zone as shown in Table 4.8.

Table 4.7 Heads (in mm) for a vertical profile in equilibrium

Depth	ψ_p	ψ_m	ψ_z	h
0	0	−1000	0	−1000
100	0	−900	−100	−1000
200	0	−800	−200	−1000
300	0	−700	−300	−1000
400	0	−600	−400	−1000
500	0	−500	−500	−1000
600	0	−400	−600	−1000
700	0	−300	−700	−1000
800	0	−200	−800	−1000
900	0	−100	−900	−1000
1000	0	0	−1000	−1000
1100	100	0	−1100	−1000
1200	200	0	−1200	−1000

Table 4.8 Heads (mm) for water evaporating from a saturated soil profile

Depth	Ψ_p	Ψ_m	Ψ_z	h
0	0	−3000	0	−3000
100	0	−2400	−100	−2500
200	0	−1900	−200	−2100
300	0	−1550	−300	−1850
400	0	−1250	−400	−1650
500	0	−1000	−500	−1500
600	0	−800	−600	−1400
700	0	−650	−700	−1350
800	0	−500	−800	−1300
900	0	−320	−900	−1250
1000	0	−205	−1000	−1205
1100	0	−101	−1100	−1201
1200	0	0	−1200	−1200
1300	100	0	−1300	−1200
1400	200	0	−1400	−1200
1500	300	0	−1500	−1200

4.4.4 Piezometers

The basic device for the measurement of hydraulic head is a tube or pipe in which the elevation of the water level can be determined. In the laboratory, the tube is a manometer; in the field the pipe is called a piezometer. It must be open at the base, so that water can flow into the pipe through a screen, and open to the atmosphere at the top.

The point of measurement is at the base of the piezometer and not at the level to which water rises inside the pipe. The open area section of the piezometer base should therefore be kept restricted to be representative of the head in the zone of interest. The simple standpipe piezometer in which the depth to water is measured by a tape has been replaced to some extent by more complex designs incorporating pressure transducers, pneumatic devices and electronic components, allowing the remote collection and recording of data. The availability of data at frequent time intervals and in a digital manner allows much improved analysis of hydrogeological parameters.

The NSW government has instigated a web-based system whereby pressure transducers installed in regional monitoring boreholes provide regular (hourly) updates on aquifer status.

4.4.5 Tensiometer

Water potential in the unsaturated zone is measured by a tensiometer. A tensiometer consists of a water-saturated porous ceramic cup connected to a manometer through a water-filled tube. The ceramic consists of very fine pores and remains water saturated even when placed in contact with soil at relatively low water potentials. Upon contact, water moves from the tube to the soil, creating a suction at the manometer end until equilibrium is reached and the total potential in the water system is balanced throughout. The principles are shown in Figure 4.10, and an example of an installation is shown in Figure 4.11 where irrigation scheduling of a vineyard is controlled by tensiometers.

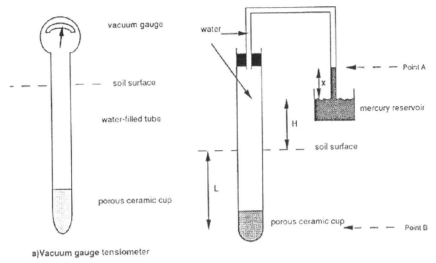

Figure 4.10 Tensiometer principle with a pressure gauge and a mercury manometer (adapted from Jury *et al.* (1991))

Figure 4.11 Soil tensiometer installation (Photo: Ian Acworth)

The tensiometer has a practical range of about -8.0 m. Below this head, gasses dissolved in the water will begin to form bubbles and the liquid column will break up.

4.4.6 Zero-flux planes

If the water potential can be measured as a profile through the unsaturated zone, as shown in Table 4.7 or Table 4.8, then the vertical gradient can be calculated. It may be that the profile is in equilibrium (Table 4.7) where the gradient in water potential is zero. Conversely, the profile may have been subject to long-term evaporation and drying at the surface and there is a constant flux of water moving from the saturated zone to the atmosphere. In this case, the gradient in water potential is negative. If rain wets the surface and starts to infiltrate downwards, the gradient is reversed and some depth there will be a location where the gradient is zero and the water is moving neither upwards or downwards. This depth is called a zero-flux plane (Wellings and Bell, 1980; Acworth, 1981; Rushton, 2003; Butler *et al.*, 2012) and has become an important concept in the calculation of groundwater recharge (See Chapter 3).

4.4.7 Equipotentials

In two dimensions, lines can be drawn connecting points of equal hydraulic head. These are similar to contours on the ground surface which define the surface topography.

In three dimensions, a surface of equipotential is created which is not possible to represent in two dimensions, other than to take a cross-section and to assume that there is not significant variation in the third dimension.

In an isotropic medium, the flow will occur at right angles to the lines of equipotential, as shown in Figure 4.12a. If the medium is anisotropic in one direction, then flow occurs at an angle to the equipotential rather than at right angles. This case is shown in Figure 4.12b.

Vertical flow nets A flow net is a two-dimensional device constructed to describe groundwater flow in either an x-y plane (horizontal) or an x-z plane (vertical).

Two sets of orthogonal curves can be constructed which describe the flow system. A set of equipotentials defines the hydraulic head distribution and a set of flow lines, at right angles to the equipotentials (at least in an isotropic domain). Flow nets are useful tools to help characterize a flow system. Examples of vertical flow nets which can be used to demonstrate the occurrence of vertical groundwater flow in local flow systems containing recharge and discharge are provided by Fetter.

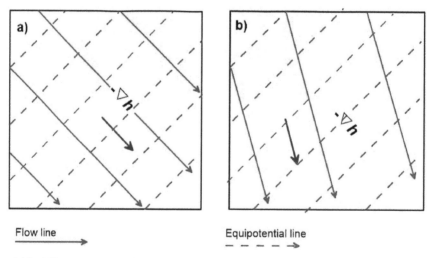

Figure 4.12 a) Flow occurs at right angles to the equipotentials; b) the flow deviates from right angles as determined by the degree of anisotropy

A vertical flow net is shown in Figure 4.13 for undulating topography (Hubbert, 1940). Figure 4.14 shows typical recharge and discharge conditions down a hill slope.

Horizontal flow nets – piezometric maps Field measurements of the hydraulic head at a number of different locations may be used to construct a piezometric map. If the flux of groundwater can be approximated as horizontal and vertical components neglected, then measurements of hydraulic head in piezometers or boreholes can be used to construct equipotentials.

Note that the construction of equipotential maps from open borehole data will produce a false picture of flow conditions in an aquifer if vertical flow components are significant. Open boreholes often provide the most efficient means for water to move through the aquifer, and there are many instances of two aquifers being unintentionally connected by a borehole sunk to investigate aquifer conditions.

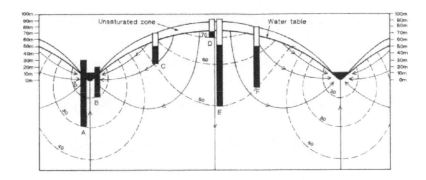

Figure 4.13 Vertical flow net showing piezometric levels beneath undulating terrain (adapted from Fetter (2001))

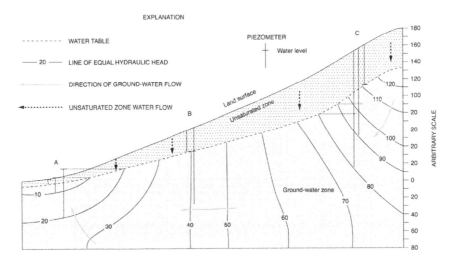

Figure 4.14 Vertical flow net showing piezometric levels beneath a hill slope (USGS)

It is important to realize that the construction of an equipotential map from well and bore-hole data implies that the conceptual model chosen represents horizontal flow only. This clearly causes problems in the representation of recharge and discharge areas.

4.4.8   Measurement of head at flowing wells

Where the hydraulic head in the aquifer is above ground surface, water will discharge under its own pressure from a well, and the well is called a flowing well or an artesian well (Figure 4.15). In Figure 4.16, the cap has been removed so that a water sample can be taken.

Figure 4.15 An extension of the piezometer pipe above ground is required to monitor an artesian head using a chart recording device (Photo: Ian Acworth)

Figure 4.16 Sampling from an aquifer with a flowing artesian head is easily arranged (Photo: Ian Acworth)

To determine the correct value of hydraulic head, the well must be capped and an accurate pressure gauge installed. The hydraulic head may then be derived from:

$$h = z + \frac{p}{\rho g} \tag{4.32}$$

where:
z is the distance to the reference plane,
ρ is the density of the water,
g is the acceleration due to gravity (9.80 m s^{-2}), and
p is the gauge pressure in Pascals.

Note that it may be necessary to correct the data for the temperature or the density of the water if the discharge is flowing up from depth in an artesian basin.

4.4.9   Hydraulic head in groundwaters of varying density

Where the density of groundwater is variable, Equation 4.30 cannot be used to calculate the hydraulic head (potentiometric head), as the density variation itself may lead to groundwater flow. This is a particular problem when dealing with the intrusion of sea water or the calculation of groundwater flow in the vicinity of saline lakes. The saline lakes problem has received much attention in Australia, with particularly useful studies conducted around Lake Tyrrell (Macumber, 1991).

In the vicinity of saline lakes or the ocean, the groundwater density may vary in piezometers completed at the same depth and at different distances from the lake shore and also in piezometers completed at the same location but at different depths.

Point-water heads

Lusczynski (1961) defines a point-water head in groundwater of variable density as the water level, referred to a given datum, in a well filled with water of the type at the point (base) sufficient to balance the existing pressure at the base of the piezometer. The point-water head is then:

$$h_{p,i} = z_i + \frac{p_i}{\rho_i g} \tag{4.33}$$

where:
i is any point in ground water of variable density,
ρ_i is density of ground water at point i,
$h_{p,i}$ is the point-water head at i,
p_i is pressure at i,
z_i is elevation of i, measured positively upward, and
g is gravitational acceleration.

Calculation of freshwater head

The freshwater head at any point i in the ground is defined (Lusczynski, 1961; Holzbecher, 1998) as the head that would be developed in a variable density system if the pressure at the piezometer intake was generated by an equivalent column of fresh water. For waters that have higher salinity contents than fresh water, the head will be higher than the point-water head. The freshwater head is then:

$$h_{f,i} = h_{p,i}\frac{\rho_i}{\rho_f} - z_i\frac{\rho_i - \rho_f}{\rho_f} \tag{4.34}$$

where ρ_f is the density of fresh water (often close to 998.3 kg m^{-3}).

Calculation of environmental-water head

Lusczynski (1961) demonstrates that freshwater heads only define the hydraulic gradient in a variable salinity medium when measured in the horizontal plane. Environmental-water heads are required to define the hydraulic gradient in the vertical plane (Acworth, 2007).

Lusczynski (1961) demonstrates that when the head measurements are made with reference to mean sea level as the datum, the environmental-water head can be defined in terms of the point-water head as:

$$h_{e,i} = h_{p,i}\frac{\rho_i}{\rho_f} - z_i\frac{\rho_i - \rho_a}{\rho_f} \tag{4.35}$$

where:
i is any point in groundwater of variable density,
ρ_a is average density of groundwater between mean sea level and point i,
ρ_i is density of water at point i,
ρ_f is density of fresh water (998.3 kg/m^3),
$h_{p,i}$ is point-water head at i,
z_i is elevation of i, measured positively upward, and
g is gravitational acceleration.

The average density between mean sea level and a depth z_i can be calculated from individual piezometers or from manometer board data (Acworth, 2007). The half-metre interval of the mini-piezometer tubes provides a good approximation to the density distribution which can be calculated from:

$$\rho_a = \frac{1}{z_i}\int_0^z \rho\,dz \tag{4.36}$$

In general, the exact density of the water is unknown. However, it can be estimated (kg m^{-3}) from the fluid EC (Stuyfzand, 1993) by:

$$\rho_s = 1000 \cdot (1 + 8.05 \times 10^{-7} \cdot RE - 6.5 \times 10^{-6} \cdot (T - 4 + 2.2 \times 10^{-4} \cdot RE)^2) \tag{4.37}$$

where:
T is the fluid temperature (°C) and
RE is the residue on evaporation at 180 °C but which can be approximated as $0.69778 \times EC_{fluid}$. The fluid EC is measured in micro-siemens/cm.
In the case that there is a vertical variation in density, sufficient information is required to establish the average density of water above the reference point. This can be estimated from

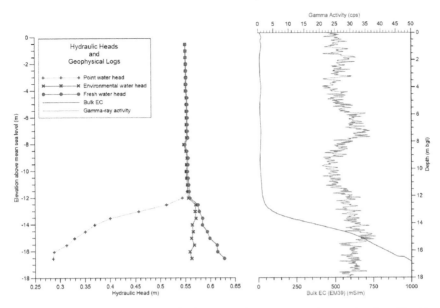

Figure 4.17 Comparison of pointwater, freshwater and environmental-water heads at Hat Head

geophysical logs (Dasey, 2010) or from taking a small sample of fluid from a bundled piezometer (Acworth, 2007).

The pointwater, freshwater and environmental-water heads have been calculated for a central part of a coastal aquifer at Hat Head (Acworth, 2007) where a multi-channel manometer board has been installed. The three plots are shown along with the gamma and bulk EC logs (Chapter 13) in Figure 4.17. If only the point-water heads had been measured, a strong downward gradient is seen below the saline interface that is found at approximately 12 m depth. If the freshwater heads are computed, the gradient seen to be reversed with upward movement of water indicated. The environmental-water head gives the true picture in that there is little or no vertical movement of water at this site. The small step in the environmental head is most likely due to a thin layer of cemented sands.

4.4.10 Factors affecting the hydraulic head

A list (not necessarily complete) of possible causes which can produce a change in water level is given below:

1 Groundwater recharge
2 Evapotranspiration
3 Tidal effects near oceans
4 Earth tide effects
5 Atmospheric pressure effects
6 Earthquakes

7 Groundwater pumping
8 Deep well injection
9 Artificial recharge
10 Agricultural irrigation
11 Geotechnical drainage

Groundwater abstraction

There is developing concern that long-term abstraction of groundwater is removing far more water from the aquifer than is being replaced by recharge. Under these conditions, groundwater pressures must fall. Data from the Liverpool Plains (Figure 4.18) illustrates this point.

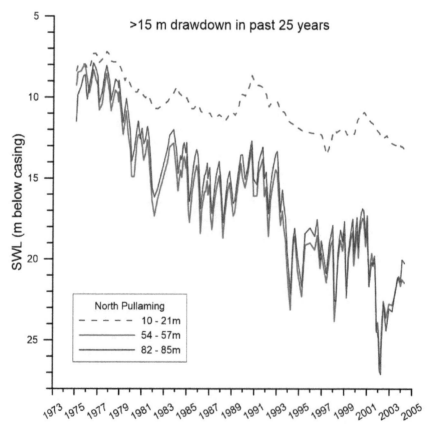

Figure 4.18 Falling water pressures in a set of nested piezometers close to Breeza on the Liverpool Plains of NSW (adapted from data available from NSW Department of Water)

4.5 SIGNAL ANALYSIS TECHNIQUES

4.5.1 Introduction

Any parameter that is measured at a fixed separation in space or time can be effectively analyzed using signal analysis techniques. An example of a measurement at a fixed separation in space would be the apparent resistivity – measured along a profile at perhaps 5 m station separation. An example of a measurement in time is the hydraulic head record obtained from a data logger measured perhaps every 1 hour. The data shown in Figure 4.19 gives an example of the possible complexity. There are clearly high frequency components (one or two cycles per day) superimposed on a much lower frequency variation that represents weekly or monthly variation in water level. Figure 4.19 shows a similar example.

The effective way to analyze this data is to use a signal analysis package such as TSoft (Van Camp and Vauterin, 2005).

4.5.2 TSoft package

The TSoft package is freeware that can be downloaded from the web. As with much freeware, the manual is not particularly extensive but the package is very useful. As an example, consider the Hat Head data set shown in Figure 4.19.

The data can be represented in the frequency domain by carrying out a Fourier transform using TSoft. Approximately half the data set shown in Figure 4.19 is analyzed below. The frequency spectra for HH3 are shown on the right hand side of Figure 4.20. Fourier analysis can be used to examine different components of the spectrum.

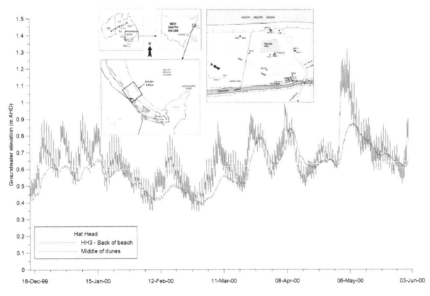

Figure 4.19 Groundwater levels at Hat Head – showing conditions at the back of the beach (HH3) that are heavily influenced by the tide, and conditions in the centre of the sand spit (HH5) where tidal influence has declined

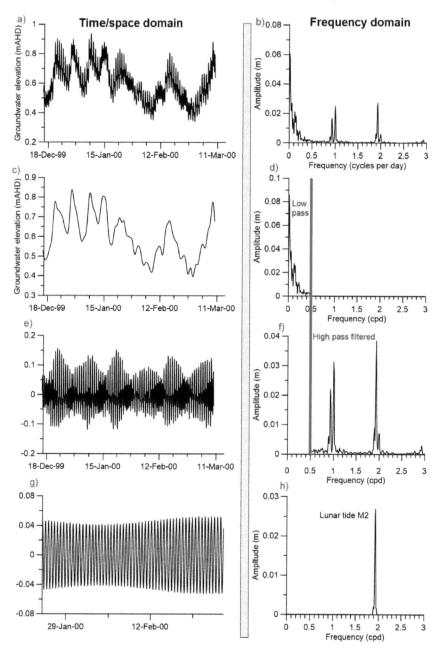

Figure 4.20 (a) A sequence of water levels from a piezometer at the top of an ocean beach; (b) the Fourier spectrum of (a); (c) the data filtered using an FFT with a 0.5 cpd low-band pass filter; (d) the Fourier spectrum of (c); (e) the data filtered using an FFT with a 0.5 cpd high-band pass filter; (f) the Fourier spectrum of (e); (g) a 1.93 band-pass filter applied to the data and (h) the Fourier spectrum

There is clearly a lot of 'energy' (high-amplitude frequency components) with a frequency of less than one cycle per day (24-hour period) in the data (Figure 4.20a). The impact of higher frequency information can be removed by applying a low-pass filter to the data. In Figure 4.20c, a filter cut off of 0.5 cps was used with a bandwidth of 0.05 cps. Conversely, the low frequency data can be removed by selecting a high-pass filter of the same characteristics. The result of this is shown in Figure 4.20e.

In Figure 4.20g, a band-pass filter is used to show the energy at a specific frequency; 1.93 cpd is used as that is the main energy frequency for the ocean tide (M_2).

Acworth *et al.* (2015a) give examples of the use of Fourier analysis in the interpretation of hydraulic head data.

The TSoft package has many useful applications for data checking and data filling using either linear or cubic spline interpolation where gaps in data are apparent. There is also a useful application that will resample your data to a different time base. For example, if you are trying to compare hourly data with 30-minute data, it is possible either to take a subset of the 30-minute data or to interpolate the hourly data.

4.6 SEA WATER INTRUSION IN COASTAL AQUIFERS

Coastal aquifers may come in contact with the ocean and, under natural conditions, fresh groundwater is discharged into the ocean. This has been recognised for very many years, and the coastal zone is an important source zone for minerals derived from groundwater that support fisheries. Estuaries, in particular, are complex environments where sea water and groundwater mix together. They can be significantly impacted by a number of factors: rising sea level; decreasing rainfall – leading to decreased recharge and freshwater head in the aquifer; increased abstraction of groundwater, etc. Fresh groundwater will discharge at the ocean beach if the aquifer is homogeneous and isotropic. However, this is rarely the case and there is frequently a reduction in permeability in the vertical direction when compared to the horizontal direction. This may cause groundwater to discharge beneath fresh water some considerable distance out to sea. The Romans recognised this and used to collect fresh water from springs beneath the ocean by lowering a funnel to the ocean bed that was connected by a leather pipe to the surface.

With the increased demand for groundwater in many coastal areas, the seaward flow of groundwater has been decreased or even reversed, allowing saltwater to enter and move inland in aquifers. This problem is called seawater intrusion. If the salt water travels inland to bore fields, underground water supplies become useless; moreover, the aquifer becomes contaminated with salt which may take years to remove even if adequate fresh groundwater is available to flush out the saline water. The importance of protecting coastal aquifers against this threat has led to investigations directed towards methods of prevention or control of seawater intrusion.

4.6.1 The Ghyben-Herzberg relationship

Two investigators, Ghyben and Herzberg, working independently from each other, became aware of the fact that in coastal aquifers, salt water existed below fresh water in locations on land close to the beach. The depth at which it is encountered is dependent on the head of fresh water above mean sea level at that point. They showed that, generally, it could be

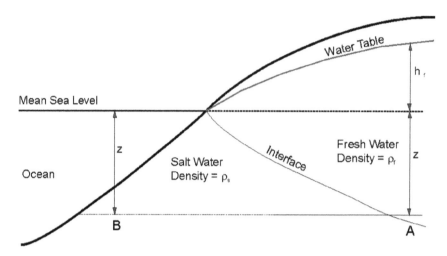

Figure 4.21 Gybenherzberg relationship for hydrostatic equilibrium between fresh water and salt water at the coast

expected to be encountered first at a depth below mean sea level, which is approximately 40 times the head of fresh water above mean sea level, i.e. for every 1 m of fresh water above mean sea level there should be approximately 40 m of fresh water below mean sea level before the salt water is encountered. The depth to the salt water/fresh water interface is based on, among other things, the densities of the two fluids. In the evaluation of the relationship above, it has been assumed that the density of fresh water is 1,000 kg m^{-3} and that of the sea water is 1025 kg m^{-3}. Ghyben and Herzberg worked their relationship assuming that they were dealing with hydrostatics. The water is assumed to be stationary. This is clearly not the case if there is groundwater discharge, and in reality, groundwater is moving both horizontally and vertically in the coastal zone and both components of movement require accounting for. However, as an introduction, the Ghyben-Herzberg relationship provides a useful starting point.

The basis behind the Ghybern-Herzberg relationship is given in Figure 4.21.

At point A (in Figure 4.21) on the interface, for equilibrium to exist, the pressure on the salt water side of the interface (pA_s) must equal the pressure on the freshwater side (pA_f).

From hydrostatics:

$$PB_s = \rho_s gz \tag{4.38}$$

and:

$$PA_f = \rho_f gz + \rho_f gh_f \tag{4.39}$$

where:
PB_s is the pressure at B on the salt water side,
PA_f is the pressure at A on the freshwater side,
ρ_f is the density of fresh water,

ρ_s is the density of salt water,
h_f is the head of fresh water above mean sea level, and
z is the depth to the interface below mean sea level.

Since the pressure at both locations at the same depth must be equal, we can set:

$$\rho_s g z = \rho_f g z + \rho_f g h_f \tag{4.40}$$

or, solving for z gives:

$$z = \frac{\rho_f}{\rho_s - \rho_f} h_f \tag{4.41}$$

The Ghyben-Herzberg concept is also applicable, with the same limitations, to confined aquifers, where the potentiometric surface replaces the water table. However, it should be remembered that the Ghyben-Hertzberg concept also assumes a static sea level. In reality, the impact of tides and storms has the result that waves run up the beach and introduce sea water above the discharging fresh water that causes significant mixing at the interface.

It should be remembered that sea water intrusion is quite natural and cannot be overcome completely. What has to be determined is the magnitude of intrusion which will be accept-able in a particular area. A freshwater flow to sea is necessary to stabilize the interface, and the position of the wedge depends on the magnitude of the flow to the sea. A reduction in the magnitude of freshwater flow will result in a movement of the toe of the wedge inland until stability is again achieved. A smaller freshwater flow is associated with a smaller gra-dient which in turn results in a flatter wedge. If the magnitude of the freshwater flow is increased, then the toe of the wedge moves seaward until stability is achieved and a steeper wedge results. The impact of sea water intrusion is then a management problem.

Chapter 5

Hydraulic conductivity and Darcy's Law

5.1 INTRODUCTION TO DARCY'S LAW

The hydraulic conductivity is a proportionality constant derived from experimental data in which the gradient in a hydraulic head is related to the water discharge through a constant cross-sectional area. The original work was carried out by Henry Darcy.

The French hydraulic engineer Henry Darcy published a report (Darcy, 1856) on the water supply in the city of Dijon in 1856. In this report, Darcy described a laboratory experiment that he had carried out to analyze the flow of water through sands. The results of this experiment can be generalized into the empirical law upon which much of the later development of the science of groundwater is based.

5.1.1 The experiment

Darcy's experimental apparatus may be generalized as shown in Figures 5.1 and 5.2.

A circular cylinder of cross-section A is filled with sand, stoppered at each end and fitted with inflow and outflow tubes and a pair of manometers. Water is allowed to flow through the cylinder until such time as all the pore space is filled with water and the inflow rate Q is equal to the outflow rate.

An arbitrary datum is selected at elevation $z = 0$; the elevations of the manometer intakes are z_1 and z_2 and the elevations of the fluid levels are h_1 and h_2. The distance between the manometer intakes is ΔL. The specific discharge (v) (volume flux, or flux per unit volume) through the cylinder is defined as:

$$v = \frac{Q}{A} \tag{5.1}$$

If the dimensions of Q are L^3/T and those of A are L^2 then v has the dimensions of velocity $[L/T]$.

Darcy's experiments show that when ΔL is held constant:

$$v \propto \Delta h \tag{5.2}$$

and when Δh is held constant:

$$v \propto \frac{1}{\Delta L} \tag{5.3}$$

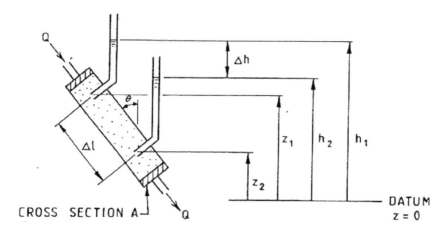

CROSS SECTION A

Figure 5.1 Darcy's concept (adapted from Freeze and Cherry (1979))

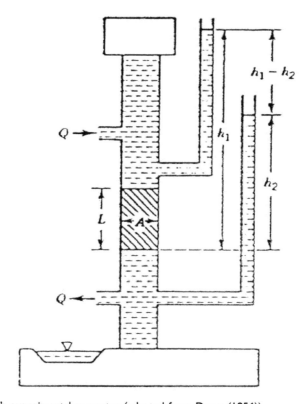

Figure 5.2 Darcy's experimental apparatus (adapted from Darcy (1856))

where:
Δh is the head difference $(h_1 - h_2)$ between the manometer intakes and
ΔL is the difference between the manometer intakes.

The proportionality constant in these relationships has been defined as the *hydraulic conductivity*. Darcy's Law can then be written as:

$$v = -K\frac{\Delta h}{\Delta L} \tag{5.4}$$

or in differential form:

$$v = -K\frac{dh}{dL} \tag{5.5}$$

where:
h is the hydraulic head and
dh/dL is the hydraulic gradient (or gradient of the water potential).

An alternative form of Darcy's Law can be obtained by substituting for specific discharge as:

$$Q = -K\frac{dh}{dL}A \tag{5.6}$$

Darcy's Law for fluid flow through a porous media has applications in many branches of science, varying from the oil and gas industry, where it is used to describe the flow of several phases through a petroleum reservoir, through the soil and agricultural sciences, where it is used to describe flow in the unsaturated zone and in plants, to medical science, where the same relationship has been used to describe fluid flow through the brain.

5.1.2 Hydraulic conductivity and intrinsic permeability

It is clear that the hydraulic conductivity must be a function of the porous medium, in that the porosity, the degree of interconnectivity of pores, the grain size and sorting of the material will all have a bearing on the value of hydraulic conductivity. It is also evident that the viscosity and density of the fluid must influence hydraulic conductivity. It is intuitively obvious that the rate of movement of a thick oil through a permeable medium will be slower than that of water, or gas, through the same medium.

The hydraulic conductivity is therefore a function of both the media and the fluid flowing through it. A dimensional analysis of Equation 5.4 indicates that hydraulic conductivity has dimensions of $[L/T]$ when the water potential is expressed in metres head of water. These units are satisfactory if dealing with clean water and a volume flux, rather than a mass flux, of water. Typical values of hydraulic conductivity are given in Table 5.1.

It is unusual to find a variable in nature which has a range of 15 orders of magnitude.

In attempts to further define hydraulic conductivity, experiments have been carried out with ideal porous media consisting of uniform glass beads of diameter d. When various fluids of density and dynamic viscosity are run through apparatus similar to

Table 5.1 Representative hydraulic conductivity values

Hydraulic Conductivity	From ms^{-1}	To ms^{-1}
Gravel	10^{-3}	1
Clean Sand	10^{-6}	10^{-2}
Silty Sand	10^{-7}	10^{-3}
Silt, Loess	10^{-9}	10^{-5}
Glacial Till	10^{-12}	10^{-6}
Marine Clay	10^{-12}	10^{-9}
Shale	10^{-13}	10^{-9}
Unfractured Basement	10^{-14}	10^{-10}
Sandstone	10^{-10}	10^{-6}
Limestone	10^{-9}	10^{-6}
Fractured Basement	10^{-8}	10^{-4}
Basalt (Interflows)	10^{-7}	10^{-3}
Limestone Karst	10^{-6}	10^{-3}

Darcy's under a constant hydraulic gradient, further proportionality relationships are observed:

$$v \propto d^2 \tag{5.7}$$

$$v \propto \rho g \tag{5.8}$$

$$v \propto \mu^{-1} \tag{5.9}$$

Together with the original observation that $v \propto -\frac{dh}{dL}$, these relationships lead to a new definition of Darcy's Law:

$$v = -\frac{Cd^2 \rho g}{\mu} \frac{dh}{dL} \tag{5.10}$$

The parameter C is another dimensionless constant of proportionality. For real soils, it must include all the influence of other media properties which affect the flux, such as grain size and distribution, grain sphericity and roundness and the nature of the packing.

Comparison with the original Darcy equation shows that:

$$K = \frac{Cd^2 \rho g}{\mu} \tag{5.11}$$

In this equation, ρ and μ are properties of the fluid and Cd^2 is a function of the medium alone. If we define lower case k as:

$$k = Cd^2 \tag{5.12}$$

then:

$$K = \frac{k \rho g}{\mu} \tag{5.13}$$

The parameter k is known as the *intrinsic permeability*. The term intrinsic permeability rather than permeability is used, as in some earlier texts; hydraulic conductivity was also referred to as the coefficient of permeability, or simply permeability. The term permeability is still in general use but restricted to the physical properties of the medium. The intrinsic (or specific) permeability is widely used in the oil industry where the existence of gas, oil and water in multiphase flow systems makes the use of a fluid-free conductance term of use. Petroleum engineers have defined the *darcy* as a unit of permeability. The unit is not in common use in the water industry.

The value of a darcy when expressed as square metres is $1\ darcy = 0.987\ ^{-12} \times m^2$.

If a material has significant permeability, it must contain pores for the fluid to flow through. A material may be porous, but if the pores are not interconnected, it is not permeable. A full description must therefore specify both porosity and intrinsic permeability.

5.1.3   Specific discharge and average linear velocity

The units of specific discharge v in Equation 5.1 were noted to be velocity. It is important to realize that the specific discharge does not represent the linear velocity at which water moves through the pore space. It is an apparent velocity, representing the velocity at which water would move through the aquifer if it were an open conduit. The cross-sectional area of the open pore space in the aquifer is much smaller than an open conduit of the same dimensions, therefore the real fluid velocity through the pore space will be significantly higher. The average linear velocity may be calculated from Equation 5.14:

$$u_L = \frac{Q}{\phi_e A} = \frac{K}{\phi_e}\frac{dh}{dL} \tag{5.14}$$

where:
u_L is the average linear velocity in the direction of L and
ϕ_e is the effective porosity.

The effective porosity is less than the total porosity, as some water remains trapped in dead-end pore space and adheres to the grain surfaces. The total porosity for a Botany Sands sample is approximately 0.40; the effective porosity is approximately 0.36.

5.1.4   Homogeneity and heterogeneity of hydraulic conductivity

When the hydraulic conductivity K is *independent of position* within a geologic formation, the formation is homogeneous. When the hydraulic conductivity K is *dependent on position* within a geologic formation, the formation is heterogeneous.

This can be expressed mathematically by examining the components of K in an x-y-z coordinate system. For a homogeneous formation, $K(x, y, z) = C$ where C is a constant; whereas in a heterogeneous formation, $K(x, y, z) \neq C$, and C is no longer constant.

Three broad classes of heterogeneity exist (Freeze and Cherry, 1979).

Layered heterogeneity – common in sedimentary formations and unconsolidated lacustrine or marine deposits. The individual beds are each homogeneous, but the entire system is heterogeneous. Layered heterogeneity can result in K contrasts of as much as 13 orders of magnitude.

Discontinuous heterogeneity – caused by the presence of faults or other discontinuities.

Trending heterogeneity – where the properties of a sand, for example, change with distance from the source. Trending heterogeneity is common in deltaic environments Fetter (2001).

5.1.5   Isotropy and anisotropy

If the hydraulic conductivity K is independent of the direction of measurement at a point in a geologic formation, the formation is *isotropic* at that point.

By contrast, if the hydraulic conductivity K varies with the direction of measurement at a point in a geologic formation, the formation is *anisotropic* at that point.

To fully describe the nature of the hydraulic conductivity in a geologic formation, it is necessary to use two adjectives, one dealing with heterogeneity and the other with anisotropy. For example, for a homogeneous, isotropic system in two dimensions:

$$K_x(x,z) = K_z(x,z) = C_1 \tag{5.15}$$

Whereas, for a homogeneous and anisotropic system:

$$K_x(x,y) = C_1 \tag{5.16}$$

for all (x, z), and:

$$K_z(x,y) = C_2 \tag{5.17}$$

for all (x, z), but:

$$C_1 \neq C_2 \tag{5.18}$$

The relationship between heterogeneity and anisotropy is shown in Figure 5.3.

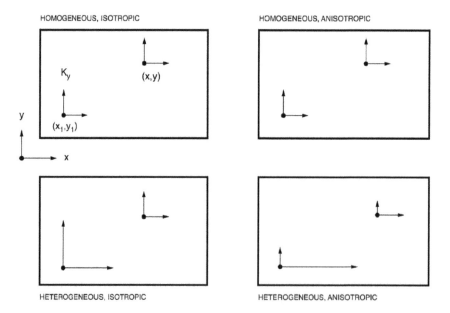

Figure 5.3 Heterogeneity and anisotropy of hydraulic conductivity (adapted from Freeze and Cherry (1979))

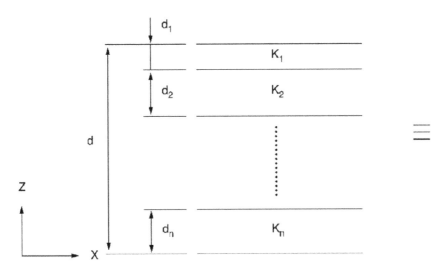

Figure 5.4 Layered heterogeneity (adapted from Freeze and Cherry (1979))

There is a relation between layered heterogeneity and anisotropy. A layered formation, as shown in Figure 5.4, with individual homogeneous and isotropic layers of hydraulic conductivity K_1, K_2, K_3, ..., K_n can be represented as a single anisotropic layer.

For flow perpendicular to the layering, the volume flux must be the same entering the system as it is leaving. Therefore, if Δh_1 is the head loss across the first layer, Δh_2 the head loss across the second layer etc., the total head loss is:

$$\Delta h = \Delta h_1 + \Delta h_2 + ... + \Delta h_n \tag{5.19}$$

From Darcy's Law we can equate the volume flux across each layer and then write:

$$v = K_1 \frac{\Delta h_1}{d_1} = v = K_2 \frac{\Delta h_2}{d_2} = ...v = ... = K_n \frac{\Delta h_n}{d_n} = K_z \frac{\Delta h}{d} \tag{5.20}$$

where K_z is an equivalent vertical hydraulic conductivity for the system of layers. It can then be shown from Equation 5.20 that:

$$K_z = \frac{d}{\sum_{i=1}^{n} d_i / K_i} \tag{5.21}$$

where $d = \sum_{i=1}^{n} d_i$, and:

$$K_x = \sum_{i=1}^{n} \frac{K_i d_i}{d} \tag{5.22}$$

In the field, it is not uncommon for layered heterogeneity to lead to regional anisotropy values of 100:1 or greater (Freeze and Cherry, 1979). This is often the result of sedimentary depositional processes which fluctuate in carrying capacity in response to climatic variation. A wet period will produce increased runoff and increasing carrying capacity in streams. This will result in coarser grained material being mobilized. A dry period will produce reduced runoff, reduced carrying capacity and silt or clay deposition on top of the coarse sand.

5.1.6 Darcy's Law in 3D

For three-dimensional flow, in a medium that may be anisotropic, it is necessary to generalize the one-dimensional form of the equation given earlier (Equation 5.4). In three dimensions the volume flux (specific discharge) is a vector with components v_x, v_y, and v_z where:

$$v_x = -K_x \frac{\partial h}{\partial x} \tag{5.23}$$

$$v_y = -K_y \frac{\partial h}{\partial y} \tag{5.24}$$

$$v_z = -K_z \frac{\partial h}{\partial z} \tag{5.25}$$

where K_x, K_y, and K_z are the hydraulic conductivity values in the x, y, and z directions respectively.

As it is limiting to restrict discussion to one of three directions, a unifying mechanism is required and can be provided by the vector notation commonly used in mathematics.

The hydraulic head is a scalar field. It can be specified anywhere in the space defined by the Cartesian coordinate directions. We then define the gradient of this field as:

$$grad\,h = \nabla h = i\frac{\partial h}{\partial x} + j\frac{\partial h}{\partial y} + k\frac{\partial h}{\partial z} \tag{5.26}$$

where:
i, j, k are the unit vectors in the x, y and z directions, respectively, and
∇h pronounced del h, is the mathematical abbreviation for grad h.

Equations 5.23, 5.24 and 5.25 can therefore be combined and written as:

$$v = -K\nabla h \tag{5.27}$$

5.1.7 Hydraulic conductivity or intrinsic permeability tensor

The hydraulic conductivity quantity as used in Equation 5.27 is a tensor quantity with components which may be written in long hand as:

$$v_x = -K_{xx}\frac{\partial h}{\partial x} - K_{xy}\frac{\partial h}{\partial y} - K_{xz}\frac{\partial h}{\partial z} \tag{5.28}$$

$$v_y = -K_{yx}\frac{\partial h}{\partial x} - K_{yy}\frac{\partial h}{\partial y} - K_{yz}\frac{\partial h}{\partial z} \tag{5.29}$$

$$v_z = -K_{zx}\frac{\partial h}{\partial x} - K_{zy}\frac{\partial h}{\partial y} - K_{zz}\frac{\partial h}{\partial z} \tag{5.30}$$

This set of equations shows that there may be nine independent components of hydraulic conductivity in the most general case. For the special case, where $K_{xy} = K_{xz} = K_{yx} = K_{yz} = K_{zx} = K_{zy} = 0$, the nine components reduce to a more manageable three.

The necessary and sufficient condition that causes the off-diagonal elements of the hydraulic conductivity tensor to be zero is that the directions of the Cartesian coordinate system be chosen to lie in the principal directions of anisotropy. It is usually possible to choose these directions accordingly, and the choice is particularly important in groundwater modelling.

5.2   LIMITATIONS OF THE DARCIAN APPROACH

Darcy's Law provides an accurate description of the flow of groundwater in almost all hydrogeological environments. In general it holds for:

- saturated flow and for unsaturated flow,
- for steady-state flow and for transient flow,
- for flow in aquifers and aquitards,
- for flow in homogeneous systems and for flow in heterogeneous systems,
- for flow in isotropic media and flow in anisotropic media, and
- for flow in both rocks and granular media.

Although the Darcy relationship does have wide applicability, it is necessary to examine the limits which do exist.

5.2.1   Darcian continuum and the representative elementary volume (REV)

Use of Darcy's Law enables the actual microscopic flow of groundwater within the matrix to be replaced by a macroscopically averaged description of the microscopic flow. Note that the dimensions of specific discharge are length/time and that the macroscopic representation of the term is as a velocity. A detailed theoretical analysis of the validity of using the macroscopic representation based upon the concept of representative elementary volume (REV) is given in several texts (Bear and Verruijt, 1987). This concept is essentially related to scale. Darcy's Law breaks down if the scale of the problem is reduced to flow around individual grains and only becomes valid when a sufficient volume of material is included that the individual details of specific flow paths can be replaced by the specific discharge.

5.2.2   Upper and lower limits of Darcy's Law

Darcy's Law is a linear law. If it were universally valid, a plot of the specific discharge v vs the hydraulic gradient dh/dL would be a straight line for all gradients between 0 and ∞. A plot of specific discharge against gradient is given for a permeable soil (I) and a clayey soil (II) in Figure 5.5 (Galperin et al., 1993). The flow through the permeable soil is shown to be linear at low gradients but to depart from linearity at high gradients. The flow through a clayey soil is shown to have a threshold below which no flow occurs. This threshold is referred to as the *incipient hydraulic gradient* or the *seepage threshold*. As the hydraulic gradient increases above the seepage threshold, the specific discharge becomes linear.

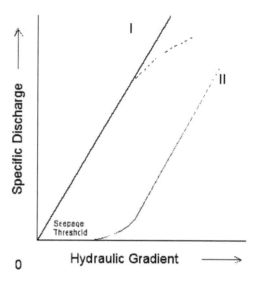

Figure 5.5 Relationships between specific discharge and hydraulic gradient for sands and clays (adapted from Freeze and Cherry (1979))

Low flow

At very low gradients, the law breaks down. There appears to be a threshold value of gradient below which no flow occurs, particularly in fine-grained materials. This threshold value is related to the energy needed to break the hydrogen bonding of water around the fine-grained (clay) particles. Experimental data (Dudgeon, 1985b) indicates that pure water behaves as a Newtonian fluid (obeying Darcy's Law) down to shear rates as low as $10^{-8} sec^{-1}$, but that water containing dissolved air acts as a thixotropic non-Newtonian fluid at shear rates below $10^{-4} sec^{-1}$. Dudgeon (1985b) also found that the viscosity of the water could be altered by the input of energy resulting from shock or vibration. These changes were attributed to changes in the molecular structure of the water. It is probable that this aspect has little practical importance in groundwater studies, although predictions of flow in large basins in which the gradient is very low and the groundwater has high values of total dissolved solids may be suspect. For this reason there may be significant over-estimates of available groundwater resources in some areas such as the Nubian sands in Libya.

High flow

At very high rates of flow, Darcy's Law also breaks down. The upper limit is usually identified in terms of the *Reynolds number R_e*, a dimensionless number that expresses the ratio of inertial to viscous forces during flow. The Reynolds number for flow through porous media is defined as:

$$R_e = \frac{\rho v d}{\mu} \qquad (5.31)$$

where:

ρ is the fluid density,

v is the specific discharge,

μ is the viscosity, and

d is a representative length dimension for the porous medium.

Darcy's Law is valid as long as the Reynolds number (based on average grain diameter) does not exceed some value between 1 and 10 (Bear, 1972). For this range of Reynolds numbers, all flow through the granular media is laminar. This subject area was researched in detail (Huyakorn, 1972a; Cox, 1977; Dudgeon, 1985b) and will be returned to later in Chapter 15.

5.3   FLOW IN FRACTURED ROCKS

The analysis of flow in fractured rocks can be carried out either with the continuum approach adopted thus far or with the non-continuum approach based upon the hydraulics of flow in individual fractures. The continuum approach considers flow on a macroscopic scale, where flow through individual pores or fractures can be averaged.

The continuum approach for fractured rocks involves the replacement of the fractured media by a representative continuum in which the spatially defined values of hydraulic conductivity, porosity and compressibility can be assigned. This approach is valid as long as the fracture spacing is sufficiently dense that the fractured media reacts in a similar fashion to granular media. The conceptualization is the same with the exception that the REV is considerably larger than the REV for a granular medium.

It may be the case that the REV can exceed the volume which contains an abstraction borehole, and that in these instances, conventional well hydraulics are not applicable, although Darcy's Law does describe satisfactorily the flow through the larger volume surrounding the abstraction site. For example, in a limestone formation, it is possible to develop a groundwater flow model which will accurately quantify the regional flow components, yet is not able to predict the abstraction rate from a borehole at any given location. In the extreme, boreholes may be dry if drilled through the limestone at a location which does not encounter a regionally connected fracture. If fracture spacings are irregular in a given direction, the media will exhibit trending heterogeneity. If the fracture spacings are different in one direction to another, the media will exhibit anisotropy (see for example Figure 4.2). If the fracture density is extremely low, it may be necessary to analyze the flow in individual fissures. This approach is often used in geotechnical investigations of slope stability.

Flow in individual fractures can be represented by the cubic law. For a given gradient, flow through a fracture is proportional to the cube of the fracture aperture. For laminar flow between smooth parallel plates, the cubic law can be expressed by Equation 5.32:

$$Q = \frac{\rho g b^2}{12\mu}(bw)\frac{\partial h}{\partial L} \tag{5.32}$$

where:

Q is the flow,

ρ is fluid density,

g is acceleration due to gravity,
μ is fluid viscosity,
b is the aperture opening, and
w is the fracture width perpendicular to b.

The form of Equation 5.32 is clearly similar to the Darcy equation.

5.4   RELATIONSHIP BETWEEN GRAIN-SIZE DISTRIBUTION AND HYDRAULIC CONDUCTIVITY

It has long been suspected that an indication of the hydraulic conductivity of sands could be established from a detailed study of the grain-size distribution, and much careful research has been undertaken (Vukovic and Soro, 1992).

5.4.1   Grain-size distributions

The distribution of different grain sizes in a sediment sample has an important impact upon the hydraulic conductivity of a sand. A system to describe the grain-size variation is therefore required. The simplest system would be a simple histogram, but for various reasons, a cumulative frequency plot is more useful. Grain-size data can be used to classify material into silt, sand or gravel. There are two standards for these subdivisions, the Wentworth size class and the phi unit. The respective subdivisions are shown in Table 5.2. Phi units can be defined as follows:

$$\phi = -log_2 \frac{d}{d_0} \tag{5.33}$$

where:

d_0 is the standard grain diameter of 1 mm and included to make the definition of the ϕ unit dimensionless. The negative sign is to make the phi unit for the common sand grain sizes, which are less than 1mm, positive and

d is the grain diameter in mm.

Table 5.2 Udden-Wentworth size classification for sediment grains

Udden–Wentworth Limits (Diameters) (mm)	Phi Notation Ø	Wentworth Size Class
4.00–2.00	−2.0 to −1.0	granule
2.00–1.00	−1.0 to 0.0	very coarse sand
1.00–0.50	0.0 to 1.0	coarse sand
0.50–0.25	1.0 to 2.0	medium sand
0.250–0.125	2.0 to 3.0	fine sand
0.125–0.0625	3.0 to 4.0	very fine sand
0.0625–0.031	4.0 to 5.0	coarse silt
0.031–0.0156	5.0 to 6.0	medium silt
0.0156–0.0078	6.0 to 7.0	fine silt
0.0078–0.0039	7.0 to 8.0	very fine silt
< 0.0039	> 8.0	clay

The series of sizes is therefore $2^3, 2^2, 2^1, 2^0, 2^{-1}, 2^{-2}, 2^{-3}$ mm respectively.

The general result for logarithms is that $log_a x = \frac{log_b x}{log_b a}$, and for a log to base 2 conversion: $log_2 x = \frac{log_{10} x}{log_{10} 2} = 3.321928 log_{10} x$.

5.4.2 Grain-size analysis

There are several methods for the estimation of grain size in a sand. The simplest is based upon a set of sieves. The sand sample is dried and weighed and then shaken through a set of sieves. The sample retained on each sieve is then weighed and a curve is plotted showing the cumulative frequency of the grain-size distribution. If significant clay material is suspected then wet sieving should be used. There is an Australian Standard Practice which fully describes the procedure. Typical grain-size data is shown in Table 5.3. The data can also be presented as a bar chart or as a frequency plot as shown in Figure 5.6.

There are a number of characteristics of the grain-size population which can be derived from the particle size distribution data, viz.:

Md_ϕ median diameter. This is equal to the d_{50} of the distribution.
M_ϕ mean diameter. This is calculated from:

$$M_\phi = \frac{\phi_{16} + \phi_{84}}{2} \tag{5.34}$$

Phi deviation measure is a measure of the degree of sorting in the sand sample and is calculated from:

$$\sigma_\phi = \frac{\phi_{84} - \phi_{16}}{2} \tag{5.35}$$

Table 5.3 Grain-size calculation for a Botany Sands sample

Total weight of sample is 26.780 gms

Sieve Size (mm)	Sample Weight Retained	Weight % Retained	% Coarser A	% Finer 100-A
0.710	0.118	0.441	0.441	99.559
0.590	0.226	0.844	1.285	98.715
0.500	0.806	3.010	4.294	95.706
0.420	3.198	11.942	16.236	83.764
0.360	4.127	15.411	31.647	68.353
0.300	5.161	19.27	50.917	49.083
0.250	6.718	25.086	76.004	23.996
0.210	3.129	11.684	87.689	12.311
0.177	2.247	8.391	96.079	3.921
0.149	0.752	2.808	98.887	1.113
0.125	0.171	0.639	99.526	0.474
0.105	0.055	0.205	99.731	0.269
0.088	0.012	0.045	99.776	0.224
0.074	0.027	0.101	99.877	0.123
0.063	0.006	0.022	99.899	0.101

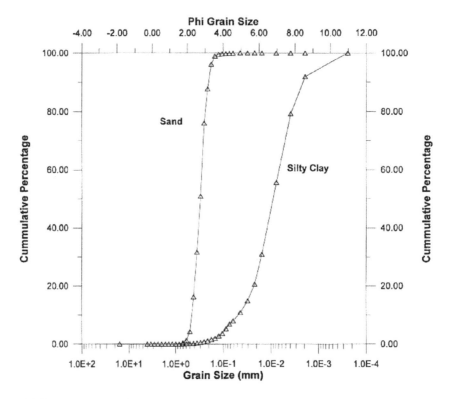

Figure 5.6 Typical grain-size distribution curve for the Botany Sands

The phi deviation is used as an indication of the sand depositional history. Dune sands are usually very well sorted with σ_ϕ values between 0.20 and 0.40.

Phi skewness is a measure of the symmetry of the grain size distribution and can be derived from:

$$\alpha_\phi = \frac{0.5 \times (\phi_{16} + \phi_{84}) - Md_\phi}{\sigma_\phi} \tag{5.36}$$

The phi skewness measure is zero for a symmetrical size distribution. If the distribution is skewed towards the smaller grain sizes then the distribution is said to be negatively skewed.

d_e Effective grain size is determined as the grain size which represents 10% of the sample. C_u The uniformity coefficient is a measure of the sand sorting expressed as:

$$C_u = \frac{d_{60}}{d_{10}} \tag{5.37}$$

A sample with a C_u less than 4 is well sorted; if the C_u is greater than 6, it is poorly sorted.

There are at least 10 methods by which the hydraulic conductivity of sandy sediments can be estimated in the laboratory (Israelsen and Hasen, 1962; Campbell, 1985; Vukovic and Soro, 1992). All these methods are based upon a semi-quantitative representation of the grain-size distribution of the sands determined from the grain-size distribution curve or from measures of the representative masses of silt, sand and clay in the sample.

5.4.3   Hazen's method for determination of hydraulic conductivity

The laboratory method which is most often quoted in the literature was developed by Hazen (1911). The method is applicable to sands where the effective grain size d_e is between approximately 0.1 and 3.0 mm.

The Hazen approximation can be written in non-homogeneous form as:

$$K = AC\tau d_e^2 \tag{5.38}$$

where:
K is the hydraulic conductivity where the dimensions depend on the coefficient A and
A is the coefficient which defines the dimension of the estimated hydraulic conductivity.
$A = 1$ for units of m/day; $A = 0.00116$ for units of cm/s.

C is an empirical coefficient which, according to Hazen, depends on the clayey fraction content of the media. For clean uniform sands, C varies between 800 and 1200. For clayey nonuniform sands, C varies between 400 and 800. The coefficient may also be determined from Lange's formula, where $ø$ is the porosity (expressed as a percentage):

$$C = 400 + 40 \times (\phi - 26) \tag{5.39}$$

d_e is the effective grain size diameter most often taken as the d_{10} passing (finer than) value from the grain-size analysis curve.

τ is the correction for temperature at the water temperature (t) determined from:

$$\tau = 0.70 + 0.03t \tag{5.40}$$

If hydraulic conductivity is expressed in cm/s, at the temperature $t = 10\ °C$, and if the coefficient $C = 860$ is used, then Equation 5.38 is simplified to:

$$K[cm/s] = d_{10}^2[mm] \tag{5.41}$$

This relationship is commonly quoted in the literature (Fetter, 2001).

Note the major changes in hydraulic conductivity estimates in what appears to be a uniform sand. This is demonstrated by the data in Figure 5.7 from the East Lakes Experimental Site in the Botany Sands Aquifer in Sydney. Multi-level piezometers were installed in five lines at 1 m distance from a source piezometer where a conductive tracer was placed as a part of a contaminant tracing experiment. Prior to that work, falling-head hydraulic conductivity measurements were made at each of the mini-piezometer locations. The variability, even in supposedly uniform aeolian sands, is apparent. Lines C, D and E were 1 m apart and orientated on a slope towards a constant-head lake boundary. Detailed

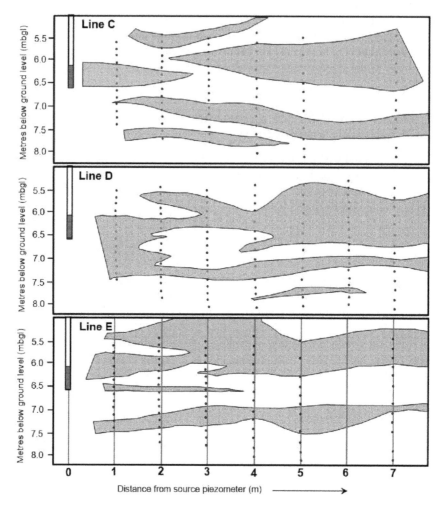

Figure 5.7 Hydraulic Conductivity Layering in 'Uniform' Sands (adapted from Jankowski and Beck (2000)). The shaded areas represent lower hydraulic conductivity regions where K<10 m/day. Each dot in the flow field represents the termination of a mini-piezometer. The hydraulic conductivity at each mini-piezometer was evaluated using a falling-head test

hydrochemical studies carried out by Jankowski and Beck (2000) demonstrated that the lower hydraulic conductivity layers contained more silt which was probably deposited from aeolian activity.

Where such a range of values is common, some authors suggest that the geometric mean of a set of hydraulic conductivity values is more representative than the arithmetic mean. The geometric mean will not be biased to the same degree by the higher values. The geometric mean is determined by taking the natural log of each value, finding the mean of the natural logs and then obtaining the exponential e^x of that value to give the geometric mean.

5.5   LABORATORY MEASUREMENT OF HYDRAULIC CONDUCTIVITY

5.5.1   Constant-head permeameter

The constant-head permeameter is used to measure the hydraulic conductivity of non-cohesive sediments. The equipment is available commercially and a schematic is shown in Figure 5.8.

The sand material is repacked into the chamber to achieve field conditions. The value of hydraulic conductivity may only approximately represent field conditions if the sands are not packed properly into the chamber.

The hydraulic conductivity is derived by application of Darcy's Law as follows:

$$K = \frac{VL}{Aht} \qquad (5.42)$$

where:
V is the volume of water discharged measured in time t,
L is the length of the sample chamber,
A is the cross-sectional area of the sample chamber,
h is the hydraulic head, and
K is the hydraulic conductivity.

The flow through the permeameter should be laminar. It is therefore important to keep the constant-head reservoir close to the top of the chamber. If the permeameter is designed for upward flow, as shown in Figure 5.8, care should be taken to ensure that quick-sand conditions are not created. The remedy for this is again to keep the hydraulic head h small.

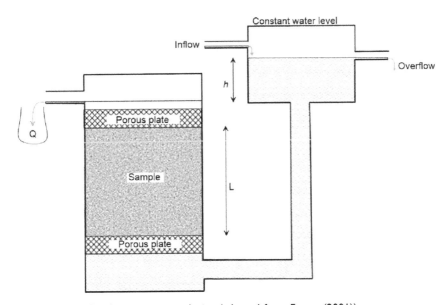

Figure 5.8 Constant-head permeameter device (adapted from Fetter (2001))

5.5.2   Falling-head permeameter

A falling-head device can be used where the hydraulic conductivity of the material is lower. A schematic of the device is shown in Figure 5.9. The initial water level above the outlet h_0 is noted. After some time, the water level h is again measured.

The hydraulic conductivity is then derived from the following:

$$K = \frac{d_t^2 L}{d_c^2 t} \log_e \left(\frac{h_0}{h} \right) \qquad\qquad (5.43)$$

where:
d_t is the inside diameter of the falling-head tube,
d_c is the diameter of the sample container,
L is the sample length, and
t is the elapsed time.

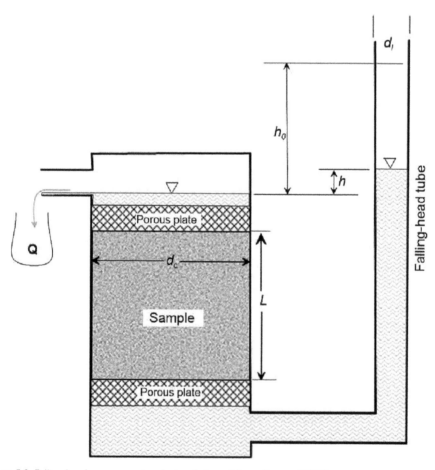

Figure 5.9 Falling-head permeameter device (adapted from Fetter (2001))

5.6   FIELD MEASUREMENT OF HYDRAULIC CONDUCTIVITY

5.6.1   Slug tests

It is possible to determine *in situ* hydraulic conductivity by means of tests carried out in a single piezometer or in an open borehole in which a section of the hole has been isolated by the use of packers (Hvorslev, 1951). The tests are carried out by causing an instantaneous change in the water level in the section by the sudden introduction or removal of a known volume of water. The recovery of the water level with time is then observed.

When water is removed, the tests are often called bail tests; when it is added, the tests are known as slug tests. It is possible to create the same effect by introducing or removing a cylinder of known volume to the piezometer.

Equipment designed at the Water Research Laboratory (WRL) uses compressed air to push the water level down in the piezometer tube. The equipment is shown in Figure 5.10. The water flows into the aquifer and a new equilibrium condition is created whereby the water surface represents the water pressure in the aquifer and the pressure exerted by the air trapped in the piezometer. The water level in the well is measured prior to the time that the air is introduced by using a pressure transducer connected to a data logger.

Immediately after the air is released by opening the quick release valve, the levels are measured frequently (1 second intervals initially) as the water level rises back toward the static water level line. Figure 5.11 shows a logger record from a field test.

The height the water level rises above the static water level immediately upon lowering the slug is h_0. The height of the water level above static water at some time, t, after the slug is lowered is h. The data is plotted by computing the ratio $\frac{h}{h_0}$ and plotting that versus time on

Figure 5.10 Slug test equipment designed at the Water Research Laboratory, UNSW, Australia (Photo: Ian Acworth)

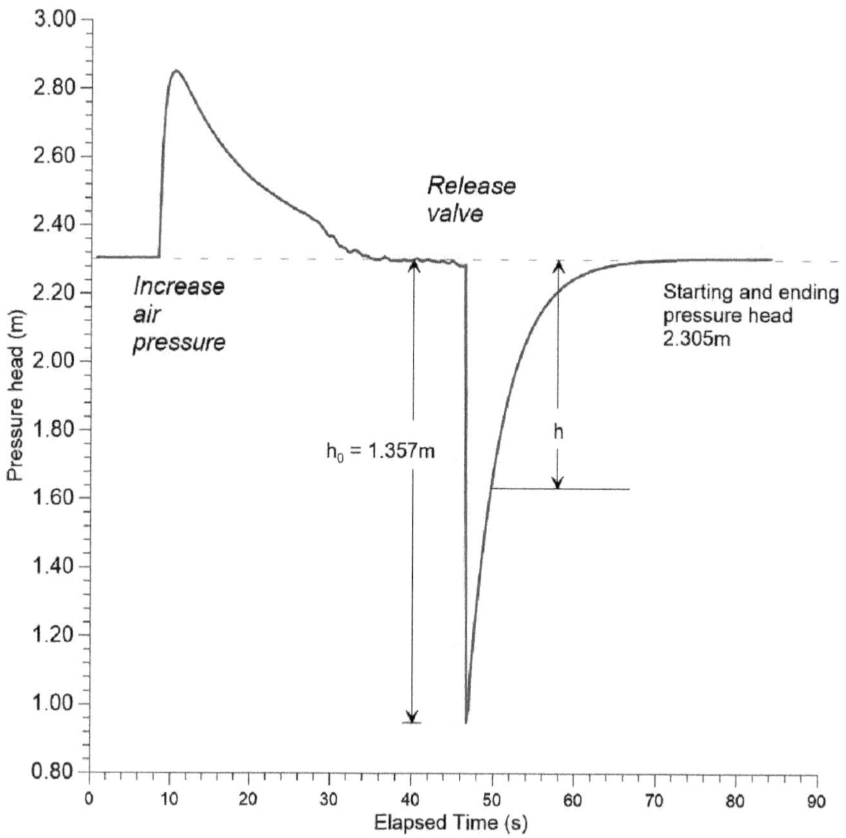

Figure 5.11 Example of field data obtained with the WRL equipment

semi-logarithmic paper. The time drawdown plot should be a straight line. The analysis of the data shown in Figure 5.11 is shown in Figure 5.12.

If the length of the piezometer is more than eight times the radius of the well screen ($L/R > 8$), the following formula applies:

$$K = \frac{r^2 log_e(L/R)}{2LT_0} \tag{5.44}$$

where:
K is hydraulic conductivity,
r is the radius of the well casing,
R is the radius of the well screen,
L is the length of the well screen, and
T_0 is the time it takes for the water level to rise or fall to 37% of the initial change.

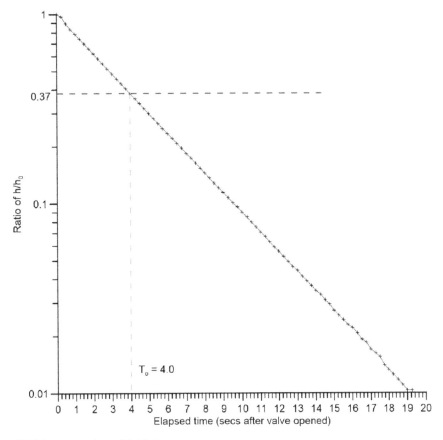

Figure 5.12 Interpretation of field data

This relationship is one of many calculated available (Hvorslev, 1951) for differing pie-zometer geometry and aquifer conditions.

Piezometers in low-permeability clays may take hours or days to recover and in these cases use of a manual dipping method is adequate. Piezometers in sands, however, will react very quickly and the use of a pressure transducer and data logger are required to obtain the data.

Chapter 6

Transport equations and steady-state flow

6.1 TRANSPORT EQUATIONS

6.1.1 Introduction

Any transport problem in which the processes are linear can be described by a differential equation of the form:

$$J = -K \frac{dC}{dL} \tag{6.1}$$

where:
J represents the flux density of energy or material,
K is a conductivity of some form, and
$\frac{dC}{dL}$ is a gradient in concentration or potential.

The flow of groundwater or the flow of heat are two processes that can be analyzed by this approach. In each case, the gradient in potential (either water or heat) is the driving force for the flux.

If the conductivity and the flux density are constant, then Equation 6.50 can be easily integrated to give:

$$J = -K \frac{C_2 - C_1}{L_2 - L_1} \tag{6.2}$$

where the subscripts indicate the values at positions L_1 and L_2. If the J and K terms are not constant, the integration can still be carried out, but some averaged value of these parameters must be used, representative of the space between L_1 and L_2. Equation 6.2 can be used to describe the steady flux density of energy or matter (water) through the subsurface soil.

If the space is divided into a number of layers, and the conductance $(K/\Delta L)$ of each layer as well as the flux density of heat or material through the layer is known, Equation 6.2 can be used to solve for the value of C for individual layers. One equation in two unknowns may be written for each layer. One of the unknowns can be eliminated between each of two equations representing adjacent layers with the exception of the top and bottom layers. These conditions at the top and bottom form the boundary conditions. Values of the boundary conditions must be supplied in order to solve the set of equations for all the layers.

The flow of water through the soil can be studied by the application of the transport equation approach, but first some description of the conductivity term in the transport equation for water flow is required.

6.2 GROUNDWATER TRANSPORT

Darcy demonstrated that the flux of water through a porous medium is directly proportional (for a Newtonian fluid) to the potential gradient.

The transport equation for flow can then be written as:

$$J_w = -K \frac{d\Phi}{dL} \tag{6.3}$$

where:
J_w is the water flux,
K is the hydraulic conductivity,
Φ is the water potential, and
L is distance.

The units of the variables in Equation 6.3 depend upon the definition of the flux and the potential.

It is this variation which leads to alternative formulations for the proportionality constant which contain the different combinations of the intrinsic permeability, fluid density and fluid viscosity shown in Table 6.1.

The flux may be defined either as:

- a volume flux of water, or
- a mass flux of water.

The mass flux is easier to understand in terms of the physical quantities and is widely used in soil physics and in theoretical analysis (Smiles, 2000; Milburn, 1979; Domenico and Schwartz, 1997), but has the disadvantage that it is more difficult to use in practice. The mass flux is first described below and then the conversion to a volume flux is presented.

Table 6.1: Conversion of terms between the mass and volume fluxes in the Darcy transport equation

Mass Flux	=	Conductivity	×	Potential Gradient
J_w	=	$-k\frac{\rho^2}{\mu}$	×	$\frac{d\Phi}{dL}$
$\left(\frac{kg}{m^2 s}\right)$		$\left(\frac{kg}{m^3} \, \frac{s}{}\right)$		$\left(\frac{J/kg}{m} = \frac{m^2}{m} \, s^{-2}\right)$
$\Downarrow$		$\Downarrow$		$\Downarrow$
$\div \rho$		$\div \rho^2$		$\times \rho$
$\left(\frac{kg}{m^3}\right)$		$\left(\frac{kg^2}{m^6}\right)$		$\left(\frac{kg}{m^3}\right)$
$\Downarrow$		$\Downarrow$		$\Downarrow$
U	=	$-\frac{k}{\mu}$	×	$\frac{dP}{dL}$
$\left(\frac{m}{s}\right)$		$\left(\frac{m^3}{kg} \, s\right)$		$\left(kg \quad m^{-1} \quad \frac{s^{-2}}{m}\right)$
Volume Flux	=	Conductivity	×	Potential Gradient

6.2.1 Mass flux of water

The units of Equation 6.3 for this case are as follows:

Mass flux J_w has units of $\frac{kg}{m^2 s^1}$ with dimensions of $\frac{M}{L^2 T^1}$.

Water potential Φ is defined as energy per unit mass with units of joules/kilogram and dimensions of $\frac{ML^2}{T^2 M}$ which further reduce to $\frac{L^2}{T^2}$.

Hydraulic conductivity has units of $\frac{kg\,s}{m^3}$, with dimensions of $\frac{MT}{L^3}$. This represents the intrinsic permeability (k) with units of m^2, multiplied by the square of the density ρ^2 with units of $\frac{kg^2}{m^6}$ and dimensions of $\frac{M^2}{L^6}$, divided by the viscosity μ with units of $Pa\,s$ with dimensions of $\frac{kg}{ms}$.

The mass flux has the advantage that the reference for the flux, the mass, is constant.

6.2.2 Volume flux of water

The units and dimensions for terms in Equation 6.3 for this case are as follows:

Volume flux v The volume flux is the specific discharge. The units are $\frac{m^3}{m^2 s}$ with dimensions of $\frac{L}{T}$. The volume flux can be derived from the mass flux by dividing by the fluid density.

Water potential P The water potential is now defined as energy per unit volume with units of $\frac{Nm}{m^3}$ and has dimensions of $\frac{ML^2}{T^2 L^3}$. The conversion from a mass basis to a volume basis for the potential is achieved by multiplying by the fluid density.

The dimensions for the volume-based potential reduce to those of pressure $\frac{M}{LT^2}$ and provide a useful insight into the nature of this potential. The ability to measure a negative gauge pressure is particularly useful for measurements of water potential in the unsaturated zone.

To convert a water potential measured as a pressure to a potential measured in metres head of water, it is necessary to divide by $\rho\,g$.

The proportionality constant (hydraulic conductivity) has units of $\frac{m^3 s}{kg}$ with dimensions of $\frac{L^3 T}{M^1}$. It represents the intrinsic permeability k, with units of m^2, divided by dynamic viscosity μ with units of $Pa\,s$ $\left(\frac{kg}{ms}\right)$

The volume flux is then seen to be proportional to the pressure difference between two gauges measuring pressure in the system, and inversely proportional to the distance between the gauges.

The formulation of the groundwater transport equation in terms of the volume flux and the pressure gradient is the basis of the SUTRA groundwater modelling approach. It has the advantage that both saturated and unsaturated conditions can be readily handled.

6.2.3 Incorporation of anisotropy and inhomogeneity in intrinsic permeability

For three-dimensional flow in a medium which may be anisotropic and/or inhomogeneous, a generalization of the one-dimensional equation is required.

In three dimensions, the specific discharge can be written:

$$v_x = -\frac{k_x}{\mu}\frac{\partial P}{\partial x} \tag{6.4}$$

$$v_y = -\frac{k_y}{\mu}\frac{\partial P}{\partial y} \tag{6.5}$$

$$v_z = -\frac{k_z}{\mu}\frac{\partial P}{\partial z} \tag{6.6}$$

where:
k_x is the intrinsic permeability in the x direction,
k_y is the intrinsic permeability in the y direction, and
k_z is the intrinsic permeability in the z direction.

The water potential, represented as a pressure, or as a head, is a scalar quantity. It can be specified anywhere in the space defined by the Cartesian coordinate system. We can define the gradient of this field as:

$$\nabla P = i\frac{\partial P}{\partial x} + j\frac{\partial P}{\partial y} + k\frac{\partial P}{\partial z} \tag{6.7}$$

The three components of the specific discharge can then also be written as:

$$v = -\frac{k}{\mu}\nabla P \tag{6.8}$$

The gradient of P is a vector that represents the direction in which the pressure changes fastest. For the condition that $k_x = k_y = k_z$, the direction of ∇P is perpendicular to lines joining points of equal pressure in the system. This will be the direction in which groundwater flows. For anisotropic conditions, the gradient is not normal to the equipotentials but varies depending upon the degree of anisotropy.

We could write a similar expression for the mass flux:

$$J_w = -k\frac{\rho_w^2}{\mu}\nabla\Phi \tag{6.9}$$

6.2.4   Conservation of fluid mass

The equations written so far allow the calculation of the flux but cannot be solved to determine the value of the hydraulic potential at any location or time. To achieve this, the principal of conservation of mass and the concepts of how much water can be stored or removed from storage as the hydraulic potential changes must be incorporated.

In words, the conservation of fluid mass can be stated as:

mass inflow rate – mass outflow rate = change in mass storage with time

in units of mass per time.

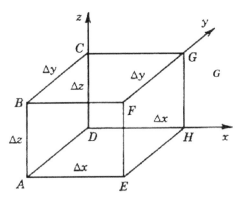

Figure 6.1 Elemental control volume

This statement can be applied to a domain of any size, but consider the unit volume of porous media as shown in Figure 6.1 where $\Delta x = \Delta y = \Delta z = 1$. Such an element is usually called an elemental control volume. The size of this control volume is chosen to be greater than the representative elementary volume wherein the macroscopic average velocity holds as an accurate representation of the fluid flow.

The law of conservation of mass for steady-state flow through a saturated porous medium requires that the fluid mass flux into any elementary control volume must be equal to the fluid mass flux out of the control volume.

From Figure 6.1 the mass *inflow* rate through the face ABCD is J_x.

In general, the mass *outflow* can be different from the mass inflow rate. The mass outflow rate through the face EFGH and can be written as:

$$\left(J_x + \frac{\partial(J_x)}{\partial x} \right) \tag{6.10}$$

The net outflow rate is then the difference between the inflow and outflow rates. The net outflow rate through EFGH is:

$$-\frac{\partial(J_x)}{\partial x} \tag{6.11}$$

Making similar calculations for the remaining faces of the cube, the net outflow through CDHG is

$$-\frac{\partial(J_y)}{\partial y} \tag{6.12}$$

and through BCGF is

$$-\frac{\partial(J_z)}{\partial z} \tag{6.13}$$

Adding these components of the mass flux in the three coordinate directions gives the net mass outflow rate through all the faces:

$$-\left(\frac{\partial(J_x)}{\partial x} + \frac{\partial(J_y)}{\partial y} + \frac{\partial(J_z)}{\partial z}\right) \qquad (6.14)$$

If the components of Equation 6.14 sum to zero – the system is said to be in *steady state*. This does not mean that there is no flow of water through the system, simply that the same amount of water that enters the REV leaves the REV. There is no change in the amount of water stored in the REV.

If the components of Equation 6.14 do not sum to zero, then there has to be a change in the mass stored in the aquifer. The system is said to be in *non-steady state*, and to solve the equation we need to consider the ways in which the water can be stored in the aquifer. This will be returned to in detail later.

6.2.5   Vector notation

The scalar, vector and tensor properties of the hydraulic head and the hydraulic conductivity can be expressed conveniently using vector notation. The following tensor terms can be defined:

- A **zero-order tensor** is also referred to as a scalar and is characterized by its size or magnitude only. The hydraulic head, temperature or chemical concentration are all examples of scalar quantities.
- A **first-order tensor** is also referred to as a vector and has both quantity and direction. Vectors require three components, each having both magnitude and direction. Heat flux, specific discharge and velocity are all vector terms.
- A **second-order tensor** (or simply – a tensor) acts as the product of two vectors and requires nine components to specify it completely. The hydraulic conductivity is an example of a tensor quantity.

The hydraulic head is a scalar quantity. If we take the gradient of the head, we get a vector quantity which has both magnitude and direction. The gradient of *h* is written mathematically as *grad h* or ∇h.

$$\nabla h = i\frac{\partial h}{\partial x} + j\frac{\partial h}{\partial y} + k\frac{\partial h}{\partial z} \qquad (6.15)$$

where:
i, j, and k are unit vectors in the *x*, *y*, and *z* directions.

The vector differential operator ∇ is equivalent to:

$$\nabla h = i\frac{\partial()}{\partial x} + j\frac{\partial()}{\partial y} + k\frac{\partial()}{\partial z} \qquad (6.16)$$

Equation 6.16 means little unless the parentheses contain either a vector or a scalar quantity.

If we place a vector quantity such as the mass flux υp into the vector differential operator, we get a divergence, written as $\nabla \cdot (p\upsilon)$ We can then write:

$$-\nabla \cdot (\rho\upsilon) = -\left(\frac{\partial(\rho\upsilon_x)}{\partial x} + \frac{\partial(\rho\upsilon_y)}{\partial y} + \frac{\partial(\rho\upsilon_z)}{\partial z}\right) \tag{6.17}$$

which is read as the "divergence of $p\upsilon$."

In this context we can understand the divergence as the difference between the mass inflow rate and the mass outflow rate per unit volume. Another useful vector operator is the Laplacian ∇^2. This can be written as:

$$\nabla^2\,() = \frac{\partial^2()}{\partial x^2} + \frac{\partial^2()}{\partial y^2} + \frac{\partial^2()}{\partial z^2} \tag{6.18}$$

or

$$\nabla^2 h \tag{6.19}$$

6.3 STEADY-STATE FLOW

If we recognise that $J = \rho\upsilon$ (from Table 6.1) then we can write Equation 6.14 for the steady-state condition as:

$$-\left(\frac{\partial(\rho\upsilon_x)}{\partial x} + \frac{\partial(\rho\upsilon_y)}{\partial y} + \frac{\partial(\rho\upsilon_z)}{\partial z}\right) = 0 \tag{6.20}$$

If we assume that the fluid density does not change spatially, the density term on the left-hand side of the equation can be taken outside of the partial differential operators so that Equation 6.20 becomes:

$$-\rho \times \left(\frac{\partial(\upsilon_x)}{\partial x} + \frac{\partial(\upsilon_y)}{\partial y} + \frac{\partial(\upsilon_z)}{\partial z}\right) = 0 \tag{6.21}$$

Clearly $\rho \neq 0$ and therefore the terms in Equation 6.21 can then be further simplified. Substitution of Darcy's Law for the υ_x, υ_y, and υ_z components yields an equation for steady-state flow through an anisotropic saturated porous medium as:

$$\frac{\partial}{\partial x}\left(K_x \frac{\partial h}{\partial x}\right) + \frac{\partial}{\partial y}\left(K_y \frac{\partial h}{\partial y}\right) + \frac{\partial}{\partial z}\left(K_z \frac{\partial h}{\partial z}\right) = 0 \tag{6.22}$$

The expression is now positive because the specific discharge terms are negative in Darcy's equation.

For an isotropic medium $K_x = K_y = K_z$, and if the medium is also homogeneous then $K(x, y, z) = $ a constant. The equation then reduces to:

$$K\left(\frac{\partial}{\partial x}\left(\frac{\partial h}{\partial x}\right) + \frac{\partial}{\partial y}\left(\frac{\partial h}{\partial y}\right) + \frac{\partial}{\partial z}\left(\frac{\partial h}{\partial z}\right)\right) = 0 \tag{6.23}$$

or:

$$K\left(\frac{\partial^2 h}{\partial x^2} + \frac{\partial^2 h}{\partial y^2} + \frac{\partial^2 h}{\partial z^2}\right) = 0 \tag{6.24}$$

Equation 6.24 reduces to Laplace's equation when the hydraulic conductivity is homogeneous and isotropic which, as $K \neq 0$, gives:

$$\frac{\partial^2 h}{\partial x^2} + \frac{\partial^2 h}{\partial y^2} + \frac{\partial^2 h}{\partial z^2} = 0 \tag{6.25}$$

The representation of the water flux as shown in Equation 6.25 forms the basis of groundwater modelling for steady-state flow. The solution to this equation describes the value of the hydraulic head at any point in the three-dimensional flow field.

The steady-state flow equation is not describing a system in which there is no flow. It simply means that the fluxes into the system are matched by the fluxes out of the system. All groundwater systems are in steady state if considered over a long enough time step. Groundwater abstraction may be matched by rainfall recharge so that, over a year or more, the average groundwater levels remain static. Another example would be that recharge by rainfall to the groundwater system is matched by discharge as baseflow to the river systems without any long-term change in head in the aquifer.

While no groundwater abstraction occurs, groundwater systems will be full and in steady-state. The R term in Equation 6.26 can be used to represent steady-state fluxes into and out of the aquifer where the hydraulic conductivity (if 3D) or the transmissivity (for a 2D example) is not isotropic or homogeneous:

$$\frac{\partial}{\partial x}\left(T_x \frac{\partial h}{\partial x}\right) + \frac{\partial}{\partial y}\left(T_y \frac{\partial h}{\partial y}\right) + R = 0 \tag{6.26}$$

Major changes occur when abstraction from the groundwater system occurs. This will be examined later.

Note that the assumption has been made in this analysis that the aquifer framework, as defined by the Cartesian coordinate system, is rigid. This assumption is significant as it means that all the pressure head is supported by the water in the system while the aquifer remains independent. A system of interconnecting pipes set in concrete matrix – or fissures in limestone – would satisfy this conceptual model. While limestones certainly exist, the majority of aquifer materials are not so rigid or incompressible, and the overlying weight of both the water and the aquifer material are jointly supported by the water in pore space and also by the formation. There is a partitioning of the overlying stress with strains in both the formation and the water. The relative components of the partitioning are determined by the ratio of the compressibilities of the aquifer and water: as described in Chapter 7.

6.4   NUMERICAL SOLUTION TO THE STEADY-STATE FLOW EQUATION

The steady-state flow equation can be solved using finite difference methods. This is a standard tool deployed in groundwater modelling and will be developed here.

The finite difference methods are fundamentally very simple. The approximation of the continuous variable leads directly to a set of simultaneous equations without the need to introduce further mathematical complications.

The geometry of a two-dimensional solution is maintained in the computer code or set out using a spreadsheet. This means that the approach is ideally suited to a two-dimensional representation of the aquifer. Two-dimensional arrays are used to store aquifer properties and heads which closely resemble the field situation. The extension to three dimensions comes naturally as additional layers of information are added.

Finite difference representations of the first order $\frac{\partial h}{\partial x}$ and second order $\frac{\partial^2 h}{\partial x^2}$ partial derivatives need to be developed. The finite difference approximation to the smooth and continuous function can best be appreciated in terms of gradients (Feynman *et al.*, 1964).

6.4.1   FD discretization of second-order partial differential

The use of the finite difference method allows the discretization of the continuous partial differential equation. Instead of values of head specified at all points in the $-x-y-z-$ space, the space is covered by a two-dimensional grid and the head only calculated for nodes of the grid. If it is necessary to examine a particular area in detail, then the mesh is increased in density in the region of interest. Conversely, if large areas of the problem domain are of little interest, then a coarse mesh can be chosen in this area.

As the mesh is orthogonal, there are some limitations in choosing the most appropriate mesh and this is one of the weaknesses of this method. Note also that the axes of the grid must be chosen to be parallel to the directions of principal anisotropy of the hydraulic conductivity tensor.

For a given portion of an aquifer, the smooth hydraulic head variation can be described by a number of straight-line segments as shown in Figure 6.2.

The accuracy with which the smooth function can be represented is determined by the number of straight-line segments into which it is divided. Approximate values of the gradients (first derivative) at intermediate points in Figure 6.2 can be calculated using the point centered grid system as:

$$\left(\frac{dh}{dx}\right)_a = \frac{h_{i+1} - h_i}{\Delta x_a} \tag{6.27}$$

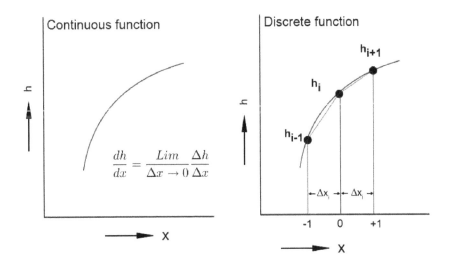

Figure 6.2 Continuous function and finite difference approximation

and:

$$\left(\frac{dh}{dx}\right)_b = \frac{h_i - h_{i-1}}{\Delta x_b} \tag{6.28}$$

where the suffixes x_a and x_b signify intermediate positions on either side of node 0. Values of the second differential can be written as:

$$\left(\frac{d^2 h}{dx^2}\right)_0 = \frac{d}{dx}\left(\frac{dh}{dx}\right) = \frac{\left(\frac{dh}{dx}\right)_a - \left(\frac{dh}{dx}\right)_b}{0.5(\Delta x_a + \Delta x_b)} = \frac{2}{\Delta x_a + \Delta x_b}\left(\frac{h_{i+1} - h_i}{\Delta x_a} - \frac{h_i - h_{i-1}}{\Delta x_b}\right) \tag{6.29}$$

If the mesh spacing is regular and $\Delta x_a = \Delta x_b$ then the expressions simplify to:

$$\frac{d^2 h}{dx^2} = \frac{h_{i+1} - 2h_i + h_{i-1}}{\Delta x^2} \tag{6.30}$$

The approximation introduced by the finite difference approach can best be examined by considering the Taylor series expansion of the terms:

$$h_{i+1} = h_i + \frac{\Delta x}{1!}\left(\frac{dh}{dx}\right)_0 + \frac{\Delta x^2}{2!}\left(\frac{d^2 h}{dx^2}\right)_0 + \frac{\Delta x^3}{3!}\left(\frac{d^3 h}{dx^3}\right)_0 + \frac{\Delta x^4}{4!}\left(\frac{d^4 h}{dx^4}\right)_0 + \cdots \tag{6.31}$$

$$h_{i-1} = h_i - \frac{\Delta x}{1!}\left(\frac{dh}{dx}\right)_0 + \frac{\Delta x^2}{2!}\left(\frac{d^2 h}{dx^2}\right)_0 - \frac{\Delta x^3}{3!}\left(\frac{d^3 h}{dx^3}\right)_0 + \frac{\Delta x^4}{4!}\left(\frac{d^4 h}{dx^4}\right)_0 - \cdots \tag{6.32}$$

Summing:

$$h_{i-1} - 2h_i + h_{i+1} = 2 \times \frac{\Delta x^2}{2!}\left(\frac{d^2 h}{dx^2}\right)_0 + 2 \times \frac{\Delta x^4}{4!}\left(\frac{d^4 h}{dx^4}\right)_0 + \cdots \tag{6.33}$$

Therefore, dividing by Δx^2 gives:

$$\left(\frac{d^2 h}{dx^2}\right)_0 = \frac{1}{\Delta x^2}(h_{i-1} - 2h_i + h_{i+1}) - \frac{\Delta x^4}{12}\left(\frac{d^4 h}{dx^4}\right)_0 + \cdots \tag{6.34}$$

The terms $-\frac{\Delta x^4}{12}\left(\frac{d^4 h}{dx^4}\right)_0$ are called the truncation error. If two solutions are obtained, the second having a mesh interval equal to half that for the first solution, the truncation error for the second solution is only one quarter of that of the first. As a result the second solution will be more accurate. For certain problems the truncation error is zero; then, the finite difference solution is identical with the theoretical solution.

6.4.2   FD representation of Poisson's equation in 2D

The equation representing horizontal flow in a two-dimensional aquifer in steady state is:

$$\frac{\partial}{\partial x}\left(T_x \frac{\partial h}{\partial x}\right) + \frac{\partial}{\partial y}\left(T_y \frac{\partial h}{\partial y}\right) = R \tag{6.35}$$

where:

T is the product of the hydraulic conductivity and the aquifer thickness. It has dimensions of $\frac{L^2}{T}$.

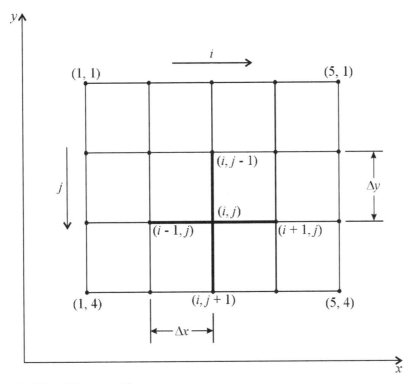

Figure 6.3 Finite difference grid

If the transmissivity is homogeneous and isotropic, it can be removed from the partial derivatives to produce the Laplace equation in 2D, *viz.*:

$$\frac{\partial^2 h}{\partial x^2} + \frac{\partial^2 h}{\partial y^2} = \frac{R}{T}$$

(6.36)

Consider a finite set of points arranged orthogonally on a regular mesh as shown in Figure 6.3.

The space between the points in the -*x*- and -*y*- directions is constant. The smaller this distance becomes, the better the accuracy of the approximation. However, as the number of mesh points and equations increases, the solution time also rapidly increases. To locate any point in the grid we specify an integer ordered pair (i, j), relative to the pair $(1,1)$ located in the upper left-hand corner of the grid. The finite difference expressions can be written in terms of the nodes $h_{i-1,j}$, $h_{i,j}$, $h_{i+1,j}$ by taking a central difference approximation to $\frac{\partial^2 h}{\partial x^2}$ at the point x_0, y_0. These are obtained by approximating the first derivative at $x_0 + \frac{\Delta x}{2}, y_0$ and at $x_0 - \frac{\Delta x}{2}, y_0$, and by then obtaining the second derivative by taking a difference between the first two differences at those points. That is:

$$\frac{\partial^2 h}{\partial x^2} = \frac{\frac{h_{i+1,j} - h_{i,j}}{\Delta x} - \frac{h_{i,j} - h_{i-1,j}}{\Delta x}}{\Delta x}$$

(6.37)

which may be written as:

$$\frac{\partial^2 h}{\partial x^2} = \frac{h_{i-1,j} - 2h_{i,j} + h_{i+1,j}}{\Delta x^2} \tag{6.38}$$

Similarly, the relationships for the approximation in the -y- Cartesian coordinate direction can be written as:

$$\frac{\partial^2 h}{\partial y^2} = \frac{h_{i,j-1} - 2h_{i,j} + h_{i,j+1}}{\Delta y^2} \tag{6.39}$$

The two components are added together to represent Laplace's equation in two dimensions. For a regular grid such that $\Delta x = \Delta y$, the finite difference expression for the point i, j simplifies to:

$$\frac{\partial^2 h}{\partial x^2} + \frac{\partial^2 h}{\partial y^2} = \frac{h_{i-1,j} - 2h_{i,j} + h_{i+1,j} + h_{i,j-1} - 2h_{i,j} + h_{i,j+1}}{\Delta x^2} = 0 \tag{6.40}$$

Multiplying through by Δx^2 gives:

$$h_{i-1,j} + h_{i,j+1} + h_{i+1,j} + h_{i,j-1} - 4h_{i,j} = 0 \tag{6.41}$$

and solving for $h_{i,j}$ gives:

$$h_{i,j} = 0.25 \times (h_{i-1,j} + h_{i,j+1} + h_{i+1,j} + h_{i,j-1}) \tag{6.42}$$

This is equivalent to saying that the head at a central node may be approximated by taking the average of the four surrounding nodes. Note that for the node i, j the diagonal nodes $h_{i-1,j-1}$, $h_{i+1,j-1}$, $h_{i-1,j+1}$ and $h_{i+1,j+1}$ do not contribute to the approximation.

The head in a three-dimensional model could be calculated in the same manner by taking the average of the six surrounding heads. Note that the simplification in Equation 6.41 is only valid if $\nabla^2 h = 0$.

6.4.3   Iterative solution methods

The simplest iteration method is to assume some starting point for each of the four unknown heads in the above problem and to apply the finite difference approximation to each unknown head in turn, repeatedly sweeping through the mesh until the change in head at each unknown is less than some predetermined acceptability criteria. This method was developed by Southwell and is often referred to as the *relaxation method*, because for each step, one of the errors is relaxed.

If the iteration index is referred to as m, then the formula to be used for a regular mesh in which $\Delta x = \Delta y$ can be written as:

$$h_{i,j}^{m+1} = \frac{h_{i-1,j}^m + h_{i,j-1}^m + h_{i+1,j}^m + h_{i,j+1}^m}{4} \tag{6.43}$$

Use of this formula at each node in succession is known as the *Jacobi iteration method*.

A more efficient method is to sweep through the mesh in an ordered manner from left to right and top to bottom, using the most recently computed heads at the left and top in the computation of the new head at the next iteration for node (i, j). This method of solution is known as the *Gauss-Seidel method*. This method allows the use of two newly computed values in the iteration formula, *viz.*:

$$h_{i,j}^{m+1} = \frac{h_{i-1,j}^{m+1} + h_{i,j-1}^{m+1} + h_{i+1,j}^{m} + h_{i,j+1}^{m}}{4} \tag{6.44}$$

The use of newly computed head values whenever possible makes the Gauss-Seidel method more efficient than Jacobi iteration.

6.4.4   Successive over-relaxation (SOR)

The change between two successive Gauss-Seidel iterations can be thought of as a residual c. The residual is defined by:

$$c = h_{i,j}^{m+1} - h_{i,j}^{m} \tag{6.45}$$

In the *SOR* method, the Gauss-Seidel residual is multiplied by a relaxation factor ω where $\omega \leq 1$, and the new value of $h_{i,j}^{m+1}$ is given by the formula:

$$h_{i,j}^{m+1} = \omega c + h_{i,j}^{m} \tag{6.46}$$

Substituting and rearranging these terms gives:

$$h_{i,j}^{m+1} = (1 - \omega)h_{i,j}^{m} + \omega \frac{h_{i-1,j}^{m+1} + h_{i,j-1}^{m+1} + h_{i+1,j}^{m} + h_{i,j+1}^{m}}{4} \tag{6.47}$$

If $\omega = 1$, the equation reduces to that for Gauss-Seidel iteration. For values of $\omega > 1$, we are, in effect, overestimating the correction for each iteration. The most efficient value of ω for a given problem depends on the geometry of the mesh and the field parameters of the equation.

6.4.5   Treatment of boundary conditions

A conceptual model and the related mathematical models of flow in a 2D aquifer are shown in Figures 6.4 and 6.5 (Wang and Anderson, 1982).

The mathematical problem is specified by Equation 6.35 inside the flow domain represented by the model. To complete the mathematical specification, the conditions around the boundary and at the start of the simulation must be specified.

Two basic conditions can exist around the boundary, *viz.*:

* a known flow (flux) across the boundary, or
* a fixed head on the boundary.

A specified flow across the boundary is referred to as a Neumann boundary condition. The most common boundary condition of this type is a no-flow boundary condition, as shown in Figure 6.4 at the left and right edges of the model domain, where the

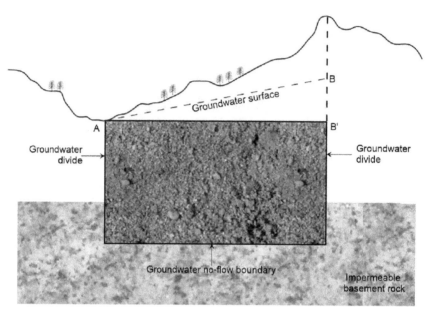

Figure 6.4 Conceptual representation of the boundary conditions (adapted from Wang and Anderson (1998))

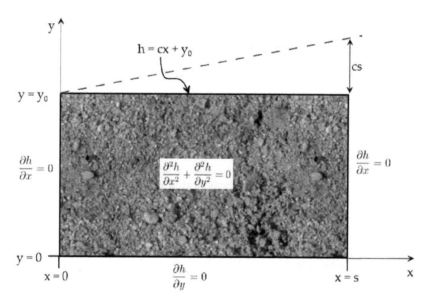

Figure 6.5 Mathematical representation of boundary conditions (adapted from Wang and Anderson (1998))

boundary is established as a groundwater divide and along the base of the domain where the aquifer is bounded by solid impermeable rock. Consideration of the basic Darcy equation indicates that no flow will occur when the gradient of the head across the boundary is zero:

$$\frac{\partial h}{\partial x} = 0 \tag{6.48}$$

A specified head along a boundary is known as a Dirichlet boundary condition. This is shown in Figure 6.4 along the top of the model domain where the head is a simple linear function of the distance away from the left top boundary. Specification of the Dirichlet condition is achieved by setting the head to a known value on the boundary and not allowing it to be changed during the solution procedure. Implementation of a fixed head condition is very significant in groundwater modelling as it creates an unlimited source of water into the aquifer, no matter what hydraulic gradient occurs. This can lead to completely unrealistic model results and should be avoided unless the quantity of water is available to sustain a fixed head.

A no-flow condition across a boundary can be implemented by introducing a fictitious row of nodes outside the boundary and setting the head inside the boundary equal to the adjacent head outside the boundary. This is equivalent to setting $h_{i-1,j} - h_{i+1,j}$ for the boundary at $h_{i,j}$.

Alternatively, if the boundary lies along $h_{i,j}$:

$$h_{i,j} = 0.25(2 \times (h_{i-1,j}) + h_{i,j+1} + h_{i,j-1}) \tag{6.49}$$

6.5   ANALOGIES TO GROUNDWATER FLOW

At the start of this chapter, it was noted that any transport problem in which the processes are linear can be described by a differential equation of the form:

$$J = -K\frac{dC}{dL} \tag{6.50}$$

where:
J represents the flux density of energy or material,
K is a conductivity of some form, and
$\frac{dC}{dL}$ is a gradient in concentration or potential.

There are two particular analogues that occur widely in groundwater studies:

Ohm's Law describing the flow of electrical current and
Fourier's Law describing the flow of heat.

6.5.1   Ohm's Law

Ohm's Law written in this form introduces the concept of electrical resistance. This will be discussed in detail in Chapter 9. Ohm's Law is often written as $I = \frac{V}{R}$, or as:

$$V = IR \tag{6.51}$$

where:

V is the voltage measured across the conductor in volts,

I is the electrical current through the conductor measured in amperes, and

R is the resistance of the conductor measured in ohms.

This formulation is a little different to the general transport equation, and the equation could be written as $I = \frac{V}{R}$ or, perhaps more sensibly, as $I = V \cdot \sigma$, where σ is the electrical conductivity (measured in Siemens) and is the reciprocal of the resistance. As noted in Chapter 9, this doubling of names for the same quantity leads to extensive confusion!

In theoretical physics, Ohm's Law is expressed as:

$$J = \sigma E \tag{6.52}$$

which is similar in form to the general transport equation and states that the current density (J) is proportional to the electric field (E) at the location and the constant of proportionality is the electrical conductivity (s).

Electrical resistivity networks were built as an analogue to groundwater flow prior to the availability of computers and computational methods (finite difference or finite element). These networks were constructed by water authorities to assist with groundwater management. Rushton and Redshaw (1979) provide a general description of these methods. Figures 6.6 and 6.7 illustrate the implementation and construction of this type of model. Although there was a disadvantage in that the model took considerable time to construct, a simple voltmeter was all that was required to read out the potential at any node – rather than examination of large paper files of data.

6.5.2 Fourier's Law

Fourier's Law describes the flow of heat in response to a gradient in temperature. It is a general transport equation described in detail in Chapter 11. Fourier's Law (Equation

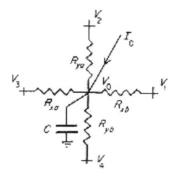

Figure 6.6 Arrangement of electrical resistors in a 2D model of transient flow. V_0 is the central voltage, surrounded in 2D by nodes with electrical voltages: V_1, V_2, V_3 and V_4, R_{y-a} and R_{y-b} are the resistances between the nodes V_2 and V_0. I_0 is the current entering or leaving the central node and C is a capacitance representing electrical storage

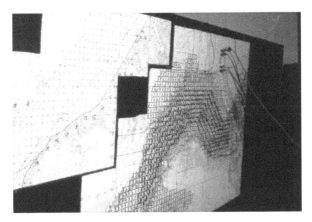

Figure 6.7 2D groundwater model development using network of resistors to represent hydraulic conductivity – built on a map representing the hydrogeology. Each node has connections as shown in Figure 6.6.

6.53) is directly comparable to Darcy's Law:

$$J_{H_c} = -\lambda \frac{dT}{dz} \tag{6.53}$$

where:

J_{Hc} is the heat flux density ($\frac{W}{m^2}$) in the z-direction due to conduction,

$-\lambda$ is the local thermal conductivity – heat moves from higher temperature to lower temperature – explaining the negative sign, and

$\frac{dT}{dz}$ is the local thermal gradient in the z-direction.

Chapter 7

Aquifer storage and abstraction impacts

7.1 WATER STORAGE IN THE UNSATURATED ZONE

7.1.1 The specific yield (S_y)

The storage term for unconfined aquifers is known as the specific yield S_y and represents the volume of water that drains from a unit cross-sectional area of the unconfined aquifer by allowing the water level (water table) to fall by a unit amount. For saturated Botany Sands with a porosity of 35%, approximately 22% of the water will drain out under gravity and the specific yield is then 22%. Note that $\frac{13}{35}$ (37%) of the original water remains in the aquifer – held there by surface tension. To put this value of S_y into perspective, a cubic metre of saturated sand will yield 220 L of water rapidly on draining and a further 130 L of water will be held in the sand by surface tension – available to support the growth of vegetation. Plants can suck out a significant proportion of the 130 L but can never completely remove all the water that is held in by surface tension.

Delayed drainage is a concept developed to account for the slower release of moisture from soils during pumping tests and is directly related to the rate at which water drains from the soil.

The specific yield will only be equal to the porosity for very clean coarse gravels. For finer material, a proportion of the water contained at saturation is held back in the soil by surface tension effects. The quantity is sometimes referred to as specific retention, where the sum of the specific yield and the specific retention is equal to the porosity:

$$\phi = S_y + S_r \tag{7.1}$$

Field capacity moisture content is analogous to the amount of water contained in the soil when the specific yield has drained away. Typical values of specific yield for a range of materials are given in Table 7.1.

7.2 CONFINED AQUIFER STORAGE

The concept of water storage below the water table is conceptually more challenging than that for draining pore space in an unconfined aquifer. Water is removed from the system but the aquifer itself is not drained. The results of removing water from confined storage have far-reaching impacts.

Table 7.1 Approximate (very) specific yields for various geologic materials

Material	Specific Yield
Coarse gravel	23
Medium gravel	24
Fine gravel	25
Coarse sand	27
Medium sand	28
Fine sand	23
Silt	8
Clay	3
Fine-grained sandstone	21
Medium-grained sandstone	27
Limestone	14
Dune sand	38
Loess	18
Peat	44
Schist	26
Siltstone	12
Tuff	21
Glacial till – silts	6
Glacial till – sands	16
Glacial till – gravels	16

7.2.1   Non-steady state flow

An analysis has been presented for steady-state flow conditions. We can now return to consider the possibility that water can be taken into storage within the REV, which will result in non-steady state conditions.

A mathematical model that is generally used to represent this possibility is:

$$-\left(\frac{\partial J_x}{\partial x} + \frac{\partial J_y}{\partial y} + \frac{\partial J_z}{\partial z}\right) = \frac{\partial(\rho\phi)}{\partial t} = \rho\frac{\partial\phi}{\partial t} + \phi\frac{\partial\rho}{\partial t} \tag{7.2}$$

Or, in volume terms, recognising that $J = \rho v$ and that ρ is independent of location:

$$-\left(\frac{\partial v_x}{\partial x} + \frac{\partial v_y}{\partial y} + \frac{\partial v_z}{\partial z}\right) = \frac{1}{\rho}\frac{\partial(\rho\phi)}{\partial t} = \frac{\partial\phi}{\partial t} + \frac{\phi}{\rho}\frac{\partial\rho}{\partial t} \tag{7.3}$$

The terms on the RHS of Equation 7.3 represent the fact that to get more water into the already saturated pore space, we can either:

- compress the skeleton – therefore increasing the porosity, or
- compress the water – achieving a higher density.

Neither mechanism looks a likely candidate to allow much increase in storage, and in reality the storage coefficient for a saturated aquifer is many times smaller than the specific yield (S_y) for an unconfined aquifer. Clearly, concepts of elasticity are involved.

There has been much debate concerning the significance of the RHS terms in Equation 7.2. The first hydrogeologist to give the problem attention was Meinzer (1928). Jacob (1940) gave a mathematical basis to the analysis and incorporated the compressibility of water into a later analysis in 1950. Terzaghi (1943), working in the area of soil mechanics, realized the effect of pore water dissipation upon the deformation of a saturated soil. The process is called *consolidation* and forms a major plank in soil mechanics theory. Unfortunately, the two streams of scientific endeavour seldom cross and only few workers acknowledge the similar basis upon which much different terminology has been established (De Wiest, 1969). The process is further complicated by the soil science discipline approaching the same ground from yet another direction using material coordinates rather than Cartesian coordinates.

7.2.2   Specific storage

The concept of *specific storage* (S_s) was introduced by Jacob (1940). In effect, the change in mass within the REV (LHS of Equation 7.3) will be proportional to the change of head with time. If the head in the aquifer remains constant, it is reasonable to assume that there is no change in the quantity of water stored in the aquifer. Mathematically, the proportionality is expressed as:

$$-\left(\frac{\partial J_x}{\partial x} + \frac{\partial J_y}{\partial y} + \frac{\partial J_z}{\partial z}\right) \approx \frac{\partial h}{\partial t} \tag{7.4}$$

The coefficient of proportionality is called the specific storage (S_s) (Freeze and Cherry, 1979):

$$-\left(\frac{\partial J_x}{\partial x} + \frac{\partial J_y}{\partial y} + \frac{\partial J_z}{\partial z}\right) = \rho S_s \frac{\partial h}{\partial t} \tag{7.5}$$

The specific storage is the volume of water released from a unit cross-sectional area of aquifer for a unit decrease in hydraulic head. The S_s is multiplied by the density to maintain the equation as a mass flux.

Jacob (1940) provided an initial analysis of the specific storage by relating it to elastic properties:

$$S_s = \rho g(\beta_p + \phi\beta) = \rho g\beta_p + \rho g\phi\beta \tag{7.6}$$

where:
β_p is the compressibility of the formation. This concept will be expanded upon later,
ϕ is the porosity, and
β is the compressibility of water.

This analysis has been the subject of some contention (Domenico and Schwartz, 1997) in the literature. While the expression is now accepted as correct, the method that Jacob used to arrive at the analysis was not. Part of the problem can be indicated by the observation that when $\phi = 0$, there will still be a component of storage as indicated by Equation 7.6 (Van Der Kamp and Gale, 1983).

The compressibility of water (β) is approximately $4.6 \ 10^{-10}$ (Table 2.8). Typical values of compressibility of soil and rock (β_p) are given in Table 7.2 (Domenico and Schwartz, 1997).

Table 7.2 Values of formation compressibility – after Domenico and Schwartz (1997)

Material	$\beta_p \ m^2/N$
Plastic clay	2.0×10^{-6} to 2.6×10^{-7}
Stiff clay	2.6×10^{-7} to 1.3×10^{-7}
Medium-hard clay	1.3×10^{-7} to 6.9×10^{-8}
Loose sand	1.0×10^{-7} to 5.2×10^{-8}
Dense sand	2.0×10^{-7} to 1.3×10^{-8}
Dense, sandy gravel	1.0×10^{-8} to 5.2×10^{-9}
Jointed rock	6.9×10^{-10} to 3.3×10^{-10}
Sound rock	$< 3.3 \times 10^{-10}$
Water at 25 °C	4.8×10^{-10}

Substituting the volume flux for the mass flux on the LHS of Equation 7.4 gives:

$$-\left(\frac{\partial \rho \upsilon_x}{\partial x} + \frac{\partial \rho \upsilon_y}{\partial y} + \frac{\partial \rho \upsilon_z}{\partial z}\right) = \rho S_s \frac{\partial h}{\partial t} \qquad (7.7)$$

The LHS of Equation 7.7 can be expanded by the chain rule to give terms of $\rho \frac{\partial \upsilon_x}{\partial x}$ and $\upsilon \frac{\partial \rho}{\partial x}$. As the density is not going to vary significantly with direction, the term $\upsilon \frac{\partial \rho}{\partial x}$ can be ignored and we are left with $\rho \frac{\partial \upsilon_x}{\partial x}$ terms, and similar terms in the other two coordinate directions.

Using Darcy's Law for the υ terms and cancelling out the density (ρ) leaves:

$$\frac{\partial}{\partial x}\left(K_x \frac{\partial h}{\partial x}\right) + \frac{\partial}{\partial y}\left(K_y \frac{\partial h}{\partial y}\right) + \frac{\partial}{\partial z}\left(K_z \frac{\partial h}{\partial z}\right) = S_s \times \frac{\partial h}{\partial t} \qquad (7.8)$$

If we make the assumption that we are dealing with an isotropic hydraulic conductivity, Equation 7.8 can be further simplified as:

$$\frac{\partial^2 h_x}{\partial x^2} + \frac{\partial^2 h_y}{\partial y^2} + \frac{\partial^2 h_z}{\partial z^2} = \frac{S_s}{K} \times \frac{\partial h}{\partial t} \qquad (7.9)$$

Equation 7.9 can be solved numerically and forms the basis of much groundwater modelling work. The inverse of the first term on the RHS of Equation 7.9 (K/S_s) is often referred to as the hydraulic diffusivity.

In analysis of vertical movement of water in soil mechanics, this term is also known as the coefficient of consolidation.

7.2.3   Compressibility of water

Returning to the second term on the RHS of Equation 7.3.

The isothermal compressibility of water is defined as (Domenico and Schwartz, 1997):

$$\beta = \frac{1}{K} = -\frac{1}{V}\left(\frac{dV_w}{dP}\right)_{T,M} \qquad (7.10)$$

where:
β is the fluid compressibility with units of reciprocal pressure,
K is the bulk modulus of compression of the fluid,

V is the volume of the fluid, and
P is the pressure.

A negative sign indicates that the volume decreases with increasing pressure.
It is worth noting that *a bulk modulus of elasticity* is also defined as:

$$E = -\frac{dP}{dV/V} \tag{7.11}$$

The compressibility of a fluid can be expressed as the bulk modulus of elasticity (E), which is the ratio of the increase in pressure to the resulting volumetric strain. The several ways of denoting the same quantity do not help!

The conservation of mass demands that the product of the fluid density and the fluid volume must be a constant as $M = \rho V$. A decrease in density that occurs as the result of a decrease in pressure must be accompanied by an increase in volume:

$$\rho\, dV + V\, d\rho = 0 \tag{7.12}$$

Taking the term dV to the left of Equations 7.10 and 7.12 and equating RHS gives:

$$\frac{V}{\rho} d\rho = V\beta dP \tag{7.13}$$

Cancelling V, dividing by dT and noting that $P = \rho gh$ gives:

$$\frac{d\rho}{dt} = \rho\beta\frac{dP}{dt} = \rho\beta\rho g\frac{dh}{dt} \tag{7.14}$$

Returning to Equation 7.3 – repeated below for clarity:

$$\frac{1}{\rho}\frac{\partial(\rho\phi)}{\partial t} = \frac{\partial\phi}{\partial t} + \frac{\phi}{\rho}\frac{\partial\rho}{\partial t} \tag{7.15}$$

If the matrix is considered to be incompressible then $\frac{\partial\phi}{\partial t} = 0$ and substituting Equation 7.14 we can write:

$$\frac{1}{\rho}\left(\phi\frac{\partial\rho}{\partial t}\right) = \rho g\phi\beta\frac{\partial h}{\partial t} \tag{7.16}$$

The quantity $\rho g\phi\beta$ is the specific storage of a rock with an incompressible matrix. This part of Jacob's analysis is correct!

What about the first term on the right-hand side of Equation 7.15? What happens if $\frac{\partial\phi}{\partial t} \neq 0$. What if the components of the aquifer/aquitard are themselves compressible? Further investigation is required into the way in which the aquifer skeleton can deform and in which coupling can occur between stress and strain in the system.

7.3 EFFECTIVE STRESS IN A SATURATED SYSTEM

7.3.1 Introduction

The initial work by Jacob assumed that only the water derived from the elastic expansion of water contributed to aquifer storage. Regional groundwater analysis carried out in the USA indicated that if this were the case then the water level in confined aquifers should have fallen more rapidly than had in fact been observed (Domenico and Schwartz, 1997). It was then realized that the aquifer skeleton could also deform – a process that would also release water. The analysis is further complicated by the possibility that the grains of material may themselves be compressible and that the resulting consolidation leads to a deformation of the REV, changing the dimensions against which the deformations are measured.

The response of water levels in rigid confined aquifers, or in saturated deformable materials such as clays, to changes in pressure can lead to interesting and important phenomena. Water levels may respond to changes in the weight (pressure) of the atmosphere or to changes in the weight of the overlying formations due to additional loading from rainfall or the reduced loading that results from evapotranspiration losses. The loading may even be the result of moving a train or plane onto an aquifer for a short period of time. The response to these transitional loads will depend very much upon the rate at which the porous medium can respond. From a soils mechanics perspective, the response can either be a drained response, where the fluid shifts away from the site of additional loading, or an undrained response. A stiffer elastic response will occur for an undrained response as compared to a drained response.

The fact that the water level in an aquifer responds to change in effective stress was well demonstrated by Jacob (1939). He plotted the hydraulic head variation as observed in a well close to a railway station. The data is shown in Figure 7.1. Similar effects can be seen in piezometers installed adjacent to the aircraft parking bays at Mascot Airport in Sydney.

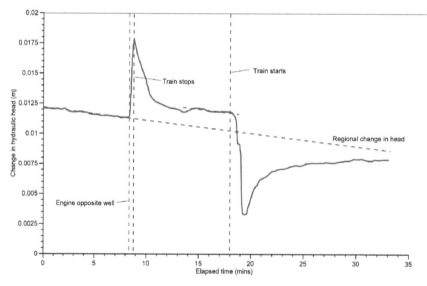

Figure 7.1 Hydraulic head variation due to external loading represented by a freight train stopping for 10 minutes adjacent to a well installed into a confined aquifer (adapted from Jacob (1940))

The piezometers are installed in a sand aquifer confined by clays. The weight of a Boeing 747–400 significantly changes the total stress (σ_T) on the aquifer!

As the train/plane approaches, the load is added to the total stress (Equation 7.17). This additional stress appears to first be carried by the groundwater as the pressure (P) increases (as indicated by the increase in head shown in Figure 7.1). The increase in pressure is then dissipated as the additional stress is transferred to the aquifer skeleton. An equilibrium is reached after some time when the aquifer skeleton appears to be carrying all the increased stress from the additional load and the excess water pressure has diminished as water flows away from the site. There has been a fairly rapid transfer of stress from the fluid to the solid, which must elastically compress.

When the train/plane leaves, there is a rapid fall in water pressure (head) observed. (The total stress is reduced; the effective stress remains constant and therefore the pore pressure decreases). However, the head shortly returns to its initial position as the stress is partially transferred from the water to the aquifer skeleton to be taken up as effective stress.

A similar but cyclic response is seen in Figure 7.2 from Hat Head in NSW. The record for a shallow piezometer at 9 m depth close to a tidal creek is shown with the tide data and the record for a deeper piezometer (18 m) at the same location. There is clearly a direct correspondence between the data sets – with no time lag in the response. There is a thin zone of cemented sand at 10.5 m, otherwise the aquifer is entirely composed of aeolian and alluvial sand.

The water level in each piezometer at the height of the tide will tend to cause a lateral drainage of water away from the location of increased stress – in the same way that the

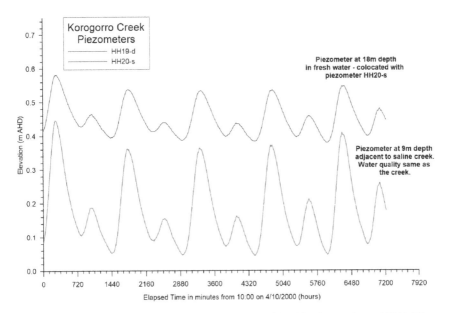

Figure 7.2 Water level response to changes in tide at Hat Head, northern NSW. The upper piezometer could be considered to be responding to advective movement of water from a tidal creek adjacent. The deeper piezometer is installed beneath a low hydraulic conductivity layer and has clean water. This response is the result of loading at the surface by the tide

excess pressure caused by the train was dissipated. The process is reversed as the tide falls again. It is of great importance to sort out whether the water level in a piezometer is moving as the result of water flowing through a rigid matrix or whether the movement is the result of changes in applied pressure. Clearly, the water level change in the deeper piezometer is not the result of sea water moving downward into the aquifer and back again with each tide. The water quality in the lower piezometer is always very good – with no saline contamination.

7.3.2   The effective stress concept

For a saturated porous material to undergo compression there must be an increase in the grain-to-grain pressures within the matrix and/or an increase in the water pressure. Without such pressure changes, no volumetric change can occur.

The total vertical stress (pressure) acting on a horizontal plane at any depth can be resolved into two components (Figure 7.3 and Equation 7.17):

$$\sigma = P + \sigma_e \qquad (7.17)$$

where:
σ is the total stress acting downwards and is the weight of all the overlying material,
σ_e is the effective stress – the component of σ supported by the grain to grain contact of the media, and
P is the neutral stress or water pressure.

Clearly, under many circumstances there is no significant change in total stress. If the water pressure is changed as a result of groundwater abstraction, then there must be a reciprocal change in effective stress:

$$P = -\sigma_e \qquad (7.18)$$

A reduction in hydraulic pressure will result in an increase in effective stress. If the formation responds elastically, then this increase will be returned when the pressure is again increased – as in the train example above. If the formation deforms as a result of the increase in effective stress, it will not be possible to recover the deformation.

The stress balance represented by Equation 7.17 is important in understanding much of what occurs in media that deform. Consider the way in which this balance would change as the total stress changes. Differentiating Equation 7.17 with respect to σ gives:

$$\frac{\partial \sigma}{\partial \sigma} = 1 = \frac{\partial P}{\partial \sigma} + \frac{\partial \sigma_e}{\partial \sigma} \qquad (7.19)$$

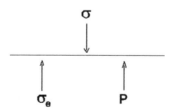

Figure 7.3 Stress balance about a horizontal plane

where:

$\frac{\partial P}{\partial \sigma}$ is called the pore pressure coefficient and represents the percentage of the load that is carried by the fluid provided that the fluid is not allowed to drain and

$\frac{\partial \sigma_e}{\partial \sigma}$ must then represent the percentage of the load carried by the matrix.

7.3.3 A flow equation for the impact of excess pressure

The normal Darcy equation is based on the flow of water in response to a gradient in hydrostatic head in a rigid matrix. But water flow can also be considered to occur as the result of a gradient of excess pressure. Following the discussion on effective stress, we can write:

$$\sigma + \Delta\sigma = \sigma_e + (P_s + P_{ex}) \tag{7.20}$$

where the total pore water pressure comprises two parts, P_s is the hydrostatic pressure and P_{ex} is a transient pore water pressure in excess of the hydrostatic pressure and is related to the increased downward total stress component ($\Delta\sigma$).

In a similar fashion, the total hydraulic head can be restated (Domenico and Schwartz, 1997) as:

$$h = z + \frac{P_s}{\rho g} + \frac{P_{ex}}{\rho g} \tag{7.21}$$

where:
h is hydraulic head,
z is distance above some datum plane – normally sea level,
ρ is the density of water, and
g is acceleration due to gravity.

For one-dimensional flow in the vertical direction, Darcy's Law can be restated as:

$$v_z = -K_z \frac{\partial h}{\partial z} = -K_z \frac{\partial}{\partial z} \left(z + \frac{P_s}{\rho_w g} + \frac{P_{ex}}{\rho_w g} \right) \tag{7.22}$$

Following the approach of Domenico and Schwartz (1997), when considering the flow that occurs in response to excess pressure change only, Equation 7.22 can be simplified to:

$$v_z = -\frac{K_z}{\rho g} \frac{\partial P_{ex}}{\partial z} \tag{7.23}$$

The derivation of a mass conservation equation for this relationship has been the subject of considerable discussion in the literature. A relationship that describes the way in which the flux (v) changes with space and time is required. Early attempts at this (e.g., Jacob (1940)) ignored the fact that the coordinate system had to deform if there was to be a change in σ_e due to compression. This was overcome by the use of material derivatives that followed the motion of the solid phase rather than using a fixed Cartesian coordinate system. The approach is described fully by Domenico and Schwartz (1997) and has been recently invoked in the soil physics literature by Smiles (2000).

Following Jacob (1940), the conservation equation could be written as shown in Equation 7.3, repeated here for convenience again as Equation 7.24. This formulation is not correct as shown by several more recent analyzes (see Domenico and Schwartz (1997)):

$$-\frac{\partial v_z}{\partial z} = \frac{\partial \phi}{\partial t} + \frac{\phi}{\rho}\frac{\partial \rho}{\partial t} \qquad (7.24)$$

The significance of incorporating the effect of deformable coordinates is to give a new equation:

$$-\frac{\partial v_z}{\partial z} = \frac{1}{1-\phi}\frac{\partial \phi}{\partial t} + \frac{\phi}{\rho}\frac{\partial \rho}{\partial t} \qquad (7.25)$$

The RHS of Equation 7.25 describes two terms. The first describes the way in which the porosity (ϕ) changes with time. The second relates to the way in which the density of water (ρ) changes with time. If water is to be added to or removed from storage, it has to either change the density of the water or the volume of the pore space.

The change in density of water is related to the pressure and the compressibility of water as derived before (Equation 7.16) and repeated here, viz.:

$$\frac{\partial \rho}{\partial t} = \rho\beta\frac{\partial P}{\partial t} \qquad (7.26)$$

The change in porosity can be written (Domenico and Schwartz, 1997) as:

$$\frac{1}{1-\phi}\frac{\partial \phi}{\partial t} = \beta_p\left(\frac{\partial P_{ex}}{\partial t} - \frac{\partial \sigma}{\partial t}\right) \qquad (7.27)$$

where:
β is the compressibility of the water and
β_p is the compressibility of the matrix.

The quantity $\partial\sigma/\partial t$ can be thought of as the vertical stress that gives rise to the excess pressure development. Substituting the relationships for density change with time (Equation 7.26) and porosity change with time (Equation 7.27) into the 1D conservation of mass equation (Equation 7.25) gives:

$$-\frac{\partial v_z}{\partial z} = \phi\beta_w\frac{\partial P_{ex}}{\partial t} + \beta_p\left(\frac{\partial P_{ex}}{\partial t} - \frac{\partial \sigma}{\partial t}\right) \qquad (7.28)$$

Substituting Darcy's Law (Equation 7.23) into Equation 7.28 and rearranging terms on the RHS gives:

$$K_z\frac{\partial^2 P_{ex}}{\partial z^2} = \rho g(\phi\beta + \beta_p)\frac{\partial P_{ex}}{\partial t} - (\rho g\beta_p)\frac{\partial \sigma}{\partial t} \qquad (7.29)$$

The term $\rho g(\phi\beta + \beta_p)$ is the specific storage already derived by Jacob, whereas the quantity $\rho g\beta_p$ is the component of specific storage due exclusively to pore compressibility.

7.3.4   Coefficient of consolidation

In some problems it can be assumed that the stress is applied rapidly and then held constant. This could occur as the result of a considerable quantity of rain added to the top of a soil column – or a vehicle moving onto the soil and staying there – or the construction of a building/embankment etc. In this case, the second term on the RHS of Equation 7.29 diminishes to zero. If we also consider that the compressibility of water is so much less than the compressibility of the soil skeleton that it can be ignored, Equation 7.29 can be collapsed to:

$$\frac{\partial^2 P_{ex}}{\partial z^2} = \frac{\rho g \beta_p}{K_z} \frac{\partial P_{ex}}{\partial t} = \left(\frac{1}{C_v}\right) \frac{\partial P_{ex}}{\partial t} \qquad (7.30)$$

This is the 1D consolidation equation of Terzaghi and Peck (1948) where the hydraulic diffusivity C_v is also known as the coefficient of consolidation. Note that there is some variability in terminology here. The hydrogeological representation of diffusivity is generally taken as $\frac{K_z}{S_s}$ and not $\frac{K_z}{\rho g \beta_p}$.

Using Equation 7.30 we can calculate the rate at which a hydraulic change will propagate through an aquifer. Rearranging we get:

$$\frac{\partial^2 P_{ex}}{\partial z^2} = \frac{S_s}{K_z} \frac{\partial P_{ex}}{\partial t} \qquad (7.31)$$

or:

$$\frac{\partial P_{ex}}{\partial t} = C_v \frac{\partial^2 P_{ex}}{\partial z^2} \qquad (7.32)$$

Equation 7.32 is analogous to Fick's 2nd Law that allows the calculation of the rate of movement of a solute front in terms of the concentration gradient, *viz.*:

$$\frac{\partial C}{\partial t} = D \frac{\partial^2 C}{\partial x^2} \qquad (7.33)$$

where:
D is the diffusion coefficient.

7.3.5   Subsidence

Subsidence occurs as a direct result of groundwater abstraction from a sequence that contains silts and clays that can deform (compress) under increased effective stress. In a sequence where Equation 7.17 holds, if groundwater pressure is decreased by abstraction, and the total stress remains the same, then effective stress must increase. In formations that do not have an elastic response, a one-way deformation occurs in which clays are compressed and the water they contain is expelled. The volume of water expressed can be a significant component of the system yield. The reduction in volume of the clay results in subsidence of the ground surface and is not reversible.

Subsidence has occurred in many groundwater abstraction basins around the world and is a particularly significant problem when it occurs in a coastal groundwater system as it can lead to seawater flooding. Bangkok and California are two well-documented examples of subsidence.

7.3.6   Case study – Mexico City

Mexico City and its metropolitan zone (MCMZ) have an estimated population of 22 million inhabitants (Huizar-Alvarez *et al.*, 2004), concentrated in an area of 2,269 km^2. The water supply comes from imported surface water (11 m^3/s) and groundwater sourced from within the watershed (53 m^3/s). Consolidation effects have been a major problem for many years. The soil consolidation rate in the historical city area (Figures 7.4, 7.5, 7.6 and 7.7) varied between 17 cm/year (1940–1970), 4.5 cm/year in the late 1970s and 9.2 cm/year from 1986 to 1991. The Chalco area has experienced a soil consolidation rate of about 40 cm/year for the last 10 years.

The aquifer is approximately a 300 m succession of alluvial and volcanoclastic material.

Figure 7.4 Subsidence in the centre of Mexico City (Photo: Ian Acworth)

Figure 7.5 Subsidence causing differential settlement (Photo: Ian Acworth)

Figure 7.6 Subsidence closing access to a building (Photo: Ian Acworth)

Figure 7.7 The Cathedral in central Mexico sinks into the soft sediment (Photo: Ian Acworth)

7.4   UNDRAINED RESPONSE TO NATURAL LOADING

7.4.1   Introduction

There are many interesting and 'strange' results as a consequence of the undrained response to natural loading. These include:

- streams and seeps that begin to flow before it rains and
- bores that make noise as the weather changes.

It is important to consider the boundary conditions under which the loading takes place. There are two possibilities:

- For field deformations that are slower than the characteristic time for diffusion of fluid through the matrix – a *drained response* occurs. This can be seen where a load is applied to a sandy aquifer. There is an initial increase in water pressure, but water then moves away and the load shifts to the grains – so the water pressure returns to the initial level.
- When the application of the load is rapid in comparison to the time taken for diffusion of the pore fluid, the local fluid mass remains essentially the same and an *undrained response* is noted.

The undrained response is elastically stiffer than the drained response so that the respective deformations are characterized by different coefficients of elasticity.

All water-level fluctuations as observed at a point that are caused by rapid loading such as tidal or atmospheric variation can be treated as constant mass phenomena. This implies that no fluid flow occurs and therefore the fluid flow term in Equation 7.29 ($K_z \frac{\partial^2 P_{ex}}{\partial z^2}$) can be set to zero leaving, after rearrangement:

$$\left(\frac{\partial P_{ex}}{\partial \sigma}\right)_M = \frac{\beta_p}{\beta_p + \phi\beta} \tag{7.34}$$

where the 'M' represents constant mass.

This equation is called the pore pressure coefficient equation and defines the change in fluid pressure at constant mass (undrained) as the stress increases. Literally speaking, it represents the proportion of an incremental load that is carried by the pore fluid assuming that the pore fluid is not allowed to drain. Jacob (1940) referred to this quantity as the *tidal efficiency (T.E.)*. As used by Jacob (1940), the *T.E.* is defined as the ratio of a piezometric level amplitude as measured in a well to the oceanic tidal amplitude, or:

$$T.E. = \frac{dP}{\rho g \Delta h} = \frac{\beta_p}{\beta_p + \phi\beta} \tag{7.35}$$

Loading of clay can occur due to flooding or to rainfall moisture induced changes in the weight of the top soil. The above theory suggests that the water level in a piezometer will respond to these changes as an additional stress is applied to the surface.

Barometric pressure has an inverse relationship with water levels in a piezometer. As the atmospheric pressure increases, the water level decreases. For a rock with a very low compressibility (lower than water), the rock matrix takes all the additional pressure and the well acts as a perfect barometer.

Returning to Equation 7.19, it can be seen that if $\partial P/\partial\sigma$ represents the tidal efficiency, the other term in this equation ($\partial\sigma_e/\partial\sigma$) represents the barometric efficiency, and that:

$$1 = TE + BE \tag{7.36}$$

so that:

$$BE = 1 - TE = 1 - \frac{\beta_p}{\beta_p + \phi\beta_w} = \frac{\phi\beta_w}{\beta_p + \phi\beta_w} \tag{7.37}$$

Thus as $\beta_p \to 0$ for a rigid aquifer, $BE \to 1$ and $TE \to 0$. Conversely, as β_p becomes larger (less small), $BE \to 0$ and $TE \to 1$. BE is closely related to the compressibility of the fluid and TE is closely related to the compressibility of the matrix.

The BE and TE can be thought of as partitioning coefficients that are determined by the compressibility of the matrix. Significantly, BE can be found fairly simply from comparing the water level change in a well to atmospheric pressure change. Using Equation 7.37 we can simply find TE. From TE we can determine β_p, and, if K_z can be estimated, the coefficient of consolidation could be found from Equation 7.30.

7.4.2 Barometric efficiency

In a rigid system, the changes in σ_e will be completely supported by change in σ_e and the value of $\frac{\partial\sigma_e}{\partial\sigma}$ will $\to 1$. As the rock matrix is rigid, it has a very low compressibility ($\beta_p \to 0$). Consider what happens in a borehole drilled into a rigid aquifer, such as a limestone, to change in atmospheric pressure. A schematic is given in Figure 7.8.

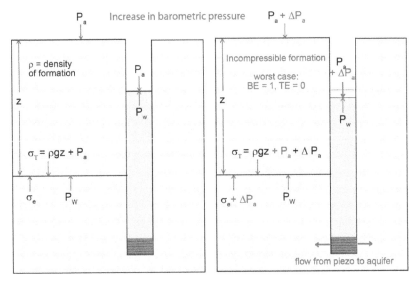

Figure 7.8 (a) Rigid and confined matrix with stresses in balance; (b) Subject to an increase in atmospheric pressure – water level falls (figure prepared by Gabriel Rau)

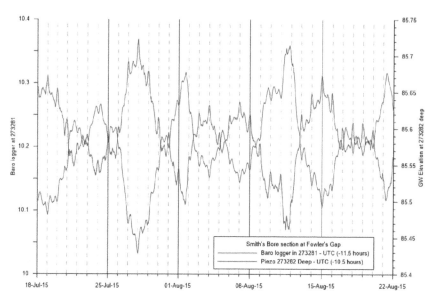

Figure 7.9 Water level response due to atmospheric pressure change. Results from a confined aquifer at Fowlers Gap, NSW

In Figure 7.8a, an increase in ΔP_a (a high pressure area approaching) is transferred to σ_e as the value of β_p is very small and the rock supports any change in weight. This leaves P_w in the formation less than the pressure in the well $P_w + \Delta P_a$ as the atmospheric pressure change is transmitted instantaneously to the water in the well. Water then moves into the formation in response to the pressure differential between the well and the formation. The water level in the piezometer falls as the atmospheric pressure increases. A barometric efficiency (*BE*) can be calculated as a ratio of the fall in water level (Δh), computed as a pressure, to the change in atmospheric pressure. $BE = \Delta h \rho g / \Delta P_a$. An example of this response is shown in Figure 7.9.

The barometric response is simply calculated as a ratio. It can be readily carried out if a detailed record of water level and atmospheric response are available. For a rock with a very small compressibility $\beta_p \ll \beta$, the water level acts as a perfect barometer. Boreholes drilled into limestone cave systems often exhale so much air in response to an approaching low pressure system that they are known to make an audible sound as the air exits and material is seen to be blown out of the bore!

In most cases, the formation compressibility is greater than that of limestone and the additional weight of air is shared (partitioned) between the formation and the water such that the barometric efficiency is less than 1. In the limiting case that all the stress is transferred to the water (the formation consolidates), there is no pressure gradient established through the intake of the piezometer and no change in water level occurs. This is also the case where an unconfined aquifer occurs as the definition of a water table is that the water pressure is at atmospheric pressure. Changes in atmospheric pressure are therefore transmitted directly to the water surface in the piezometer and to the water table.

It follows from the above discussion that, if a piezometer shows some barometric efficiency, it must be confined.

7.4.3   Tidal efficiency

In a system that can deform, where the $\beta_p \gg \beta$, increases in total stress are shared differently. Figure 7.10b indicates the response to an increase in surface loading indicated by ΔP_L. The formation compresses elastically and leaves the water to take the load. As the surface of the water in the piezometer is not subject to the increased load, a gradient in water pressure exists around the piezometer with water moving into the well in response to this gradient. A tidal efficiency (*TE*) can be defined (Jacob, 1940) as:

$$T.E. = \frac{dP}{\rho g \Delta h} \tag{7.38}$$

An example of this response is shown in the Hat Head data (Figure 7.2) for the piezometer at 18 m. The water level does not rise as the result of flow of salt water to this depth – the water in this piezometer is of very low salinity – but as a result of the additional loading caused by the extra water in the creek at high tide. Further data is shown in Figure 7.11 from Timms and Acworth (2005), where the impact of additional rainfall is transmitted rapidly to a piezometer at 60 m depth. This is through 40 m of clay and could not have occurred by advective groundwater flow.

7.4.4   Transmission of pressure

The transmission of pressure changes to a specific depth within a matrix depends on the frequency of pressure changes in addition to the hydraulic diffusivity of the matrix. The fluctuation is transmitted downward into the aquitard and is delayed and damped out as

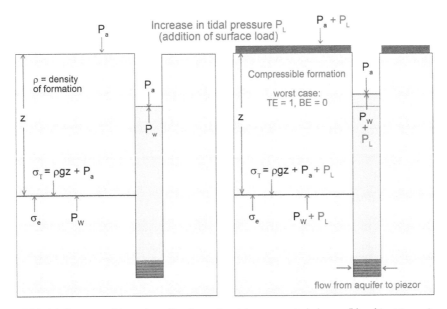

Figure 7.10 (a) Compressible and confined matrix with stresses in balance; (b) subject to an increase in surface loading – water level rises (figure prepared by Gabriel Rau)

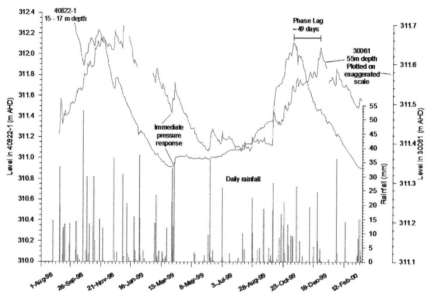

Figure 7.11 Impact of additional loading due to surface rainfall at Yarramanbah on the Liverpool Plains, NSW (adapted from Timms and Acworth (2005))

it moves (van der Kamp and Maathuis, 1991). If a sinusoidal fluctuation is assumed, as would approximately result from the annual wetting and drying of a soil in a semi-arid climate, the amplitude of the wave diminishes with depth according to Keller *et al.* (1989):

$$a(z) = a_0 exp(-z\pi/C) \tag{7.39}$$

where:
$a(z)$ is the amplitude of the wave at depth z and
a_0 is the amplitude of the wave at the top of the aquitard.

The relative penetration depth, C, is defined by Equation 7.40:

$$C = (\pi TD)^{1/2} \tag{7.40}$$

where:
T is the period of the sinusoidal pressure change and
D is hydraulic diffusivity $\left(\frac{K_z}{S_s}\right)$.

From Equation 7.39, at a depth z equal to C, the amplitude of the wave equals only 4% of the initial amplitude and there is a phase lag δ of π radians (or half the period of fluctuation). From this relationship it is evident that the relative penetration depth of short-term pressure changes (e.g., atmospheric pressure) is much less than for long-term pressure changes (e.g., seasonal loading by soil moisture).

The relative amplitude and phase lags of pressure responses in piezometers at specific depths may be used to derive hydraulic diffusivity. Given S_s derived from atmospheric pressure response, K_v may then be determined over a specific depth range, z, using Keller *et al.* (1989):

$$K_v = \frac{S_s \pi}{T}(z/\delta)^2 \tag{7.41}$$

where the phase lag δ is expressed in radians.

Alternatively, the change in amplitude of a pressure wave transmitted to a specific depth can be used to calculate K_v. Substitution of Equation 7.40 into Equation 7.39 and solving for K_z gives:

$$K_v = \frac{z^2 S_s \pi}{T}\left(\ln\frac{a_0}{a_z}\right)^{-2} \tag{7.42}$$

where a_0 is the initial amplitude and a_z is the amplitude at a specific depth (amplitude equivalent to half the peak height measured from trough to trough). This approach was used by Timms and Acworth (2005) to analyze the Yarramanbah data shown in Figure 7.11.

Note that this response to the transmission of pressure where the pressure is applied approximately as a sinusoid is similar to the analysis of heat flow in response to diurnal heating and cooling. There is an amplitude decay and a phase lag as the energy moves down through the system.

7.5   DERIVATION OF SPECIFIC STORAGE FROM RECORDS OF HYDRAULIC HEAD IN CONFINED AQUIFERS

7.5.1   Frequency components of atmospheric pressure variation

Cartwright (1999) notes that the 24-hourly (S_1) and 12-hourly (S_2) oscillations in atmospheric pressure were first noted by a Frenchman (Robert de Paul, Chevalier de Lamanon) during exploratory voyages in the southern oceans in 1785. The cause of these pressure oscillations is linked to heating of the upper atmosphere and can be explained by the theory of internal tides in a stratified atmosphere (Volland, 1996; Palumbo, 1998). The S_1 tide shows distinct seasonal variability while the S_2 tide is fairly uniform throughout the year (Acworth *et al.*, 2015a). The S_2 tide is considered to be a harmonic of the S_1 tide that surprisingly has a similar amplitude to the primary S_1. This is probably because several S_1 tides exist related to the differential heating of gasses in the atmosphere and that while these tides have the same periodicity, they have different phases which in some cases act in opposition to each other (destructively).

The meso-scale variation in atmospheric pressure that brings different weather systems to bear is a familiar concept. It was perhaps because there was no immediate use of the observation of 12- and 24-hour variations in atmospheric pressure that they were forgotten about for so long. Figure 7.12 is a Fourier transform pair showing these variations. The general variation in atmospheric pressure that varies over days and weeks is clearly seen at lower frequencies in Figure 7.12. Annual variation is also seen, but the more regular sub-diel variation is hidden. Fourier analysis reveals these variations clearly.

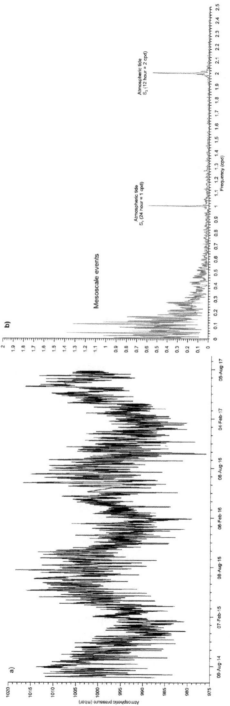

Figure 7.12 (a) Record of atmospheric pressure variation for a 3-year period at Fowlers Gap, NSW. Data frequency was 10 minutes. (b) The Fourier spectrum of (a) showing the discrete response to the atmospheric tides at 1 and 2 cps

7.5.2 Groundwater response in a confined aquifer to atmospheric tides

Despite the regularity of the variation in atmospheric pressure, it was not until much later that it was realized that the atmospheric pressure variation was also seen in confined aquifers. If the size of the atmospheric pressure variation could be established and compared with the response in the confined aquifer, the *BE* could be established.

Figure 7.13 shows data where the groundwater elevation is compared to the atmospheric pressure and a value of *BE* calculated using the method of Gonthier (2007) and Acworth and Brain (2008) evaluated. In this approach, the groundwater data is plotted against the atmospheric data measured at the same time and the slope of the series calculated to give a value of *BE*. To check that this *BE* value is correct, the groundwater data is recalculated by adding or subtracting a head quantity calculated as *BE* times the atmospheric pressure. A trial and error approach will find that value of *BE* which reduces the slope to zero. The method works well for a data set where the *BE* is close to unity as in this example from Fowlers Gap in western NSW. However, where *BE* is much less than unity and where recharge or abstraction impact the groundwater level, it is much more difficult to achieve an acceptable value of *BE*. This is clearly seen in Figure 7.14 where data for a low *BE* at Cattle Lane (Acworth *et al.*, 2017) and an intermediate *BE* at Baldry are shown next to the high *BE* data from Fowlers Gap.

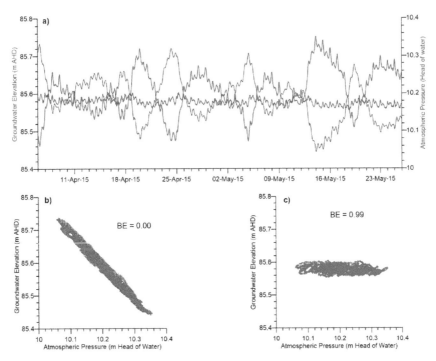

Figure 7.13 (a) Groundwater level elevation data (green) and atmospheric pressure data (red). The blue data is the groundwater level data corrected for atmospheric pressure. (b) Plot of uncorrected groundwater elevation data vs atmospheric pressure data and (c) *BE* corrected groundwater level data (Gonthier, 2007)

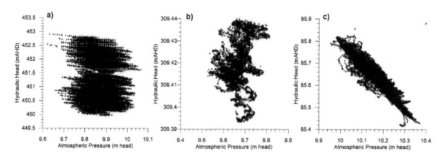

Figure 7.14 Scatter plots of atmospheric pressure vs groundwater level for (a) intermediate *BE* on fractured granite at Baldry, (b) a low *BE* at Cattle Lane from unconsolidated clays and sands and (c) a high *BE* from a confined aquifer at Fowlers Gap (adapted from Acworth *et al.* (2016a))

Table 7.3 Earth tides at diel and semi-diel frequencies

Constituent	Period (hours)	Frequency (cycles per day)	Vertical amplitude (mm)	Horizontal amplitude (mm)
M_2	12.421	1.932	384.83	53.84
S_2 Solar semi-diel	12.000	2.000	179.05	25.05
N_2	12.658	1.896	73.69	10.31
K_2	11.967	2.006	48.72	6.82
K_1	23.934	1.003	191.78	32.01
O_1	25.819	0.930	158.11	22.05
S_1 (Solar diel)	24.000	1.000	1.65	0.25
P_1	24.066	0.997	70.88	10.36

7.5.3 Incorporating the earth tide at a specific frequency

Acworth *et al.* (2016a) demonstrated that it was possible to calculate the *BE* in the frequency domain where the impact of the atmospheric tide as 2 cpd (S_2) was very constant. A simple ratio of the amplitude of the groundwater response at 2 cpd to the atmospheric pressure response, expressed in the same units and at the same frequency, is an approximation to *BE* (Acworth and Brain, 2008). Further work (Acworth *et al.*, 2015a) demonstrated that the *BE* values were a slight underestimate of *BE* because there was another factor involved. The groundwater level response to atmospheric pressure at 2 cpd is contaminated by a response to earth tides at the same frequency. The earth tide acts to diminish the response to atmospheric pressure and requires backing out before the ratio is calculated to estimate *BE*.

Fortunately, the earth tide can be predicted accurately at any location and elevation using software that is freely available (Van Camp and Vauterin, 2005). The earth tide amplitudes at the various frequencies are given in Table 7.3. The amplitude of the earth tide at S_2 can be estimated from taking the ratio of $\frac{S_2^{ET}}{M_2^{ET}}$ as this ratio in the earth tides is a constant. If this ratio is applied to the M_2^{GW} term, the earth tide contribution to the S_2^{GW} term can be calculated.

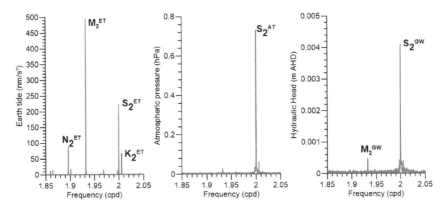

Figure 7.15 Sub-diel frequency components of the earth tide, the atmospheric pressure and the groundwater from a site at Fowlers Gap, NSW

This contribution is added back into the S_2^{GW} value as the earth tide acts to oppose the atmospheric tide. Also, recognising that there may be a phase difference between M_2^{GW} and M_2^{ET} the BE can then be calculated from Equation 7.43, as described by Acworth *et al.* (2016a) with components as labelled in Figure 7.15:

$$BE = \frac{S_2^{GW} + S_2^{ET}\cos(\Delta\phi)\frac{M_2^{GW}}{M_2^{ET}}}{S_2^{AT}} \tag{7.43}$$

If M_2^{GW} is very small, then Equation 7.43 approaches:

$$BE = \frac{S_2^{GW}}{S_2^{AT}} \tag{7.44}$$

Using this approach, the BE for the data sets shown in Figure 7.14 can be calculated as 0.059 for the unconsolidated silts and clays at Cattle Lane; 0.563 for the weathered granite at Baldry and 0.976 for the confined aquifer at Fowlers Gap (Acworth *et al.*, 2016a).

Acworth *et al.* (2017) described the use of this method in a study of the BE variation in a nest of piezometers installed in silty clay at Cattle Lane, NSW.

With accurate values of BE available, and recognising that:

$$BE = 1 - TE \tag{7.45}$$

and that:

$$TE = \gamma = \frac{\beta_p}{\theta\beta + \beta_p} \tag{7.46}$$

where:
α is the formation compressibility (Pa^{-1}),
β is fluid compressibility (PA^{-1}) (4.59×10^{-10} Pa^{-1} at 20 °C), and
θ is porosity.

This equation can be solved for β_p. Then, with ρ, β_p, β_w and θ either known or assumed, the value of specific storage for the confined formation is (Cooper, 1966):

$$S_s = \rho g (\beta_p + \theta \beta_w) \tag{7.47}$$

The specific storage values for the three data sets in Figure 7.14 are then 2.663×10^5 at Cattle Lane; 2.094×10^{-6} at Baldry and 1.586×10^{-6} at Fowlers Gap (Acworth et al., 2016a).

7.6 DERIVATION OF SPECIFIC STORAGE – A LINEAR POROELASTIC APPROACH

Specific storage is one of the fundamental elastic parameters in the development of the theory of linear poroelasticity. This has seen separate development in the areas of geomechanics, petroleum engineering and hydrogeology (Wang, 2000) that has caused a wide variety of definition and terminology.

The elastic coefficients involved in poroelastic coupling vary depending upon the time taken for a load to be applied and stress to dissipate (Domenico and Schwartz, 1997; Wang, 2000). For rapid loading, as occurs with the passage of a seismic wave or the response to atmospheric tides at sub-daily frequency, there is insufficient time for water to flow in response to the increased stress and pore pressure. Therefore, the loading occurs at constant mass ($d\zeta/dt = 0$ where ζ is the mass of fluid) and poroelastic coefficients represent *undrained* conditions. By contrast, if the loading occurs slowly and fluid has the opportunity to redistribute, the loading occurs at constant pore pressure ($dp/dt = 0$ where p is pore pressure) and represents *drained* conditions. Note here that the term *drained* should not be confused with the interpretation that subsurface pores are drained of water, i.e., the pressure head in a confined aquifer is lowered below its upper limit causing unconfined conditions, as is a common interpretation in hydrogeology.

Wang (2000) gave a detailed analysis of poroelastic theory for both drained and undrained conditions, and Van Der Kamp and Gale (1983) developed expressions for the analysis of atmospheric and earth tides, which are undrained phenomena. The developments build on the coupled equations for stress and pore pressure derived by Biot (1941) for very small deformations, typical of those that occur with the passage of seismic waves or in response to atmospheric tides. In the most general case, it is necessary to consider a fully deformable medium in which all components are compressible. Besides the bulk formation compressibility, two more components require consideration. The water compressibility:

$$\beta_w = \frac{1}{K_w} \approx 4.58 \cdot 10^{-10} \, Pa^{-1}. \tag{7.48}$$

The solid grain (or unjacketed) compressibility:

$$\beta_s = \frac{1}{K_s} \tag{7.49}$$

assumes homogeneous solids and is not well defined for mixtures of different grain types (Wang, 2000).

The volume of water squeezed from a sediment is always less than the change in bulk volume when grain compressibility is included (Domenico and Schwartz, 1997).

To take account of this change, the *Biot-Willis* coefficient is given as (Biot, 1941; Wang, 2000):

$$\alpha = 1 - \frac{\beta_s}{\beta} = 1 - \frac{K}{K_s}. \tag{7.50}$$

Note that if $\beta_s \ll \beta$ then there is relatively little, if any, change in volume of the grains when compared to the total volume change and therefore $\alpha \approx 1$.

Van Der Kamp and Gale (1983) and Green and Wang (1990) presented a comprehensive relationship for specific storage that assumes only uniaxial (vertical) deformation (zero horizontal stress) and includes solid grain compressibility:

$$S_s = \rho_w g \left[\left(\frac{1}{K} - \frac{1}{K_s} \right)(1 - \lambda) + \theta \left(\frac{1}{K_w} - \frac{1}{K_s} \right) \right], \tag{7.51}$$

where the density of water $\rho_w = 998\ kg/m^3$, the gravitational constant is $g = 9.81\ m/s^2$, θ is total porosity and:

$$\lambda = \alpha \frac{2}{3} \frac{(1 - 2\mu)}{(1 - \mu)} = \alpha \frac{4G}{3K\upsilon}. \tag{7.52}$$

Here, K_υ is the drained vertical (or confined) bulk modulus and expressed as (Green and Wang, 1990; Wang, 2000):

$$\frac{1}{K_\upsilon^{(u)}} = \beta_\upsilon^{(u)} = \frac{1 + \mu^{(u)}}{3K^{(u)}(1 - \mu^{(u)})} = \left(K^{(u)} + \frac{4}{3}G \right)^{-1}. \tag{7.53}$$

where:
G is the shear modulus that can be measured in the field using cross-hole seismic methods (see Chapter 8) and does not vary between drained and undrained conditions (Wang, 2000) and
$\mu^{(u)}$ is the undrained Poisson's ratio that can also be measured with cross-hole geophysics.

If the solids are incompressible ($\beta_s = 1/K_s \to 0$) then Equation 7.51 reduces to the well-known formulation (Jacob, 1940; Cooper, 1966):

$$S_s = \rho_w g \left(\frac{1}{K_v} + \frac{\theta}{K_w} \right) = \rho_w g (\beta_p + \theta \beta_w), \tag{7.54}$$

We note that if $\mu^{(u)} = 0.5$ then it can be seen from Equation 7.53 that $K_\upsilon^{(u)} = K^{(u)}$. Note, however, that this will only be the case for very unconsolidated silts or clays.

To summarize, specific storage values derived from Equations 7.51 and 7.54 represent vertical and isotropic stress only and are therefore smaller compared to the case where horizontal stress and strain is allowed to occur (Wang, 2000). However, this is a reasonable and common assumption which suffices to represent the conditions encountered in a hydrogeological setting. For example, Equation 7.54 is widely used in hydrogeology (Van Der Kamp and Gale, 1983), particularly for the analysis of head measurements obtained from pump testing (e.g., Kruseman and de Ridder, 1990; Verruijt, 2016).

Poroelastic theory can be used to determine minimum and maximum values of specific storage that can exist. Using the values determined by Acworth *et al.* (2016a), it is clear that even for a soft clay, a maximum of $S_s \approx 2 \cdot 10^{-5}\ m^{-1}$ exists. This should be remembered when a groundwater model is calibrated. If values of $S_s \gg 2 \cdot 10^{-5}$ have been used during the calibration, there is a clear indication that the conceptual model is not correct because an impossible amount of water is shown as being released by the formation when, in fact, it can only be coming across the boundary of the model by downward, upward or lateral leakage. This observation is further demonstrated and confirmed by Rau *et al.* (2018), who combined poroelastic data derived from cross-hole seismics (Chapter 8) with loading efficiency values derived from observations of head response to atmospheric tides (Chapter 4) to determine maximal possible values of specific storage. The theoretical analysis developed by Rau *et al.* (2018) demonstrates that there is a maximum value of specific storage of $\approx 1.3 \times 10^{-5} m^{-1}$.

Geophysical investigation techniques: seismic

8.1 INTRODUCTION TO SEISMIC METHODS

The seismic method is one of the most commonly used geophysical methods in exploration. There are three types of application, *viz*.:

* borehole seismics, including borehole to borehole and surface to borehole methods,
* seismic refraction, and
* seismic reflection.

Borehole to borehole and surface to borehole methods are extensively used in geotechnical investigation work to provide profiles of Poisson's ratio, bulk, shear and Young's modulii that are particularly useful in foundation design or in tunneling work. These data can also be used to investigate specific storage in a confined aquifer.

The seismic reflection method has been developed by the oil industry to the point that no exploration wells are drilled for oil without a very detailed and expensive seismic reflection survey preceding the decision to drill. More recently, and with the availability of extensive computing power in the field, this method has been adapted to shallow investigations (Hill, 1992; King, 1992). In general however, this method is not suitable for depths of less than 50 m and has not been widely used in groundwater studies. By contrast, the seismic refraction method is suitable for shallow investigation and is the method which will be described in detail. The refraction method works most effectively if there is a clear velocity contrast between different rock types.

8.1.1 Seismic relationships

Rocks of very low porosity, including most igneous and metamorphic rocks and evaporites, have seismic velocities which are controlled mainly by their composition and which can be well predicted from a knowledge of the velocities of their component minerals. Depth of burial is important even in such low porosity rocks, however, since microcracks may easily reduce the velocity measured at ground surface to little more than half of its expected value. A confining pressure of about 1 kbar is sufficient to close the cracks, but this corresponds to a depth of burial of 3000 m.

In rocks of medium to high porosity, the velocity (more particularly the P-wave velocity) will depend on the nature of the fluid (normally air or water) filling the pore space. Over the commonly occurring porosity range of 20%–40%, the velocity-porosity relationship is quite well approximated by the so-called time-average relationship (Equation 8.1):

$$\frac{1}{V} = \frac{\phi}{V_f} + \frac{1-\phi}{V_m}$$
(8.1)

where:
V_f is the velocity of the fluid filling the pore space and
V_m is the velocity of the matrix of granular material.

Rearranging Equation 8.1 and solving for porosity gives:

$$\phi = \frac{\frac{1}{V} V_m V_f - V_f}{V_m - V_f}$$
(8.2)

This equation is useful for estimating porosity ϕ from velocity or vice versa, using values of 300 m/s (air) or 1500 m/s (water) for dry and saturated rocks respectively and an appropriate value of V_m, which will rarely lie outside the range 5000–6000 m/s. Partially saturated rocks are found to have velocities close to the *dry* value up to 80% – 90% saturation, when a rapid increase to the *wet* value occurs.

8.1.2   Wave terminology

The wavelength (λ) is the distance between two adjacent points on a wave front which have the same displacement or phase. For the case of the ripple on the surface of a pond, the wavelength is the distance between two successive peaks. The amplitude (A) of the wave is the maximum displacement of a particle associated with the passage of the wave. The pattern of particle displacement in space (Figure 8.1) and time (Figure 8.2) for a simple sine wave are shown.

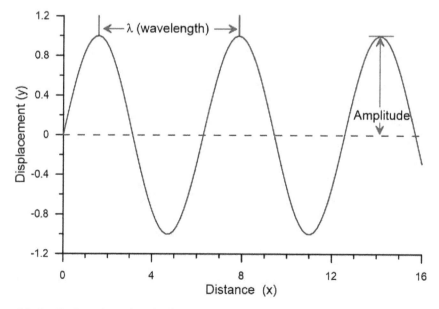

Figure 8.1 Amplitude and wavelength of a wave in space domain

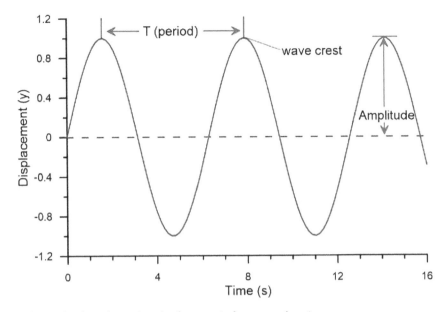

Figure 8.2 Amplitude and wavelength of a wave in frequency domain

The repetition rate of the waveform, or the time taken for two successive wave crests to pass by a fixed observation point, gives the period (T) of the wave. The number of repetitions or cycles per second is defined as the frequency (f) of the wave.

$$f = \frac{1}{T} \tag{8.3}$$

where:
f is the frequency [Hz] (s^{-1}) and
T is the period [s].

The velocity V of a wave is clearly:

$$V = \frac{distance}{time} \tag{8.4}$$

or:

$$V = f\lambda \tag{8.5}$$

8.1.3 Elastic coefficients

There is an unfortunate overlap here between the Greek letters used to denote quantities in seismic theory and those used in electrical theory. The context of the use should always be clear, but be warned!

If a perfectly elastic (Hookean) material is subjected to a uniaxial compression or tension, a linear relationship exists between the applied stress σ and the resulting strain ϵ:

$$\sigma = E\epsilon \tag{8.6}$$

where:

E is a constant of proportionality known as Young's modulus.

The value of the strain ϵ is the ratio of the change in line length in its deformed state (l_f) to its initial state (l_o):

$$\epsilon = \frac{l_f - l_o}{l_o} = \frac{\Delta l}{l_o} \tag{8.7}$$

If a Hookean solid is subject to uniaxial compression, it will shorten in the direction of compression and expand in the plane at right angles to the direction of compression.

If ϵ_1 represents the shortening in the direction of compression and ϵ_3 represents the expansion in the plane at right angles to the compression, then the ratio of these two quantities is referred to as Poisson's ratio (μ):

$$\mu = \frac{\epsilon_1}{\epsilon_3} \tag{8.8}$$

where $\mu < 0.5$.

There are two further elastic constants which are also important. In an isotropic material subject to a general change in pressure a change in volume will occur. The change from the original volume (V_o) to a final volume (V_f), when compared to the pressure change, is called the bulk modulus (K). It can be defined as:

$$K = \frac{\Delta p}{\Delta V/V_o} = \beta^{-1} \tag{8.9}$$

where:

Δp is the pressure change,
ΔV is the volume change $V_f - V_o$,
V_o is the initial volume,
V_f is the final volume, and
β is the compressibility of the material.

The bulk modulus is the reciprocal of the compressibility and is therefore also referred to as the incompressibility!

It is also possible to deform a solid by means of a simple shear. A shear strain γ is induced in response to a shear stress σ_s. The ratio of these quantities is the rigidity modulus (G):

$$G = \frac{\sigma_s}{\gamma} \tag{8.10}$$

Some representative values for these quantities are given in Table 8.1 (Burger, 1992). Note that each rock type will cover a range of values and that the data is indicative only.

Table 8.1 Indicative elastic coefficients and seismic velocities for selected rock types

Rock Type	Density ρ kg/m^3	Young's modulus E GN/m^2	Poisson's ratio μ $-$	P-wave velocity V_p m/s	S-wave velocity V_s m/s
Shale	2670	12.0	0.040	2124	1470
Siltstone	2500	13.0	0.120	2319	1524
Limestone	2710	33.7	0.156	3633	2319
Limestone	2440	17.0	0.180	2750	1718
Quartzite	2660	63.6	0.115	4965	3274
Sandstone	2280	14.0	0.060	2488	1702
Slate	2670	48.7	0.115	4336	2860
Schist	2700	54.4	0.180	4680	2921
Gneiss	2640	25.5	0.146	3189	2053
Marble	2870	71.7	0.270	5587	3136
Granite	2660	41.6	0.055	3967	2722
Gabbro	3050	72.7	0.162	5053	3203
Basalt	2740	63.0	0.220	5124	3070
Andesite	2570	54.0	0.180	4776	2984
Tuff	1450	1.40	0.110	996	659

Note: Units for Young's modulus are GNm^{-2} or GPa.

The rigidity and bulk modulii are not independent quantities and can be expressed in terms of the Young's modulus and Poisson's ratio as:

$$G = \frac{E}{2(1 + \mu)} \tag{8.11}$$

and:

$$K = \frac{E}{3(1 - 2\mu)} \tag{8.12}$$

Poisson's ratio for shale is generally > 0.3; for tuff about 0.11; for silts and sands between 0.2 and 0.4. Poisson's ratio for clays is between 0.4 and 0.5.

Young's modulus for shale is approximately 10 GPa; for tuff 1.5 GPa; for silty clay between 0.007 and 0.008 GPa and 0.2 GPa for hard clay.

8.1.4   Seismic waves

Seismic energy can pass through the air with a velocity typically dependent upon air density (humidity) and temperature. The velocity of sound in air is V_{air} = 343 m/s at 20 °C. The distance travelled by the air wave is then:

$$x = V_{air} \times t \tag{8.13}$$

The speed of sound passing through a medium is given by the Newton-Laplace equation:

$$V = \sqrt{\frac{K}{\rho}} \tag{8.14}$$

where:

K is the bulk modulus and

ρ is the material density.

The Rayleigh wave velocity is often approximated as $V_R \approx 0.9V_s$. The distance time equation is then simply:

$$x = V_R \times t \tag{8.15}$$

There are two types of body wave which occur within a uniform homogeneous rock. One type of wave is transmitted by particle movements back and forth along the direction of propagation of the wave. These waves are referred to as longitudinal waves. As these waves have the highest velocity, they are the first waves to arrive at a receiver and are also referred to as primary waves (P-waves). Longitudinal waves are also sometimes referred to as compressional waves. The particle motions which produce P-waves can be thought of as a series of compressions and expansions which occur in the direction of wave propagation. Sound moves through air in this manner.

The second wave type is the transverse wave, in which particle motion is in the plane at right angles to the direction of motion. These waves are referred to as secondary waves (S-waves) as they arrive at a receiver after the primary wave. They are also called shear waves, as the wave motion through the rock causes the particles to be subject to a shearing motion. A common analogy for shear wave motion is the transmission of wave energy along a rope. A flick of the end will send a wave along the rope, but the only motion of the particles is in a direction at right angles to the wave propagation direction. Using this analogy we can also explain the existence of vertical and horizontal polarization of wave energy. If the initial flick of the rope is in a horizontal plane, the wave is said to be horizontally polarized. Similarly, if the initial energy is given in the vertical direction, vertically polarized waves result. When seismic waves pass through interfaces or boundaries, there is often a transfer of energy between P-waves and S-waves which result in new wave formation. The study of the polarization of this energy can often be diagnostic of soil structure.

In areas which are not homogeneous, two other wave types are generated. These are the Rayleigh and Love waves. These travel at the interface or the surface of the earth and cause ground roll. They are not generally used for seismic analysis but can cause difficulties in the field by swamping the more useful P- and S-wave energy.

The surface wave travels more slowly than either body wave and is generally complex. In the special case of a homogeneous ground, the surface disturbance is caused entirely by the wave known as a Rayleigh wave, in which both vertical and horizontal components of ground motion are present. The horizontal ground motion is of rather smaller amplitude than the vertical and is 90° out of phase with it, so that the resultant path of an element of the medium during the passage of a Rayleigh wave cycle follows an ellipse lying in the plane of propagation. The ground motion becomes negligibly small within a distance from the free surface of the same order as the wavelength of the disturbance. The velocity V_r of the Rayleigh wave is only about 10% less than the body shear wave velocity, the ratio $\frac{V_r}{V_s}$ depending only on Poisson's ratio.

If the ground were perfectly elastic, the amplitudes of the waves would decrease with range from their source simply as a result of the spreading of the wave energy (proportional to the square of the amplitude) over the increasing wave-front area. This is analogous with the amplitude of the wave in the pool decreasing as the diameter of the wave front increases. This area will vary as the square of the range for spherically spreading body waves and linearly with range r for cylindrically spreading surface waves. Body and surface waves would thus have amplitudes varying as $\frac{1}{r}$ and $\frac{1}{\sqrt{r}}$ respectively.

The form of the pulse is determined by the relative amplitudes of the different frequency components which make up its spectrum. Since these spectral components are equally affected by geometrical spreading, one would not expect the pulse form to be changed during its propagation. Real earth materials, however, are imperfectly elastic, leading to energy loss and attenuation of the seismic waves, that is, to an amplitude reduction more rapid than would be expected from geometrical spreading alone. This attenuation is more pronounced for less consolidated rocks and is also greater for higher frequencies. The selective loss of the high-frequency components in a pulse leads to a progressive broadening in time as the pulse propagates.

Rayleigh waves are a type of surface wave confined to the near surface of the ground. Rayleigh waves include both longitudinal and transverse motions that decrease exponentially in amplitude as distance from the surface increases. There is also a phase difference between these component motions. The speed of the Rayleigh wave is a little less than the shear wave and can be approximated from Equation 8.16:

$$V_{Rayleigh} = \frac{0.862 + 1.14\mu}{1 + \mu} \tag{8.16}$$

Since Rayleigh waves are confined near the surface, their in-plane amplitude when generated by a point source decays only as $\frac{1}{\sqrt{r}}$, where r is the radial distance. Surface waves therefore decay more slowly with distance than do bulk waves, which spread out in three dimensions from a point source. This slow decay is one reason why they are of particular interest to seismologists; Rayleigh waves can circle the globe multiple times after a large earthquake and still be measurably large. It is the surface wave that does the most damage in an earthquake.

Seismic wave velocities expressed in elastic coefficients

The velocities of the P- and S-waves may be expressed in terms of the elastic constants:

$$V_p = \sqrt{\frac{K + 4/3G}{\rho}} = \sqrt{\frac{E}{\rho} \frac{(1 - \mu)}{(1 - 2\mu)(1 + \mu)}} \tag{8.17}$$

and:

$$V_s = \sqrt{\frac{G}{\rho}} = \sqrt{\frac{E}{\rho} \frac{1}{2(1 + \mu)}} \tag{8.18}$$

As the bulk modulus (K) and the rigidity modulus (G) are always positive and Poisson's ratio is less than or equal to a half, then the shear wave velocity must always be less than the primary wave velocity.

The ratio of the primary and shear wave velocities can be used to determine Poisson's ratio directly by solution of Equation 8.19:

$$\frac{V_p}{V_S} = \sqrt{\frac{1 - \mu}{0.5 - \mu}} \tag{8.19}$$

Because the rigidity modulus (G) is zero for liquids, S-waves cannot propagate through liquids. Representative primary wave velocities for a variety of materials are given in Table 8.2.

It is apparent that only ranges of P-wave velocity such as those quoted in Table 8.2 can be expected to correspond to a given lithological type, and that the overlap between these ranges is such that a velocity of 4000 m/s, for example, might correspond to anything from a well-cemented sandstone or limestone to a fractured igneous or metamorphic rock. Velocities within one formation, however, normally show a much smaller variability than these figures suggest, and provided that some geological control is available, the identification of such a formation from its seismic velocity in a local context is often quite practicable.

A guide to seismic P-wave velocities would be 500 m/s for dry unconsolidated material, 1500 m/s for saturated unconsolidated material, 4000 m/s for sedimentary rocks and 6000 m/s for unweathered igneous rocks. The following generalizations can be made:

*   seismic velocity increases as saturation increases,
*   seismic velocity increases with consolidation of the sediments,
*   velocities are all similar in saturated unconsolidated sediments,
*   seismic velocity decreases as weathering increases, and
*   seismic velocity decreases as fracturing in bedrock increases.

Energy sources (hammer or explosives) used to generate seismic waves typically produce frequencies in the range 10 Hz to 100 Hz. The wavelength of a P-wave traveling through saturated unconsolidated material will then be between 150 m and 15 m.

Table 8.2 Representative primary wave velocities

Material	Velocity m/s
Air	331.5
Water	1400–1600
Weathered layer	300–900
Soil	250–600
Alluvium	500–2000
Clay	1100–2500
Unsaturated sand	200–1000
Saturated sand	800–2200
Unsaturated sand & gravel	400–500
Saturated sand & gravel	500–1500
Granite	5000–6000
Basalt	5400–6400
Metamorphic rocks	3500–7000
Sandstone and shale	2000–4500
Limestone	2000–6000

The ratio between the P-wave and the S-wave velocities can be useful. In general, the following apply:

- $V_s \approx 0.6V_p$ for crystalline rocks,
- $V_s \approx 0.5V_p$ for sedimentary rocks, and
- $V_s \approx 0.4V_p$ for soils and unconsolidated materials.

A rule of thumb estimate for Rayleigh (surface) waves is $V_r = 0.9V_s$.

Different soil characteristics are also listed by the French Oil Institute as:

Table 8.3 Cross-hole derived characteristics

Material	P Velocity (m/s)	S Velocity (m/s)	Density (kg/m^3)
Vegetated soil	300–1700	100–300	1700–2400
Dry sand	400–1200	100–500	1500–1700
Wet sand	1500–4000	400–1200	1900–2100
Clay	1100–2500	800–1800	1800–2200
Siltstone	2000–3000	750–1500	2100–2600
Sandstone	3000–4000	1200–2800	2100–2400
Limestone	3500–6000	2000–3300	2400–2700
Coal	2200–2700	1000–1400	1300–1800

Seismic wave attenuation and amplitude

As seismic waves move through the ground, they cause small distortions of the mineral grains which are the strains caused by the stress represented by the passage of the wave. The ground is not perfectly elastic, and some energy is lost due to the production of heat.

The reduction in energy of the wave is called the absorption. This can represented as:

$$I = I_0 e^{-qr} \tag{8.20}$$

where:
I is the intensity of the energy,
q is the absorption coefficient, and
r is distance.

If I_0 is the energy at some point, then this will have decayed to I at a distance r. In general, losses due to absorption are greater at high frequencies than at low frequencies. For this reason, it is the lower frequency energy which propagates the furthest through the earth.

8.1.5 Seismic equipment

Energy sources

Two seismic sources are commonly used in shallow seismic work:

- A weight drop, or hammer blow, and
- an explosive.

Figure 8.3 A hammer seismic source in use (Photo: Ian Acworth)

Figure 8.4 BISON Betsy shotgun seismic source. Shotgun shells are loaded into the seismic gun and fired into the ground. Coupling is improved by standing on the tyre forming the top of the source (Photo: Ian Acworth)

The simplest and most common form of a weight drop is a hammer. A hammer struck against a metal plate lying on the ground surface can effectively generate sufficient energy to enable the seismic wave to be detected at 100 m distance (Figure 8.3). The hammer has an inertial switch taped to the shaft that closes when the hammer strikes the ground, and the signal is used to initiate the collection of seismic data.

If a hammer does not impart sufficient energy, then a variety of guided weight drops can be used. The complexity of these is often a function of the ease of mobilizing the weight drop to the required site. A wide range of shallow seismic work can be carried out with a hammer striking against a metal plate laid on the ground. The plate is used to get good seismic coupling between the hammer and the ground. Clearly, seismic work on soft ground becomes a problem, as the plate is knocked into the ground and has to be dug out!

Explosives are used for cases where the energy imparted by a weight drop is insufficient. These may be shotgun cartridges, as in the Betsy shotgun source (Figure 8.4), or gelignite placed in prepared drill holes. The explosive is placed in a prepared drill hole and tamped down to improve the coupling between the explosive charge and the ground. There are

Figure 8.5 ABEM-Terraloc digital seismograph (Photo: Ian Acworth)

various types of explosive designed specifically for seismic work. Permission to use explosives on a site may be a problem in some areas, and permission to travel with explosives is becoming more and more difficult.

Signal recording

Seismic equipment consists of the receiver or geophone and a means of recording the output from the geophone. Multi-channel devices which are typically 12 or 24 channels are typically available.

The seismograph accepts a signal produced by the seismic source which initiates a recording of the output from the geophones. The seismograph recording used to be directly onto a chart recorder. The analogue signal from the geophone is digitized directly and recorded on a computer.

The seismograph display often has the capability of several forms of data presentation. The seismogram can be displayed as a simple wiggle trace or as an amplitude shaded signal as shown in Figure 8.6. It is also possible to vary the amplification on each channel individually so that the signal from distant geophones can be boosted in favour of those from geophones close to the source. Filters can also generally be applied to each signal. In this way, the effects of transmission line interference (high-amplitude filter) or ground roll (low-amplitude filter) can be removed.

Signal stacking

Seismographs have the capacity to stack the seismograms derived from repeated signals. In this way, the effect of cultural noise can be overcome. Noise is a major problem in surveys and is created by traffic, industrial processes, farm equipment and even trees. The best time to carry out a seismic survey in non-urban areas is early in the morning when there is little wind. The wind blowing through trees can set up low frequency noise in the ground which

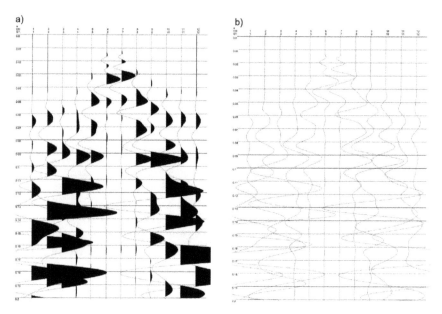

Figure 8.6 A variable area trace (a) and a simple wiggle trace (b) of the same data set

can swamp the seismic response from a hammer blow. Even the wind blowing through grass and causing the grass blades to knock against the geophones can be a problem. For these reasons, the best time to measure seismic data is early in the day before local winds build up.

Geophones

The geophone monitors the ground motion produced by the passage of the seismic wave train. The geophone design basically consists of a cylindrical coil of fine wire suspended in a cylindrical cavity in a magnet. The coil moves as a result of the ground motion and induces an electric current in the coil which is proportional to the ground motion. The natural frequency of the geophone is damped to avoid the output from these oscillations obscuring the ground motion.

A schematic of a geophone is given in Figure 8.7.

The standard geophone used in refraction work is shown in Figure 8.8a. The characteristic response frequency of the inertial mass and leafspring system can be constructed so that the geophone is more sensitive to higher frequency sources used in reflection seismic work. The coil and inertial mass can be located horizontally to produce a geophone sensitive to shear waves as seen in Figure 8.8b or can be jointly mounted with three components allowing the complete characterization of the waveform in x-y-z space (Figure 8.8c).

A special multi-core cable is used in seismic work, often with 24 pairs of connectors for the connection of 24 geophones. It is worth noting that multi-channel systems are not absolutely necessary – they are just more efficient. It is quite possible to achieve the same results by using one geophone and moving the source position each time. This is time consuming for any but the most simple of shallow problems. However, it does illustrate the principle of

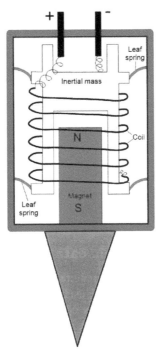

Figure 8.7 Schematic of a typical geophone (adapted from Burger (1992)). The inertial mass is supported on fine leaf springs that allow the mass to move with respect to the fixed magnet and in response to the ground movement transmitted to the body of the geophone through the spike at the base. A current is produced inside the coil that is wrapped around the inertial mass and the received signal transmitted to the seismic recorder

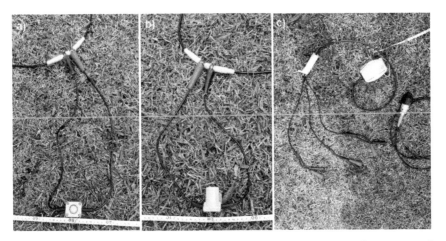

Figure 8.8 (a) Standard vertically mounted inertial mass geophone used in refraction or reflection studies; (b) a geophone with the inertial mass on its side to make it sensitive to horizontally moving shear waves; and (c) a three-component geophone with three orthogonally arranged components (Photos: Ian Acworth)

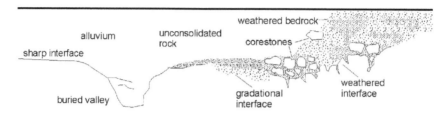

Figure 8.9 Varying form of the bedrock interface (adapted from a sketch by Ron Barker)

reciprocity that is used significantly in the delay-time methods described below. The travel time from a source to a geophone must be the same if the source and geophone locations are switched. This element of reciprocity is also used in the interpretation of seismic refraction data described below.

8.1.6 Interpretation of seismic data

As a general rule, any geophysical method requires a lateral or vertical contrast in the physical property being measured to allow some useful deductions to be made. Discontinuities or changes in the physical property normally coincide with some geological or formation boundary, although different geological formations may have the same physical properties and therefore the reverse is not always true. To interpret seismic data, information on rock density must be available. In some cases, differing sedimentary units have clearly different densities and, as a general rule, the density tends to increase with depth. Perhaps the most difficult problem for seismic interpretation is where gradational changes in density exist such as in the variation of weathering depicted in Figure 8.9. A seismic interpretation based upon data that was collected to the left of Figure 8.9 will be more accurate than an interpretation based on data to the right. Unfortunately, reality frequently lies to the right and not the left in Figure 8.9.

Surface geophysical measurements give bulk averaged estimates of the subsurface properties. Further discrimination can be provided by making measurements in boreholes or on materials retrieved from boreholes. Geophysical methods are best used in collaboration with a limited drilling program. Consider a 1 ha field in which five boreholes are drilled. Some would say that the uncertainty about ground conditions is eliminated by having as many as five bores per hectare. What about reality? Each bore may sample a 50 mm diameter hole. This is an area of 0.00196 m^2, or 0.0000196% of the total area per borehole! In reality we know *nothing* about the remaining 99.99998% of the area, and drilling many other exploratory boreholes will not help that much and are expensive. The problem is *severely* exacerbated by including the third dimension.

8.2 CROSS-HOLE SEISMIC METHOD

8.2.1 Introduction

There is an increasing need for definitive values of dynamically determined elastic constants of foundation materials such as bulk modulus, Young's modulus, rigidity (shear) modulus and Poisson's ratio. The construction of large structures, such as nuclear power stations or off-shore oil platforms, requires the input of such data at the design stage to account for dynamic loading from earthquakes, ocean waves or machinery. The information is also required for the study of wave propagation and the calculation of pile performance using the wave equation solution (Davis and Taylor-Smith, 1980). Clayton (2011) provides a detailed account of cross-hole seismic work in the geotechnical area.

There are two levels at which the interpretation can be carried out, *viz.*:

*   measurements between boreholes which provide a single value of the compressional and shear wave velocities, and
*   full tomography in the plane between bores. This is mainly restricted to compressional wave velocities.

It has been suggested that the determination of elastic modulii at low strains (Clayton, 2011) provides values which are more representative of the strains encountered in the environment than measurements made on samples in the laboratory. In either case, cross-hole measurements provide an *in situ* measurement technique which provides a bulk averaged measurement. This is in contrast to *ex situ* measurements on discrete samples.

8.2.2   Survey configuration

A number of configurations are possible.

*   Measurements can be made between two boreholes where the source and receiver are lowered to the same depth. This configuration is shown in Figures 8.10 and 8.11. Figure 8.11 shows the arrangement of three bores with the source in the closest bore and the two down-hole geophones in the further bores.

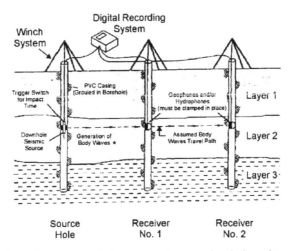

Figure 8.10 Schematic to show a cross-hole survey using one shot hole and two receiver holes (US EPA Archives)

Figure 8.11 Cross-hole seismic survey with a source bore and two receiver bores (Photo: Ian Acworth)

- If more than one receiver is available, measurements can be made at multiple depths to
  record the signal generated by a single source. The collection of this type of data leads
  to full-scale tomography.
- Measurements can be made from a surface source and a sensor lowered to different
  depths or to a string of sensors. This type of survey is known as vertical seismic
  profiling. A hammer source is shown in Figure 8.17 next to a borehole with a single
  down-hole geophone.

Shear waves can be generated by imparting a shear stress to the ground surface. The most
common method used is to hit the side of a plate fixed into the ground, or the side of a
timber placed on the ground surface and held in place by parking a vehicle with its
wheels on the timber (Figure 8.16). Opposing polarity shear waves can be generated by
striking either end of the timber.

8.2.3   Field instrumentation

The borehole environment requires the use of specialized equipment to provide both the
source of seismic energy (compressional and shear wave) and the sensing elements to
detect the energy. The requirement of good acoustic coupling between the source (and
sensors) and the formation beyond the borehole also needs to be addressed.

Borehole construction

Two boreholes are required for the measurement of seismic velocities. The source is placed
in the first bore and the receiver in the second bore. Problems with the establishment of an
accurate start time for the seismic waves (inaccurate triggering) mean that these velocities
may not be sufficiently accurate and the problem is sometimes overcome by drilling a third
borehole. The seismic source is placed in the first bore, and geophones are placed in each of
the second and third bores. The difference in travel time between the second and third bore
is used as an accurate measure of seismic velocity.

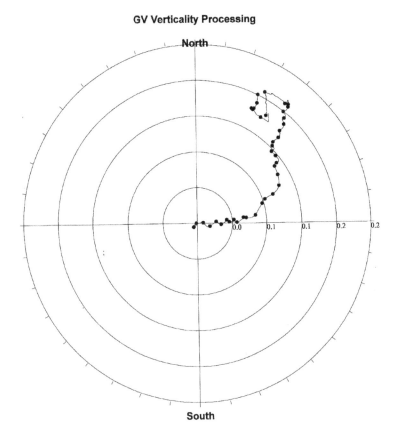

Figure 8.12 Example of a borehole verticality survey

The bores are drilled using conventional methods and thin-walled PVC casing installed and grouted in place to ensure good seismic coupling between the formation and the casing.

Cross-hole seismics can be carried out to any depth, but depths of 50 m to 70 m represent a typical maximum. Spacing between bores is often between 5 m and 10 m, depending upon the anticipated velocities and absorption of the material.

It is often assumed that boreholes are vertical. This is seldom the case, and a verticality survey is required on each borehole so that an accurate distance between source and receiver can be used in the calculation of velocities. Borehole verticality surveys can be a significant element in the overall cost of cross-hole work. A verticality survey returns a file containing x-y-z data for locations down the bore and a plot indicating deviation from verticality. An example of this type of survey is given in Figure 8.12. Figure 8.13 shows the change in horizontal distance between two boreholes initially 7.85 m apart at the surface.

The borehole deviation is measured using a three-axis flux gate magnetometer and can be mounted in a logging tool with a probe diameter of approximately 40 mm and a probe length of 1.5 m. In circumstances where there is a large natural magnetic field, a gyroscopic system can also be used to measure the verticality.

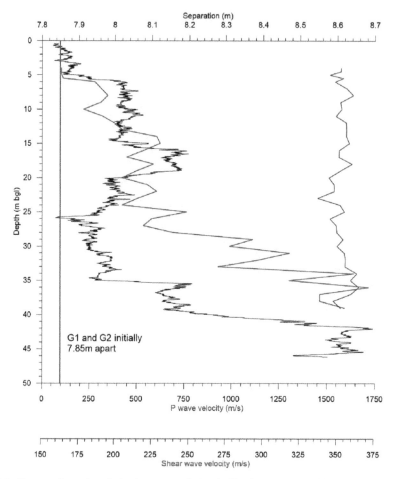

Figure 8.13 Change of interborehole distance with depth. The bores were 7.85 m apart at the surface. Deviations with depth determined by verticality survey

Borehole seismic sources

The borehole sparker is used as a compressional energy source. This equipment causes a spark to be created by a high-voltage discharge from a sonde lowered into the water-filled borehole. The spark causes vaporization of water and creation of a small bubble. The bubble collapses back in on itself and generates a shock wave which propagates radially out from the source. The sparker source does not need to be in contact with the borehole wall, as the energy passes from the fluid, through the casing and into the formation.

Shear waves are more difficult to generate. Vertical shear waves can be generated by a shear wave hammer. This is basically a frame which is lowered into the bore and clamped into position by inflation of a rubber diaphragm using either compressed air (Figure 8.14) or hydraulic fluid pumped from the surface. Shear waves are generated by letting a striker fall against the frame to produce vertical shear waves with an initial downward motion, or pulling the striker sharply upwards to generate vertical shear waves with an

Figure 8.14 Shear wave hammer source manufactured by Ballard. A foot pump is used to expand the rubber bladder on the side of the sonde until it is locked inside the casing. The weight on the end of the sonde is then sharply pulled upwards or dropped downwards to generate opposite polarity vertical shear energy (Photo: Ian Acworth)

initial upward motion. Recognition of the shear wave energy is facilitated by observing the difference in shear wave motion on the seismogram.

Borehole geophones

The detection of compressional wave energy can readily be achieved by use of hydrophones. These are completely sealed sensor elements which can pick up the arrival of the seismic energy through water. A string of hydrophones is often used to record the arrival of compressional wave energy at regular depths down the bore from a single source signal.

The arrival of shear wave energy requires a sensor to be held against the casing wall so that shear wave energy, which is not transmitted through fluids, can be detected. A three-component sensor can be used. A vertical component and two orthogonal horizontal components can be recorded in a unit held against the bore wall by a pneumatic element which is expanded from the surface (Figure 8.15). The orientation of the horizontal components can be maintained by using rigid PVC pipe to lower the geophone assembly into the borehole.

8.2.4   Shear waves

Shear wave detection

The detection of vertically polarized shear waves, generated in a borehole, is facilitated by comparing seismograms obtained from striking the hammer downwards and then upwards. If two separate seismograms are collected, then the shear wave energy can be separated from the other seismic energy by comparing the opposing polarity of the waves. Alternatively, subsequent seismograms can be added together by reversing the polarity of the signal each time an opposing strike of the hammer is made. In this way, other energy will tend to cancel out, and the shear wave energy will stack up and re-enforce.

An example of seismograms collected in a cross-hole survey using the technique described is shown in Figure 8.18. The arrival of the shear wave is clearly seen from the

Figure 8.15 Borehole geophones (manufactured by GEOSTUFF) with the mechanically operated leaf spring that locks the geophone assembly against the side of the bore. The 3-component geo-phone assembly inside the bore can be rotated remotely so that the first component is lined up with the source borehole. The surface controller used to select components of the signal and an ABEM 12-channel seismograph are also shown (Photo: Ian Acworth)

Figure 8.16 Shear waves being generated for a down-hole survey by hitting the end of a railway sleeper anchored to the ground by a vehicle parked onto the sleeper (Photo: Anna Greve)

Figure 8.17 Surface to down-hole seismic investigation with a hammer source used to generate P waves. The geophone assembly is clamped into place inside the borehole (Photo: Anna Greve)

commencement of trace diversion when the two traces from the vertically aligned geophone are presented together. This occurs because the energy is imparted at the source in the vertical direction. Note also that the P-wave arrival is much more clearly seen in the horizontal component geophone where the horizontal-2 component (Figure 8.18) is aligned with the source borehole. Both upward and downward blows (red and blue) have the same polarity – as expected for the compressional P-wave.

The detection of horizontally polarized shear waves, generated from the surface, can be assisted by comparing the seismogram obtained by first hitting a surface shear wave plate from one side and then comparing this seismogram with a second seismogram obtained by hitting the shear wave plate from the other side. In this case, the shear waves will be horizontally polarized rather than vertically polarized, as is the case for the borehole source.

8.2.5   Survey results

A profile of shear waves from a borehole at Cattle Lane is shown in Figure 8.19. The shear wave amplitude varies considerably. It has the greatest values in solid clay (16 to 20 m depth) and the lowest values in a very loose layer (14 to 15 m) that contained a high concentration of fine sand that was unstable when drilled and needed to be cased out (Acworth *et al.*, 2015b).

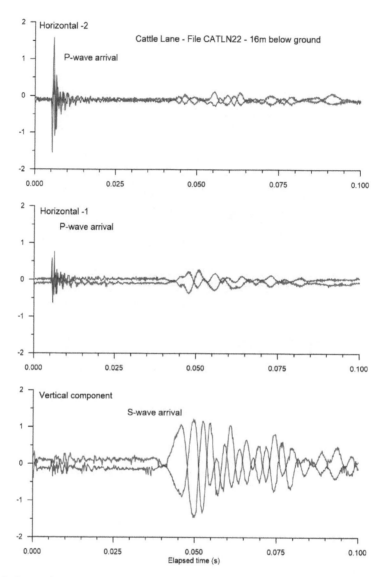

Figure 8.18 Output from a three-component geophone installed at 16 m depth in a silty clay sequence at Cattle Lane on the Liverpool Planes of NSW. The saturation depth was approximately 3 m. Blue signifies the weight dropped down; red signifies the weight pulled up. Five blows were stacked for each direction. A Ballard down-hole source and Geostuff borehole geophones were used to collect the data. The vertical component best represents the opposing shear waves while the P-wave arrival is best seen in the 'horizontal-2' geophone

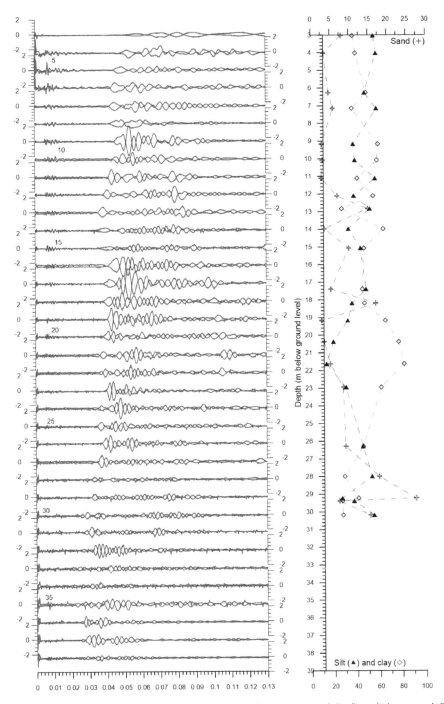

Figure 8.19 40 m profile of shear wave seismograms showing upward (red) and downward (blue) blows by the borehole hammer source. The seismograms are all scaled to the same amplitude showing significant vertical variation. Results from grain-size determination from cores are shown for comparison (Acworth *et al.*, 2015b)

Shear wave anisotropy

The down-hole hammer produces dominantly vertically polarized shear wave energy which does not show much anisotropy.

Many sediments are anisotropic to the passage of shear wave energy. The anisotropy may relate to a dominant fracture field, to faulting or to other geological reasons. By measuring the shear wave energy in the horizontal plane which results from vertically propagating surface sources and comparing this with energy from borehole sources, the anisotropy of the formation can be determined.

8.3　SURFACE REFRACTION METHODS

8.3.1　Critical refraction at an interface

Seismic wave energy is reflected or refracted when it passes across an interface. The relative amounts of reflected and refracted energy depend upon the angle of incidence of the seismic energy to the interface and upon the seismic velocity contrasts between the two layers. The relationship between the incident angle and the refracted angle is known as Snell's Law and is shown in Equation 8.21:

$$\frac{\sin \theta_i}{\sin \theta_r} = \frac{V_1}{V_2} \tag{8.21}$$

where:
θ_i is the angle of incidence,
θ_r is the angle of refraction,
V_1 is the seismic velocities of the upper layer, and
V_2 is the seismic velocity of the lower layer.

A P-wave incident on an interface that separates materials of different seismic velocity creates a disturbance which gives rise to both reflected and refracted P-waves and S-waves, as shown in Figure 8.20.

The percentage of energy which is transferred from one wave type to another is a function of the seismic velocity contrast between the media and the angle of the incident wave

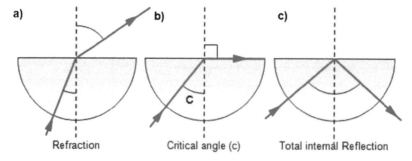

Refraction　　　Critical angle (c)　　　Total internal Reflection

Figure 8.20 Seismic energy passes into the lower layer (a) and is critically refracted along the interface (b) or reflected at the interface (c)

front. The only case for which an incident S-wave does not give rise to a reflected or refracted P-wave is when the S-wave is entirely horizontally polarized. The incident energy is partitioned in the process of refraction where the energy in each of the refracted primary or shear waves sums to equal the energy in the incident wave train.

The ratio of reflected to refracted energy is a function of the angle of incidence of the seismic wave front. The angle of refraction increases as the angle of incidence increases until a point is reached when all the energy is refracted along the horizontal interface. Further increase in incident angle causes complete internal reflection of energy at the interface with no energy penetration into the underlying medium. This is the property used in the analysis of seismic reflection data.

The angle at which no further seismic energy penetrates a deeper layer, and all energy is refracted along the interface creating a head wave travelling at the velocity of the deeper layer, is called the critical angle. This critical angle is given by Equation 8.22:

$$\theta_{ic} = \sin^{-1} \frac{V_1}{V_2} \tag{8.22}$$

8.3.2 Two-layer problem

The geophones are placed in a line at regular intervals and connected to a seismograph. A hammer is used to strike the ground, and the signals from the geophones are all recorded at the seismograph. The hammer has an inertial switch connected to it which sends a zero-time signal to the seismograph and causes the recording of the geophone response to begin.

The time for the wave front to reach each geophone is measured from a seismograph trace and plotted against distance. The time to reach each geophone is given by Equation 8.23:

$$t = \frac{x}{V_1} \tag{8.23}$$

where:
t is the time for the wave to travel to the geophone,
x is the distance from the shot (hammer) point to the geophone, and
V_1 is the velocity of seismic energy through the ground.

The first derivative of Equation 8.23 with respect to distance gives the reciprocal of the velocity. A plot of arrival times (y-axis) against distance (x-axis) produces a straight line, whose slope is then the reciprocal of the velocity:

$$\frac{dt}{dx} = \frac{1}{V_1} \tag{8.24}$$

The geometry for a single subsurface layer is shown in Figure 8.21.

The total travel time for a seismic wave following the path shown in Figure 8.21 can be easily calculated. The total travel time (T_{EG}) along the path E-M-N-G in Figure 8.21 is:

$$t_{EG} = \frac{EM}{V_1} + \frac{MN}{V_2} + \frac{NG}{V_1} \tag{8.25}$$

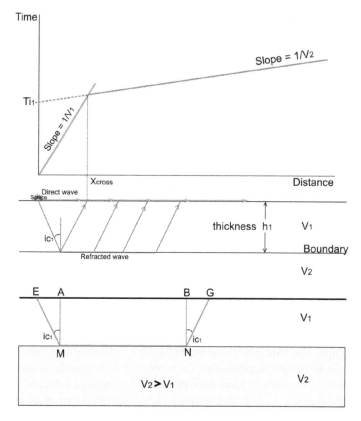

Figure 8.21 Schematic to show the elements of seismic refraction from a single horizontal layer

Since:

$$\cos \theta_{ic} = \frac{h_1}{EM} \tag{8.26}$$

and $EM = NG$, then:

$$EM = NG = \frac{h_1}{\cos \theta_{ic}} \tag{8.27}$$

Noting that $EA = BG = h_1 \tan \theta_{ic}$ and that $MN = x - 2h_1 \tan \theta_{ic}$, we can rearrange Equation 8.25 with the objective of writing the equation only in terms of the layer velocities and depths. This will allow a more readily derived solution:

$$t_{EG} = \frac{h_1}{V_1 \cos \theta_{ic}} + \frac{x - 2h_1 \tan \theta_{ic}}{V_2} + \frac{h_1}{V_1 \cos \theta_{ic}} \tag{8.28}$$

The first and third terms in Equation 8.28 ($h_1/V_1 \cos_{\theta c}$) represent the travel time expended in the top layer as the wave first moves down to the interface and then back to the surface from the interface. They will be referred to as *delay times* in later analyzes.

Utilizing the relationship for the critical angle that:

$$\sin \theta_{ic} = \frac{V_1}{V_2} \tag{8.29}$$

and:

$$\tan \theta_{ic} = \frac{\sin \theta_{ic}}{\cos \theta_{ic}} \tag{8.30}$$

and substituting in Equation 8.28 we get:

$$t_{EG} = \frac{2h_1}{V_1 \cos \theta_{ic}} - \frac{2h_1 \sin^2 \theta_{ic}}{V_1 \cos \theta_{ic}} + \frac{x}{V_2} \tag{8.31}$$

Rearranging Equation 8.31 produces:

$$t_{EG} = \frac{2h_1(1 - \sin^2 \theta_{ic})}{V_1 \cos \theta_{ic}} + \frac{x}{V_2} \tag{8.32}$$

Using the relationship that $\sin^2 \theta_{ic} + \cos^2 \theta_{ic} = 1$ and the relationships listed previously, we can simplify Equation 8.32 as follows:

$$t_{EG} = \frac{2h_1 \cos \theta_{ic}}{V_1} + \frac{x}{V_2} \tag{8.33}$$

$$t_{EG} = \frac{2h_1(1 - \sqrt{V_1^2/V_2^2})}{V_1} + \frac{x}{V_2} \tag{8.34}$$

$$t_{EG} = \frac{2h_1 \sqrt{V_2^2 - V_1^2}}{V_2 V_1} + \frac{x}{V_2} \tag{8.35}$$

After all that analysis, if we take the first derivative of the time with respect to distance we get:

$$\frac{dt}{dx} = \frac{1}{V_2} \tag{8.36}$$

A plot of the time of arrivals against distance from a source, for a seismic wave refracted from an interface, will give a straight line with a slope value equal to the reciprocal of the refractor velocity (Figure 8.21).

The depth to the interface between the first and second layers can be derived from Equation 8.35. If the line defining the second layer velocity is extrapolated back to zero distance, a value of the intercept time t_i is produced. Setting $x = 0$ in Equation 8.35 and using the intercept time for a value of t_{EG} allows a solution for h_1, as shown in Equation 8.37:

$$h_1 = \frac{t_i}{2} \frac{V_2 V_1}{\sqrt{V_2^2 - V_1^2}} \tag{8.37}$$

The distance from the shot point at which the refracted wave becomes the first arrival is known as the *crossover distance*. The crossover value is a function of the velocity contrast and the depth to the refractor and is important, as it determines the distance from the shot point that measurements have to be made to accurately determine the thickness of the first layer and the velocity of the second layer. A value of the crossover time can be derived by equating Equation 8.23 with Equation 8.35 and solving for the distance:

$$h_1 = \frac{x_{co}}{2} \sqrt{\frac{V_2 - V_1}{V_2 + V_1}}$$
(8.38)

Note that it is only the arrival time of the first energy at the geophone that is used in refraction analysis. The direct wave is the first arrival (Equation 8.24 and as seen in Figure 8.21) until the refracted wave arrives after the critical distance from the source is reached, whereby the refracted wave (Equation 8.35) overtakes the direct wave and is received first at the geophone. The arrival of the direct wave and the refracted wave can be seen in Figure 8.21 or in Figure 8.22.

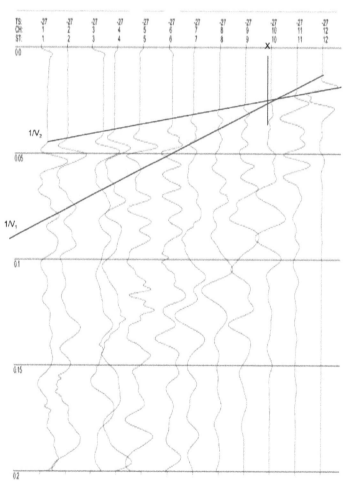

Figure 8.22 Seismograph wiggle trace of a simple spread showing the first arrival with a velocity V_1 being overtaken by a faster (V_2) refracted wave

Reflected wave equation

It is not intended to review seismic reflection methods in any detail, as they are rarely used in groundwater studies. There is considerable difficulty involved in identifying the reflected wave arrival from the refracted and direct wave arrivals as shown in Figure 8.23. Note also the half-parabolic shape of the reflection wave front. This complexity necessitates the recording of the complete waveform for every geophone along with considerable post processing. Nevertheless, for problems where the depth of interest is greater than 40 m, reflection techniques should be investigated (Steeples and Miller, 1990; Burger, 1992). Good success with a hammer, 14 Hz geophones and 100 Hz low-cut filters to resolve bedrock depth at 50 m have been reported (Burger, 1992). However, better data has been obtained using a 10-gauge shotgun source, 100 Hz geophones and a 200 Hz low-cut filter. Studies have been able to delineate structure within the overlying cover material as well as the depth to bedrock.

The travel-time equation for the reflected wave is the time taken for the wave to pass from the source to the interface (distance EA) and back again to a receiver (distance AG) as seen in Figure 8.24. As the distance between source and receiver increases, the travel time has a parabolic character.

The travel time for a reflected wave is:

$$time = \frac{EA + AG}{V_1} \qquad (8.39)$$

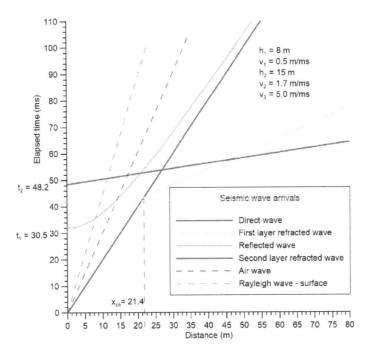

Figure 8.23 Calculated first arrivals for the different wave, sets (direct, refracted, reflected, air and Rayleigh waves) using a simple spreadsheet

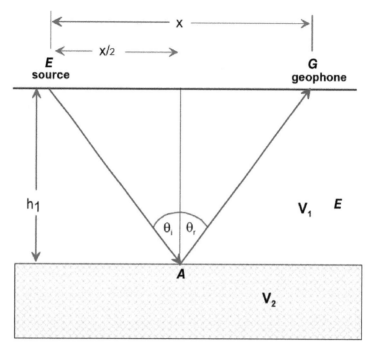

Figure 8.24 Geometry for a simple reflected wave case from a horizontal interface

Using a simple geometry:

$$EA = AG = \sqrt{\left(\frac{x}{2}\right)^2 + h_1^2}$$ (8.40)

and then:

$$time = \frac{\sqrt{x^2 + 4h_1^2}}{V_1}$$ (8.41)

The first derivative of Equation 8.41 is parabolic in shape and there is no possibility of a velocity determination from a straight line fit on a simple t-x plot.

The generation of a parabolic shape to the first arrival leads to the concept of *normal move-out (NMO)*. NMO is defined as the difference in reflection travel times from a horizontal reflecting surface due to variations in the source to geophone distance. NMO is determined by subtracting the travel times *t* the source, where $x = 0$ (T_0), from the travel time at the receiver distance t_x. One of the goals in reflection seismic data processing is to remove the effect of NMO from the profiles as it complicates the data interpretation considerably.

Data recovered over a dipping interface would produce a parabola skewed to one or other sides of the shot point, depending upon the angle of dip.

The depth to the reflector can be determined by taking the square of Equation 8.41 as:

$$t^2 = \frac{x^2 + 4h_1^2}{V_1^2} = \frac{1}{V_1^2}x^2 + \frac{4h_1^2}{V_1^2}$$ (8.42)

A plot of t versus x^2 then gives a straight line with a slope of $\frac{1}{V_1^2}$ and an intercept of:

$$t_0^2 = \frac{4h_1^2}{V_1^2} \tag{8.43}$$

or:

$$t_0 = \frac{2h_1}{V_1} \tag{8.44}$$

and:

$$h_1 = \frac{t_0 V_1}{2} \tag{8.45}$$

The recovery of depths to multiple interfaces is more difficult than in refraction seismics and tends to be handled by data processing methods rather than analytical derivations. It should be clear that there is a significant problem in separating out the clear reflection data from all the other arrivals at the geophone (Figure 8.23). In reality, there is an optimum window for reflection data acquisition where the reflected wavelet is not obscured by other arrivals. The selection of this optimum window depends upon the depths to the reflector and source geophone locations.

8.3.3   Simple refraction models

The equations for the direct wave (Equation 8.24), the refracted wave (Equation 8.35) and the reflected wave (Equation 8.41) can be used in a simple spreadsheet to calculate the arrival times for these waves and how they vary for different velocities and depth to the interface. For comparison, the air wave (Equation 8.13) and the surface (or Rayleigh wave) (Equation 8.15) can also be included as shown in Figure 8.23. If the velocities (V_1 and V_2) and the depth to the interface are set as variables in the spreadsheet, a range of problems can be investigated; specifically, the line length that will be required to recover refracted data for the second layer. A hammer source may work to a distance of 100 m under very quiet conditions. Beyond that, assisted weight drops, a gun source (Bison or similar) or explosives will be required. A rule of thumb for the seismic line length would be five times the depth to the main refractor (Kearey and Brooks, 1991).

The geometry for two horizontal refractors is shown in Figure 8.25. The travel time for a two-layer model is given in Equation 8.46:

$$t_{EG} = \frac{EP}{V_1} + \frac{PR}{V_2} + \frac{RS}{V_3} + \frac{SQ}{V_2} + \frac{QG}{V_1} \tag{8.46}$$

Using a similar approach to that used previously, it can be shown that:

$$t_{EG} = \frac{x}{V_3} + \frac{2h_1\sqrt{V_3^2 - V_1^2}}{V_3 V_1} + \frac{2h_2\sqrt{V_3^2 - V_2^2}}{V_3 V_2} \tag{8.47}$$

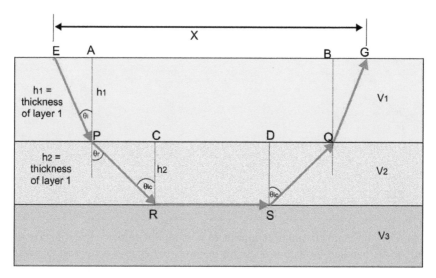

Figure 8.25 Seismic wave path through two horizontal layers

This can be equivalently expressed as:

$$t_{EG} = \frac{x}{V_3} + \frac{2h_1 \cos\theta_1}{V_1} + \frac{2h_2 \cos\theta_2}{V_2} \tag{8.48}$$

where: $\theta_1 = \sin^{-1}\frac{V_1}{V_3}$ and $\theta_2 = \sin^{-1}\frac{V_2}{V_3}$.

The first derivative of these equations again yields a relationship which gives the velocity:

$$\frac{dt}{dx} = \frac{1}{V_3} \tag{8.49}$$

A plot of travel-time data over two interfaces will yield three segments. The reciprocal of the slope for each segment gives the velocity of each segment. The second and third segments can both be extrapolated back to zero distance to produce intercept times (ti_1 and ti_2). The thickness of the second layer can be established from the equation:

$$h_2 = \left(t_{i_2} - \frac{2h_1\sqrt{V_3^2 - V_1^2}}{V_3 V_1} \right) \frac{V_3 V_2}{2\sqrt{V_3^2 - V_2^2}} \tag{8.50}$$

The critical distance for a two-interface model can be calculated from:

$$x_{crit} = 2\left(h_1 \frac{V_1}{\sqrt{V_3^2 - V_1^2}} + h_2 \frac{V_2}{\sqrt{V_3^2 - V_2^2}} \right) \tag{8.51}$$

The critical distance is of significant help when trying to recognize the onset of a refracted wave.

8.4 SEISMIC REFRACTION: DELAY-TIME OR PLUS-MINUS METHOD

8.4.1   Introduction

There are many solutions available for multiple layers or multiple dipping layers, but they are rarely used and are not presented here. Burger (1992) gives a good summary. However, many (most) geological problems involve the calculation of depths to an irregular interface. A typical groundwater problem would be to shoot across a valley and to determine the location of maximum weathering for the placing of an abstraction bore. The weathered profile cannot be expected to be either dipping or horizontal, so a more complete interpretation method is required. The method for solving this problem was developed by Hagedoorn (1959) and reviewed by van Overmeeren (2001). It is also considered to be an example of the generalized reciprocal method (Palmer, 1980).

When the interface is not horizontal, ambiguity in the interpretation arises with results from a dipping layer being possible to interpret in terms of a different velocity and depth combinations. To resolve this ambiguity, the seismic spread is shot from both ends. The reciprocal time should be identical. If the geophone and shot points are interchanged, the travel time should be the same. This is the basis of the plus-minus method.

The solution can be developed from the geometry shown in Figure 8.26.

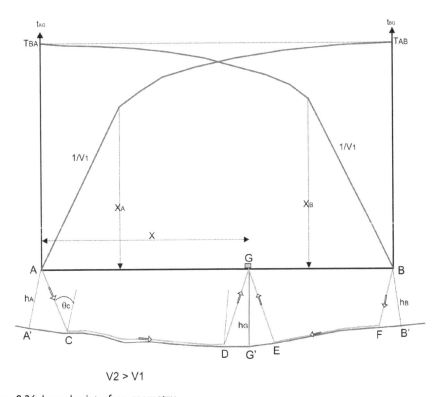

Figure 8.26 Irregular interface geometry

8.4.2 Mathematical basis

We have to make the assumption that the V_2 velocity will be greater than the V_1 velocity. This is normally a reliable assumption as the lower material is both more consolidated and probably saturated.

Consider a geometry such as that shown in Figure 8.26, where a single interface separates an upper lower velocity layer from a deeper higher velocity layer.

A travel-time equation for this model is then:

$$t_{AG} = \frac{AC}{V_1} + \frac{CD}{V_2} + \frac{DG}{V_1} \tag{8.52}$$

where:

t_{AG} is the *travel time* for the wave energy to pass down through the top layer (A to C) with a velocity of V_1; then along the interface (C to D) with a velocity of V_2 and then back up to the geophone (D to G), again with a velocity of V_1.

If the interface is approximately planar over the distance between A' and C then – as shown in Figure 8.26:

$$AC \approx \frac{h_A}{\cos \theta_c} \tag{8.53}$$

Similarly, if the interface is approximately planar over the distance between D and G', then:

$$DG \approx \frac{h_G}{\cos \theta_c} \tag{8.54}$$

where:
h_A is the perpendicular thickness of overburden beneath A and
h_G is the perpendicular thickness of overburden beneath G.

Similarly, if the overall dip on the interface is small, $x \approx A'C + CD + DG'$, then:

$$CD \approx A'G' - h_A \tan \theta_c - h_G \tan \theta_c \approx x - h_A \tan \theta_c - h_G \tan \theta_c \tag{8.55}$$

Thus, the travel-time equation can be written (using the trigonometric identities developed previously) as:

$$t_{AG} \approx \frac{x}{V_2} + \frac{h_A}{V_1} \left(\frac{1}{\cos \theta_c} - \frac{V_1}{V_2} \tan \theta_c \right) + \frac{h_G}{V_1} \left(\frac{1}{\cos \theta_c} - \frac{V_1}{V_2} \tan \theta_c \right) \tag{8.56}$$

We can interpret Equation 8.56 as the travel time t_{CD} in the V_2 layer plus the times taken to get from the surface to the V_2 layer (the delay time t_A) and then back up again – the delay time t_G. We refer to these times taken for the energy to pass down and then up through the V_1 layer (of overburden) as the delay times. The first is t_A and represents the time taken passing down to the V_2 layer below A. The second is t_G and represents the time taken passing from the V_2

layer back up to the geophone at G. Then, simplifying Equation 8.56 gives:

$$t_{AG} \approx \frac{x}{V_2} + t_A + t_G \qquad (8.57)$$

where:

t_A is $\frac{h_A \cos \theta_c}{V_1}$ – the *delay time* at A and

t_G is $\frac{h_G \cos \theta_c}{V_1}$ – the *delay time* at G.

It is easy to show, using the basic trigonometric identities developed above, that:

$$\frac{h_A}{V_1} \left(\frac{1}{\cos \theta_c} - \frac{V_1}{V_2} \tan \theta_c \right) = \frac{h_A \cos \theta_c}{V_1} \qquad (8.58)$$

and similarly for t_G.

It is worth noting here that if $h_A = h_B$ (the interface is flat), then Equation 8.57 becomes the same as Equation 8.33 produced for the simple two-layer interpretation problem. The delay times are the same.

In a similar manner to Equation 8.57 for the forward path of A to G, we can account for the reversed path length (B to G) as:

$$t_{BG} = \frac{BF}{V_1} + \frac{FE}{V_2} + \frac{EG}{V_1} \approx \frac{L-x}{V_2} + t_B + t_G \qquad (8.59)$$

where:

L is the spread length AB and

t_B is the *delay time* at B $\left(t_B = \frac{h_B \cos \theta_c}{V_1} \right)$.

Minus times (t_)

The travel times for the forward (t_{AG}) and the reversed (t_{BG}) paths can be measured in the field. These travel times have been shown to be given by Equation 8.57 for the forward time and Equation 8.59 for the reverse time. We can use these relationships to give a profile of depths to the interface.

If we subtract the reverse time from the forward time for all first arrivals between for $X_A < x < X_B$, we get:

$$t_- = t_{AG} - t_{BG} = \frac{2x}{V_2} + \left(t_A - t_B - \frac{L}{V_2} \right) \qquad (8.60)$$

which produces:

$$t_- = \frac{2x}{V_2} + constant \qquad (8.61)$$

A plot of *minus times* against distance gives a straight line with a slope of $\frac{2}{V_2}$ and an intercept of T_0, as shown in Figure 8.27. Note that a variation in the V_2 velocity can be detected

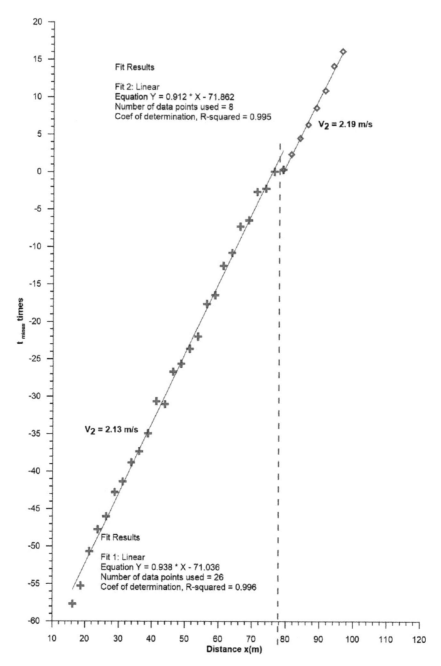

Figure 8.27 Derivation of V_2 from t minus data. The T_0 time can be read at the interception of the minus time plot at $x = 0$. Two segments of refractor velocity are shown with a slightly higher velocity in the far right (blue). The different values of V_2 can be incorporated into the analysis

from this plot. It is not uncommon to pick up a decrease in the second layer velocity which is often related to fracturing in the refractor. This would be a good place to site an abstraction borehole.

Plus times (t₊)

Using the same arrival data for T_{AG} and T_{BG} but adding the arrivals gives:

$$t_+ = t_{AG} + t_{BG} = 2t_G + t_A + t_B + \frac{L}{V_2} \tag{8.62}$$

Then, if the total travel time (T) from one end of the spread to the other (A to B) can be expressed as the total spread length (L) divided by the refractor velocity (V_2), plus the delay time to get from A to C (t_A) and the delay time to get back from F to B (t_B), we can then write:

$$T = \frac{L}{V_2} + t_A + t_B \tag{8.63}$$

Substituting Equation 8.63 into Equation 8.62 and rearranging gives:

$$t_+ = 2t_G + T \tag{8.64}$$

where:
T is the end-to-end time ($T_{AB} = T_{BA}$) and has to be true as the result of reciprocity (the path time from source to geophone has to be the same as the same path – geophone to source).

The *plus times* give the delay time (t_G from Equation 8.64) and then the depths (h_G) beneath each of the geophones in the section of the t-x plot between the crossovers ($X_a < x < X_B$) from:

$$h_G = \frac{V_1 t_G}{\cos \theta_c} \tag{8.65}$$

Note that when more than two layers are present, the value of T is obtained by projecting the V_2 slope line to the start or end of the line. This means that T will be greater than the end-to-end times. This important point is illustrated in Figure 8.28. If the end-to-end time is used rather than the correct value of T, the predicted depths to the interface using Equation 8.64 will be too great.

8.4.3   Summary of the interpretation procedure for the delay-time or plus-minus method

The interpretation procedure can be summarized as follows:

1   Plot t-x data (t_{AG}, t_{BG}, versus x),
2   calculate V_1,
3   calculate the t_- and t_+ times for $X_A < x < X_B$,

4   plot t_- versus x and compute V_2 (Equation 8.61),
5   compute the critical angle $\theta_c = \sin^{-1}\frac{V_1}{V_2}$,
6   compute the delay times t_G for $x_A < x < x_B$ from the relationship $\frac{1}{2}(t_+ - T)$, and
7   compute thicknesses from delay times, $hG = \frac{V_1 t_G}{\cos\theta_c}$.

Steps 1 to 7 above allow the depths to the interface to be calculated for parts of the time-distance data set where you have coverage of returns from the interface ($x_A < x < x_B$ – i.e., between 20 m and 80 m in Figure 8.28).

It is also possible to get depths beneath the shot points (h_A and h_B) by considering that:

$$T = t_A + t_B + \frac{L}{V_2} \tag{8.66}$$

and:

$$T_0 = t_A - t_B - \frac{L}{V_2} \tag{8.67}$$

where:
T_0 is the intercept time from the T_- versus distance (x) plot (Figure 8.27).

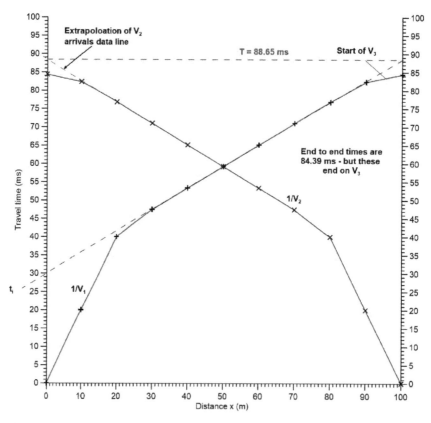

Figure 8.28 The selection of an end-to-end travel-time value (T) when an additional layer is present is based on an extrapolation of the second-layer arrivals

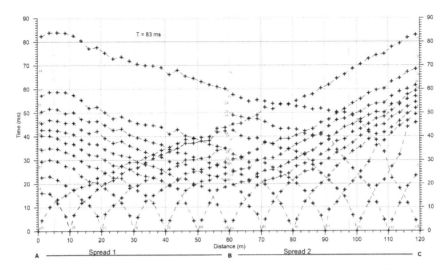

Figure 8.29 An example showing coverage of two spreads of 60 m each with five additional shots into the spread that allow more confidence in the establishment of refractors

The depth to the interface beneath the shot points can then be calculated by solving the simultaneous equations (Equations 8.66 and 8.67).

Note that this analysis is based upon the following assumptions:

1   The velocities are assumed constant,
2   the irregularity of the interface is not extreme, and
3   the overall dip of the interface is small.

Uncertainty concerning the first-layer velocity has prompted many surveys to increase the number of shots into a spread. The example in Figure 8.30 has only one mid-spread location. It would be possible to collect further data as shown in Figure 8.29. With the additional data, automated interpretation routines are used to invert the data. In Figure 8.29, showing a 60 m spread length collected with a 24-channel hammer system with geophones every 2.5 m, there was an end shot at A (0 m); five intra-spread shots (10 m, 20 m, 30 m, 40 m and 50 m); an end shot at B (60 m); an off-end shot at 120 m and a middle off-end shot at 90 m.

8.4.4   Extension to three layers

The extension of the method to three layers requires overlapping coverage of each velocity layer along the complete profile. The method follows almost exactly the same approach to that given for a single irregular interface. Plus and minus times are calculated for the overlapping sectors on the second refractor (clearly seen in Figure 8.30). The V_3 velocity is calculated from the minus times. The second-layer plot of minus times is shown in Figure 8.30.

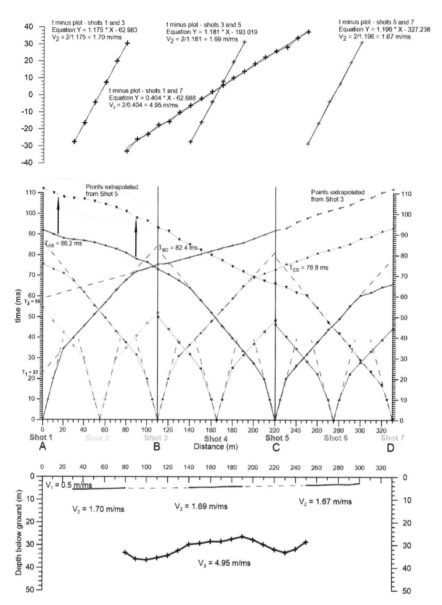

Figure 8.30 Example of the plus-minus method for two irregular interfaces. In this example a 12-geo-phone cable is used with a 12-channel seismograph. The cable is first laid out between A and B and arrivals from each geophone recorded for shots at A (Shot 1), the mid-point of AB (Shot 2), at B (Shot 3) and C (Shot 5). The cable is then moved to between B and C and the geophones are planted again. Shots are recorded at A (Shot 1), B (Shot 3), the mid point of BC (Shot 4), C (Shot 5) and D (Shot 7). Data for the far end of the line is then collected by moving the cable to CD and shooting at B (Shot 3), C (Shot 5), the mid-point of CD (Shot 6) and D (Shot 7)

The t_G times for the second layer need some modification, as they comprise the delay through both the first layer and the second layer. The t_G times are first calculated in the normal manner from the end-to-end time (T) and the plus times. Calculation of h_g using these values will greatly overestimate the depths to the bottom layer.

A correction needs to then be made that represents the delay in the top layer that can be subtracted from the t_G times. This correction is $\frac{h_1 \cos \theta_{c_{1,3}}}{V_1}$. The h_G depths are then calculated from: $\frac{h_2 t_g^{corrected}}{\cos \theta_{c_{2,3}}}$.

Survey data should be collected with the objective of obtaining complete coverage over the refractor of interest. With a survey combining forward and reverse directions with continuous subsurface coverage of each target refractor, high-confidence interpretations can be made using the plus-minus method giving the velocity and depth of the various refractors. An example is given in Figure 8.30.

Decisions have to be made concerning the distance between geophones, the length of line which is shot, the location of shot points, etc. The length of the spreads and the geophone separation are chosen based upon some simple modelling of the crossover expected from a consideration of planar two- or three-layer models. As one of the major areas of uncertainty is the V_1 refractor velocity, sufficient data should be collected to establish this accurately. One method of doing this is to have a mid-spread shot point in addition to the shot-points at either end of the line. Collection of reversed travel-time data over deeper refractors may also (usually will) require an off-end shot at each end of the spread. Figure 8.30 shows a typical data collection. There are seven shots in this example where each spread length (AB, BC, CD) contains 12 geophones each separated by 10 m. Note that, in this example, where the shot is to be made coincident with a geophone location, the geophone is shifted by 5 m along the line. This has the advantage of not damaging the geophone (!) but also allowing better data coverage of the low V_1 layer. This is important as inaccuracy in the determination of V_1 has a large impact on depth calculation to deeper layers. After the shot is made, the geophone is returned to its original position.

In Figure 8.30, shot 1 is only made twice into the line of geophones between A and B. The shot is made once as this is the beginning of the line and a second time from the end of spread BC. Most shot locations are occupied at least three times, and practical considerations need to be considered that the shots, if made by explosives, do not destroy the ground so that the geophone cannot be relocated. For this reason, a shot location may be offset to the side and the data corrected accordingly for the different path lengths.

To get better coverage of the variability in V_1 along the line, a mid-spread shot is also made (shot 2 at 55 m). Shot 3 is made firstly into spread AB and then repeated after the geophones are shifted to spread BC. Before moving the geophones from spread AB to spread BC, a shot is made into line AB from the location for shot 3 (at C in Figure 8.30). The geophones are then moved to spread BC and shot 3 is repeated, but now into spread BC. A mid-spread shot is made (shot 4) as is an end shot (shot 5) at the far end of spread BC. While the geophones are located for spread BC, a shot is also made into this spread from shot 7 location and a shot at the first location is also made. The data in Figure 8.30 is colour coded to indicate the shot location. The geophones are then moved to spread CD and the shot process repeated, firstly at shot 3 location, then shot 5 location, then shot 6 location for the mid-spread data and finally, shot 7 location. If the survey were to be continued, data would also be collected for a shot 9 location.

The choice of geophone distance and spread length will be determined by the initial analysis of the site geology in the pre-program planning stage. It is at this time that critical decisions will be taken, such as whether to rely on a hammer energy source or to use explosives. This decision will effect project costing significantly, as the use of explosives requires specially qualified personnel, the means to dig shot holes and increased safety on site (care of detonators). These additional requirements require a more costly mobilization and longer time in the field. However, many problems can only be solved with explosives, especially if the top layer does not conduct seismic energy efficiently.

Surveys across sand dunes, areas of marsh or deep clay soils are all difficult (or impossible) using hammer energy sources. Surveys through thick bush or across rough ground also increase field times significantly and need to be accounted for at the project planning stage. Problems can be generated by overhead power lines or surveys close to major sources of noise, such as roads or railways, or through thick vegetation. The geophysicist is commonly the first person to occupy a site and is the first to discover problems such as unmapped deep channels, bogs, impenetrable undergrowth or just the favorite haunt of big snakes!

8.4.5   Corrections to travel-time data

Shot offset correction

It is frequently the case that a geophone spread cannot be arranged so that all the geophones are at the same elevation or in the same plane. Placing of shot holes at an offset (right angle) from the geophone line has the advantage that data can be recorded from a regular array of geophones with a number of shot holes prepared adjacent to the line. This is particularly useful if the shot destroys the ground in the near vicinity and a further hole must be drilled for an off-end shot or an end shot. Recall that in the three-layer delay-time example (Figure 8.30), shot positions 3 and 5 were occupied three times, requiring three separate blast holes if explosives were used. The geometry of the problem is shown in Figure 8.31.

It is also necessary to correct t-x data for the fact that the shot is buried beneath the ground. This may make a significant difference to calculation of times for nearby geophones.

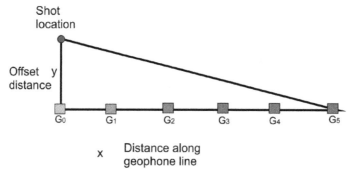

Figure 8.31 A correction (Pythagoras) to the distance along the line is required for geophones close to a shot set offset from the main line

Elevation corrections

In undulating ground, it is necessary to correct all the travel times such as they then represent a flat surface. If this is not done, then the proportion of the observed travel times which are due to the varying elevation will be erroneously interpreted as changes to the depth of layers, or velocities. This is particularly a problem as seismic wave travel through the low-velocity first layer takes a disproportional percentage of the total travel time.

8.4.6   Interpretation packages

The methods described above provide a general introduction to the basics of seismic refraction. More sophisticated interpretation packages are available that make use of ray tracing and approach the refraction interpretation task more as a modeling exercise. The ray tracing method (Cerveny *et al.*, 1974) or the generalized reciprocal method (GRM) (Palmer, 1980) is a popular approach to seismic refraction interpretation. The GRM is based upon the delay-time method shown above, however Palmer (1980) extended the method to deal especially well with irregular refractors with dip angles up to 20 degrees. The GRM also indicates the presence of hidden layers and their probable nature (velocity inversion or thin layer). Commercial software packages are available. REFLEXW (info@sandmeier-geo. de) is one such package; Interpex are another company selling software (info@interpex. com).

8.4.7   Ambiguities in seismic refraction interpretation

Each of the interpretation methods described above has been based on the assumption that the seismic velocity increases with depth. This is most often the case with shallow investigation, as saturation increases with depth and weathering decreases with depth. However, there are important exceptions to this generalization.

Discontinuities such as faults give rise to t-x plots with specific characteristics. These can often be seen in the t-x data sets, as long as reversed data has been collected.

Blind-zone problem

Sometimes a deeper layer will have a lower velocity than a shallower layer with both layers underlain by solid rock of a much higher velocity ($V_2 < V_1 < V_3$). This may be caused by several conditions:

- if laterite is present developed as a resistate close to the ground surface or
- if the ground is frozen at the surface (permafrost problem that is common at higher latitudes), or
- if the ground is part of an alluvial fill sequence where clay has a higher velocity than underlying sand layers overlying bedrock.

These conditions lead to what is referred to as a *blind-zone problem*.

The *t-x* data for an example of a blind-zone problem is shown in Figure 8.32. Arrivals for the direct wave and the refracted waves are calculated using Equations 8.24 and 8.35.

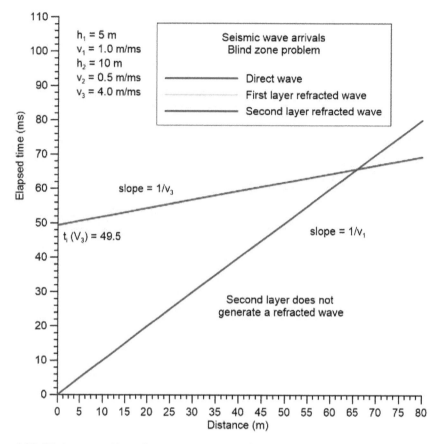

Figure 8.32 Blind zone problem: there are never any refractions generated from layer 2

Critical refraction does not occur at the interface between the first and second layers because, according to Snell's Law, the ray is bent deeper (and is never seen as a refraction) rather than towards the interface.

A plot of possible arrival times is shown in Figure 8.32 where a spreadsheet is used to calculate arrival times for the model shown. The t-x data has the appearance of a simple two-layer problem. The depth to the second layer can be calculated by Equation 8.37. Using the velocities shown in Figure 8.32 and the indicated intercept time, the h_1 depth is 25.56 m. This is a significant *overestimate* of the true depth. This occurs as the result of the assumption that all the material to the top of the third layer has a velocity equal to V_1 and does not include the higher V_2. The only way in which this problem can be overcome is to establish the presence of the second layer by some independent means such as electric imaging, if there are satisfactory contrasts in electrical properties between the layers, or by drilling.

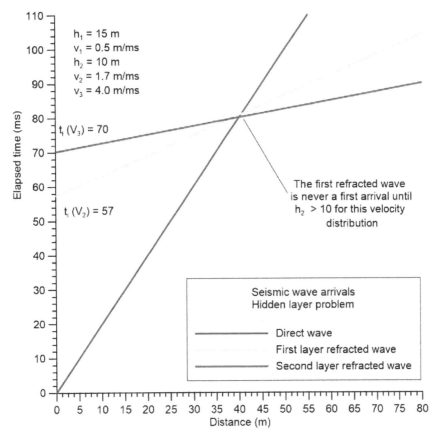

Figure 8.33 Hidden-layer problem: layer 2 is not sufficiently thick to generate significant refractions that arrive before deeper refractions

Hidden-layer problem

Thin layers, or layers which only have a low velocity contrast with underlying layers, will also escape detection. The problem is illustrated in Figure 8.33.

The intermediate velocity layer does produce a head wave (in contrast to the blind-zone problem) but it is overtaken by the head wave from the deeper layer before it can be detected at the geophones. The predicted depth to the third layer can be estimated using the model data shown in Figure 8.33. In this case, a depth to the interface of 17.6 m is calculated. This is a significant *underestimate* of the 25 m used in the model.

The hidden-layer problem is more common than the blind-layer problem. An example of a hidden-layer problem occurs where a thin layer of groundwater is located at the base of unconsolidated material and overlying rock. It may not constitute a useful aquifer but could form a significant geotechnical problem and will cause problems with interpretation of other data sets.

Chapter 9

Geophysical investigation techniques: electrical

9.1 ELECTRICAL METHODS

9.1.1 Introduction

Electrical methods using four electrodes were first introduced by Wenner (1912) in the USA. Schlumberger introduced the first expanding array system to investigate change of conductivity with depth in 1934. The Megger earth tester was introduced as a resistance measuring device sometime during the Second World War and was probably used to investigate for groundwater resources in the North African desert campaigns. This was augmented by the Geophysical Megger comprising two boxes on tripods and was in extensive use in Africa for groundwater investigation. Dr G. L. Paver designed a system used in Iran in the 1940s, and Robin Hazell was using similar equipment in Northern Nigeria in the 1950s.

BRGM/Schlumberger used portable generator sets to produce sufficient current to probe deep for groundwater in Algeria and French West Africa in the 1950s. By the 1960s, the Swedish company ABEM was manufacturing a two-box system called a Terrameter. Derivatives of this equipment remain on sale.

9.1.2 Electrical conductivity or resistivity

In electrical circuit theory, it is possible to accurately measure the conductance of a component, because the component has only limited dimensions. In geophysical surveys, many different components of the ground may add to the total conductance, and we need to normalize the conductance in terms of the dimensions of the sample. For example, if we take a tank as shown in Figure 9.1, connect a battery and ammeter in series, and fill it with sand and water, we could show that the conductance of the material in the tank is determined as the ratio of the current flow to the voltage. This is Ohm's Law written in terms of conductance instead of resistance and can be written as:

$$G = \frac{I}{V} \tag{9.1}$$

If we take a number of tanks of differing dimensions and fill them with the same mixture of sand and water, we could show that the conductance was proportional to the area (A) of the end plates and inversely proportional to the distance (L) between the plates.

The constant of proportionality is termed the *conductivity* and is a property of the sand and water mix alone, or whatever else is placed between the plates. The conductance can

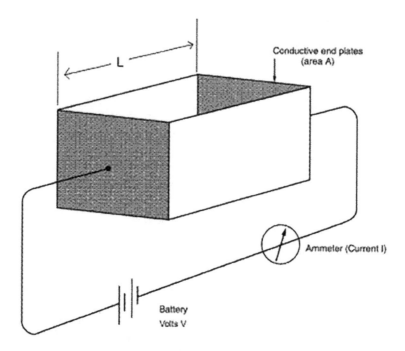

Figure 9.1 Tank apparatus to demonstrate conductivity or resistivity relationships

then be written as:

$$G = \sigma \frac{A}{L} \tag{9.2}$$

where:
σ is the conductivity with units of S/m,
G is the conductance (S),
A is the area of the tank end plates (m^2), and
L is the distance between the plates (m).

Confusion is often caused by the use of both the conductance (and the conductivity) at the same time as the resistance (and the resistivity). The resistivity is defined as the proportionality constant shown in Equation 9.3:

$$R = \rho \frac{L}{A} \tag{9.3}$$

where:
ρ is the resistivity with units of $\Omega \cdot$ m,
R is the resistance in ohms (Ω),
A is the area of the tank end plates, and
L is the distance between the plates.

The resistivity is simply the inverse of the conductivity and does not have any separate special significance or physical properties.

Table 9.1 Conversion factors

Resistivity	Conductivity
1 $\Omega \cdot$ m	1 S/m
1 $\Omega \cdot$ m	1000 mS/m
1000 $\Omega \cdot$ m	1 mS/m
1 $\Omega \cdot$ m	10 dS/m
10 $\Omega \cdot$ m	1000 μS/cm
100 $\Omega \cdot$ m	10 mS/m
10 $\Omega \cdot$ m	100 mS/m
100 $\Omega \cdot$ m	0.1 dS/m

The conversion factors shown in Table 9.1 are provided. Note that to derive a conductivity reading in mS/m from a resistivity reading in ohm metres, the following relationship is used:

$$\sigma = \frac{1000}{\rho} \tag{9.4}$$

The SI unit of measurement for conductivity is siemens per metre (S/m), but most groundwater measurements are reported in μS/cm; electromagnetic surveys tend to be recorded in mS/m and soil science measurements are often made in dS/m.

Electric current flow can occur by:

- electronic,
- electrolytic, or
- dielectric conduction processes.

Electronic conduction can occur only through a conductor such as copper or graphite, where the atomic structure of the material allows direct movement of loosely bound electrons within the crystal lattice to act as charge carriers. Dielectric conduction becomes important at high frequencies. Electrolytic conduction is the dominant process at low frequencies (0.3–3.0 Hz) where the electrical conductivity of the substance predominates.

9.1.3 Factors influencing the bulk electrical conductivity

The electrical conductivity of a material is impacted by several factors, including the fluid electrical conductivity of water held in the pore space. To differentiate these terms, it is common to refer to fluid electrical conductivity as σ_w and the conductivity of the complete mixture as σ_b – the bulk conductivity.

Typical ranges for the conductivity of earth materials (Palacky, 1987) are given in Figure 9.2.

The bulk electrical conductivity (σ_b) of soil and rock is dominantly controlled by the following factors:

- the porosity available to be filled by fluid,
- the electrical conductivity of the pore-filling solution,
- the type and quantity of clay material present,
- the amount of fluid filling the porosity (degree of saturation), and
- the frequency (electrical) at which the measurement is made.

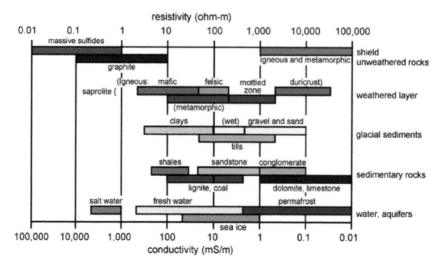

Figure 9.2 Conductivities of typical earth material (adapted from Palacky (1987))

Porosity available to be filled by fluid

The most widely used relationship to determine porosity from electrical measurements is Archie's Law. This has been developed in the oil industry where it was used to predict porosity (and therefore oil content) from the ratio of the fluid resistivity and the bulk resistivity:

$$\sigma_b = a\sigma_w\phi^m \qquad (9.5)$$

where:
σ_b is the observed electrical conductivity at the surface,
σ_w is the conductivity of the water present,
ϕ is the porosity,
m is a factor related to grain shape, and
a is a constant.

Archie's Law does not work well for fluid conductivities of less than 100 mS/m², for sediments containing clay, or for partially saturated sediments. Modifications have been suggested to extend the relationship as shown in Equation 9.6:

$$\sigma_b = a\sigma_w\phi^m S^n + \sigma_s \qquad (9.6)$$

where:
σ_b is the observed electrical conductivity at the surface,
σ_w is the fluid conductivity of the water present,
σ_s is the conductivity associated with the movement of surface charge along clay surfaces,
ϕ is the porosity, and
S is the saturation.

Archie's Law is not the only relationship (see Acworth and Jorstad (2006), for example), and in fact physically derived expressions are often more useful in defining a range of acceptable porosity.

Saturation of the pores

The following empirical relationship for partially saturated rock has been given (Archie, 1942, 1950):

$$\rho = \rho_b S^{-n} \tag{9.7}$$

where:
S is the fraction of the pore filled with water,
ρ_b is the resistivity of the partially saturated rock, and
n is an empirical constant for each rock, now believed to be related to the pore distribution.

Electrical conductivity of the pore-filling solution

Water conductivity is controlled by the presence of ions in solution (Helander, 1983) from the dissociation of salts; for example, NaCl dissociates into Na^+ and Cl^-; where salts are dissolved in water the solution is called an electrolyte.

Current flow is created by subjecting the ions in the electrolyte to an electric potential. Since each ion carries a specific electric charge, the greater the number of ions in solution, the greater the current carrying capacity; therefore, the greater the salinity of the water, the greater its fluid electrical conductivity.

Temperature influences the electrolytic conductivity of a solution, as it greatly affects the movement of ions in solution. Ions are also subject to a frictional resistance (viscous drag) as they move through the solution. The higher the water viscosity, the greater the frictional resistance or viscous drag on the ion, and the slower it moves. Water viscosity is a function of temperature, decreasing with increasing temperature. Therefore, when the temperature of the electrolyte is increased, the ions can move faster.

For salinity ranges of interest, there is for most salts a linear relationship between salt content and electrolyte conductivity. It should also be appreciated that it is common practice to measure fluid conductivity in the field, corrected to 25 °C. This may considerably over-estimate actual field conductivity. The correction factor for a potassium chloride solution is 2.2% per °C. A more complete relationship is:

$$\sigma_{water} = \sigma_{water\ at\ 25°C}\left(1 - 0.022(25 - T)\right) \tag{9.8}$$

where T is the water sample temperature.

Since ions deposit as salts on evaporation of water, there is clearly a direct relationship between the water conductivity and the total dissolved solids content. This is expressed by the empirical Equation 9.9:

$$c\sigma_w = TDS \tag{9.9}$$

where:
c is a constant of proportionality and ranges from 0.54 to 0.96, but lies mostly between 0.55
    and 0.75 (Hem, 1985),
σ_w is water conductivity (μS/cm), and
TDS is total dissolved solids (mg/L).

Table 9.2 Salinity limits for drinking water

	Conductivity mS/per/metre	Resistivity Ωm
humans	250	4.00
poultry	580	1.72
pigs	660	1.52
horses	1160	0.86
beef cattle	1720	0.60
sheep	2300	0.43

This is only a very approximate relationship and should be established for samples within limited hydrogeological areas and salinity ranges. Note that it is not possible to use the value of σ_w to predict the concentration of a particular ionic species for waters which are mixtures of several salts in solution.

The following limits for drinking water have been suggested (Wolfgram and Karlik, 1995) and are given in Table 9.2.

9.2 COMPLEX CONDUCTIVITY

9.2.1 Impact of clay type

When sediment samples have a high clay content or the pore water salinity is low (<50 mS/m), surface conduction becomes a significant component of the total bulk conductivity. The origin of this surface conduction is due to the surface charge and the formation of the electrical double layer.

The most useful description of the ion distribution at the interface is often called the Gouy-Stern double layer model shown in Figure 9.3. Here the surface of the clay is coated with water and the ions are surrounded by hydration water. The two parts of the electrical double layer (Levitskaya and Sternberg, 1994) are:

The Stern Layer – An immobile region of strongly bound water is composed of fixed ions, or oriented water dipoles, which are strongly connected to the solid surface by electrostatic forces. The water in this region has a greater density (1.2 to 2.0 kg/m^3) than free water and has greater viscosity, shear strength and elasticity and less conductivity compared to that of pure (distilled) water (Galperin *et al.*, 1993). The freezing point of water in this region is also much lower than free water and depends upon the pore size and its specific surface. For the two common clays it is:
Kaolinite −20 °C
Montmorillonite −193 °C
Water in this region does not dissolve either salt or sugar.
When contact adsorption of ions occurs, the locus of the centre of these partially unsolvated ions is called the *inner Helmholtz plane (IHP)*.
Ion, or water molecule size, determines the thickness of this dense layer. The electrical potential decreases linearly in the Stern layer.
The Diffuse Layer – Hydrated ions are attracted to the layer of contact adsorbed ions and from a second layer of closest approach to the surface of the solid (Levitskaya and

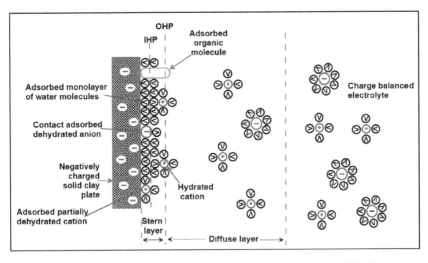

Figure 9.3 Conceptual model of an electrified interface (adapted from Kelly (1994))

Sternberg, 1994). The locus of the centres of these second layer ions is called the *outer Helmoltz plane (OPH)*. The OHP is a boundary of approach for other hydrated ions involved in thermal motion. They form a diffuse layer where ion density decreases deeper into the liquid phase. These ions are relatively mobile and may continuously exchange with other ions in the free solution. The thickness of the diffuse layer may be considerable (up to hundreds of micrometres).

The reduction in electrical potential in the diffuse layer is exponential. The potential in the liquid at the boundary between the Stern layer and the diffuse layer is called the zeta-potential (ζ). As electrolyte concentration increases, thermal molecular motion is unable to tear ions and water dipoles from the surface and diffuse them into the electrolyte. The electrical double layer contracts as a result and, if the concentration is high enough, all counterions are within the dense Stern layer and the zeta-potential reduces to zero.

For charged surfaces on which isomorphous substitution has occurred, the density of charge is independent of the electrolyte concentration; therefore, the surface conductance σ_s changes only if the average mobility of the ions changes. For surfaces of constant potential (i.e., sand), the density of charge increases with increasing electrolyte concentration. Hence, an increase in surface conductance with increasing electrolyte concentration can be expected, although this increase may be partly offset by a decreased average ion mobility (van Olphen, 1963).

Where theoretical models for electrical conduction have been attempted, they treat the surface and double layer conduction in one of the following ways:

• No conduction of the Stern layer counter ions which are fixed to the surface; conduction occurs only in the diffuse zone.
• Conduction in the Stern layer and the diffuse zone; ions are fixed to the Stern layer.
• Conduction in the diffuse zone and the Stern layer; ions are free to move.

9.2.2 Complex conductivity response

The electrical properties of water are relatively well known (Olhoeft, 1985). Pure water is an insulating material with an electrical conductivity that is strongly temperature dependent but independent of frequency below several GHz. However, small quantities of solutes can dramatically increase the conductivity.

Systems comprised of clean saturated sand have a linear response to the application of an electrical current. An irreversible migration of the electric charge occurs in response to an applied electric field. This simple system does not exist as soon as some clay is added to the mixture. Depending upon the frequency of the driving current, it takes a small amount of time for the electrons in the Stern and diffuse layers to respond to the driving current, as there is some material resistance involved in getting the electrons to move their position in the clay lattice. Figure 9.3 indicates the location and binding of the electrons and ions that result in a phase lagged response. Importantly, this movement of electric charge is reversible.

The combination of irreversible charge migration (associated with electrolytes away from charged surfaces) and frequency-dependent reversible charge migration (electric charge associated with charged surfaces) means that the electrical conductivity needs to be described by both the amplitude of the response and also the phase of the response (Brandes, 2005).

The electrical conductivity of systems containing clay becomes complex (in the mathematical sense) with a both a real component and an imaginary component – or an in-phase component and a phase-lagged component.

Both components of the electrical conductivity can be represented by Equation 9.10:

$$\sigma^* = \sigma' + i\sigma'' \tag{9.10}$$

where:
σ^* is the complex conductivity (S/m) – also referred to as the admittance,
σ' is the real part of the electrical conductivity representing the steady-state condition, and
σ'' is the imaginary (out-of-phase) component of the electrical conductivity representing relaxation processes
$i\sqrt{-1}$

When alternating currents (AC) are measured, the expression becomes:

$$\sigma^*(\omega) = \sigma'(\omega) + i\sigma''(\omega) \tag{9.11}$$

where:
$\omega = 2\pi f$ and
f = the AC frequency.

Note that in-phase conductivity (σ') is also called conductance (G) and out of phase conductivity (σ'') is also called susceptance (B).

Complex conductivity can be converted to polar coordinates, where the phase angle (Ω) can be expressed as:

$$\Omega = tan^{-1}\frac{\sigma''}{\sigma'} \tag{9.12}$$

and:

$$\sigma^* = \sqrt{(\sigma')^2 + (\sigma'')^2}$$
(9.13)

9.2.3   Electrical impedance spectroscopy

In complex conductivity processes, the effect of charge-polarization is to make the magnitude and phase of electrical impedance measurements frequency dependent. The rate at which a polarization process occurs is reflected in the frequency at which the electrical impedance changes. This property is exploited in electrical impedance spectroscopy (EIS) measurements, where the electrical impedance magnitude and phase are recorded over a range of frequencies (Brandes, 2005).

The SOLARTRON 1260 EIS equipment was used to measure the complex conductivity of clay samples from the Liverpool Plains as reported by Acworth and Jankowski (1997) and shown in Figure 9.4. These samples were from shallow cores taken at a site at Breeza on the Liverpool Plains. Note the strong frequency dependence of the smectite clay–dominated

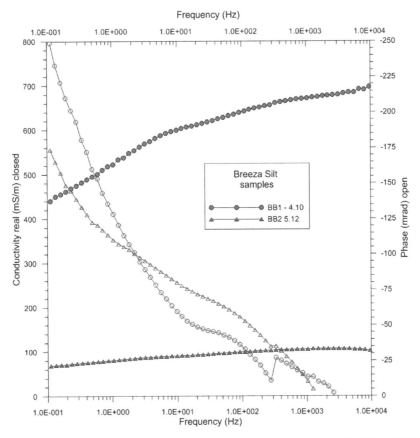

Figure 9.4 Complex conductivity spectra for a clay (BB1 4-10) and a sandy silt (BB2 5-12) from the Liverpool Plains

sample (BB1 4-10). Considerable error will be caused by the assumption that the electrical conductivity of these clays is independent of the frequency of the imposed field.

Complex conductivity over a range of frequencies are generally too time consuming to measure during field studies. Each measurement location would take several minutes to collect the necessary data. Instead, a time-domain method is used in which the decay of voltage as the current is either turned on or turned off is recorded. This is described in more detail below.

9.2.4 Electrical conductivity models

Archie (1942) showed that for high-salinity clay-free materials, there was a good relationship dependent upon porosity (Equation 9.5). However, if clay is present in the sample then non-linearity is produced into the relationship between σ_f and σ_b for low values of σ_w. This is indicated in Figures 9.5 and 9.6.

Modelling the non-linear part of the relationship has long remained a challenge, particularly in the agricultural industry (Rhoades *et al.*, 1989a–c).

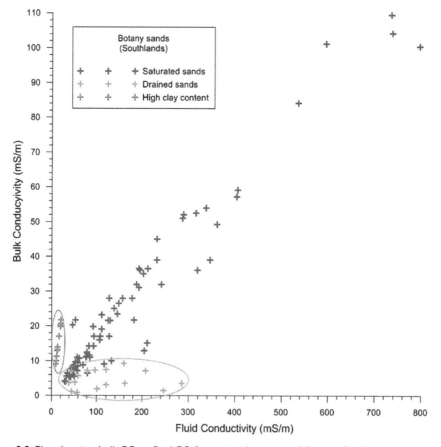

Figure 9.5 Plot showing bulk EC vs fluid EC for variously saturated Botany Sand samples

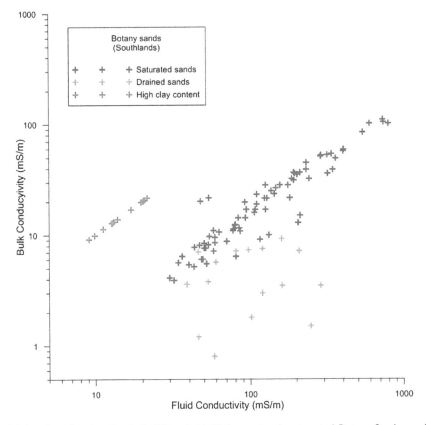

Figure 9.6 Log-log plot showing bulk EC vs fluid EC for variously saturated Botany Sand samples

The Hanai-Bruggeman effective medium theory (EMT) (Bussian, 1983) has a strong physical basis and has been used for fully saturated conditions (Kelly, 1994; Acworth and Jorstad, 2006). This model is:

$$\sigma_b = \sigma_w \phi^m \left(\frac{1 - \sigma_r/\sigma_w}{1 - \sigma_r/\sigma_b} \right)^m \qquad (9.14)$$

where σ_r is the grain surface conduction. Kelly (1994) carried out extensive tests in the Botany Sands in Sydney and showed that the values in the EMT could be represented by a porosity (ϕ) of 0.36, a geometry factor (m) of 1.51 and grain surface conduction σ_r of 0.56 mS/m.

However, this model does not perform well under unsaturated conditions and has been further modified as described by Revil (2013) and Greve *et al.* (2013). These authors modify an original expression proposed by Waxman and Smits (1968):

$$\sigma_b = \frac{S_w^n}{F} \left(\sigma_w + \frac{B Q_v}{S_w} \right) \qquad (9.15)$$

where:

S_w^n is the fraction of water occupying the pore space and n is an empirical parameter called the saturation exponent,

F is ϕ^{-m}, ϕ is porosity and m is the cementation factor,

B $(S \cdot cm^2/meq)$, and

Q_υ is usually expressed in meq/mL (1 meq/mL = 96.32×10^6 Cm^{-3} in SI units).

9.2.5 Complex conductivity processes

A full understanding of the causes of chargeability is beyond the scope of this book. Suffice to say that the chargeability can be both positive and usually is when caused by ionization in mineral deposits, but may also be negative in some expanding clays that contain moisture levels close to the liquid limit. Brandes (2005) reviews processes with particular emphasis on the negative chargeability in clays. Extensive studies by Olhoeft (1985) and Coster and Chilcott (1999) are also valuable.

The presence of a significant chargeability (in IP measurements) or complex-conductivity indicating a frequency dependence (from EIS measurements) can be used to distinguish between salty water in sand (no IP) and clays. This can be particularly valuable in a coastal environment.

9.3 ELECTRICAL POTENTIAL THEORY

Electric fields may be set up by a variety of natural processes, some of which are not as yet well understood. In many instances, the flow of groundwater forms an essential component of the physical system which generates the electrical field. The study of naturally occurring potentials forms the basis for the self-potential method of investigation (Corwin, 1987).

There are three principal mechanisms associated with these processes, two are electro-chemical and the third mechanical.

9.3.1 Liquid junction (diffusion) potentials

This potential is created by the differences in mobility of various ions in solutions of different concentrations:

$$\phi_d = -\frac{R\theta_A(I_a - I_c)}{Fn(I_a + I_c)}\log_e(C_1/C_2) \tag{9.16}$$

where:

R is the gas constant (8.31 J/°C),

F is the Faraday constant (9.65×10^4 C/mol),

θ_A is the absolute temperature,

n is valence,

I_a is the mobility of anions,

I_c is the mobility of cations, and

C_1, C_2 are solution concentrations.

The liquid junction potential for a sodium chloride solution where $I_a/I_c = 1.49$ at 25 °C is (Telford et al., 1976):

$$\phi_d = -11.6 \quad log_{10}\left(\frac{C_1}{C_2}\right) \tag{9.17}$$

where ϕ_d is given in millivolts.

When two identical metal electrodes are placed in a single solution, no potential is created. However, if solution concentrations around the electrodes change, then a potential difference is created, similar to the diffusion potential. This additional potential is called the *Nernst* potential and may act to increase the diffusion potential significantly.

9.3.2   Mineralization potentials

If two different metal electrodes are immersed in a solution, a potential is set up between them. This process, while not well understood, is thought to be the basis for the potential observed around zones of mineralization. These potentials are particularly well developed around zones containing sulphides, graphite or magnetite and are much larger than the diffusion potential described above.

Currents associated with bacteriological activity are also noted in the literature. Negative potentials of 100 mV associated with passing from a forest area to a cleared area have been noted (Telford et al., 1976). These potentials are probably related to plant root activity.

9.3.3   Electrokinetic (streaming) potential

This potential is observed when a solution of conductivity σ_w and viscosity μ is forced to flow through a porous medium. The resultant electrical potential between the ends of the flow path is given by Equation 9.18:

$$\phi_k = -\frac{\zeta \Delta P \kappa}{4\pi\mu\sigma_w} \tag{9.18}$$

where:
ζ is the adsorption (zeta) potential,
ΔP is the pressure difference,
κ is the dielectric permittivity of the solution, and
σ_w is the fluid electrical conductivity.

The quantity ζ is the potential at the double layer between the liquid and the solid (see Figure 9.3).

Research is currently underway to relate the size of the streaming potential to hydraulic conductivity. Seismic energy from a surface hammer blow or weight drop passes downward into the formation. As the seismic wave passes, there is a relative movement of fluid with respect to the porous medium. This is sufficient to set up a streaming potential (electrokinetic effect) which can be sensed at electrodes placed at the surface. Although the method appears to be theoretically possible, detailed work so far in several countries has failed to reveal a unique relationship between hydraulic conductivity and the streaming potential.

Applications

The major environmental and engineering applications of the presence of natural potentials (self-potentials) has been in the investigation of groundwater flow associated with leakage from dams; along dikes; leakage from containment structures; and delineation of flow patterns in the vicinity of landslides, wells, faults, shafts, tunnels and sinkholes (Corwin, 1987). Field data acquired around contaminant plumes indicates that chemical concentration gradients also produce significant self-potential (SP) anomalies.

High noise levels associated with human-made structures can mask natural anomalies, especially at old industrial sites often associated with groundwater contamination.

Interpretation of SP data is qualitative, using visual correlation between the observed profile or contour patterns and known or suspected seepage flow paths or other sources (Corwin, 1987).

The equipment for an SP survey includes a measuring instrument, electrodes, wire and a tool for digging electrode positions. The choice of electrode is significant if a polarization associated with the electrode-to-ground interface is to be avoided. The use of copper to copper sulphate or silver to silver chloride electrodes is required.

9.3.4 Potential due to a point source of current

The development of many electrical field techniques in geophysics has occurred based upon two assumptions: that the imaginary component of the conductivity is zero and that there is no variation of bulk electrical conductivity with the frequency of the driving current.

Consider a DC source of current inside a homogeneous space. Current will flow radially from the source outwards such that at any spherical surface S, a distance r from the source (origin) current of strength I, the current density will be:

$$J_r = \frac{I}{4\pi r^2} \tag{9.19}$$

We may write Ohm's Law in terms of the resistivity rather than the conductivity as follows:

$$E_r = J_r \rho = \frac{I\rho}{4\pi r^2} \tag{9.20}$$

The potential ϕ (or voltage V) at a distance r from the electrode is given by the integral of E_r between r and infinity:

$$\phi = \int_r^\infty E_r dr = \frac{I\rho}{4\pi r} \tag{9.21}$$

If the current electrode is at the ground surface (the surface of the half-space), the terms on the right-hand side of Equations 9.19 to 9.21 are doubled because the area of the hemisphere appears in the denominator. Equation 9.22 provides the fundamental relationship for all electrical prospecting performed at the surface of the earth:

$$\phi = \int_r^\infty E_r = \frac{I\rho}{2\pi r} \tag{9.22}$$

Field measurement of potentials

In practice, four electrodes are required to measure a potential difference at the surface of the earth.

*   A current source electrode C_1 (+),
*   a current sink electrode C_2 (−), and
*   and two potential electrodes, P_1 and P_2, to measure a potential difference between them.

The distribution of current flow between two adjacent current electrodes is shown in Figure 9.7. Note that this pattern is identical to the equipotentials in hydraulic head formed between a point source of water flowing into the ground at an injection well and water being recovered at an abstraction well. The analogue in well hydraulics will be returned to in Chapter 15.

If we adopt the convention that $\phi_{x,y}$ is the potential at the electrode position -x- due to a current source or sink at the position -y- we can write an expression for the total potential at electrode position 1, due to a source current at position 0 and a sink term at position 3, shown for the generalized configuration of Figure 9.8, as:

$$\phi_{1,0} - \phi_{1,3} = \frac{I\rho}{2\pi}\left(\frac{1}{r_1} - \frac{1}{r_2}\right) \tag{9.23}$$

where r_1 is the distance between potential electrode 1 and current source electrode 0 and r_2 is the distance between potential electrode 1 and current sink electrode 3 (as shown in Figure 9.8). The sign of the second term is negative, as the potential due to a current sink will be negative.

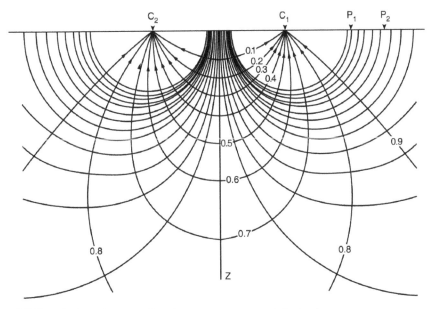

Figure 9.7 Distribution of current flow between two current electrodes

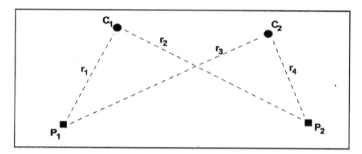

Figure 9.8 The four-electrode array

We then introduce a second potential electrode at position 2. The potential at electrode 2 due to a current source term at electrode 0 and a current sink at electrode 3 is then:

$$\phi_{2,0} - \phi_{2,3} = \frac{I\rho}{2\pi}\left(\frac{1}{r_3} - \frac{1}{r_4}\right)$$ (9.24)

where r_3 is the distance between potential electrode 2 and current source electrode 0 and r_4 is the distance between potential electrode 2 and current sink electrode 3 (also as shown in Figure 9.8). We measure a potential difference rather than trying to measure an absolute value of either of the potentials given in Equations 9.23 and 9.24. We can write an expression for the potential difference between the two potential electrodes as:

$$\Delta\Phi = (\phi_{1,0} - \phi_{1,3}) - (\phi_{2,0} - \phi_{2,3})$$ (9.25)

Substituting for the values of the potential (Equation 9.22) and gathering terms in Equation 9.25 gives:

$$\Delta\Phi = \frac{I\rho}{2\pi}\left(\frac{1}{r_1} - \frac{1}{r_2} - \frac{1}{r_3} + \frac{1}{r_4}\right)$$ (9.26)

Rearranging Equation 9.26 for the resistivity (ρ) gives:

$$\rho = \frac{\Delta\Phi}{I} \times K = R \times K$$ (9.27)

Where the K term is called the geometric factor and R is the resistance $\frac{\Delta\Phi}{I}$:

$$K = \frac{2\pi}{\left(\frac{1}{r_1} - \frac{1}{r_2} - \frac{1}{r_3} + \frac{1}{r_4}\right)}$$ (9.28)

The geometric factor is a function of the distance between the electrodes and therefore varies for each specific array configuration. The R is resistance. Modern field equipment holds the current constant and varies the voltage. The ratio of these two (resistance) is presented directly to the observer.

Apparent resistivity or conductivity

The true resistivity (conductivity) of an homogeneous half-space is given by Equation 9.27. However, the ground is very rarely homogeneous and volumes with different electrical conductivity will impact on the measured resistance, so it will vary depending upon the separation of the electrodes and the location of the electrodes with respect to the anomalous areas. In practice, an apparent resistivity will be measured over inhomogeneous ground and is normally written as ρ_a.

Ward (1990) comments as follows,

> Each such measurement (of apparent resistivity), is the apparent resistivity of an equivalent homogeneous isotropic half-space. It is not representative, necessarily, of the true resistivity of any element in the earth nor is it necessarily representative of a simple average of true resistivities but it is a useful normalisation for any given array.

The apparent resistivity does vary in a systematic manner from which an interpretation of true resistivities can often be derived. However, it is important to realize that if Archie's Law or some derivative thereof is to be used, the apparent resistivity must not be simply substituted for the true resistivity. Deriving the values of true resistivity from the measurements of apparent resistivity is a problem for inversion. It is analogous to the problem of deriving hydraulic conductivity values from pumping test results.

9.3.5 Four-electrode field methods

There are three basic methods:

- continuous separation traversing (CST),
- vertical electrical sounding (VES), and
- electrical imaging – which is essentially a combination of CST and VES.

These methods are described in the following sections.

9.4 CONTINUOUS SEPARATION TRAVERSING (CST)

CST work was carried out extensively in Africa prior to the mid-1980s where it was used to locate linear features of interest for borehole location on fractured basement rocks. Typically, four equally spaced electrodes were used as shown in Figure 9.9.

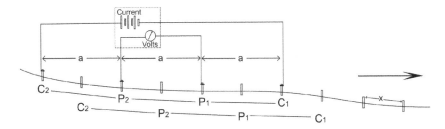

Figure 9.9 CST using the Wenner alpha array

Many different electrode array combinations have been tried for CST work. Wenner (1912) determined that there were only three ways in which the four electrodes could be connected. These were the alpha (CPPC), beta (CCPP) and gamma (CPCP) configurations shown in Table 9.3. The depth of penetration for an array (Acworth and Griffiths, 1985) is defined as the depth at which 50% of the contribution to the electrical potential arises from ground above this depth and 50% below. These depths have been determined by numerical analysis of homogeneous ground. The depth of penetration and the geometric factor for each of the tripotential arrays is given in Table 9.3.

An important benefit arises from measuring the alpha, beta and gamma resistances in the field. It can be shown that the sum of the beta and gamma resistances will be equal to the alpha resistance for any electrode configuration over any inhomogeneous or anisotropic ground. This property can then be used as a field check on data quality and is an important provision of field quality control.

An example of a profile over weathered basement rock collected using the tripotential configuration is shown in Figure 9.10.

Table 9.3 Depth of penetration and geometric factors for linear tripotential arrays where the electrodes are separated by the distance 'a'

Array	Configuration	Depth of penetration as a function of a	Geometric factor
alpha	C1-P1-P2-C2	0.519	$2 \pi a$
beta	C1-C2-P2-P1	0.416	$6 \pi a$
gamma	C1-P1-C2-P2	0.594	$3 \pi a$

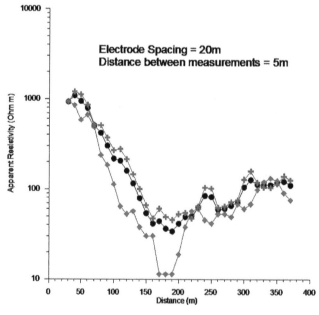

Figure 9.10 Tripotential profile data over weathered basement. Alpha (circle), beta (diamond) and gamma (cross) data is shown

The difference in depth of investigation of the tripotential array configurations produces definitive results when the array is passed over a zone of lateral inhomogeneity. A characteristic crossover occurs of the beta and gamma resistivities as shown in Figure 9.10. Examination of the full tripotential data set can therefore reveal significantly more information than measurement of the simple alpha data alone; however, it takes three times as long to acquire the data set! CST provides information on the presence of lateral variations in subsurface bulk electrical conductivity. The data can be collected using the CST method (Figure 9.9), but the data is more readily acquired using EM profiling equipment such as that produced by GEONICS (McNeil, 1980). The main reason for this is that there is no requirement for electrodes connected to the ground. EM profiling has developed into a significant field in its own right and is described later.

It is important to realize that neither CST nor EM profile work can be quantitatively interpreted. There is no method available to invert the data into zones of true conductivity. CST/EM profiling are excellent tools to identify areas where inhomogeneity exists, but further field work is required to acquire data capable of inversion.

9.5 VERTICAL ELECTRICAL SOUNDING (VES)

It should be clear that CST/EM profiling over uniform or homogenous ground will return resistance values that are all equal. The apparent resistivity can be calculated using the geometric factor. As electrode spacing doubles, the resistance will decrease by half so that the apparent resistivity remains the same. In this case, as noted above, the apparent resistivity will be the same as the true resistivity. CST in this environment is not beneficial! If the ground consists of horizontal plane layers (Figure 9.11), then CST or EM profiling will return apparent resistivities that are all equal as long as the electrode separation is constant.

If the electrode separation is changed, but the layers remain horizontal, then sufficient apparent resistivity data can be collected to allow the apparent resistivity data to be inverted. In this way, the original thickness of the layer and the true resistivity of each layer can be recovered. The apparent resistivity data is collected at a succession of different electrode separations to form a vertical electrical sounding (VES). In reality, it is only possible to determine three or four layers at a maximum because there is a problem with equivalence.

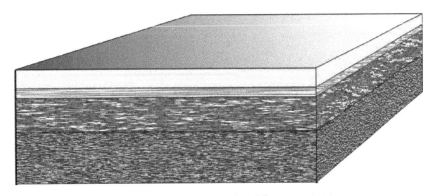

Figure 9.11 A set of horizontal plane layers, each with a different resistivity

Variation from the condition of plane layering will significantly affect the accuracy of the interpretation.

A VES is conducted using four electrodes in much the same manner as electrical profiling. Instead of moving the four electrodes along a line and keeping the separation of the electrodes constant, in VES the spacing between the electrodes is expanded about a fixed centre point. Measurements of apparent resistivity are frequently made at 0.5, 1, 2, 4, 8, 16, 32, 64, 128 and 256 m separation. At 256 m electrode separation, the total distance between the two outer electrodes is 768 m. Clearly, the larger the electrode separation, the greater the chance of deviation from the condition of plane layering. For this reason, the accuracy of the method for determining conditions at depth reduces substantially.

The depth of penetration for a four-electrode array is approximately equal to half the electrode spacing (Table 9.3). The field data is plotted in the same manner as Figure 9.12.

In Figure 9.12, a forward modelling routine has been used to calculate what the apparent resistivity would be for a particular model of plane layers and depths. The forward modelling method uses a technique established by Ghosh [1971a, b]. It is important to recognize that the apparent resistivity values are only the same as the true resistivity values for very small electrode spacings. For this reason, use of apparent resistivity field data must not be

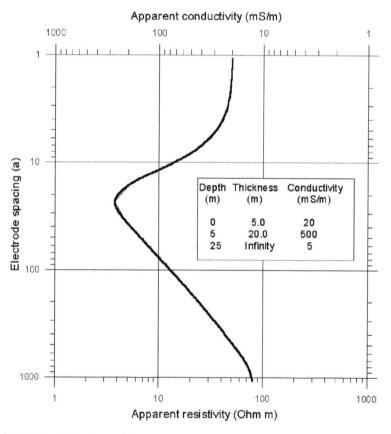

Figure 9.12 Vertical electrical sounding curve

Figure 9.13 Plane layering demonstrated by an electrical conductivity image measured over uniform ground over acid-sulphate soils in Hexham, NSW

used as an indication of true resistivity without a detailed and well-controlled (by drilling or sampling) inversion. Even when this is the case, the problems of equivalence are still difficult to overcome.

The presence of lateral variation in resistivity using VES methods was described by Barker (1981) in his offset sounding method. If VES data collection is to be carried out, then this approach to field collection and interpretation should be followed.

It is important to appreciate that the presence of an anomaly detected with a profiling device (CST or EM) is strong evidence that the earth is not comprised of plane layers. If that is the case, then it means that a VES should not be attempted in this area. It is unfortunate that there have been many groundwater resource surveys on fractured basement rocks in Africa that appear to ignore this dictum. A VES arranged along the axis of a fracture zone mapped out with CST cannot indicate plane layering, and therefore a VES should not be attempted.

Data that provide a good VES interpretation can only be collected in areas where CST give constant values. This point is demonstrated by the electrical conductivity image (Figure 9.13) collected over layered recent sediments north of Newcastle, NSW. A CST at any given spacing will produce the same value of apparent resistivity, and the interpreted image data gives a model similar to Figure 9.11.

9.6   ELECTRICAL CONDUCTIVITY IMAGING (ECI)

The results obtained with CST give an averaged 1D representation of bulk resistivity variation at a single electrode separation and approximately constant depth of penetration. VES interpretation will allow a model of plane layers (no lateral variation) and depths to be established. A combination of techniques is required to interpret more real-world problems.

Electrical conductivity imaging (ECI) (Acworth and Griffiths, 1985; Acworth, 1999, 2001a) is one way to approach this problem. From the perspective of fieldwork sequence, it is usual to carry out electromagnetic surveys to provide a quick map of apparent bulk electrical conductivity so that the optimum location for an ECI can be chosen. However, rather than break the description of electrical measurements made with electrodes and cables (galvanic techniques), ECI is described here, and a full description of electromagnetic techniques is delayed to Section 9.13.

The ECI approach uses repeated profiles at increasing electrode separation across an area to build up an apparent electrical conductivity image section. In effect, CST data is collected at multiples of the base electrode separation using multi-core cables and computer-controlled resistance measuring equipment. This routinely used technique has now replaced CST/VES as the most efficient method of electrical resistivity investigation.

The ECI method can be used in 2D or 3D. The most common application is for 2D collection and interpretation. The 2D image (in the x-z domain) of apparent resistivities can be inverted to produce a section of true resistivity where the resistivity in the third dimension (y) is assumed to be constant. The ground model is thus one of infinitely long (y direction) prisms of known resistivity. These values of known resistivity can then be interpreted using Archie's Law, or the other models derived from Archie, to investigate clay content, salinity and porosity.

9.6.1   Data acquisition equipment

Electrical resistivity imaging equipment is available from a number of suppliers including ABEM in Sweden, OYO in Japan and Geometrics in the USA. All systems make use of multicore cables with a minimum number of 21 nodes. Some systems can address as many as 128 nodes. There are two basic methods for equipment design. Either the required switching is carried out in the main instrument and the cables are connected to each electrode, or the switching is done at the electrode, making use of a small electronics pad at each electrode. There are advantages and disadvantages to either approach.

The ABEM Lund system uses four cables of 21 active electrodes. The outer cable set is joined to the inner cable set by a connector (this can be seen in Figure 9.16) that allows every other electrode on the far cable and all electrodes on the near cable to be addressed at the same time.

ABEM Lund electrical imaging equipment is shown deployed in Figure 9.14.

The current is applied to the electrodes using a modified square wave as shown in Figure 9.15. The duration of each phase of the signal is set as input at the beginning of a survey, allowing great control over the data acquisition.

Figure 9.14 Cables, electrodes and a resistivity measuring device (ABEM Terrameter) in use (Photo: Ian Acworth)

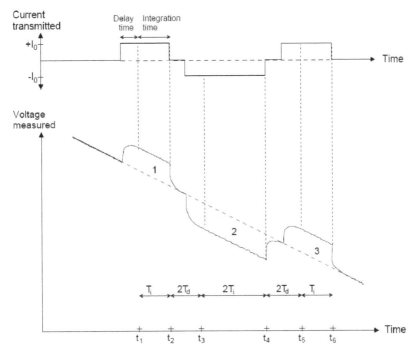

Figure 9.15 The response to a variable input of current to the ground which is switched between positive and negative to avoid electrode polarization. In this example, a spontaneous potential variation requires backing out. This is carried out automatically

Figure 9.16 Small amount of drilling mud placed around the electrodes at the start and end of the two cables used in the ABEM LUND system. Drilling mud has been prepared and is dispensed from the plastic bottle. The electrode connector is shown on the left – although not connected in this image (Photo: Ian Acworth)

9.6.2   Electrode resistance

All galvanic methods require that there is good electrical contact between the electrode and the earth. The current injected into the earth is limited by the resistance of the ground (contact resistance) in the immediate vicinity of the electrodes and by the power supply available. There is a rapidly expanding volume of earth material available for the current to flow in as the current moves out from the electrodes. The resistance to current flow therefore decreases markedly away from the electrodes.

The efficiency of the current injection can be improved by careful preparation of the electrode sites. There are three methods available to achieve this.

1   Lowering the ground resistance in the immediate vicinity of the electrode by placing a wet clay paste around the electrode. The mud can be prepared using driller's supplies and added salt. A mixture that can be dispensed using a 'squeezy' bottle is the most effective in the field. Two or three bottles are frequently sufficient for a survey, but it is advisable to add the necessary components to the field equipment so that further supplies can be prepared if required. It is recommended that mud be used unless the ground is wet from recent rain, as even clay-dominated soils can dry out to such an extent that the electrode does not make sufficient contact with the ground for low-noise data to be collected.
2   By using multiple electrodes in parallel. It may be necessary to use as many as nine electrodes wired together in a small box configuration to significantly reduce the resistance. A saline solution would also be added. This method of approach can only be adopted for electrodes which are widely separated from each other (<10 m).
3   By the use of a sheet of metal buried in saline-soaked soil or in a clay paste.
4   In dry sand, it may be necessary to use long electrodes to penetrate to the deeper damper sand. This can only be used for the electrodes at distances of 32 m and above. Use at small spacings violates the assumption that the electrodes are point sources of current. They become line sources.

9.6.3   ECI protocols

For many years there were only 3 electrode configurations that were used routinely. These were:

- Wenner – equispaced array C1-P1-P2-C2
- Schlumberger C1- – -P1-P2- – -C2
- Dipole dipole C1-C2 – - – - – P1-P2

This situation has evolved with the availability of equipment that can measure four or more potential pairs for each current pair. Loke (2016) provides a complete review of all these options and is also a valuable general reference for electrical field work and interpretation.

The gradient array has been optimized for use with the Terrameter LS system (ABEM) for use with sets of four multicore cables (Dahlin and Zhou, 2006). This provides a very rapid method for profiling by completing all measurements on the first of four cables before working on the remaining measurements at the other three cables. The first cable can be disconnected and laid out in front of the line while the remaining measurements

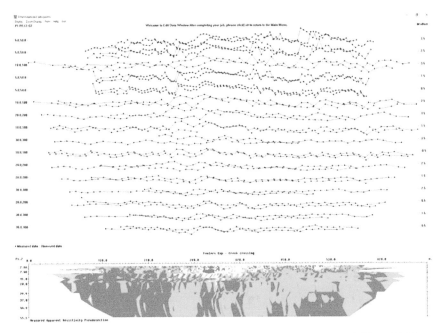

Figure 9.17 Plot of field data collected using the gradient array (Dahlin and Zhou, 2006) with the apparent resistivity section displayed beneath (Loke, 2016)

are completed, thus minimizing down time. The gradient protocol addresses four potential pairs simultaneously while a single current pair is activated. This increases survey output by a factor of four!

An example of a gradient array data set is given in Figure 9.17.

9.7 INDUCED POTENTIAL (IP) MEASUREMENTS

Chargeability is a time-domain complex-conductivity measurement. A variety of names are used for the technique but, in general, it is called the *induced potential* method. It has the advantage over a frequency-domain method as it is less prone to electromagnetic coupling effects, because there is no incident current during impedance decay measurements. Time-domain complex conductivity can be transformed into the frequency-domain using a Fourier transform (Telford *et al.*, 1976).

The time-domain complex-conductivity measurements can be made by the same multi-core cables and equipment as electrical images (Figure 9.14). Electrodes can be stainless steel, and the same arrangements of electrodes can be used for collecting data. Resistivity data and induced potential data can be collected at the same time, if the sampling times used for the resistivity data collection are modified to enable the decay of the potential when the current is switched off to be accurately sampled. Induced potential data can be inverted using similar software approaches (Loke, 2016) to those for resistivity.

The chargeability is calculated using Equation 9.29. Chargeability is analogous to capacitance where a capacitor is used to store electrical energy:

$$m = \frac{1}{V_0} \int_1^2 V(t)dt \tag{9.29}$$

The process is illustrated in Figure 9.18, where the driving voltage and the transmitted current are shown for an experiment over fractured and mineralized metamorphosed lavas and ignimbrites. The driving voltage and the potential measured between three potential electrode pairs are shown in Figure 9.19.

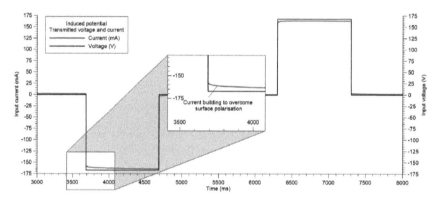

Figure 9.18 Driving voltage and the transmitted current for an induced potential signal

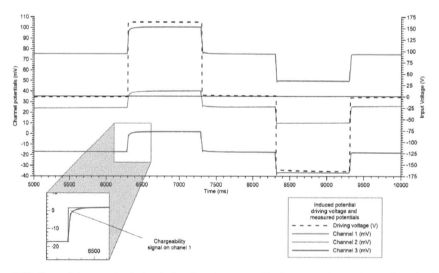

Figure 9.19 Part of the received signal showing the potentials between three electrode pairs for a current cycle. The chargeability is highlighted

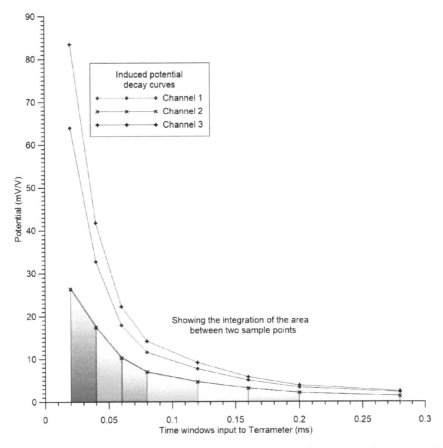

Figure 9.20 The recorded induced potential data at eight time slices for the same three potential pairs shown in Figure 9.19. The green shaded area represent the integrated areas beneath the curve for Channel 2. These values, summed together, give the total chargeability (Equation 9.29 for this electrode configuration.)

The IP signal measured between the same three electrode pairs is shown in Figure 9.20. There were eight time windows selected and the green shaded segments in Figure 9.20 depict the integrated IP signal between the times selected. Appropriate time steps can be determined by experimentation and examination of the recorded signal (Figure 9.18).

9.8 MODELLING OF ECI DATA

A typical set of field data and the interpreted output is shown in Figure 9.21. The field work took approximately 3 hours to acquire and comprised 350 m of image line at a base electrode separation of 5 m. The contoured apparent conductivity field data is shown in the top section of Figure 9.21. The middle and lower sections form a part of the data inversion method. Surface conductivity profiling using an electromagnetic device across the same area is shown in Figure 9.45. Data collected for resistivity inversion or induced potential

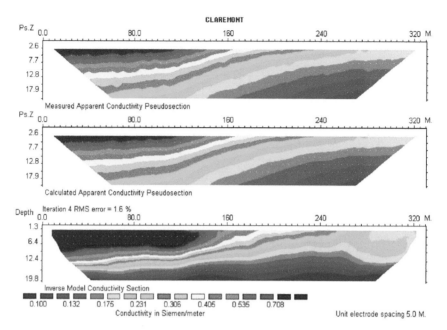

Figure 9.21 Electrical conductivity image close to Claremont off Cattle Lane on the Liverpool Plains

inversion can be interpreted at the same time by the RES2DINV or RES3DINV packages developed by Loke (2016).

9.8.1 Introduction

Full three-dimensional modelling of electrical conductivity data can be carried out using finite difference methods. This approach is analogous to the forward modelling of VES data using a convolution of the resistivity transform with filters as reported by Koefoed (1979), and allows detailed investigation of three-dimensional conductivity distributions.

9.8.2 Three-dimensional finite difference model

The details of a three-dimensional finite difference model were published by Dey and Morrison (1979). Their derivation starts with Ohm's Law:

$$J = \sigma E \tag{9.30}$$

where:

J is the current density Am^{-2},

σ is an isotropic conductivity distribution S/m, and

E is the electric field intensity $V\ m^{-1}$.

As stationary electric fields are conservative, the electric field is the gradient of the scalar potential ϕ, therefore:

$$E = -\nabla\phi \tag{9.31}$$

where ϕ is the electric potential in volts. Thus we get:

$$J = -\sigma\nabla\phi \tag{9.32}$$

If the assumption is made that electric charge is conserved within the volume of interest, then applying the continuity equation, we can write:

$$\nabla \cdot J = \frac{\partial\rho}{\partial t}\delta(x)\delta(y)\delta(z) \tag{9.33}$$

where ρ is the charge density specified at a single location in Cartesian x-y-z space by the dirac delta function (δ).

Substituting Equation 9.30 and Equation 9.31 into the left-hand side of Equation 9.33, we can write, for a generalized 3D space:

$$-\nabla \cdot (\sigma_{(x,y,z)}\nabla\phi_{(x,y,z)}) = \frac{\partial\rho}{\partial t} \cdot \delta(x - x_s)\delta(y - y_s)\delta(z - z_s) \tag{9.34}$$

where (x_s, y_s, z_s) are the coordinates of the current electrode.

Over a volume element containing a source term, the right-hand side of Equation 9.34 can be written as:

$$\frac{\partial\rho}{\partial t} \cdot \delta(x - x_s)\delta(y - y_s)\delta(z - z_s) = \frac{I}{\Delta\Phi} \cdot \delta(x - x_s)\delta(y - y_s)\delta(z - z_s) \tag{9.35}$$

Equations 9.34 and 9.35 can then be combined to give:

$$-\nabla \cdot (\sigma_{(x,y,z)}\nabla\phi_{(x,y,z)}) = \frac{I}{\Delta\Phi} \cdot \delta(x - x_s)\delta(y - y_s)\delta(z - z_s) \tag{9.36}$$

Equation 9.36 can be solved for an inhomogeneous conductivity distribution by standard finite difference application. A three-dimensional representation for an interior node is shown in Figure 9.22.

The finite difference representation of the partial differential terms arising from an expansion of Equation 9.36 for a homogeneous electrical conductivity distribution follows the form shown in Equation 9.37. The subscript '0' represents the approximation about node '0' in the mesh shown in Figure 9.22.

$$\left(\frac{\partial^2\phi}{\partial x^2}\right)_0 = \frac{\frac{[\phi_1 - \phi_0]}{\Delta x_{01}} - \frac{[\phi_0 - \phi_3]}{\Delta x_{03}}}{\frac{[\Delta x_{01} + \Delta x_{03}]}{2}} \tag{9.37}$$

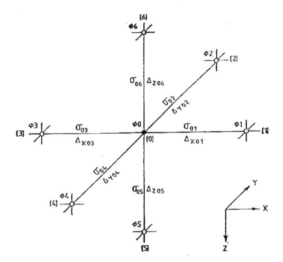

Figure 9.22 Three-dimensional discretization of an inhomogeneous conductivity distribution

A more generalized representation in three coordinates for an inhomogeneous mesh and incorporating a source term is shown in Equation 9.39.

$$\frac{2}{\Delta x_{01} + \Delta x_{03}} \left(\sigma_{x01} \frac{\phi_1 - \phi_0}{\Delta_{x01}} - \sigma_{x03} \frac{\phi_0 - \phi_3}{\Delta x_{03}} \right)$$

$$+ \frac{2}{\Delta y_{02} + \Delta y_{04}} \left(\sigma_{y02} \frac{\phi_2 - \phi_0}{\Delta_{y02}} - \sigma_{y04} \frac{\phi_0 - \phi_4}{\Delta y_{04}} \right)$$

$$+ \frac{2}{\Delta z_{05} + \Delta z_{06}} \left(\sigma_{z05} \frac{\phi_5 - \phi_0}{\Delta_{z05}} - \sigma_{z06} \frac{\phi_0 - \phi_6}{\Delta z_{06}} \right) + S_{(x,y,z)}$$

$$= 0 \qquad\qquad\qquad\qquad\qquad\qquad\qquad\qquad\qquad\qquad (9.38)$$

This equation can be rearranged and the terms collected to form an N by N matrix equation in the N unknown potentials. The boundary conditions require special consideration but can be implemented within the matrix representation. The method presented is similar to that used for a simple groundwater model and uses a node-centered mesh. The block centered-mesh implementation (Dey and Morrison, 1979) is more efficient in terms of computer memory.

Inversion of the matrix and back substitution with a single source term (Equation 9.39) at a time allows the potential to be calculated throughout the model domain for a source term at any position. The source term is then moved to the next position and the potentials that result for this current source calculated.

Although all potentials are calculated for all parts of the mesh, only those potentials that represent the position of electrodes are stored in an $n \times n$ array for later processing. The required potential differences are then calculated by forming the potential sums as described by Equation 9.25 for the alpha array, for example.

The full three-dimensional forward modelling requires major amounts of computer storage for the coefficient matrix which represents any sensible problem. For this reason, Dey and Morrison (1979) also investigated a pseudo three-dimensional approach.

9.8.3   Two-dimensional finite difference model

In this model, the degree of conductivity variation was limited to two dimensions, -x- and -z-. This is equivalent to considering models which do not change along the strike direction with a conductivity variation as shown in Equation 9.39:

$$\frac{\partial}{\partial y}(\sigma_{(x,y,z)}) = 0 \tag{9.39}$$

The potential formulation has to remain three-dimensional, as it is not possible to represent a point source of current in 2D.

Equation 9.40 represents the two-dimensional formulation:

$$-\nabla \cdot (\sigma_{(x,z)}\nabla\phi_{(x,y,z)}) = I_{x,y,z} \tag{9.40}$$

The algorithm in general use (Dey and Morrison, 1979) uses a Fourier transform of Equation 9.40 and solves for the potentials in transform space. The three-dimensional potential distribution $\phi_{(x,y,z)}$ due to a point source of current at (x, y, z) over a two-dimensional conductivity distribution $\sigma_{(x,z)}$ is reduced to a 2D transformed potential $\phi_{(x,Ky,z)}$. The Fourier transform of Equation 9.40 subject to these constraints is given in Equation 9.41:

$$-\nabla \cdot (\sigma_{(x,z)}\nabla\phi_{(x,K_y,z)}) + K_y\sigma_{(x,z)}\phi_{(x,K_y,z)} = I_{(x,K_y,z)} \tag{9.41}$$

Equation 9.41 is solved for a number of values of K_y in transform space. It is normally possible to use only six or seven values before performing the reverse Fourier transform back to -x-y-z- space. This is carried out as an integration.

The method adopted is to invert the coefficient matrix formed by the two-dimensional finite difference distribution and to back substitute the inverted matrix with the required current source terms one at a time. The potential distribution for each source term is then saved and converted to an -x-y-z- value by the reverse Fourier transformation. The individual terms $\phi_{(0,1)}$, $\phi_{(0,2)}$, $\phi_{(1,2)}$ are thus created and combined using the equations developed above.

Considerable savings in computer time are therefore achieved as the matrix of transform potentials is only 2D rather than 3D and only has to be inverted for the six or seven values of the transform variable for each source term.

9.8.4   Forward modelling

Any 2D conductivity distribution can be modelled using this approach and the apparent conductivity image created for a particular electrode array. The Wenner equispaced electrode method is readily modelled, as the constant electrode separation can easily be incorporated into the finite difference distribution. Modelling has a significant advantage using

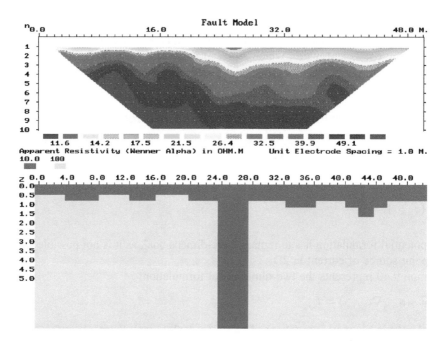

Figure 9.23 Model conductivity results produced by a 2D forward model using the RES2DMOD code written by Loke (2016)

this approach in that the infinite boundaries can be represented by using a stretched mesh on the boundaries of the modelled area.

Free software for DOS applications is available at http://www.abem.se/. The software is called RES2DMOD. An example of the forward model results is shown in Figure 9.23.

9.8.5   ECI inversion methods

One method for inversion of the ECI data is to trial a large number of forward models to find the model bulk electrical resistivity distribution that best matches the field data. This can be very time consuming, and some method of converging on a solution is required. The problem was solved by Loke (1995), who first developed a method for transforming the apparent resistivity field results using a filter based upon the signal contribution sections. This transformed data set gives a good estimate of the real distribution of bulk resistivity and acts as the starting model for the 2D forward model. A least squares fitting procedure is then used to converge the model data to the field data (Loke, 1995; Loke and Barker, 1996). A limited use software program called RES2DINV is also available for free download at http://www.abem.se/.

The output from the forward model shown in Figure 9.23 has been used as the input to the inversion software (RES2DINV) to produce the results shown in Figure 9.24. The recovery of the initial model is quite good. Further refinement can be made to the inversion to even better determine the model.

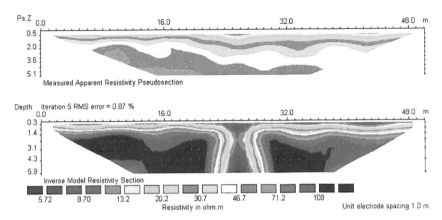

Figure 9.24 Inversion results for a dike model using the RES2DINV software (Loke, 2016)

Loke (1999) provides a detailed manual on the interpretation of electrical images using his software. This can be downloaded from the ABEM site as well. In general, the interpretation method provides three sections as shown in Figure 9.21.

1 The first (top) image is the contoured field data showing the apparent resistivity or conductivity data.
2 The bottom image is a distribution of true bulk electrical resistivity or conductivity distribution derived from the optimization. It is this lower section that provides a real indication of the ground conditions beneath a site and from which values of true resistivity can be used for further interpretation using the various models of electrical conductivity.
3 The central image is derived by forward modelling based upon the conductivity distribution shown in the lower image. This is an image of apparent conductivity or resistivity and should be compared to the field data in the top image. The inversion method seeks to produce a forward model (central image) that most closely resembles the field data (top image). In Figure 9.21, the method produced a fit after only four iterations, where the root mean square (RMS) error between the model apparent conductivity data and the field apparent conductivity data was only 1.6%. The model distribution (bottom image) is therefore considered to be a good solution to the field data (top image).

9.9 EXAMPLES OF ECI APPLICATION: BLACK COTTON SOILS

The moisture and salt content held in the deep black cotton soils, that have been derived from weathering calc-alkali basalts volcanics on the Liverpool Ranges of NSW, are of great significance for agricultural management.

9.9.1 Cattle Lane studies

Long-term monitoring has been carried out in many areas but frequently with little understanding of the processes occurring. Extensive work at Cattle Lane was published by

Figure 9.25 General location of the ECI line along Cattle Lane adjacent to a dryland cotton crop (Photo: Ian Acworth)

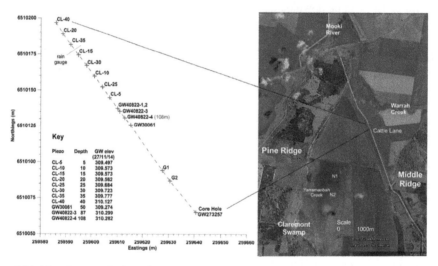

Figure 9.26 Map showing the bore locations

Acworth and Jankowski (1997), Acworth (1999), Acworth and Jankowski (2001), Acworth *et al.* (2015b), and Timms *et al.* (2018) that frequently used the ECI as a building block in the establishment of a conceptual model for the site.

Acworth *et al.* (2015b) presented a very detailed comparison of electrical image data with down-hole geophysics and the physical properties of recovered core from the Liverpool Plains. The electrical image is presented in Figure 9.27. In this investigation, the high bulk electrical conductivity proved to be due to the inclusion of salt with wind-blown (aeolian) silt during the last ice age. The recovery of sand grains from the core allowed optically stimulated luminescence measurements to be made from which the core could be dated. The base of the core at 30.5 m proved to be about 112 ka old. The general location of the ECI is shown in Figure 9.25.

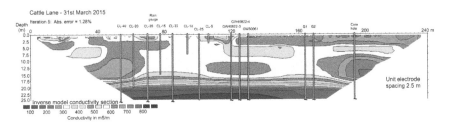

Figure 9.27 Inverted ECI data for a part of the Liverpool Plains, NSW, where silts with a high bulk electrical conductivity overlie lower EC material. Individual piezometers were drilled (shown) to check the interpretation

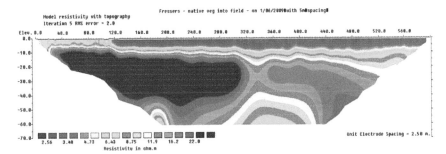

Figure 9.28 Electrical image of the ground adjacent to and beneath a flood irrigated field. Data collected by Anna Greve and interpreted using RES2DINV (Loke, 2016)

9.9.2   Change in soil moisture beneath a growing crop using time-lapse ECI

Electrical image techniques can be seen in Figure 9.28 to very clearly demarcate areas that have been wetted up by flood irrigation beneath an irrigated field. The edge of the ECI between 0 m and 120 m very clearly has a higher resistivity than the ground beneath the field. Importantly, the depth to which the irrigation water has penetrated, indicated by the low resistivity values, extends to approximately 8 m. This indicates significant downward leakage below the flood-irrigated field. The ECI has proven to be a very valuable investigation technique in irrigated agriculture.

Acworth *et al.* (2005) used time-lapse surveys at the Hudson agricultural trial site on the southern Liverpool Plains. ECI data were collected over the site using an electrode separation of 1.25 m and the measurements repeated six times during the 2000–2001 growing season. Lucerne was planted into the central plot (Plot 6) when the entire soil profile was wetted up (having a low electrical resistivity) in October 2000. The plots on either side were left unchanged with soil at the left of the line occupied by natural grass and soil at the right planted to grass. As the crop grew, the soil moisture was depleted and the soil compressed in the plot beneath the lucerne. It is significant that the lucerne dried the profile to nearly 3.5 m depth (Figure 9.29) and that the depression in the soil profile caused by the

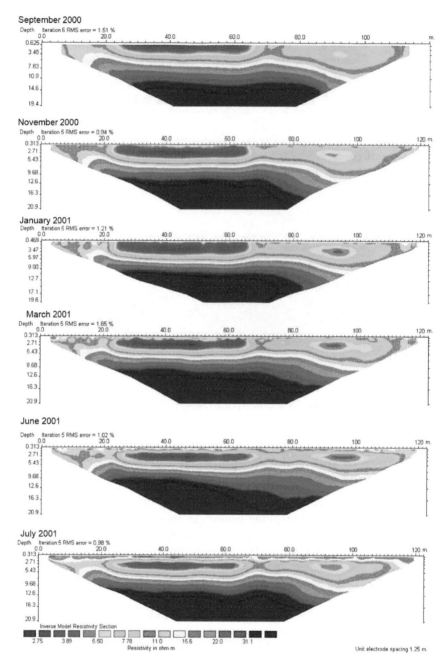

Figure 9.29 Time-lapse ECI data beneath an agricultural trial plot growing the deep-rooted crop lucerne (adapted from Acworth *et al.* (2005))

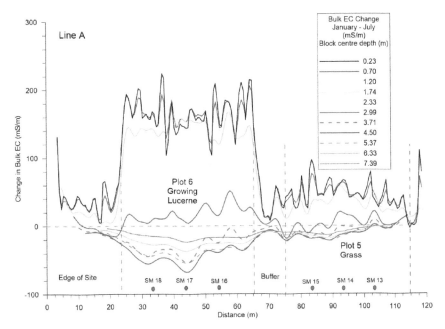

Figure 9.30 Changes in electrical resistivity beneath the profile (Figure 9.29) derived from inversion results (adapted from Acworth *et al.* (2005))

compaction of the soil in response to moisture removal took several years to disappear. The detailed changes in soil resistance along the profile were extracted from the resistivity inversion results and are presented in Figure 9.30; details of the site can also been seen in this figure.

9.9.3   Electrical image analysis of soil moisture change in 3D

Greve *et al.* (2011) have shown how 3D electrical imaging techniques can be used to map moisture change in swelling and cracking clays beneath an irrigation field. A time-lapse sequence of 3D images are shown in Figures 9.31 and 9.32. Four bundles of electrodes were prepared and placed in augered holes around a square of 1 m. Time-lapse measurements were made over a field with deep self-mulching soil containing smectite silt that was initially without cracking and the data (Figure 9.31) inverted using the RES3DINVERT program (Loke, 1999). The experiment was then repeated over a similar soil that had developed deep cracks. The influence of the cracks and the way in which the flood irrigation water fills the cracks is clearly seen in Figure 9.32.

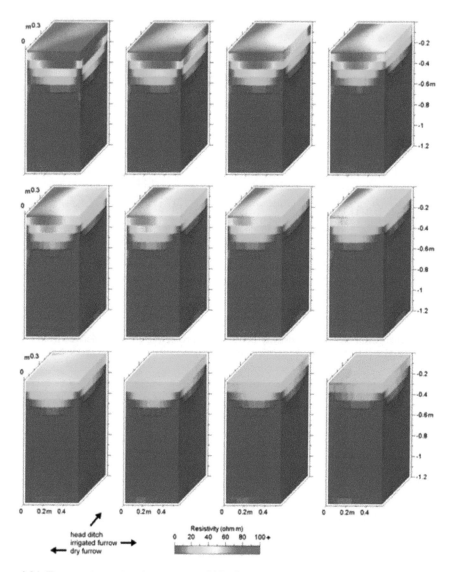

Figure 9.31 Twenty-minute time-lapse series of 3D electrical images during an irrigation event over a soil that had no cracks before the irrigation (data collected and interpreted by Anna Greve using RES3DINV (Loke, 2016)

9.10   COASTAL GROUNDWATER STUDIES

9.10.1   Change in bulk electrical conductivity in a coastal sand unit traversed by a tidal creek

In this example, an electrical image line was carried out across a tidal creek at Hat Head in northern New South Wales. The electrodes were suspended from a cable suspended across

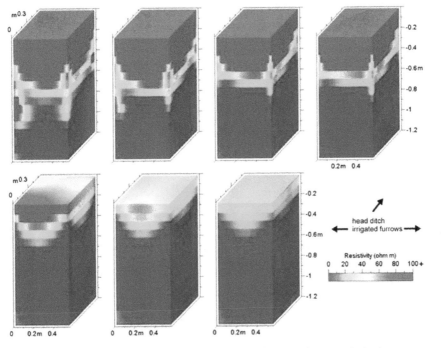

Figure 9.32 Twenty-minute time-lapse series of 3D electrical images during an irrigation event over a cracked clay soil. Water at first runs down the cracks (Greve *et al.*, 2010)

the creek (Acworth and Dasey, 2003). The tidal variation was approximately 1.8 m during flood tides. In Figure 9.33, the top image is a contoured section of the observed field apparent resistivity distribution. The central image is the apparent resistivity produced by the distribution of bulk resistivity shown in the bottom image. The 2D inversion routine was used for this data set (RES2DINV) as there was a reasonable assumption that the electrical conductivity did not vary significantly parallel to the creek. The option to show the exact boundary (in the x-z domain) of the resistivity distribution is presented as a true indication of the model distribution. The use of a contoured section for the lower plot produces a visually more appealing result but obscures the accurate values and boundaries.

The data collection is shown in progress in Figure 9.34. The extension of the work Dasey (2010) using buried electrodes (Figure 9.35) allowed a bulk electrical cross-section to be measured at the top of the beach using two strings of buried electrodes. Figure 9.36 shows the drilling preparation of the bore for installation of the string of electrodes and Figure 9.37 shows the 18 m long string being installed.

9.10.2 Cross-borehole imaging

An example of the data collected at the top of the beach between two bundles of electrodes is shown in Figure 9.38. The interpretation was cross-checked against the fluid electrical conductivity of water samples recovered from bundled piezometer tubes installed in the

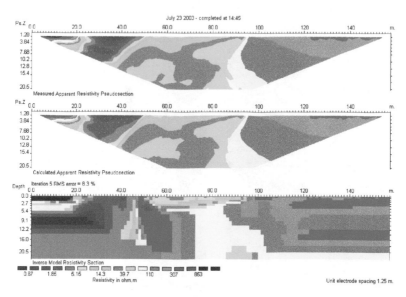

Figure 9.33 ECI data over a saline creek in unconsolidated sands at Hat Head, Northern NSW

Figure 9.34 ECI data collection across a tidal creek (Photo: Ian Acworth)

Figure 9.35 String of sample tubes and electrodes prepared for installation in a borehole (Photo: Ian Acworth)

Figure 9.36 Drilling with hollow-stem augers in coastal sands to prepare for installation of a bundle of electrodes and sample tubes (Photo: Ian Acworth)

Figure 9.37 Installation of the 18 m long bundle into the borehole (Photo: Ian Acworth)

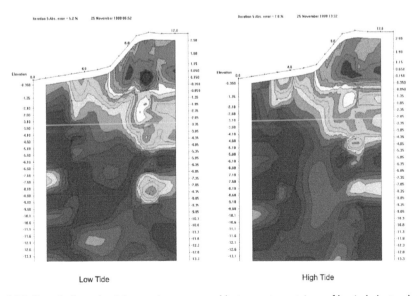

Low Tide

High Tide

Figure 9.38 Electrical conductivity section measured between two strings of buried electrodes at the top of an ocean beach. High-tide and low-tide conditions are shown

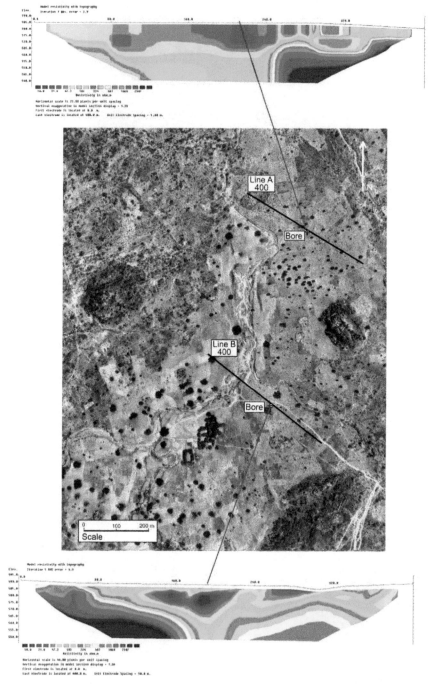

Figure 9.39 Location of two ECI at Bauchi, Northern Nigeria

same bores. The saline interface is clearly seen extending inland beneath a tongue of fresh water discharging at the beach.

9.11   ECI OVER FRACTURED ROCK

Resistivity has been widely used to determine the optimum locations to drill for water in fractured rock (Acworth, 1981, 2001a). Two ECI data sets are shown in Figure 9.39 overlain on an aerial photograph of the area. The locations of the bores shown were first determined using a number of CST lines and then hand contouring the data set. A linear low resistivity feature was discovered that served as the locus for the bores. ECI were then carried out to better determine the structure of the lineament. It is of interest to note that the lineament did not relate to the present surface drainage or to the orientation of major fracture sets in the surrounding inselbergs. This feature could only be detected by electrical resistivity surveys.

9.12   IP MEASUREMENTS

Data from an investigation using an ABEM Terrameter LS with two 32-takeout cables, steel electrodes and the gradient array were collected at the Wellington Research Station of UNSW where copper mineralization has occurred in Devonian metasediments and volcanics. An example combined resistivity and induced potential interpretation is shown in Figure 9.40.

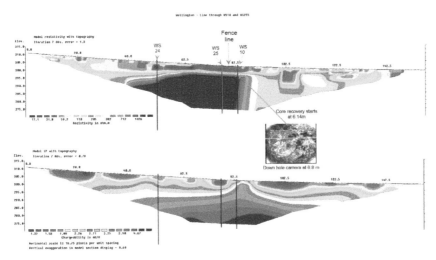

Figure 9.40 Joint inversion of resistivity and induced potential data from Devonian metasediments at Wellington, NSW. The bores WS 10 and WS 25 were cored and a substantial thickness of copper and iron sulphides was located at a shallow depth. Bore WS 10 was logged using a down-hole colour camera (inset)

9.13   ELECTROMAGNETIC PROFILING

Ground conductivity meters (such as the GEONICS EM38, EM31 and EM34-3) operate at low frequencies (called low induction numbers, or LIN) (see Figure 9.41 after McNeill (1990)). In these cases, only the quadrature phase is important and is linearly proportional to the ground apparent conductivity σ_a. To obtain information on structures at different depths, different coil spacings are used. However, it is not possible to interpret these different data sets into an accurate model of true conductivities and depths. Ground conductivity meters can be used in both horizontal and vertical dipole modes, with different penetration depths in each case. Switching from an operation mode where the plane of the coils is vertical to one where the plane of the coils is horizontal will double the penetration depth.

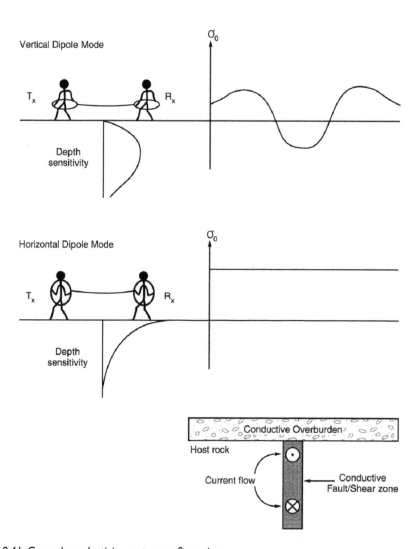

Figure 9.41 Ground conductivity meter configurations

A shallow EM survey instrument (EM38) is shown in Figure 9.42. An EM31 being carried on a survey is shown in Figure 9.43. An EM31 attached to a QUADCYCLE and used for extensive salinity surveys is shown in Figure 9.44.

The electrical frequency at which each of the GEONICS equipment works is shown in Table 9.4. There is a considerable frequency variation between the EM34-3 used at the 40 m separation and the EM38. These instruments will give different values of apparent conductivity over *uniform ground* if the material has a complex conductivity. This will be problematic with clay material and makes the comparison of data acquired with the various instruments possible only in a qualitative manner.

Eastern Australia has significant quantities of the expandable lattice mineral smectite in the clays derived from the weathering of extensive Tertiary basalt flows. These clays in particular have a complex conductivity (Acworth and Jankowski, 1997) and are not suitable for comparison of EM EC_a values derived from differing equipment coil configurations.

Figure 9.42 GEONICS EM38 (Photo: Ian Acworth)

Figure 9.43 EM31 shoulder carried for a salinity survey (Acworth and Beasley, 1998)

Figure 9.44 EM31 mounted on a QUADCYCLE (Photo: R. Beazely)

Table 9.4 Frequency ranges of GEONICS equipment

Instrument	Intercoil Spacing (m)	Dipole Coil Orientation	Exploration Depth (m)	Operating Frequency (Hz)
EM38	1.0	Horizontal	0.75	13,200
		Vertical	1.5	13,200
EM31	3.7	Horizontal	3.0	9,600
		Vertical	6.0	9,600
EM-34	10	Horizontal	7.5	6,400
		Vertical	15	6,400
EM-34	20	Horizontal	15	1,600
		Vertical	30	1,600
EM-34	40	Horizontal	30	400
		Vertical	60	400
EM-39	0.5		0.45	39,200

9.13.1 Theory of operation

The general theory of operation is explained in the technical note by McNeil (1980). In general terms, a transmitter coil (Tx) is energized with an alternating frequency current operating at audio frequencies (see Table 9.4). A receiver coil (Rx) is located a short distance (s) away from the transmitter coil. The time-varying magnetic field arising from the alternating current in the transmitter coil induces very small eddy currents in the earth.

The eddy currents generate a secondary magnetic field (H_z), which is sensed, together with the primary magnetic field (H_p), by the receiver coil. In general, the secondary magnetic field is a complicated function of the coil spacing (s), the operating frequency (f) and the ground conductivity (σ). GEONICS McNeil (1980) designed equipment such that, when operated at low induction numbers, the secondary magnetic field is a simple function of the primary magnetic field:

$$\frac{H_s}{H_p} = \frac{i\omega\mu_o\sigma s^2}{4} \qquad (9.42)$$

where:
H_s is the secondary magnetic field at the receiver coil,
H_p is the primary magnetic field at the receiver coil,
$i = \sqrt{-1}$,
$\omega = 2\pi f$,
f is the frequency,
μ_o is the permeability of free space,
σ is the bulk conductivity, and
s is inter-coil spacing.

The ratio of the secondary magnetic field to the primary magnetic field is then linearly proportional to the bulk apparent electrical conductivity (σ_a) such that:

$$\sigma_a = \frac{4}{i\omega\mu_o s^2}\frac{H_s}{H_p} \qquad (9.43)$$

The GEONICS equipment is excellent for rapid reconnaissance of ground conductivity, and there are many published examples of the successful use of this equipment. The instruments do not require any ground contact and can therefore be walked over a site to rapidly acquire data. It was this characteristic that caused the use of CST to rapidly diminish when the GEONICS range of equipment became available. The EM38 and EM31 can be used by a single operator, whereas the EM34-3 requires a two operators to profile along at separations of 10 m, 20 m or 40 m.

An example of a survey for salinity on the Liverpool Plains (Acworth and Beasley, 1998) is shown in Figure 9.45. Four borehole locations are indicated in Figure 9.45. The data was collected using the farm bike apparatus shown in Figure 9.44. The ECI shown in Figure 9.21 was measured centered on the second borehole down from the top of the image.

Two approaches to the interpretation of the EC_a data are possible. The first is to derive a statistical relationship between EC_a and EC along a particular profile line. The correlation is frequently improved by including measures of clay content, or moisture content (Cook et al., 1992). A disadvantage of this approach is that the relationships derived are valid for one area and soil type alone and frequently produce misleading results if applied in other areas without the detailed investigation required to produce the initial correlation. The second approach is to carry out a more complete sampling of the EC_a distribution using ECI and to invert this data set to produce a true two-dimensional, or preferably three-dimensional, model of the EC variation within each zone. Factors that influence the value of EC within each zone can then be derived.

The GEONICS EM-34 system has been used to sample the EC_a at different penetration depths using repeated passes over the same profile at different distances between the transmitter and receiver coils. McNeil (1980) introduced functions to interpret this profile data in terms of different layer depths and EC values, but the analytical functions employed do not allow a ready inversion of an extensive data set over inhomogeneous ground.

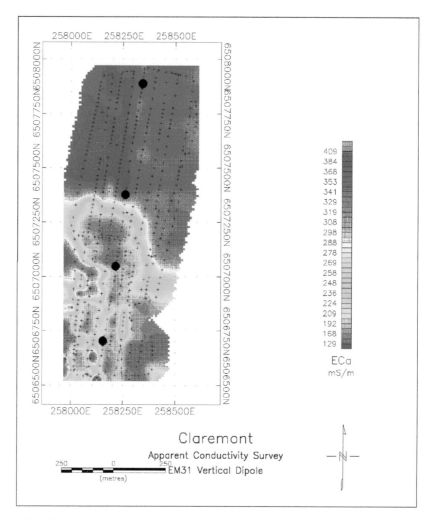

Figure 9.45 EM31 survey over clays containing variable salt concentrations. Four bores are shown (black circles); the second from the top was the centre point of the ECI (adapted from worth and Beasley (1998))

Finally, a comparison of CST and EM profiling from fractured basement is given in Figure 9.46. It can be seen that there is very little difference between the results obtained by the two different methods. CST with an electrode separation of approximately 10 m produces similar results to the Geonics EM31 with the plane of the coils horizontal (normal operational method).

A further example is shown in Figure 9.47.

9.13.2   Airborne EM methods

Airborne surveys can cover large areas efficiently. However, to be financially viable, they must also be used to cover large areas. This tends to mean that they are commissioned

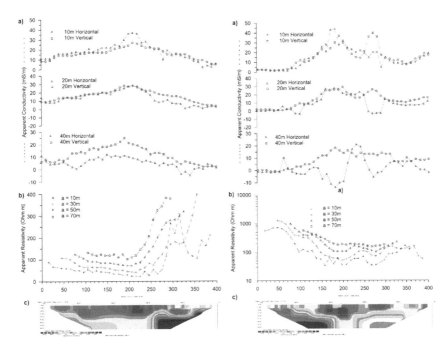

Figure 9.46 Comparisons between EM profiling and the results of ECI at two sites from Bauchi in Northern Nigeria (Acworth, 1981). The results of EM-34 (both coil orientations (a); the CST resistivity results at similar spacings (b); and (c) the ECI model of the CST data

by state agencies rather than individuals (Duncan *et al.*, 1992). Geophysical methods can provide invaluable information to assist with the overall interpretation of an area. Google maps are a good indication of the quantum leap in understanding that is possible with a well-calibrated remote sensing product. However, there has been a tendency to over-promote airborne geophysics in the same way that ground-based geophysics has often been oversold. There is no escaping the fact that geophysical methods require calibration based upon actual observation of ground conditions.

Having said that, geophysical methods provide an invaluable method to extend from, or between, known conditions established by drilling and sampling. Geophysics cannot be used to map groundwater or to monitor regional groundwater movement. The federal government financed an extensive evaluation of airborne electromagnetic surveys for the detection and monitoring of dryland salinity. The results were mixed, with the methods working well in parts of WA but less effectively on the Liverpool Plains of NSW. It is probable that this failure on the Liverpool Plains was due to the complex conductivity of the smectite clays (Acworth and Jankowski, 1997) where the bulk conductivity varies with frequency. Frequencies of 400 Hz to 56,000 Hz were used in the airborne work (Figures 9.48 to 9.50).

The appropriate use of airborne geophysics is that employed by English *et al.* (2004) where a fully integrated approach to a specific problem on the south-eastern part of the Riverine Plains of the Murray Basin has been undertaken. Airborne EM, radiometrics (gamma-ray spectrometry) and magnetics data were collected and calibrated by an extensive program of drilling, sampling and down-hole logging.

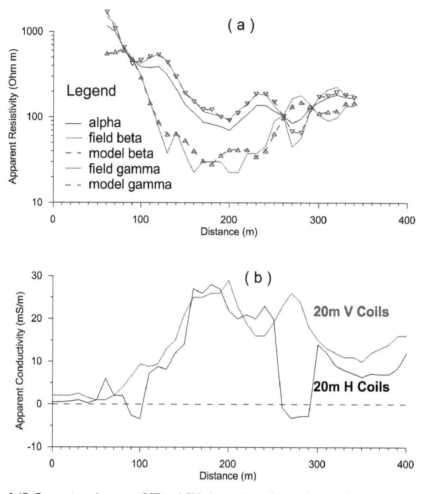

Figure 9.47 Comparison between CST and EM data collected over fractured basement ground in Nigeria. Tripotential data is shown (a) along with the modelled beta and gamma responses, while electromagnetic profile data with vertical and horizontal coils is shown in (b)

Extensive development work is continuing in the effort to improve the inversion of airborne EM work (Liu and Yin, 2016) and the interpreted bulk conductivity distribution is being incorporated into groundwater flow models.

9.13.3   Example survey data

The federal government carried out a comparison study by different companies offering airborne electromagnetic surveys in 1995. The purpose of this project was to evaluate methods to rapidly map dryland salinity. The results of the survey at five different catchments were reported in detail by George *et al.* (1998).

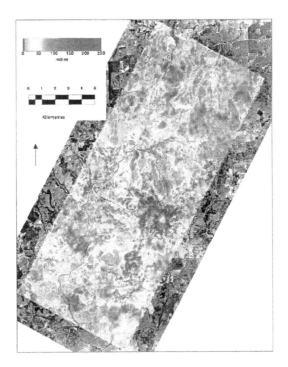

Figure 9.48 Shallow penetration using 56,000 Hz airborne EM system. National Dryland Salinity Investigation Program – Liverpool Plains (George *et al.*, 1998)

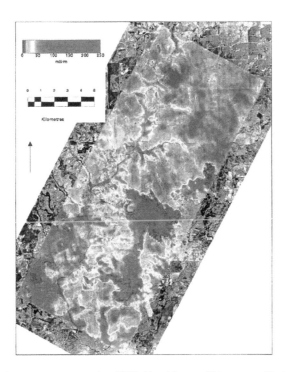

Figure 9.49 Intermediate penetration using 7200 Hz airborne EM system. National Dryland Salinity Investigation Program – Liverpool Plains (George *et al.*, 1998)

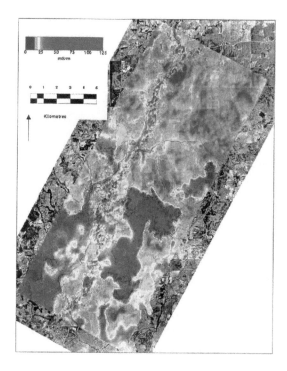

Figure 9.50 Deep ground penetration using 400 Hz airborne EM system. Note the change in scale between this data set and the earlier data sets. National Dryland Salinity Investigation Program – Liverpool Plains (George *et al.*, 1998)

The Liverpool Plains was one area where there was considered to be sufficient supplementary data that the survey products could be evaluated. An area of the Plains close to Cattle Lane and incorporating the EM profiling shown in Figure 9.45 was selected. The results for frequencies of 56,000 Hz (shallow), 7200 Hz (intermediate) and 420 Hz (deep) are shown in Figures 9.48 to 9.50.

There are clear and significant anomalies detected in these surveys with the apparent conductivity increasing with depth. However, it was not possible to resolve this data with the available information from boreholes and other geophysical methods, and while the survey methods appeared to provide good results on the weathered basement in Western Australia, the same could not be said for the Liverpool Plains (George *et al.*, 1998).

Chapter 10

Geophysical investigation techniques: gravity

10.1 GRAVITY THEORY

10.1.1 Introduction

It has long been recognized that gravity is not constant but varies with latitude, elevation and the distance from significant changes in mass – such as mountains. One of the early steps in the developments of plate tectonic theory was made by Prof. Don Griffiths at Birmingham University, who made gravity measurements in a boat along a transect across the Celtic Sea in the 1950s and began to untangle the deep geological structure of that part of the continental margin. On a smaller scale, microgravity measurements can be used to detect a well shaft covered by a concrete slab. Gravity measurements can be made from a stationary sensor to investigate changes in gravity with time; from a moveable sensor to measure gravity changes with location; installed in an aeroplane to measure changes with location on a large scale; or from space to measure even larger scale changes.

The significance of small perturbations in the gravity field of the earth has been realized as the number of satellites and associated space junk has increased. The only way to accurately predict their location is to have a very accurate map of the gravity field. More recently, global positioning systems (GPS) rely entirely upon knowing the correct value of gravity, and the satellite net that supports global communication must have a very accurate map of satellite location for efficient communication. If the value of g is known accurately, the position of the satellite can also be known. Repeated measurements of the distance between the satellite and the earth surface (satellite altimetry) can be used to detect changes in the surface elevation of the earth. These changes occur frequently in response to groundwater abstraction (Chapter 7) and have been used to highlight the impact of groundwater abstraction in California (Figure 10.1) in 17 months. The use of satellite altimetry is an important component to studies of change in mass made possible by the gravity recovery and climate experiment (GCACE) data.

The GRACE satellites were launched in March 2002 as a joint NASA–German project to map the gravity field of the earth. The GRACE experiment consists of two identical satellites placed into orbit (semi-major axis is 700 km with an inclination of 89.0° and a period of 91 min) close to each other. The two identical satellites orbit one behind the other in the same orbital plane at an approximate distance of 220 km. The distance between the satellites is measured by laser interferometry. As the satellites pass over a region of varying density, they move slightly closer together or further apart. The difference in distance is directly

Figure 10.1 Total subsidence in the San Joaquin Valley between 7 May 2015 and 10 September 2016. The data was derived from the ESA Sentinel-1A satellite and processed by the Jet Propulsion Laboratory by NASA

proportional to the change in gravity at that location. Maps of the gravity field can be made repeatedly and changes in gravity reconciled with changes in mass as water moves about the planet. For example, the end of the millennial drought in Australia was marked by very extensive rains and associated flooding. The low elevation of much of Australia meant that the floods remained for many months and that a significant addition to the mass of Australia occurred as a result. This was recorded by the GRACE data.

The gravity method is used in a broad range of applications which range from mapping geological structures such as buried valleys or variable basement (aquifer) depths to searching for voids. The gravity method provides the possibility for a rapid reconnaissance of an area and seldom provides all the required data.

The major steps in a gravity survey can be summarized as:

* acquire data in appropriate detail and precision,
* reduce data to an interpretable form,
* identify and isolate anomalies of interest,
* interpret the source characteristics of an anomaly, and
* translate the physical anomaly into a geological source.

10.1.2 Fundamental relationships

The gravity method is based on the measurement of small perturbations in the planetary fields caused by subsurface anomalous sources. The gravity field is a potential field which obeys Laplace's equations.

Newton's law of gravitation can be expressed as Equation 10.1:

$$F = G \frac{m_1 m_2}{r^2} \tag{10.1}$$

where:
m_1 is mass 1 [kg],
m_2 is mass 2 [kg],
r is the distance between the two masses [m], and
G is the universal gravitational constant 6.673×10^{-11} N m^2 kg^{-2}.

If we assume that the earth is spherical, the force exerted on the earth on a spherical body with mass m resting on the earth's surface is shown by Equation 10.2:

$$F = G \frac{mM}{R^2} \tag{10.2}$$

where: M is the earth's mass. This is estimated to be 5.9736×10^{24} kg in astrometric and geodetic data tables, but an estimate of 5.963278×10^{24} kg fits better with other estimates of the universal gravity constant and the radius of the earth as shown here.
R is the earth's radius. The earth is not spherical. The radius at the equator is 6.378140×10^6 m, while the radius at the pole is 6.356752×10^6 m, a difference of 21.388 km.

Newton's second law of motion is given by Equation 10.3:

$$F = ma \tag{10.3}$$

where:
F is force [(Newton]) (kg m/s^2),
m is mass (kg), and
a is an acceleration (m s^{-2}).

If we define g as the acceleration due to gravity, we can write:

$$F = mg = G \frac{mM}{R^2} \tag{10.4}$$

and:

$$g = \frac{GM}{R^2} \tag{10.5}$$

Table 10.1 Density of the main rock-forming minerals

Constituent	Density kg m^{-3}	Rock	Density kg m^{-3}
Quartz	2650	Saturated sand	1950–2080
Calcite	2700	Dry sand	1400–1650
Water	1000	Soft limestone	1900–2100
Oil	900	Hard limestone	2400–2700
		Sandstone	1800–2700
		Clay	2000–2500
		Granite	2500–2700
		Basalt	2800–3200

The value of g is expressed in m/s^2 in SI units. Historically, gravity has been measured in units of Gal which is $1\frac{cm}{s^2}$, although units of milliGal (mGal) are often used in exploration studies. The gravity unit gu is sometimes used instead of milliGals where 1 gu is 1 μm s^{-2}. Note that 1 mGal = 10 gu. Changes in the gravity field on the surface of the earth are due to the variable density of earth materials. Changes can be as large as 1000 mGal, although also as small as a few μGal.

10.1.3   Density

As gravity depends upon mass, it follows that it also is a function of density. The density (ρ) of rocks generally varies from 1.6 kg m^{-3} to 3.2 kg m^{-3}, and it depends upon porosity, saturation and mineral content. The densities of the important rock-forming minerals are given in Table 10.1. Together with air, which can be assumed to have zero density in these comparisons, these components make up the majority of rock types. The bulk density of rocks can be calculated from the density of the individual components as:

$$\rho_0 = f_1\rho_1 + f_2\rho_2 + f_3\rho_3 + \cdots + \cdots \tag{10.6}$$

Where ρ_0 is the bulk density, f_1 is the fraction of mineral 1 and density ρ_1. For a rock consisting of one mineral plus water, Equation 10.6 can be written as:

$$\rho_b = \phi \times \rho_w + (1.0 - \phi)\rho_m \tag{10.7}$$

Where ϕ is porosity and ρ_m is the mineral density.

10.1.4   Measurement of gravity

The classic instrument for measuring gravity is the pendulum. An ideal pendulum suspends a mass of zero dimensions from a string which does not stretch and has zero mass. If these conditions could be realized, the acceleration due to gravity can be derived from Equation 10.8:

$$T = 2\pi\sqrt{\frac{l}{g}} \tag{10.8}$$

In reality, Equation 10.9 is used:

$$T = 2\pi\sqrt{\frac{K}{g}} \tag{10.9}$$

where K is a constant that represents the characteristics of the system. Relative measurements of gravity can then be made by making measurements at two stations. Clearly, carrying around a pendulum is hardly convenient, and modern systems use the principle of an extending spring to measure gravity.

The instruments used for gravity measurement are called gravity meters or gravimeters. Gravity can be measured in the time domain to investigate temporal changes in gravity or in the space domain by moving a gravimeter about the country. The most widely available is a spring balance system in which a spring supports a mass as shown in the basic form in Figure 10.2. It is important to protect the spring from damage and changes in heat. In the CG-5 system (Scintrex Autograv – Figure 10.3), the spring is contained within two ovens held at a constant temperature.

The CG-5 instrument is nulled by turning a calibrated dial until a beam produced by a light pointer is brought to the centre of a magnifying eyepiece. The instrument is then moved to a new location and the movement required to re-center the eyepiece noted. The amount of movement necessary to bring the instrument back to balance is proportional to the change in g between the locations. Calibration on manufacture produces a dial constant (e.g., 0.0869 mGal/dial division). This constant is multiplied by the change in dial reading to produce a reading in mGal. A precision of 0.01 mGal can be obtained.

The major problem with gravimeters is that there is slight extension of the spring mechanism known as *drift*. To achieve accurate data, stations have to be re-established several times during a day of field work to check the drift that day.

Absolute measurements of gravity can be made by several techniques. One such technique involves throwing a mass up into the air and timing its descent! Gravity can be

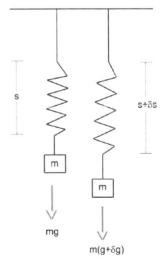

Figure 10.2 Principle of a spring balance

Figure 10.3 Scintrex CG-5 Autograv instrument for recording change of gravity with location

Figure 10.4 LaCoste Romberg gPhone total gravity station for recording gravity changes with time

established from Equation 10.10:

$$z = \frac{gt^2}{2} \tag{10.10}$$

An *International Gravity Standardization Net – IGSN71* was established in 1971. If absolute measurements of gravity are required, a survey should include one of the IGSN71 data points.

The gravitational field as measured by the GRACE system from earth orbit is shown in Figures 10.5 and 10.6.

The gravity field can also be observed from an aircraft and an airborne gravity map determined. The data from northern NSW is shown after the NSW Geological Survey in Figure 10.7. Very clear linear gravity anomalies are seen that relate to deep structure, characterizing continental accretion. The rocks of the Tablelands Complex (New England Plateau) to the east of the Peel Fault (Figure 10.7) have a negative gravity anomaly (less mass). The Boggabri High shows an area where Permian volcanics lie at shallow depth to the west of the Mooki Thrust – a Carboniferous age continental suture line.

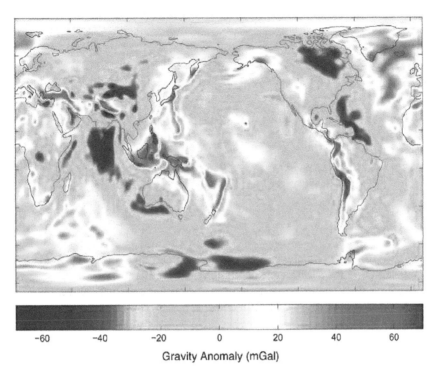

Figure 10.5 Gravity map of the earth derived from GRACE measurements

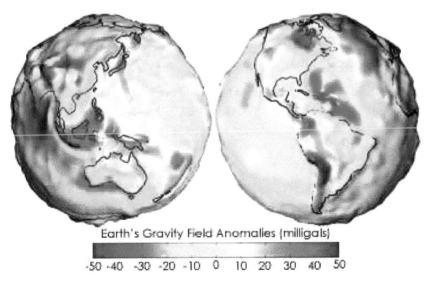

Earth's Gravity Field Anomalies (milligals)

Figure 10.6 Alternative (to Figure 10.5) map of the gravity field published by NASA from the GRACE data set

Table 10.2 Principal earth tides

Constituent	Period (hours)	Vertical amplitude (mm)	Horizontal amplitude (mm)
M_2	12.421	384.83	53.84
S_2 Solar semi-diel	12.000	179.05	25.05
N_2	12.658	73.69	10.31
K_2	11.967	48.72	6.82
K_1	23.934	191.78	32.01
O_1	25.819	158.11	22.05
S_1 (Solar diel)	24.000	1.65	0.25
P_1	24.066	70.88	10.36
M_f	13.661 days	40.36	5.59
M_m	27.555 days	21.33	2.96

10.2 EARTH TIDES

The earth exists in the gravitational fields of the sun and the moon. As the moon rotates around the earth and the earth rotates in an elliptical orbit around the sun, the changes in the gravitational fields experienced at a location on the earth vary throughout the day, with seasons and with latitude. The main components are shown in Table 10.2.

Earth tides can be seen as small changes in hydraulic head as described in Chapter 7.

The lunar tide (M_2) is almost double the size of the solar semi-diel tide (S_2). The effects of the earth tides are directly seen in a variation of gravity when measured over time. The components of the tides are also well known and constant. Their effect can be modelled anywhere around the world using a package such as TSoft (Van Camp and Vauterin, 2005) or measured using a gravity meter such as that seen in Figure 10.4. A comparison of measured and modelled tides is shown in Figure 10.8.

10.3 GENERAL FIELD PROCEDURES FOR A GRAVITY SURVEY OF AN AREA

10.3.1 Introduction

A gravimeter placed at a single location and observed every hour for a 24 hours would show a systematic variation. The variation occurs for two reasons:

- earth tide impacts and
- instrument drift.

The first of these is entirely predictable if the location of the gravimeter is known and can be calculated using public domain software such as TSoft (Van Camp and Vauterin, 2005). The calculated earth tide can be subtracted from the gravity measurements leaving the second variation due to drift. Instrument drift occurs due to non-elastic movements of the gravity system. It can be measured by making an observation at a location where gravity is known and then returning to that location during the day to repeat the measurement.

Instrument drift is clearly shown in the gPhone data in Figure 10.9 over a month-long data set where readings were taken at 1 s intervals and then re-sampled to 1 min data.

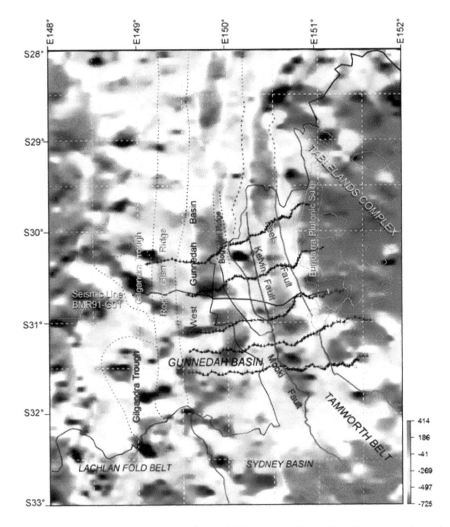

Figure 10.7 Airborne gravity map for the Gunnedah Basin area (Australian Geoscience data adapted by Guo *et al.* (2007) and published with the permission of the Geological Society of Australia)

The instrument was installed in a purpose-built hut powered by a solar array and located in a field on the Liverpool Plains of NSW at Breeza. The soil at Breeza is a deep cracking clay (smectite) and supports an extensive cotton growing industry. There was very little rain at Breeza during July 2015. The data in Figure 10.9 is shown in two panels. The lower panel shows the field measurement, initially taken at 1 s intervals and then resampled to 1 min samples using TSoft. The earth tide for the location is also shown and the result of subtracting the earth tide from the observed data (dark green line in Figure 10.9). It can be seen that there is still a linear change in gravity that is not related to a process within the earth. There was no rain, and no crops were growing in July (winter). The

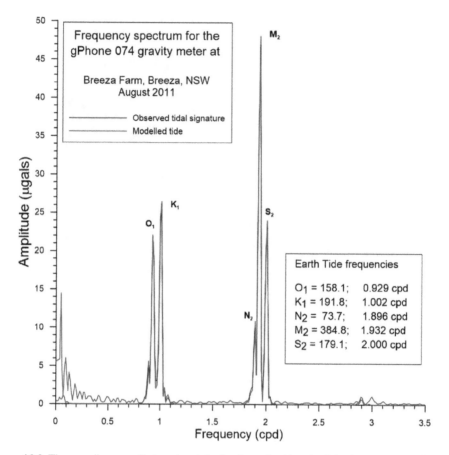

Figure 10.8 The excellent prediction (modelled) of earth tides (red line) as compared by the measured earth tide (blue line). The comparison is made in the frequency spectrum for clarity

regional groundwater level was at approximately 1 m depth with no local or regional groundwater abstraction occurring.

The linear change in gravity is therefore an instrument drift and can be removed by carrying out a linear regression to identify the slope of the best fit and removing that slope from the data. The drift in this example was approximately 50 µGal/day. The TSoft program has a function that can be used to make the calculation. The data in Figure 10.9 illustrates the process as applied during adjustment of gPhone data collected at a single location. For measurements with the CG-5 system that changes location, it is necessary to establish the drift of the instrument by returning to a set location or base station on a regular basis. If possible, the base station should be one of the IGSN71 net stations so that your readings can be related to the global measurement system. Care must be taken not to confuse the earth tide response with the drift, so the earth tide should first be calculated and subtracted from the data before the drift is identified and removed. The earth tide does not vary significantly over areas of several thousands of square kilometres.

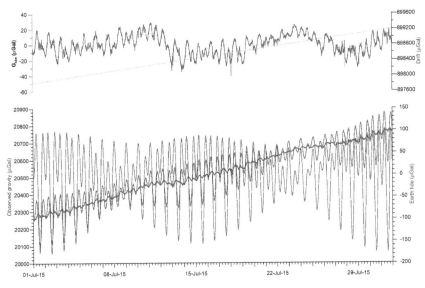

Figure 10.9 Blue line is instrument value of g. Red line is the calculated earth tide at the Breeza location. Dark green line is the observed data corrected for the earth tide. In the upper plot, the dark green line is the observed data (g_{obs}) with the instrument drift (pink line in the upper plot) removed

10.3.2   Elevation and position determination

The accuracy of measurement station elevation to better than 250 mm is required. This used to be established by cadastral surveying procedures and was an expensive task. Survey benchmarks were located within the study area and the elevation of each gravity station was established with reference to the net of benchmarks.

With the ready availability of GPS equipment with an absolute accuracy of better than 250 mm, the necessity to survey using cadastral techniques is removed.

10.4   GRAVITY CORRECTIONS TO G_{obs}

If the earth was an exact spheroid, did not rotate and was flat, then the field measurements of gravity would need no correction. In reality, a number of corrections are necessary, *viz.*:

- latitude correction,
- free air correction,
- Bouguer correction, and
- terrain correction.

10.4.1   Latitude correction

As a direct impact of the rotation of the earth, the value of *g* is greater at the poles than at the equator by 3400 mGal.

As the radius of the earth is greater at the equator than at the pole, a further decrease in g of 4800 mGal occurs from equator to pole. However, since radius length is greater at the equator, more mass is positioned between the surface and the earth's centre at the equator than at the poles.

The effect of rotation, the radius variation factor and the mass factor result in a net increase in gravity by 5.2 Gal as you travel from the equator to the pole. This accounts for the variation in gravity from 9.780 $\frac{m}{s^2}$ at the equator to 9.832 $\frac{m}{s^2}$ at the poles. A correction for field data to account for latitude is clearly required if we are to achieve the required survey accuracy of 0.01 mGal. Equation 10.11 produces the required correction:

$$g_n = 9.7803185(1 + 0.005278895 \sin^2\theta - 0.000023462 \sin^4\theta)\frac{m}{s^2} \tag{10.11}$$

where:
θ is the latitude of the observation station.

Normal gravity g_n represents the value of gravitational acceleration we should observe in the absence of any anomaly. The gravity anomaly is then $g_{obs} - g_n$.

10.4.2   Elevation correction 1 – the free-air correction

The gravitation acceleration imparted by the earth is given in Equation 10.5. To calculate the variation of g with height, we can take the first differential of Equation 10.5 with respect to distance as shown in Equation 10.12:

$$\frac{dg}{dR} = -2\frac{GM}{R^3} = -g\frac{2}{R} \tag{10.12}$$

Equation 10.12 is an approximation because the radius of the earth is not constant as noted above. A full derivation is not appropriate here, but Equation 10.12 is modified as reported by Grant and West (1965) and shown in Equation 10.13. The units are mGal/m:

$$\frac{dg}{dR} = -0.3086 - 0.00023 \cos 2\theta + 0.00000002z \tag{10.13}$$

where:
θ is latitude and
z is elevation.

For most surveys, the latitude adjustment (the second term in Equation 10.13) is small. Similarly, except in mountainous regions, the correction to the vertical gradient due to distance above the geoid (third term in Equation 10.13) is small and the elevation correction is limited to the first term. The correction does not take account of the increase in mass represented by elevations above the geoid and for this reason it is called the *free-air correction*.

The normal datum for gravity measurements is sea level. Because gravity decreases by 0.3086 mGal for every metre above sea level, the free-air correction is added to the observed gravity. To maintain an accuracy level of 0.1 mGal in the final survey, it is necessary that the station elevation be known to an accuracy of 0.33 m.

The gravity value obtained after adding the free-air correction and subtracting normal gravity from observed gravity is termed the *free-air anomaly* as shown in Equation 10.14:

$$\Delta gFA = g_{obs} - g_n + FA_{corr} \tag{10.14}$$

10.4.3   Elevation correction 2 – the Bouguer correction

The free-air correction takes no account of the presence of a slab of mass lying above the geoid surface. This additional mass has a gravitational correction of its own and needs accounting for. The correction is referred to as the *Bouguer correction* after the Swiss geophysicist who first tried to calculate the lateral gravitational attraction of the Swiss Alps. Burger (1992) gives an account of the derivation for this correction.

The Bouguer effect due to a slab of infinite extent and thickness z is $2\pi G(\rho z)$. Substituting known values gives a correction factor of $0.04193\rho z$ mGal/m. The Bouguer correction removes the effect of extra mass and therefore is subtracted from the other measurements. The residual gravity anomaly remaining after the Bouguer correction is subtracted is referred to as the *Bouguer anomaly* and is given in Equation 10.15:

$$\Delta g_b = g_{obbs} - g_n + FA_{corr} - B_{corr} \tag{10.15}$$

The most difficult problem in many gravity surveys is deciding what value of density to substitute in Equation 10.15. In the absence of local information, a value of 2670 kg/m^3 is used as this is the average density of crustal rocks.

10.4.4   Elevation correction 3 – the terrain correction

The Bouguer correction assumes that the measurement station lies on an infinite slab at a certain elevation above the geoid surface. Clearly this is not the case, and terrain corrections are required to account for the variation in topography around a site. The significance of topography basically depends upon how close it is to the measurement site. Gravity measurements are much easier, for example, around Bourke in the middle of the Darling Basin than they are in the Megalong Valley in the Blue Mountains NSW, where high cliffs of sandstone require extensive terrain corrections.

When Bouguer corrections are made, they are subtracted from the observed gravity because they represent additional mass above a plane. Areas which are below the datum chosen for the Bouguer correction are over-corrected for and the terrain correction needs to be added back into the anomaly. The mass lying above the datum plane exerts a gravitational attraction which reduces the value at the measuring point. For this reason the correction for zones above the datum plane are also added into the observed data. If *TC* represents the terrain correction, the Bouguer formula becomes:

$$\Delta g_b = g_{obbs} - g_n + FA_{corr} - B_{corr} + TC \tag{10.16}$$

Topographic corrections vary around a measuring station. It is difficult to calculate how much correction should be made for different topographic variation in different locations around the measuring point. A solution to this problem was proposed by Hammer (1939) in which a terrain-correction template is laid over the topographic map with the centre of the template at the measuring station.

10.5 MODELING GRAVITY DATA

10.5.1 Anomalies due to buried spheres and cylinders

The impact of a density variation which is associated with a change in lithology can be simply studied analytically. The sphere is the simplest geometry and can be analyzed as follows.

If the density contrast between the surrounding material and the sphere is g_c, the excess mass due to the sphere is $\frac{4}{3}\pi R^3 \rho_c$. The gravitational attraction due to a sphere of radius R in the generalized direction r is then:

$$g_{sphere(r)} = \frac{Gm}{r^2} = \frac{G4\pi R^3 \rho_c}{3r^2} = \frac{G4\pi R^3 \rho_c}{3(x^2 + z^2)} \tag{10.17}$$

Only the vertical component of this attraction is measured by the gravimeter, therefore we require Equation 10.17 to be modified as shown in Equation 10.18:

$$g_{sphere} = \frac{G4\pi R^3 \rho_c}{3(x^2 + z^2)}\cos\theta = \frac{G4\pi r^3 \rho_c}{3}\frac{z}{(x^2 + z^2)^{1.5}} \tag{10.18}$$

A similar analysis can be given for a buried cylinder. Equation 10.19 gives the anomaly:

$$g_{cylinder} = G2\pi R^2 \rho_c \frac{z}{x^2 + z^2} \tag{10.19}$$

A useful exercise is to put these two equations into a spreadsheet and to calculate the anomalies for various combinations of depth, density contrast and radii. Analytical equations exist for a variety of buried rods, sheets and faults. In practice, a more general approach is required.

10.6 ANALYSING GRAVITY ANOMALIES

10.6.1 Gravity models

Gravity models can calculate the two-dimensional anomalies created by any combination of polygons of infinite width. The approach is similar to the interpretation of 2.5 dimensional resistivity data. Danis *et al.* (2011) describe the 3D geology of the Sydney Basin using a combination of gravity, seismic reflection and borehole data that demonstrates the presence of a deep basin of high-density basic lavas that form a proto-rift structure. This work is a continuation of detailed surveys and modelling by Danis *et al.* (2010) of the Gunnedah Basin.

An example of a gravity model for a section through Figure 10.7 at Tamworth, NSW, is shown in Figure 10.10 (Guo *et al.*, 2007).

10.6.2 Regionals and residuals

Large features produce anomalies which are smooth over considerable distances. These are going to be caused by deep variations in lithology which are seldom the target for environmental and engineering surveys. These large features are referred to as *regionals* or *regional*

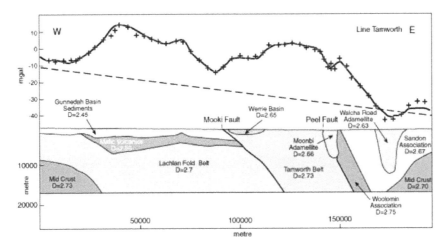

Figure 10.10 A gravity model for the Gunnedah Basin through Tamworth showing sedimentary units, density and depth (Guo *et al.*, 2007). The crosses represent gravity observations and the solid line represents the calculated profile based upon the distribution of different density blocks as shown below. (Published with the permission of the Geological Society of Australia and GEMOC, Macquarie University.)

trends. If we are interested in the small-scale variation, the effect of the regionals can be subtracted from the data set to leave behind the *residual anomaly.*

There are many techniques for calculating the regional anomaly using gravity modelling software.

10.6.3 Derivatives

The presence of a small-scale anomaly can often also be analyzed by taking a derivative of the anomaly. The first derivative is a measure of the slope of an anomaly. The second derivative is a measure of the rate of change in slope. In areas where strong regionals exist, the use of a second derivative can expose local significant variation.

All these interpretation techniques assume that the original field data and the corrections have been carried out to a sufficient degree of accuracy to enable the subsidiary data to be extracted.

10.7 APPLICATIONS OF THE GRAVITY METHOD

One of the most common applications of the gravity method is establishing the depth to bedrock. This is important in groundwater surveys to determine the transmissivity of an aquifer or to determine how far dense non-aqueous phase liquids (DNAPL) may have sunk into an aquifer.

An application of the gravity technique known as *microgravity* also exists. Microgravity surveys are particularly useful for finding buried shafts. This is a problem often encountered in the redevelopment of urban land. Microgravity can also be used to find sink holes in limestone terrain. Burger (1992) cites an example of gravity used to find the extent of the cone of depression caused by a pumping test in an unconfined aquifer. Finally, micro-gravity can be used to determine the extent and depth of a completed landfill, some time after the records have been lost!

10.7.1 Example field data

A typical set of readings for a fixed station (gPhone) on the Liverpool Plains, NSW, is shown in Figure 10.11. The drift during the time taken to collect this data was not significant.

10.7.2 Bedrock depth surveys

This is the most common application of gravity in environmental and engineering geophysics. Gravity methods can often be employed in urban environments where the use of other techniques is not possible. The presence of buildings, roads, pipes and cables may make seismic, electromagnetic and resistivity methods impossible. Careful elevation control, position control and density measurements can produce excellent results if a sufficiently accurate survey is undertaken. Gravity readings can also be taken inside buildings so that minimum impact of the survey is caused.

10.7.3 Gravity measurement variation with time

A gPhone gravimeter was installed on the Liverpool Plains and left to monitor change in gravity for a period of 10 weeks. The gPhone was set on the top of a 10 m concrete plinth that was sealed into the ground for the lower 5 m and set in a PVC casing (sleeve) for the

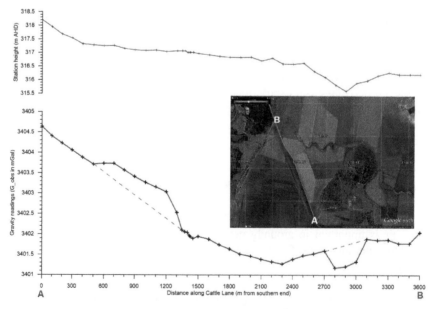

Figure 10.11 Gravity survey readings along Cattle Lane in New South Wales showing the gravity observations and the station elevations

Figure 10.12 Top of the 10 m long concrete plinth that will form the base for the gPhone (Photo: Wendy Timms)

Figure 10.13 Purpose-designed gravity hut being lowered onto the plinth (Photo: Wendy Timms)

top 5 m so that the ground surface could rise and fall with the expansive smectite soils at Breeza. The gravimeter was powered by a solar array and installed inside a purpose-designed hut. Details are shown in Figures 10.12 and 10.13. Rainfall and pressure were recorded at a Campbell Scientific weather station at a site immediately adjacent.

The results of the gravity monitoring are shown in Figure 10.14. The rainfall event on 9 September, 2011, when approximately 40 mm of rain fell, shows an increase in gravity of 116 µGal. Note that the gravity scale in Figure 10.14 is inverted. Smaller rainfall

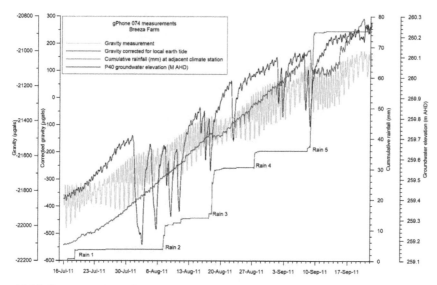

Figure 10.14 Gravity results at Breeza for July 2011. Significant rainfall events are shown

events appeared to have no impact, suggesting that evaporation may have occurred soon after the event.

10.7.4   Measuring aquifer storage change (specific yield)

Pool and Eychaner (1995) describe the use of repeated gravity surveys across a 518 km^2 alluvial drainage basin in central Arizona. They conducted repeat measurements at two base stations on basement rock outside the alluvial basin and at six borehole sites within the basin. Gravity increased by as much as 158 μGal (± 6 μGal) in response to a change in water level of 17.6 m. The average specific yield at the wells varied between 0.16 and 0.21. It is important to note that this successful survey was applied to determine specific yield and not specific storage.

10.7.5   GRACE results

The GRACE satellite data provides an excellent way to monitor the change in mass stored with time. The changes caused by surface flooding or by the shrinking of ice caps (Figure 10.15) is well established.

There has been much interest in using GRACE to monitor groundwater level changes, with work in California and India receiving much prominence. However, Alley and Konikow (2015) and Long *et al.* (2016) provide warnings against the over-interpretation of GRACE data and recommend much greater monitoring of the actual situation on the

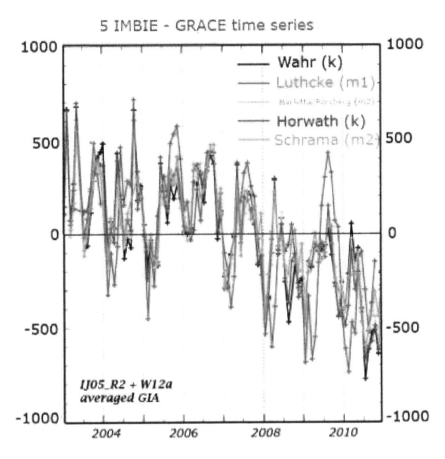

Figure 10.15 Monthly changes in Antarctic ice mass, in gigatonnes, as measured by NASA's Gravity Recovery and Climate Experiment (GRACE) satellites from 2003 to 2011

ground. Alley and Konikow (2015) in particular point out the difficulty of relating mass change to piezometric (hydraulic) head in a confined aquifer where the specific storage is a very small. This means that a small mass change will produce a very large change in hydraulic head.

Chapter 11

Geophysical investigation techniques: heat

11.1 INTRODUCTION

Heat is a form of energy expressed in the SI system as Joules (Nm), with units of $m^2 \, kg \, s^{-2}$. Temperature is a measure of the average heat or thermal energy of the particles in a substance. Heat flows by a combination of convection and conduction.

Of the four areas of geophysics currently in use in groundwater investigation, the study of heat flow has had the most recent development and is still not a mature area, with significant research results regularly being published. A major advantage of heat as an investigation technique is that reliable and stable thermometers are readily available, can be very accurate and are not expensive. All groundwater level data loggers measure heat to help correct the pressure measurement made by a transducer. Most systems allow a download of temperature at the same time as the hydraulic head (water level) is downloaded. Measurements of temperature above the ground surface, in the unsaturated zone and in the saturated zone are equally possible and allow the characterization of many processes.

An understanding of heat flow is required to study geothermal energy and remote sensing and to help identify the complexities of surface water and groundwater exchange.

11.2 GEOTHERMAL HEAT FLOW

Heat flows from the earth as the result of radioactive decay of elements within the earth's mantle. The isotopes ^{238}U, ^{235}U, ^{232}Th and ^{40}K provide the main source of heat now as shorter half-life isotopes have decayed away since the earth was formed. Heat is produced as gamma radiation is released from atomic nuclei during radioactive decay. Table 11.1 lists the thermal properties of the four major unstable isotopes involved.

Heat flows constantly from within the earth to the surface. The mean heat flow is $65/m^2$ over the continental crust and $101/m^2$ over oceanic crust. This is $0.087 \, W/m^2$ on average and compares to the $1367 \, W/m^2$ received by the earth at the top of the atmosphere as a result of solar radiation (see Section 2.3). The heat flow is greater in areas where the lithosphere is thin, such as along mid-ocean ridges and over mantle plumes (hot spots). The lithosphere acts as a thermal blanket that prevents heat escape, acting to concentrate heat flow by convection through volcanoes. The remainder of the heat loss is through the lithosphere by conduction. The heat of the earth is replenished by radioactive decay at a rate of 30 TW. The global geothermal flow rates are more than twice the rate of human energy consumption from all primary sources (Rybach, 2007).

Table 11.1 Major heat–producing isotopes

Isotope	Heat release [W/kg]	Half-life [years]	Mean mantle concentration [kg isotope/kg mantle]	Heat release [W/kg mantle]
^{238}U	9.46×10^{-5}	4.47×10^9	30.8×10^{-9}	2.91×10^{-12}
^{235}U	5.69×10^{-4}	7.04×10^8	0.22×10^{-9}	1.25×10^{-13}
^{232}Th	2.64×10^{-5}	1.40×10^{10}	124×10^{-9}	3.27×10^{-12}
^{40}K	2.92×10^{-5}	1.25×10^9	36.9×10^{-9}	1.08×10^{-12}

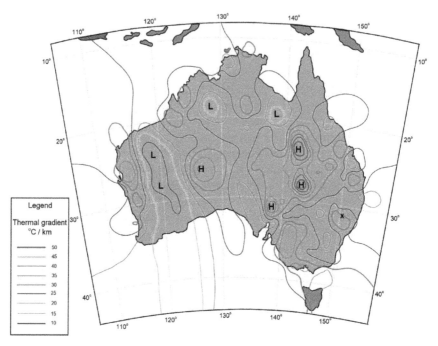

Figure 11.1 Geothermal gradient data for Australia (adapted from Cull and Conley (1983)). The cross marks the location of the site shown in Figure 11.2 which is just to the north of the Liverpool Ranges – a site of significant basalt lava flows

Temperature can be measured at the base of boreholes and a data set compiled that allows identification of different crustal temperature regimes. Three different regimes are identified in Australia, broadly outlining three major crustal elements (Cull and Conley, 1983). These three broad blocks can be seen in the geothermal gradient data shown in Figure 11.1. The maximum geothermal gradient is 50 °C/km in the central west of Queensland. The minimum of 10 °C/km occurs over the stable cratonic block in Western Australia.

Groundwater assumes the temperature of the surrounding rock. A profile of groundwater temperature carried out down a borehole is therefore equivalent to measuring the geothermal gradient. Groundwater temperatures have been extensively measured across the Great Artesian Basin (Cull and Conley, 1983). The temperature is normally established a year or so after the borehole is completed to avoid the impact of disturbances caused by drilling. The temperature of water in wells tapping aquifers in Lower Cretaceous and Jurassic

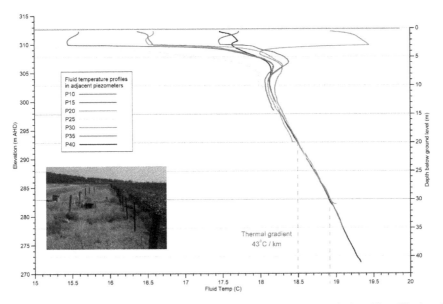

Figure 11.2 Fluid temperature logs showing a uniform geothermal gradient below 15 m. The logs have been measured in individual piezometers drilled into alluvial sands and clays on the Liverpool Plains. The inset image shows some of the piezometers. The geothermal gradient of 43° C/km is consistent with the country wide data shown in Figure 11.1.

sediments generally ranges from about 30 °C to 50 °C, but, in many areas of the basin, temperatures at the well head are as much as 100 °C. Water is discharged from mound springs along the western margin at temperatures ranging from about 20 °C to 40 °C. The average geothermal gradient in the GAB is 48 °C/km, which exceeds the global average of 33 °C/km. Polac and Horsfall (1979) noted a range from 15.4° to 102.6 °C/km with the wide variation taken as indicating significant movement of water vertically along fracture zones (Polac and Horsfall, 1979).

The fluid temperature logs in Figure 11.2 are assembled from seven piezometers installed 8 m apart into alluvial clay and gravel at Cattle Lane of the Liverpool Plains (Acworth *et al.*, 2015b). The logs of all the piezometers converge on a gradient of 43 °C/km. This is higher than shown in Figure 11.1 of about the average of 35 °C/km, but shows good broad agreement between two completely different scale investigations. The greater heat flow indicated at the Liverpool Plains may be due to deeper igneous activity associated with the Liverpool Plains basalt flows or due to the higher than normal [232]Th occurrence at the site (described further in Chapter 13). The thermal profiles show different diel heating and cooling effects to 10 m depth. Below 15 m, all logs converge on the same geothermal profile with a gradient of 43 °C/km. The difference in surface temperatures is related to the order the logs were made – starting on a cold morning. The water level was approximately 3.0 m below ground in all sites.

The temperature of groundwater can provide an indication of the source of groundwater, particularly if flowing from a spring into a warmer or cooler environment. The temperature of spring water will be the long-term temperature of the rock that forms the aquifer. If the water is flowing from depth, the spring will be a warm or hot spring as the earth warms by approximately 1 °C/100 m. Mapping the temperature of spring water can therefore indicate the source of water.

The thermal conductivity of rocks is low and heat is slow to move. The near surface may then contain information about recent climate change. Sea levels have oscillated over the past million years. They are at a maximum during the interglacial warm periods that generally last for 10 thousand years or so and $\approx$ 130 m above their lows reached during the glacial maxima. As ice ages typically last for 100,000 years, the current sea level is far above the long-term average. This factor is particularly important in limestone systems which may have developed flow systems that discharge at the sea level characteristic of average climate conditions over a glacial cycle. These discharge points are mostly now below the current higher sea level. These processes are readily observable around the Mediterranean, where fresh water has been harvested (by the Romans) from beneath the sea by placing inverted funnels over a submerged spring. The discharge of warmer water into a colder ocean can be readily observed by thermal imaging.

11.3   UNDERGROUND HEAT STORAGE IN THE SOIL

The Netherlands have been at the forefront of the design of heat storage and recovery systems that are becoming standard for new constructions. The ground strata can be anything from sand to crystalline bedrock, and depending on engineering factors, the depth can be from 50 to 300 m. Thermal energy storage systems can be subdivided into:

ATES (aquifer thermal energy storage) An ATES store is composed of a doublet, totaling two or more wells into a deep aquifer that is contained between impermeable geological layers above and below. One half of the doublet is for water extraction and the other half for re-injection, so the aquifer is kept in hydrological balance, with no net extraction. The heat (or cold) storage medium is the water and the substrate it occupies. ATES systems are now quite common.

BTES (borehole thermal energy storage) BTES stores can be constructed wherever boreholes can be drilled and are composed of one to hundreds of vertical boreholes, typically 155 mm in diameter. Systems of all sizes have been built.

Seasonal heat stores can be created using borehole fields to store surplus heat captured in summer to actively raise the temperature of large thermal banks of soil. A ground source heat pump is used in winter to extract the warmth to provide space heating. Boreholes can be either grout- or water-filled, depending, on geological conditions, and usually have a life expectancy in excess of 100 years. This area is recently reviewed by Stauffer *et al.* (2017).

11.4   THERMAL IMAGERY

Thermal imaging cameras are readily available and can be mounted on a range of platforms. The comparison between imagery from the visible spectrum and imagery from the thermal spectrum can be informative, as seen in Figures 11.3 and 11.4 (images from the USGS web site).

Every body emits radiation at a frequency determined by the surface temperature of the body (Stefan-Boltzmann Law – described in Chapter 2). The Landsat 8 satellite has

Figure 11.3 Colder (blue colour) water from the streambank storage mixing with warmer water (orange) in the stream (USGS image from a FLIR hand-held camera system).

Figure 11.4 Stream water much warmer than the stream bank foliage (USGS image from a FLIR hand-held camera system).

a thermal infrared sensor (TIRS) that records radiation in a frequency band particularly sensitive to vegetation. Data is available at https://landsat.gsfc.nasa.gov/data/where-to-get-data/.

Thermal infrared images from Landsat satellites provided a wealth of information about probable groundwater discharge locations. This is because groundwater discharge leads to springs and wetlands that support vegetation growth that is clearly seen in the thermal infra-red energy band.

The Landsat 8 satellite provides imagery with a resolution of 30 m. The remarkable level of detail is shown by the two images taken 30 years apart shown in Figure 11.5. The top figure (Figure 11.5a) shows a location in the north of Saudi Arabia taken in February 1986. There are a few circles of red – indicating actively growing vegetation associated with pivot style irrigation of wheat. These early developments were deemed successful, and the decision was taken to exploit fossil groundwater resources in the area. The second image (Figure 11.5b) is from February 2016 (30 years later) showing massive development

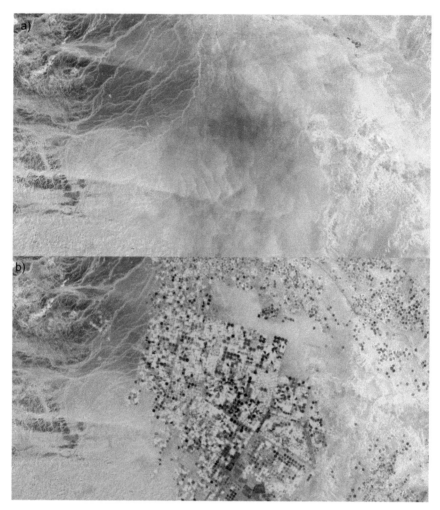

Figure 11.5 (a) The Saudi desert before the commencement of groundwater development – date: February 1986; (b) the same area of Saudi as seen in (a). The green circles are the wheat fields each fed by a rotating irrigator supplied from a groundwater abstraction bore – date: February 2016 (NASA JPL image archive)

of agriculture. Groundwater resource analysis has indicated that the resource has a lifetime of 30 years.

Figure 11.6 is another example of this technique from a limestone block close to Mpongwe in northern Zambia. This image used a combination of visible and near infrared bands from Landsat 3 to form what was referred to as a false colour composite (FCC). From field inspection, it was understood that actively growing vegetation occurred along the banks of perennial streams fed by groundwater discharge. These areas presented as a dark red colour on the FCC and can be seen occurring where groundwater discharges at springs from the limestone aquifer. The limestone pavement outcrop appears as a light blue on the FCC. In some areas, a deep loam occurs on top of the limestone and appears

Figure 11.6 Landsat false colour composite in 1978 showing GDEs (bright red tones) around a limestone block (blue and brown tones) at Mpongwe, northern Zambia (Photo: Ian Acworth)

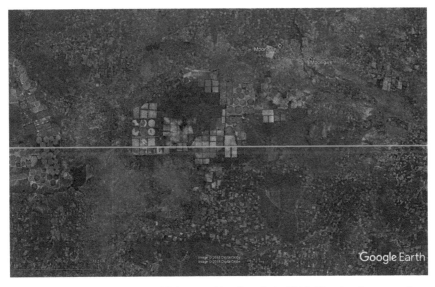

Figure 11.7 The same area as Figure 11.6 imaged by Google in 2018. The development of extensive agriculture is clearly evident. Discharge from the western end of the limestone block is captured by a dam with pumping used to irrigate immediately downstream.

as brown on the FCC. The identification of groundwater supplies at Mpongwe led to a recommendation for extensive irrigation development using overhead rotary irrigation. The success of that process can be seen on current Google Earth images, where the green circles represent crops irrigated from central pivots similar to those in the Saudi example and the square fields indicate rain-fed agriculture.

11.5 FIELD MEASUREMENT OF TEMPERATURE

11.5.1 Data loggers

Many logging systems designed for recording hydraulic head also measure and record the temperature at the same time. This information can often be of use when it indicates movement of groundwater or recharge from rainfall. Figure 11.8 shows a sequence of five HOBO temperature loggers installed inside PVC pressure pipe and the sequence of installation steps at a site in Maules Creek, northern NSW (Rau *et al.*, 2010).

11.5.2 Fibre-optic systems

Selker *et al.* (2006a) and Selker *et al.* (2006b) introduced the concept of using fibre-optic distributed temperature sensing. The temperature around a fibre-optic cable can locally change the characteristics of light transmission in the fibre. The damping of the light in the quartz glass through scattering can allow the location of the external temperature anomaly to be detected. Optical fibre can therefore be used as a linear sensor of temperature. Using these methods, it is possible to measure temperature along a cable of more than 30 km length with a temperature resolution of < 0.01 °C. Figure 11.9 from the USGS shows the detection of cooler groundwater flowing into a surface channel.

Figure 11.8 Example temperature array and field installation techniques (Rau *et al.*, 2010)

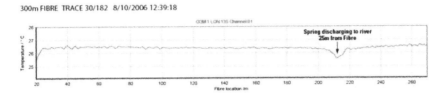

Figure 11.9 Temperature anomaly detected with a fibre-optic cable laid in the bed of a stream (USGS)

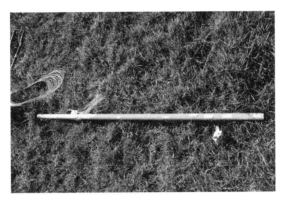

Figure 11.10 Coil of fibre-optic cable wrapped around a support prior to installation in the ground (Photo: Landon Halloran)

Figure 11.11 Fibre-optic installations at Hat Head (Photo: Landon Halloran)

There are some practical issues with laying cable out in the natural environment. Fibre-optic cable is delicate and easily fractured by, for example, boulders moving in a flood! The spatial resolution is also only accurate to a few metres over long cable lengths. Notwithstanding the advantages of the method, there are significant difficulties associated with calibration and data collection.

Halloran *et al.* (2016b) reported good results for the analysis of groundwater flow close to a pumping bore by using fibre-optic cable. They tightly wound the cable into a coil that was placed into the ground. This system allowed far greater spatial temperature resolution with data accurate to 9 mm reported.

11.6   THERMAL PROPERTIES

11.6.1   Specific heat

Before soil heat fluxes and temperature profiles can be calculated, the specific heat and thermal conductivity of the soil must be known. The volumetric specific heat of the soil is the sum of the specific heats of all soil constituents:

$$C_h = C_m \phi_m + C_w \theta_v + C_a \phi_a + C_0 \phi_0 \tag{11.1}$$

where ϕ is the volume fraction of the component indicated by the subscript, and the subscripts m, w, a and o indicate mineral, water, air and organic constituents. Table 11.2 lists values of these components. The contribution from soil air is often ignored, as C_a is so low (Table 11.2). In mineral soils, the organic fraction is also very small and therefore Equation 11.1 can often be simplified to:

$$C_h = C_m(1 - \phi_f) + C_w \theta_v \tag{11.2}$$

where:
ϕ_f is the total porosity and
v is the volumetric water content.

11.6.2   Thermal conductivity

The thermal conductivity of a soil depends upon the bulk density, water content, quartz content and the amount of organic matter (Campbell, 1985; Abu-Hamdeh and Reeder, 2000). Typical values are given in Table 11.2. As a result of the large differences in thermal conductivity of the individual components, it is clear that conductivity of a soil will depend upon the intimacy of packing and the extent to which air is displaced from the soil. Increasing the bulk density of the soil lowers the porosity and improves the thermal contact between the solid phases. But the increase caused by this process is small compared with the impact of adding water to the soil. The presence of a film of water surrounding the mineral grains improves the thermal contact and reduces the porosity.

Table 11.2 Thermal properties of soil materials

Material	Density Mg/m³	Specific heat J/g/Kelvin	Thermal conductivity W/m/Kelvin	Volumetric specific heat MJ/m³/Kelvin
Quartz	2.66	0.80	8.80	2.13
Clay minerals	2.65	0.90	2.92	2.39
Organic matter	1.30	1.92	0.25	2.50
Water	1.00	4.18	0.57	4.18
Air (20 °C)	0.0012	1.01	0.025	0.0012
Ice	0.92	1.88	2.18	1.73

11.7 FOURIER'S LAW AND DIFFERENTIAL EQUATIONS FOR HEAT FLOW

11.7.1 Mathematical basis

All the physical, biological and chemical processes which occur in the unsaturated zone are impacted by the soil temperature. Rates of some processes double as the temperature increases by 10 °C. In some cases growth of plants is better correlated with soil temperature than air temperature. Heat energy can be transported through the soil by a number of different mechanisms:

- conduction,
- advection of heat by flowing water,
- advection of heat by flowing air,
- radiation, and
- convection of latent heat.

The two most important processes under normal conditions are conduction and the advection of latent either by flowing water or flowing air.

The relation between the vertical flux of heat by conduction and soil temperature is given by another type of transport equation – the Fourier Law:

$$J_{H_c} = -\lambda \frac{dT}{dz} \tag{11.3}$$

where:
J_{Hc} is the heat flux density by conduction (W m^{-2}),
λ is the thermal conductivity (W m^{-1} K^{-1}),
T is temperature (degrees Kelvin K), and
z is vertical distance.

The negative sign indicates that heat flow is in the opposite direction to the temperature gradient. Equation 11.3 has the same form as the transport equations for water flow (Chapter 7). The solution methods used in solving equations for groundwater flow can therefore be applied to heat flow problems.

The Fourier equation (Equation 11.3) can be combined with the continuity equation to obtain the time dependent differential equation that describes heat flow. The approach is identical to the consideration of an elemental volume used when deriving the equation of groundwater flow:

$$\frac{\partial J_{H_c}}{\partial z} = \frac{\partial}{\partial z}\left(\lambda \frac{\partial T}{\partial z}\right) = C_h \frac{\partial T}{\partial t} \tag{11.4}$$

where:
C_h is the volumetric specific heat of the soil [J/m^3/K].

Solutions to Equation 11.4 describe the temperature as a function of depth and time.

If the thermal conductivity (λ) is both independent of depth in the soil and of time, we can rearrange Equation 11.4 and write:

$$\frac{\partial^2 T}{\partial z^2} = \frac{C_h}{\lambda}\frac{\partial T}{\partial t} = \frac{1}{D_h}\frac{\partial T}{\partial t} \tag{11.5}$$

where:
D_h is the soil thermal diffusivity (λ/C_h).

For a soil column that has a surface temperature distribution given by a regular function:

$$T_{0,t} = T_a + a_0 \sin \omega t \tag{11.6}$$

and which is infinitely deep, the temperature at any depth and time is given by:

$$T_{z,t} = T_a + a_0 \exp\left(-z/z_d\right)\sin\left(\omega t - z/z_d\right) \tag{11.7}$$

where:
T_a is the mean soil temperature,
a_0 is the amplitude of the temperature wave at the soil surface – the difference between the maximum and minimum temperatures,
w is the angular frequency of the oscillation (2π divided by the period in seconds), and
z_d is the sampling depth:

$$z_d = (2D_h/\omega)^{0.5} \tag{11.8}$$

The amplitude of the thermal wave is attenuated exponentially with depth and the phase of the wave is shifted with depth. At $z = z_d$, the amplitude is $e^{-1} = 0.37$ times its value at the surface. The temperature at three to four sampling depths would therefore be expected to be about the mean temperature for the cycle.

11.7.2   Thermal dispersivity

In contrast to the movement of water through a porous medium, heat can move by conduction through both the solid matrix and the fluid, in addition to the movement by advection as heat is carried by the moving fluid. Although there are similarities between the movement of heat and the movement of water and the mathematical formulations are similar, there are real differences between the physical processes involved. Mathematically, this is incorporated in the definition of the diffusivity term (Equation 11.5). It is noted that conductive heat transport should be more rapid than fluid transport as the result of the additional conductive pathway via the solid matrix. In contrast, the advective transport of heat is slower than solute transport since the heat capacity of the solids will retard the advance of the thermal front. These complexities were investigated by Rau *et al.* (2012).

Rau *et al.* (2012) constructed a hydraulic tank containing well-sorted sand that allowed precise monitoring of both heat and solute tracer movement from a point source of heat. The use of joint monitoring allowed identical conditions to be maintained to facilitate their comparison. The flow rates were chosen to replicate conductive movement only by keeping the Reynold's number to <3.0 (this threshold is discussed further in Chapter 15).

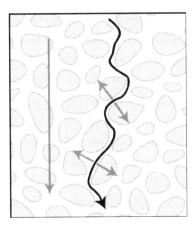

Figure 11.12 Heat movement by convection (black) and conduction (grey) in a porous medium (figure prepared by Gabriel Rau)

The experimental results demonstrated that heat transport with natural groundwater flow velocities can reach a transition zone between conduction and advection where the Peclet number lies between 0.5 < Pet < 2.5. They observed that Darcy velocities independently derived from heat and solute experimentation show a systematic discrepancy of up to 20%, and that experimental thermal dispersion results contain significant scatter.

11.7.3   Numerical solution to the heat flow equation

In a porous medium, such as streambed sediments, heat is transported by both conduction and advection, as shown in Figure 11.12. Heat conduction occurs through the bulk medium while heat advection is caused by fluid moving in the voids between sediment grains. Both processes can be superimposed.

These processes can be mathematically described using the conduction-convection equation:

$$\frac{\partial T}{\partial t} = \lambda \frac{\partial^2 T}{\partial z^2} - \frac{\phi v_f \rho_f c_f}{\rho c} \frac{\partial T}{\partial z} \tag{11.9}$$

where:
T is temperature which varies with time (t) and depth (z),
λ is effective thermal diffusivity,
ϕ is porosity,
v_f is vertical fluid velocity,
ρ_f is fluid density,
c_f is heat capacity of the fluid,
ρ is density of the saturated sediment-fluid system, and
c is the heat capacity of the sediment-fluid system.

The atmospheric temperature fluctuates due to daily, seasonal and annual changes in solar radiation. These fluctuations also affect the temperature of shallow surface waters which are then transferred into the subsurface by conduction and convection.

The temperature fluctuations can be used to derive fluxes between surface water and groundwater using two different approaches.

Approach I – forward model

If the temperature signal at the surface of the sediments is recorded, it is possible to predict the temperature signal at any depth within the sediments, using:

$$T_s^n(z,t) = T_0(z) + \Sigma_{i=1}^n \Delta T_s^{i,i-1}(z,\tau) \tag{11.10}$$

where:

$\tau = t - t_1$

and:

$$\Delta T_s^{i,i-1}(z,\tau) = \frac{\Delta T_w^{i,i-1}}{2}\left[erfc\left(\frac{z-C\tau}{2\sqrt{D\tau}}\right) + \exp\left(\frac{Cz}{D}\right)erfc\left(\frac{z+C\tau}{2\sqrt{D\tau}}\right)\right] \tag{11.11}$$

where:

C is $\frac{\rho_f c_f}{\rho c}$ and
D is $\frac{\kappa_e}{\rho c}$.

In the equations above $\Delta T_s^{i,i-1}(z,\tau)$ is the temperature at any depth (z) and time (t), T is the temperature at depth z when $t = 0$, $\Delta T_w^{i,i-1}$ is the change in temperature at depth $z = 0$, ρ_f and c_f are bulk density and heat capacity of the fluid, ρ and c are effective bulk density and heat capacity, ϕ is porosity, v_f is vertical fluid velocity and κ_e is the effective thermal diffusivity.

The forward modelling approach involves comparing the predicted temperature signal against a recorded temperature signal at some depth and adjusting the flux until both signals are the same, thereby deriving the flux. This approach is shown in Figure 11.13.

Approach 2 – transient model

A second approach is to record one temperature signal at the surface of the sediments and another at some depth within the sediments. The temperature signal at depth will display an amplitude drop and phase shift, as shown in Figure 11.14.

These measurements can then be used to derive the flux using:

$$v = \frac{2\kappa_e}{\Delta z}\ln(A_y) + \sqrt{\frac{\alpha_i + v^2}{2}} \tag{11.12}$$

and:

$$v = \sqrt{\alpha_i + \left(\frac{\Delta\phi 4\pi\kappa_e}{P\Delta z}\right)^2} \tag{11.13}$$

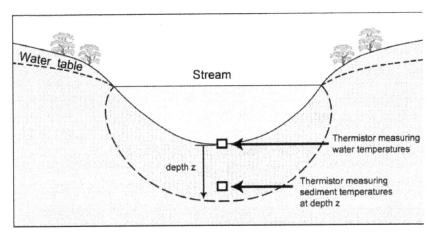

Figure 11.13 Cross-section of a stream aquifer section showing thermistor recording locations Stone-strom and Constanz (2003) (Circular, US Geological Survey)

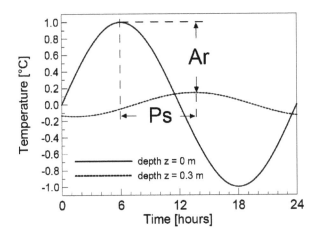

Figure 11.14 Example temperature signals showing how a signal at 0.3 m depth will have an amplitude drop and phase shift compared to a signal at 0 m depth (i.e., just above the sediments) (figure prepared by Gabriel Rau)

and:

$$\alpha_i = \sqrt{v^4 + \left(\frac{8\pi\kappa_e}{P_i}\right)^2} \qquad\qquad (11.14)$$

where:

$$v = v_f \frac{\rho_f c_f}{\rho c} \qquad\qquad (11.15)$$

where other terms are as described above, and:

P_i is the period of the temperature fluctuation and
A_y is the difference in amplitude between temperature signals.

Rau *et al.* (2010) used both solution methods to match their experimental data. Note that if groundwater pressure is similarly subject to a regular loading, then the groundwater pressure down through the aquifer will develop a time lag and a phase shift in the same way that the heat signal is modified. Timms and Acworth (2005) demonstrated this approach for a saturated clay profile on the Liverpool Plains.

11.8   HEAT AS A TRACER

Stonestrom and Constanz (2003) provide a detailed reference on the use of heat as a tracer in the saturated zone. This work was later extended by Rau *et al.* (2014). Halloran *et al.* (2016a) review the use of heat as a tracer in the unsaturated zone.

11.8.1   Saturated zone heat flow

Field application

Rau *et al.* (2010) describe field results (Figure 11.15) from a stream bed at Maules Creek in northern NSW. The upper graph shows two temperature signals: one from 0 m depth and the other from 0.6 m depth. The lower graph shows the same data but is transformed by a filtering process from true temperatures to relative temperatures. This filtering process allows the daily fluctuations in temperature to be more clearly seen. Using the data shown in the upper graph in Figure 11.15, it was possible to apply the first numerical approach to derive a flux (see Figure 11.16). For this location at Elfin Crossing, the calculated velocity was approximately 0.6 m/day downwards.

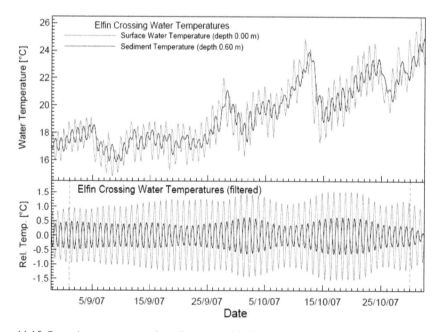

Figure 11.15 Example temperature data (Rau *et al.* (2010))

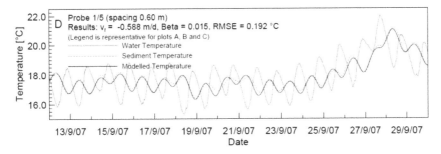

Figure 11.16 Vertical flux prediction using a forward model solution (Rau *et al.* (2010))

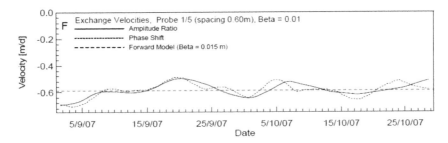

Figure 11.17 Vertical flux prediction using a transient model (Rau *et al.* (2010))

The second approach (transient model) was then applied to the data shown in the lower graph in Figure 11.15 to derive a flux (see Figure 11.17). Using this method, it was shown that the velocity varied through time but was also approximately 0.6 m/day downwards.

Rau *et al.* (2017) describe results from the same area but using a far larger (12 km) spatial separation between pairs of sensors installed at different depths. They use the heat data to better characterize ephemeral and intermittent flow processes. The approach exploited the fact that the downward propagation of the diel temperature fluctuation from the surface depends on the sediment thermal diffusivity. This is controlled by time-varying fractions of air and water contained in streambed sediments causing a contrast in thermal properties. They show how amplitude ratios of the diel component in temperature-time series measured at two vertical locations in shallow streambeds can be used to detect the presence of water in sediments and to characterize transitory flow conditions.

11.8.2   Unsaturated zone heat flow

Halloran *et al.* (2016a) review the development of heat sensing methods with particular emphasis on the unsaturated zone. They emphasize the importance of joint hydraulic head and temperature measurement in the field to better understand processes. While software for joint modelling of hydraulic head and temperature has been available for many years in computer packages produced by the US Geological Survey (SUTRA for example (Hughes and Sanford, 2004)), field data sets are uncommon.

Daily measurements of temperature at 10 cm, 20 cm and 50 cm in a clay loam soil at Narrabri in northern NSW are shown in Figure 11.19 for summer conditions and

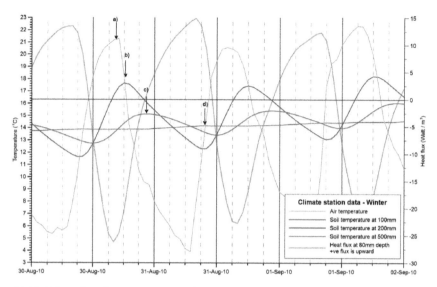

Figure 11.18 Winter hourly measurements of air temperature, soil temperatures at 10, 20 and 50 cm and the heat flux beneath a clay loam field at Maules Creek, NSW

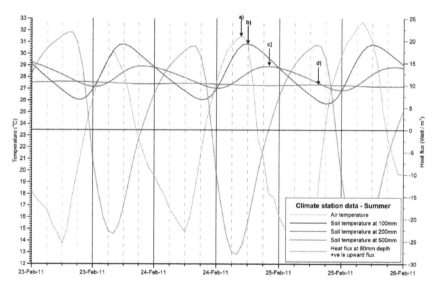

Figure 11.19 Summer hourly measurements of air temperature, soil temperatures at 10, 20 and 50 cm and the heat flux beneath a clay loam field at Maules Creek, NSW

Figure 11.18 for winter conditions. Included in these figure are the air temperature and heat flux profiles. The increasing phase lag and amplitude decay between 100 mm and 500 mm are clearly seen as the diurnal heat signal propagates down into the ground. The amplitude peaks are marked in these figures, and it is seen that there is approximately a 1.5 h lag between the maximum air temperature and the maximum ground temperature at 100 mm

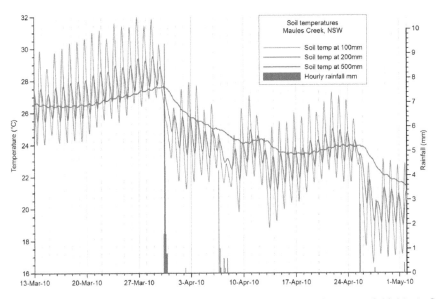

Figure 11.20 Soil temperature measured at three depths at the edge of a cotton field, Mauls Creek, NSW. Note soil cooled by rainfall recharge.

depth. This increases to 2.3 hr at 200 mm and 15 hr at 500 mm depth for summer conditions. The last two lags are greater for winter conditions, being 3.0 hr and 17 hr. The heat flux shows upward movement of heat peaking at approximately 08:00 in winter and an hour later in summer.

The difference between summer and winter conditions can be clearly seen by comparing Figure 11.19 with Figure 11.18.

The temperature of the ground can also be quite rapidly changed by a rainfall event. Figure 11.20 shows the impact of a 4-day period of rain on the temperature profile beneath a field. In periods not affected by rainfall, the diel change in temperature occurs as the result of conduction through the soil. The heat flux oscillates as the air temperature drives the change, but there is a significant phase lag and amplitude decay caused by the conduction process. However, when a rainfall event occurs, conduction gives way to advection as the heat is moved downward by the percolating rain water. The change in temperature is also a clear indication of movement of rainfall beneath the root zone at this location and may be taken as evidence of recharge. The temperature is reduced at 500 mm depth indicating rapid water movement beneath the root zone of the crops. The process was also observed in the 3D electrical images measured in the same area (Chapter 9). The cooling of the ground would be expected to reduce vegetation growth until the soil profile warms again.

Chapter 12

Drilling and sampling techniques

12.1 CABLE-TOOL (PERCUSSION) DRILLING

The Chinese are credited with the invention of the drilling rig. The ancient Chinese writings contain occasional references to well-drilling machines used on the borders of the Gobi Desert. These machines were based on the percussion drilling principle, which consists of repeatedly hitting the base of the well with a heavy weight to break the rock mass into small pieces. The rock fragments were removed from the well by bailing. The reciprocating movement at the surface was created by a large bow. This general type of device has been in use for about 4000 years and has been credited with drilling to depths in excess of 1000 m. The drill line was made of strips of bamboo spliced together with hemp.

The drilling of the Passy well in Paris, completed in 1875, added considerably to the knowledge base. The Passy well was completed at a diameter of 700 mm to a depth of 587 m. The yield from this bore was 21,152 m³/day. A 1.676 m diameter well was shortly after sunk to a depth of greater than 300 m at La Chapelle using a drill bit weighing 3628 kg operated by a steam winch which delivered 20 blows per minute.

The percussion method has changed little over the centuries. It is still slow but can handle much larger diameter constructions than more recent rotary-drill methods. With better engineering and manufacturing skills, the materials used in cable-tool drilling have evolved, but the basic method of raising and letting fall a heavy weight suspended by a cable has remained. Although this is a less popular method than in times past, it may still be appropriate if ground conditions make other methods of drilling difficult. This is the case even for investigation wells wherever the presence of large cobbles at depth occur. This situation is still best handled by the cable-tool method. It is sometimes appropriate to use cable-tool techniques to drill and set a larger diameter casing and then to switch to faster mud-based rotary methods.

In cable-tool drilling, the drill bit is attached to the lower portion of the weighted drill stem that, in turn, is attached by means of a rope socket to the rope or cable. The cable and drill stem are suspended from the mast of the drill rig through a pulley.

The cable runs through another pulley that is attached to a spudding beam. The beam is powered by the rig engine and moves up and down; the bit is alternately raised and dropped by this action. Given time, this method can penetrate all geological formations. A full string of cable-tool drilling equipment consists of five components:

1   drill bit,
2   drill stem,

3   drilling jars,
4   swivel socket, and
5   cable.

Each of the tools is an integral part of the cable-tool system. The drill bit is usually a massive steel bit with a chisel built into the base. Water flow channels are formed on either side of the bit to allow the slurry to pass away from the cutting face. The drill stem adds weight to the bit and also increases the length of the assembly. This increased length helps to keep the hole straight.

The drilling jars consist of a pair of linked steel bars. When the bit is struck, it can be freed most of the time by upward blows of the free-sliding jars. The stoke of the drilling jars is 230 mm to 450 mm. This distinguishes them from fishing jars, which have a stroke of 450 mm to 920 mm. The swivel socket connects the string of tools to the cable. The socket also transmits the rotation of the cable to the drill bit. In this way, each subsequent stoke of the drill bit is rotated from the previous stroke, keeping the hole circular.

The wire cable that carries and rotates the drilling tool is called the drill line. It is 16 mm to 25 mm in diameter and is a left-hand lay cable that twists the tool joint on each upstroke to prevent it from unscrewing. The cable is suspended over the top of the mast, down to the spudding sheave on the walking boom.

The cuttings are removed from the bottom of the hole by a bailer. This is generally a hollow metal cylinder fitted with a flap valve at the base. The drilling tools are removed from the hole and the bailer run in on a separate line.

In unconsolidated materials, a combination of pushing a casing into the ground and bailing the cuttings from the inside can be used.

Tools for fishing lost materials or components of the drill string have been developed for cable-tool methods. An initial pilot hole can be reamed out to allow construction and installation of an abstraction bore assembly. An assortment of reamers and other equipment can be seen in Figure 12.3. The long mast in Figure 12.1 allows casing runs of 12 m to be installed in one pass.

12.2   ROTARY METHODS

12.2.1   Introduction

The principle of rotary drilling is based on the controlled rotation and feed of hollow drill pipes attached to a drill bit, by means of a power driven rotary table or the more versatile hydraulic top-drive unit. Cuttings are removed from the bit and flushed up the hole by circulating fluid or air under pressure. The cuttings are delivered to the surface where they can be washed and collected for description. The conventional rotary drilling fluid is mud, used particularly for deep water wells and petroleum wells. For both it is necessary to use bore hole geophysics to determine the lithology, groundwater salinity and exact boundary of units prior to placing casing and screens.

Rotary drilling methods were developed for the oil and minerals industries and have been adapted for the groundwater industry. There are a number of possible variations, including:

*   direct mud rotary,
*   reverse circulation mud rotary,

Figure 12.1 Cable-tool rig installing an abstraction bore on the Liverpool Plains (Photo: Ian Acworth)

Figure 12.2 Cable-tool rig drilling soft limestone (chalk) in Kent, UK (Photo: Ian Acworth)

Figure 12.3 Reamers for use with a cable-tool rig (Photo: Ian Acworth)

- air-flush rotary, and
- down-hole air hammer.

The drill rig can often be equipped to drill using several of these methods. The manual produced by ADITC (2015) provides detailed instruction on the drilling process.

In direct mud rotary drilling, mud is pumped down the inside of the drill pipe and recovered up the annulus between the drill pipe and the formation. The annulus is created by using a drill bit, attached to the bottom end of the drill pipe, with a larger diameter than the drill pipe. Compressed air can also be used as a drilling fluid in the place of water. This may be a particularly good solution in areas where there is no initial water supply available to mix a drilling mud.

Down-hole hammer rigs use a pneumatic hammer drill bit to cut through solid rock. An air compressor is required to drive the drill bit, and drill cuttings are blown back to the surface via the annulus.

Reverse circulation rigs are larger than direct circulation rigs. They are used in formations where high up-hole velocities are required to recover the cuttings to the surface. The drilling fluid is pumped down the outside of the drill string and recovered up the inside of the drill pipe. The larger volumes of mud and the larger pipe diameters require the use of a heavier drill rig as an operating platform.

Excellent reviews of rotary drilling methods are provided by Driscoll (1986) and more recently by Misstear *et al.* (2006).

Figure 12.4 shows the setup of a water well rotary rig. The rig is mounted on the back of a lorry and the mast raised using hydraulics. The rotational drive to the drill string is provided by either:

- a top-drive unit where a hydraulic motor is located on top of the drill string, as in Figure 12.4, or

Figure 12.4 Rotary rig about to spud in. Note that this standard of dress would no longer meet current OHS guidelines in many countries (Photo: Ian Acworth)

Figure 12.5 Air lifting a bore to improve yield (Photo: Ian Acworth)

Figure 12.6 Water well drilling site, showing mud pits and rig tilted while geophysical logging in progress to determine design parameters (Photo: Ian Acworth)

- a rotary table where the weight of a top-drive assembly would be unstable, as shown in Figure 12.10.

Note the use of cement to allow the mud returned to the surface to be directed straight to the mud pits, shown in Figure 12.7, and the tri-cone roller drill bit, shown in closeup in Figure 12.8.

Top-drive rigs are the most common rig used in water well drilling. They have been adapted from technology used in the minerals industry for quarrying.

Figure 12.9 shows the setup of a rotary rig capable of drilling to several thousand metres. In this configuration, the rotary-table drive is provided by a rotary table as shown in Figure 12.9.

Mobile mud pits are often used and contain shaker systems for the geologist to extract cuttings from, as shown in Figure 12.11. Mobile mud pits in use at Centennial Park are shown in Figure 12.12.

Formation samples are normally taken by the driller for inspection by the geologist and for a grain-size analysis. The frequency of sampling will be determined by the drilling contract, but is often every metre. As a minimum, samples should be taken at every formation change. Minimum requirements for formation sampling in Australia are set out in *Minimum Construction Requirements for Water Bores in Australia* (Agriculture *et al.*, 1997).

12.2.2   Drill bits

Various types of drill bit have been designed for drilling in different lithologies. A drag bit is used for fast drilling in unconsolidated or semi-consolidated formations. Roller cone bits

Figure 12.7 Mud pits for rotary drilling (Photo: Ian Acworth)

Figure 12.8 Tri-cone roller drill bits for rotary drilling (Photo: Ian Acworth)

(tri-cone) are used in consolidated sedimentary formations, as shown in Figure 12.8. Pneumatic button bits are used for air drilling (Figure 12.13).

12.2.3 Drilling fluids

The use of drilling fluids in the construction of bores is a large field in its own right. The subject is covered in detail by Driscoll (1986).

The first drilling fluid utilized in rotary drilling was water. The entrainment of natural clay particles produced a mud and gave rise to the term *drilling mud*. Later on, specific clay compounds were deliberately added to the mud to control the mud properties. The technology has

Figure 12.9 Rotary rig setup for deep drilling (Photo: Ian Acworth)

Figure 12.10 Rotary table drive for rotating the drill string (Photo: Ian Acworth)

Figure 12.11 Mud shaker system to allow cuttings to be separated from the drilling mud for geological logging (Photo: Ian Acworth)

developed to a point where many different compounds can be added to the fluid to control its physical and chemical characteristics.

The control of the drilling fluid is fundamental to successful drilling using the rotary method. If the drill cuttings are not removed to the surface from the cutting faces of the

Figure 12.12 Self-contained mud pit assembly to minimize surface impact at Centennial Park, Sydney (Photo: Ian Acworth)

Figure 12.13 Air hammer drill bit (Photo: Ian Acworth)

Figure 12.14 Drag bit for use in soft formations (Photo: Ian Acworth)

drill bit, the likely outcome is that the drill string and drill bit will become lodged in the hole. Recovery of both is then a long (costly) process and may result in the bore being abandoned with loss of the equipment in the bore. If the return circulation of the fluid fails, drilling will stop while efforts are made to thicken the mud; increase the mud flow; increase the density of the mud; or any other available technique to re-establish the flow. Rotary drilling is often a 24-hour process for this reason.

Types of drilling fluids

Water based fluids are used in the water well drilling industry. Water-based drilling fluids consist of the following three phases:

1   a liquid phase,
2   a suspended-solid (colloidal) phase, and
3   cuttings entrained during the drilling.

Air-based drilling fluids may consist of only air, but more commonly have a surfactant added to create a foam.

There are many different additives commonly used, including:

- flocculants to precipitate clay particles,
- dispersants (thinning agents),
- weighting materials to increase the fluid density,
- corrosion inhibitors,
- lubricants,
- bactericides, and
- lost-circulation materials.

The exact drilling fluid system chosen for a given project will depend upon the geology of the area and the type of materials to be drilled. In general, air-based systems are used with rotary-hammer rigs to drill through hard-rock formations, while water-based fluids with clay or polymer additives are used in unconsolidated or soft sedimentary formations. Clean water is used for reverse circulation drilling equipment.

Functions of a drilling fluid

The primary functions of a drilling fluid are as follows.

- The removal of drill cuttings from beneath the drill bit to the surface. The rate at which cuttings are removed depends upon the viscosity, density and up-hole velocity of the drilling fluid and the size, shape and density of the cuttings.
- The drilling fluid preserves the stability of the borehole wall and prevents expansion of swelling clays. Water-based fluids must provide a positive pressure to the borehole walls to prevent caving and collapse of the formation. If collapse does occurs, the drill bit and string can become trapped and the hole abandoned.
- The drilling fluid serves to both cool and lubricate the drill bit.
- The correct drilling fluid can control losses of fluid to the formation by creating a highly impermeable cake of drill mud on the formation wall. The invasion of drilling fluids

into the formation must, however, be minimized, as this material has to be removed from the formation during the well development process.

- The physical characteristics of the drilling fluid require adjustment so that the drill cuttings drop out readily at the surface and are not returned down the hole by the fluid circulation.
- The drill fluid returns the cuttings to the surface where the geologist may be trying to form a geological log of the formations penetrated. The fluid should act to disguise the true nature of the cuttings as little as possible.
- Lastly, the fluid should be able to suspend the cuttings in the borehole if the circulation of fluid is halted for any reason. This ability is important, as the settling to the bottom of the hole of the cuttings in a deep bore can easily trap the drill string.

In many cases the drilling fluid will be unable to satisfy all the above criteria, and continuous monitoring of the physical properties of the fluid is advisable. The drilling fluids are commonly mixed in mud pits dug close to the drill rig. Typical configurations are shown in Figure 12.8 and Figure 12.10.

The viscosity of a water-based fluid is measured with a marsh funnel. This is an inverted cone into which a specified volume of fluid is placed. The viscosity of the fluid is related to the time taken to drain through the bottom of the funnel. Typical viscosities, as measured by a marsh funnel, for various unconsolidated formations are shown in Table 12.1 (Driscoll, 1986).

The density of the fluid can be checked by weighing a sample of the fluid in a simple balance. In general, the density of the drilling fluid must be high enough to counter any hydrostatic pressures encountered in the bore. Groundwater encountered in an artesian aquifer under pressure can suddenly cause a blow out of the drilling fluid unless the fluid density is increased by adding barite or some other high-density material.

It is worth remembering that the location of a sufficient supply of water to make up the drilling mud is often one of the first problems to overcome at a new site. The water does not have to be absolutely clean but should have a TDS of less than 5000 mg/L to allow efficient mud operation. A water tanker is frequently part of the drilling operation with the task of hauling water to the site.

12.2.4 Air drilling

The use of air as the drilling fluid has become increasingly popular for water well work. A borehole completed using air flush requires much less development time compared to the work carried out to remove a drill cake from a mud drilled hole. The compressor is

Table 12.1 Typical mash funnel viscosities for unconsolidated formation drilling

Materials Drilled	Appropriate Marsh Funnel Viscosities (seconds)
Fine sand	35–45
Medium sand	45–55
Coarse sand	55–65
Gravel	65–75
Coarse gravel	75–85
Lost circulation zones	85–120

usually the key to successful air drilling because insufficient air volume and pressure are the main causes of problems with this drilling technique. Compressors are rated to deliver a given air volume at a certain operating pressure. The mechanics of air drilling are more complex than those of fluid drilling because the drilling fluid (air) is compressible and also contains other fluids (such as water) which may be incompressible. For example, 21.2 m^3 of air at atmospheric pressure occupies 1.2 m^3 at 1720 kPa pressure.

Typical compressor ratings are given in Table 12.2 (Driscoll, 1986)).

The compressors are rated for sea-level pressures and for air temperatures of approximately 15 °C. Correction factors are required for operation at different temperatures and pressures.

Air drilling is relatively simple and effective. High-pressure, high-volume compressors are attached to drilling rigs and used to clean out the cuttings. The circulation system is the same as with a fluid system direct rotary rig. The air is passed through the drill column to the bit. All cuttings are quickly removed up the annulus at a high velocity of 900 to 1500 m/min. Compressors are rated to deliver a given volume at a certain operating pressure. Compressors connected in series will improve the pressure in an air system; a parallel arrangement will increase the volume.

Up-hole circulation velocity

The drilling fluid is used to remove cuttings from the bore. The viscosity of the fluid and the up-hole velocity will determine how efficient this process is. Viscosities and recommended up-hole velocities are given for different fluids in Table 12.3 (Agriculture *et al.*, 1997). Conventional circulation air drilling is performed where stable formations exist. Penetration rates are fast, owing to the speed at which the cuttings are removed from the bit and up the hole. Setup time is also quicker, as there is no necessity for mud tanks or pits. Sampling from an air-drilled hole is easier and the sample is more representative and recognisable.

Air drilling, however, loses its advantages when drilling unstable, unconsolidated formations, as the hole is continually prone to collapse, or where drilling under high water inflows

Table 12.2 Typical pressure and volume capacities for water well rigs

Number of Stages	Pressure kPa	Volume m^3/s
1	862	0.28–0.42
1	1030	0.35–0.54
2	1900	0.35–0.42
2	2410	0.42–0.50
2	3450	0.50

Table 12.3 Recommended minimum up-hole circulation velocities

Circulating Fluid	Marsh Funnel Viscosity (s)	Recommended Up-hole Velocity (m/s)
Air or mist	15 to 25	
Water	26	0.6
Normal mud	32–40	0.4
Thick mud	50–80	0.2

Figure 12.15 Tertiary river channel cobbles that are very difficult to drill through using conventional mudflush or air rotary techniques – particularly when dry (Photo: Ian Acworth)

Figure 12.16 (a) Raising the outer casing with air-return attached; (b) lowering casing into the bore; (c) detail of outer casing being lowered around drill rods (Photos: Ian Acworth)

or under a considerable depth of water. Water requires a minimum of 10 kPa pressure per metre depth to remove it from the bore hole.

The circulation system is the most important part of any drilling operation. If this system is poorly designed or malfunctions, all other components of the operation will deteriorate or fail.

12.2.5 Tubex drilling

ODEX is an option when unconsolidated formations are too dense or contain too many cobbles for auger drilling. ODEX is a down-hole air hammer system that is designed to advance casing during drilling. Once a desired depth is reached the eccentric bit can be retrieved, leaving the casing in place for sampling or installations.

Figure 12.17 Joining two of the outer casing sections together using welding. This could not be carried out during very dry conditions due to the risk of starting a fire (Photo: Ian Acworth)

Figure 12.18 Checking the verticality of the welds. The welding has to be cut again for withdrawal of the casing (Photo: Ian Acworth)

In the absence of an eccentric bit, casing can still be advanced outside the drill tube and air flow recovered inside the outer casing to recover cuttings. This system was used at Wellington (western NSW) where investigation bores were drilled into a Tertiary river channel now occupied by the Macquarie River. The cobble-sized alluvial deposits occurred on the surface (Figure 12.15) but also at greater depths. The rig used at the site is shown in Figure 12.16.

12.2.6   Auger drilling

A top-drive rig can be used with hollow-stem augers to advance an investigation hole in unconsolidated formations. The hollow stem is used to support the formation while samples are taken or a piezometer is installed. The typical components of a hollow-stem auger are shown in Figures 12.19 and 12.20.

Sampling can be achieved at any depth if a pilot bit is used inside the hollow stem. The pilot bit and rods are withdrawn and a sampling device inserted. This can be either a split-spoon sampler driven into the formation beneath the bottom of the augers to obtain a

Figure 12.19 Hollow-stem auger rig showing the augers on the ground and the top-head drive auger rig (Photo: Ian Acworth)

Figure 12.20 Small bobcat-type tracked drill rig equipped with augers for access over dry sandy ground (Photo: Ian Acworth)

disturbed core or a thin wall tube pushed beyond the base of the augers to collect an undisturbed sample.

When drilling into an aquifer that is under even low to moderate confining pressure, or drilling far beneath the water table, the sand and gravel will frequently "heave" upward into the hollow stem. If a center plug is used during drilling, then flow of the aquifer material frequently occurs as the rods are pulled back. When this problem occurs, the bottom of the augers must be cleaned out before further sampling or progress can occur. Unfortunately, the removal of material from the base will normally induce further material to flow up into the augers thus compounding the problem. These problems can usually be overcome by one or more of the following:

- adding water into the hollow stem in an attempt to maintain sufficient hydraulic head at the bottom of the auger flights that the flow of material is prevented;
- adding drilling mud additives to increase the viscosity and density of the water inside the auger stem;
- drilling with the first auger flight constructed around a wire wound screen. This allows formation water to enter the hollow flights and to equalize the hydraulic pressure and prevent the formation entering the base of the augers;
- drilling with a pilot bit, a knock-out plug or winged clam shell to physically prevent the formation entering the hollow-stem flights.

Rigs can be obtained with sufficient power to drive the augers to depths of 30 m to 40 m in clayey/silty/sand profiles. Because of the relative ease of sample collection, hollow-stem augers are the recommended technique for the majority of monitoring well constructions.

12.2.7 Sonic drilling

Sonic drilling is a relatively recent development that is capable of drilling through overburden or rock with excellent core recovery. The drill head contains an oscillator that causes a high-frequency oscillation to be superimposed on the drill string. The drill bit is physically vibrated up and down in addition to being gently pushed down and rotated. These three combined forces allow drilling to proceed rapidly through most geological formations.

In overburden, the vibratory action causes the surrounding soil particles to fluidize, thereby allowing the drill to penetrate the formation. In unstable formations, the same mechanism is used to vibrate a construction casing down over the core tube, allowing the core to be withdrawn on a wire line. In rock, the drill bit causes fractures at the rock face facilitating advancement of the drill bit. Drilling fluids are not required in sonic drilling. Hole verticality is also very good, as the drill string is not used to push the drill bit downwards.

The oscillator is driven by a hydraulic motor and uses out-of-balance weights to generate high sinusoidal forces that are transmitted to the drill bit. An air spring is also incorporated in order to confine the alternating forces to the drill string. The frequency can be varied to suit operation conditions and is generally between 50 and 160 Hz.

Sonic drilling can core to depths over 150 m. The depth required will determine the rig size. Figures 12.21 illustrates the use of sonic drilling on the small tracked vehicles manufactured by GEO-PROBE. In this application, the rig maintains the ability to carry out a full suite of rotary drilling using augers or rotary percussion drilling using a head-mounted hammer. Figure 12.22 provides more detail of the sonic drill head equipment.

The samples collected via sonic drilling tend to be far more disturbed than those collected using a triple-tube method, but they are far superior to those collected by conventional mud-flush methods, as depicted in Figure 12.23.

12.3 SAMPLING METHODS

12.3.1 Introduction

Obtaining samples from a bore is normally one of the main reasons for drilling an investigation well for a groundwater abstraction scheme. The least favorable types of sampling are

Figure 12.21 GEOPROBE tracked vehicle equipped with sonic drilling head (Photo: GEO-PROBE web page accessed 5/10/17)

Figure 12.22 Detail of the sonic head equipment on the GEOPROBE rig (Photo: GEOPROBE web page accessed 5/10/17)

the inspection of bailed samples from a long bailer or the inspection of drill cuttings from the returns of a rotary drill operation. It is not possible to say where these samples have come from with any degree of accuracy or how they have been disturbed on their journey to the surface. Needless to say, these methods still frequently form the most extensively used methods! Wherever possible, the bore should be logged using down-hole geophysical techniques as described in Chapter 13.

Figure 12.23 Drill samples laid out at a rotary mudflush rig site (Photo: Ian Acworth)

Figure 12.24 Drill samples collected from air return (Photo: Ian Acworth)

When air rotary techniques are used, the samples can be collected in bags as shown in Figure 12.24. In unconsolidated formations, better sample can be obtained by either:

- split-spoon driven sample, or
- a thin-walled pushed sample.

These are the primary techniques used to provide sample data for contaminated site sampling. Samples from consolidated formations can be obtained by the use of wire-line core barrels, such as the Treifus system described in Section 12.3.5.

12.3.2   Split-spoon sampling

The split-spoon sampler was developed for geotechnical investigations and provides an indication of formation density by conducting a standard penetration test (SPT).

The split-spoon sampler is attached to the end of the drill rods and lowered to the bottom of the hole. The common practice in formation evaluation (geotechnical work) is to collect 18-inch samples every 5 feet as the bore is advanced. This is only possible in a stabilized bore. The procedure for an SPT is to record the number of blows necessary to drive a sample of standard dimensions through the base of the hole using a 140-pound weight dropping

Figure 12.25 Split-spoon sample being slid out of the core tube and opened after drilling (Photo: Ian Acworth)

Figure 12.26 Sample from a split-spoon sampling tube showing the junction between clean sands and an organic rich pond sediment (peat) beneath the sand (Photo: Ian Acworth)

through a 30-inch interval. The number of blows to drive the sample provides a qualitative measure of the formation density. A disturbed sample can be obtained from the split-spoon sampler by opening it at the surface, as shown in Figure 12.25. The sample can be laid out and logged in the field (Figure 12.26) before transfer to a sample bag for later analysis.

12.3.3   Modified triple-tube methods

To obtain good quality minimally disturbed core samples in sands, an HQ triple-tube coring system can be modified for use within augers. This can be used to recover core from the interior of 250 mm hollow flight augers using a wire-line system. The inner sample barrel from the HQ3 drilling system can be adapted to engage a casing advance latch mechanism installed in the lead auger. The inner barrel fits inside and protrudes through the bottom cutting face of the lead auger. The latch mechanism locates the sample barrel inside the lead auger and enables the sample barrel to remain free from the rotation of the auger. This has been achieved by the use of a thrust bearing and ball race incorporated into the latch mechanism. The latch mechanism allows for easy recovery of the 1.5 m long sample barrel via the use of standard wire line recovery techniques.

The sample barrel is lined with a clear PET liner and the core is acquired inside the liner. To prevent cross contamination, the use of water is minimized to core barrel withdrawal. The sample barrel containing the liner and core are recovered to the surface using the wire line. The core and liner are then expelled from the sample barrel and the ends of the PET liner capped with PET caps. This method of core recovery achieved minimal disruption of the

Figure 12.27 Core catcher installed in the base of the core. As the core rises up inside the barrel, the stainless steel arms open up. With reverse movement, the arms close down preventing core and fluids from escaping (Photo: Ian Acworth)

core and maintains the pore fluids in place. A new PET liner is inserted and the sample barrel lowered back into the bore until the latch mechanism engages the barrel at the correct depth. As the system comprises three components to the core barrel – the outer steel tube that spins with the core-cutting assembly at the base, an inner tube that does not spin and a clear uPVC tube fitted inside this inner barrel – the system is referred to as a triple-tube system. The sample is held in the base of the core tube by use of a core catcher. Examples are shown in Figure 12.27.

Examples of core acquired by this system are shown in Figure 12.28.

12.3.4   Diamond-core drilling using a double barrel

The conventional coring method in consolidated formations is to use a diamond-core barrel consisting of a double-tube system. The system is widely used in core holes for coal exploration or mining investigation. This is a wire-line system where the core bit rotates around the core barrel and has a diamond-impregnated cutting face that is cooled using a mud flush or air flow. When the core barrel is full, the core inside the barrel is returned to the surface using a wire-line as described for the modified auger system. Cores are seldom cut in solid rock for groundwater investigation as groundwater, if present, is located in fractures that are difficult to obtain core. Contamination studies and detailed research into groundwater flow in fractured rocks requires careful mapping of fractures (width and orientation) that diamond-core drilling can facilitate.

12.3.5   Treifus barrel

A Treifus wire-line core barrel with rotary mud flush was used on the Liverpool Plains to recover 100 mm core in clear uPVC tubes from a 40 m sequence of unconsolidated clays and sands (Acworth *et al.*, 2015b). The system delivered almost 100% recovery of

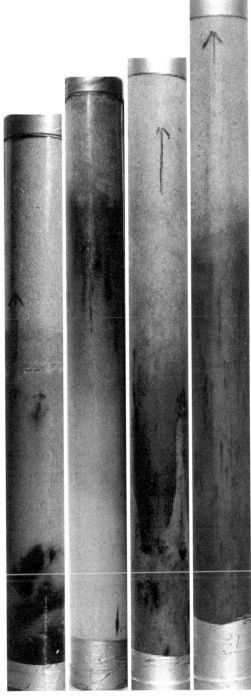

Figure 12.28 Cores recovered from unconsolidated sands at the Southlands DNAPL contaminated site in Sydney using the wire-line core barrel inside a modified lead auger to contain the ball race and core barrel latching system. A core catcher similar to Figure 12.27 was used at the base of the clear uPVC tube. This method collected core with pore fluids intact. The cores are seen here after extraction and capping. X-ray images of these cores revealed significantly more structure in the core (Acworth, 2001b)

Figure 12.29 The inner cutting face of the core tube (nose) extends proud of the rotating outer cutting face of the barrel by 40 mm (Photo: Wendy Timms)

Figure 12.30 Base of the core inside the bottom of the core nose prior to removing the drill head. This material was analyzed on site for moisture content and bulk density and bagged for later grain-size determination (Acworth *et al.*, 2015b) (Photo: Wendy Timms)

the core with excellent sample integrity. The components of the Treifus system, similar to other triple-tube systems, were originally developed during the Snowy Mountains investigation and are now available from a number of manufacturers.

The core is cut by direct pushing of the inner core barrel that extends beyond the rotating outer core nose by as much as 40 mm (Figure 12.29). In this way, the core rises up inside the clear uPVC inner tube with no contact with the circulating drill fluids. The core barrel has a catch facility at the top and can be recovered inside the outer casing using a wire line. A ball race assembly is built into the top of the core barrel that allows the outer barrel to be driven by rotation of the drill rods but prevents the inner barrel from rotating. In this way, the outer core bit removes the formation after the undisturbed core has passed into the core barrel.

Coring is carried out until the barrel is full or until refusal. Figure 12.30 shows the base of the freshly cut core before removal of the core nose assembly. Barrels can be supplied in a

Figure 12.31 Using stilsons to unscrew the core nose assembly (Photo: Wendy Timms)

Figure 12.32 Using compressed air to force the core tube from the core barrel after the core nose was removed (Photo: Wendy Timms)

number of lengths, but it was found that a 1500 mm barrel worked well in unconsolidated clays. The core nose assembly is removed and contains core material that occurs below the base of the core barrel (Figure 12.31). Extrusion of the core contained inside the uPVC tube using compressed air into the top of the core barrel is seen in Figure 12.32. Material held in the core nose was retained for analysis on site and bagged for later grain-size determination. The cores were transferred to a mobile fridge to better preserve hydrochemical parameters *in situ.* After removal of the core, the assembly is cleaned and a new uPVC tube installed and the core nose attached before the complete assembly is lowered on the wire line back to the base of the hole, where a latching facility engages the barrel at the base of the bore prior to pushing down to collect the next core.

Running sands had to be cased off by installing temporary casing to protect the core barrel. This is conceptually the same approach as used in the Tubex system for drilling through cobbles. Even with this precaution however, core material was lost, as it proved impossible to retain the material in the barrel.

Chapter 13

Geophysical logging

13.1 INTRODUCTION

Borehole geophysics may be defined as the science of recording and analysing continuous or point measurements of physical properties made in bores or cased wells. These measurements are used to identify the lithology and structure of formations and to make inferences regarding water quality, porosity and permeability of formations. Borehole geophysics has the great advantage that measurements can be made almost in the formation, whereas surface measurements can only be used to infer subsurface properties. Geophysical logging is required to determine the design of an abstraction bore when drilled using mud systems. The geophysical logs indicate the aquifer horizons that will be screened and the clay zones that will have blank casing placed against them. Geophysical logging results are also widely used for aquifer correlation in the large sedimentary basins such as the GAB and the Murray, Ottway and Perth Basins.

A wide range of borehole geophysical sondes have been developed, including oil field tools, coal investigation tools, and specialized engineering geophysics tools (Keys, 1989).

13.1.1 Logging equipment

Early logging systems comprised an analogue chart recorder that monitored the output from a down-hole probe, with the chart recorder motor activated by a pulse provided from a pulley shaft encoder (Figure 13.1 or 13.2). These early systems used a very simple probe that often required multiple and heavy cables. Log analysis had to be carried out on the paper traces from the flat-bed chart recorder or had to be later digitized.

The availability of frequency-modulated signal transmission capabilities has allowed the evolution of single core cable equipment with a simple winch and data acquisition unit on the surface controlled by a field laptop and down-hole sondes. The equipment is much lighter and often battery operated. The disadvantage is that the sondes have become much more expensive.

Equipment is available from a number of American and European suppliers. A Welsh company (GeoVista) manufactures sondes that can be chained together in such a way that as many as 15 parameters can be measured independently at the same time. The GeoVista equipment shown in Figure 13.3 is being used to investigate fracture flow in sandstone beneath the Nepean River. The fractures were considered to be due to disturbance from long-wall mining beneath the sandstone.

Figure 13.1 Eastman 1000 logger in use Nigeria in 1977 (Photo: Ian Acworth)

Figure 13.2 Early Robertson equipment in use in 1983 (Photo: Ian Acworth)

Figure 13.3 GeoVista logging equipment set up in a river bed (Photo: Wendy Timms)

13.1.2 Borehole effects

The method by which the bore was drilled, completed and tested may have a pronounced effect on the geophysical logs obtained in the bore. The objective of geophysical logging of a bore is to obtain representative measurements of physical properties in the formation or of fluids within the bore. The drilling process disturbs the rocks near the borehole to varying degrees and may open up hydraulic pathways through a layered succession which would not otherwise exist.

The use of drilling muds will significantly alter the response of logs carried out in holes drilled in this manner. Monitoring of the drilling mud properties is often required to decouple the effects of the mud from the formation. This aspect has been developed in detail by workers in the oil industry and was one of the influences behind the development of resistivity logs. Deep water bores drilled using mud also require a similar analysis.

The major effect upon many geophysical logs is the diameter of the borehole. This may vary significantly depending upon the drilling technology employed and the care and experience of the driller. It should also be remembered that it is the internal diameter (id) of the final casing installed in the bore that will determine the size of logging sonde that can be accommodated.

Two other factors impact upon the logs:

*   the depth to the water table and
*   the type, diameter and length of bore casing installed.

Accurate interpretation of the geophysical logs requires detailed information to be available on these matters.

13.1.3 Available sondes

The sondes available for borehole logging equipment include the following:

*   Gamma,
*   High-sensitivity gamma,
*   Gamma spectrometry,
*   Gamma-gamma (density tool),
*   Spontaneous potential,
*   Single-point resistivity,
*   Long (64″) and short (16″) resistivity,
*   Caliper (one, three and four arms),
*   Fluid temperature,
*   Fluid conductivity,
*   Fluid flow,
*   Magnetic susceptibility,
*   Electromagnetic induction,
*   Neutron,
*   Neutron-neutron,
*   Verticality,
*   Water sample,
*   Hydrolab water quality, and
*   Sonic,

Table 13.1 Response of logs to porosity

Log	Property measured	Response to total porosity	Response to effective porosity	Response to secondary porosity
Resistivity	Both resistivity and volume of fluid in interconnected pores	No current flow through isolated pores	Responds only to effective porosity	Detects secondary porosity, affected by the shape of pores
Gamma-gamma	Electron density	Both response to rocks with substantial porosity	Does not distinguish	Does not distinguish from primary porosity
Neutron	Hydrogen content	Best response to rocks with minimal porosity	Does not distinguish	Does not distinguish from primary porosity
Acoustic velocity	Average compressional–wave transit time	Only related to total porosity when it is primary and intergranular	Does not distinguish	Does not respond to secondary porosity under most conditions

- Acoustic televiewer, and
- Down-hole TV.

13.1.4 Quality assurance

The subject of quality assurance in borehole geophysical logging is receiving more attention. It should of course be possible to carry out logs with the same equipment at different times in the same bore, or with different equipment at the same time in the same bore and obtain the same result. This has unfortunately not been the case, and more attention is needed to ensure accurate calibration of logs and standardization of equipment.

13.1.5 Porosity measurement

There is no one log which is capable of actually measuring the porosity of a formation; however, the porosity affects the response of many geophysical logs. The neutron, gamma-gamma and acoustic-velocity logs are all sometimes incorrectly referred to as porosity logs. Resistivity logs are significantly affected by porosity changes. The various logs respond differently to changes in total, effective, primary and secondary porosity. Table 13.1 (Keys, 1989) shows some of these relationships.

13.1.6 Log selection

A list of criteria for the selection of logging tools is given in Table 13.2 (Keys, 1989).

13.2 CALIPER LOGS

13.2.1 Introduction

Caliper logs provide a continuous record of borehole diameter and are used extensively in hydrogeology. These logs are often required for the interpretation of other logs and are

Table 13.2 Selection criteria for logging tools

Log Type	Properties measured	Potential applications	Required borehole conditions	Other limitations
Spontaneous potential	Electric potential caused by salinity differences in borehole and interstitial fluids	Lithology, shale content and water quality	Uncased borehole filled with conductive fluid	Salinity difference needed between borehole fluid and interstitial fluids; correct only for NaCl fluids
Single-point resistance	Resistance of rock, saturating fluid and borehole fluid	High-resolution lithology; fracture location by differential probe	Uncased borehole filled with conductive fluid	Not quantitative; hole diameter effects substantial
Multi-electrode	Resistivity of rock and saturating fluids	Quantitative data on salinity of interstitial water; lithology	Uncased borehole filled with conductive fluid	Normal logs provide incorrect values in thin beds
Gamma	Gamma radiation from natural or artificial radio-isotopes	Lithology – may be related to clay and silt content and permeability; spectral identifies radioisotopes	Any borehole condition, except large diameter, or several strings of casing and cement	None
Gamma-gamma	Electron density	Bulk density, porosity, moisture content, lithology	Optimum results in uncased borehole; qualitative through casing or drill stem	Severe borehole diameter effects
Neutron	Hydrogen content	Saturated porosity, moisture content, activation analysis and lithology	Optimum results in uncased borehole; can be calibrated for casing	Borehole-diameter and chemical effects
Acoustic-velocity	Compressional wave velocity	Porosity, lithology, fracture location, and character, cement bond	Fluid-filled and uncased, except cement bond	Does not detect secondary porosity; cement bond and waveform require expert analysis
Acoustic televiewer	Acoustic reflectivity of borehole wall	Location, orientation, and character of fractures and solution openings, strike and dip of bedding, and casing inspection	Fluid-filled and 3- to 16-inch diameter	Heavy mud or mud cake attenuate signal, slow logging
Caliper	Borehole or casing diameter	Borehole-diameter corrections to	Any conditions	Deviated holes limit some probes;

(Continued)

Table 13.2   (Continued)

Log Type	Properties measured	Potential applications	Required borehole conditions	Other limitations
		other logs, lithology, fractures, and borehole volume for cementing		significant resolution difference between tools
Temperature	Temperature of fluid near sensor	Geothermal gradient, flow within borehole, location of injected water, correction of other logs and curing of cement	Fluid-filled	Accuracy and resolution of probe varies
Electrical conductivity	Most measure resistivity of fluid in borehole	Quality of borehole fluid, flow within borehole and location of contaminant plumes	Fluid-filled	Accuracy varied, requires temperature correction
Flow	Flow of fluid in borehole	Flow within borehole, location and apparent hydraulic conductivity of permeable interval	Fluid-filled	Spinners require faster velocities; needs to be centralized

also useful in obtaining information on bore construction, the development of secondary porosity related to fracturing and on lithology. Caliper logs may consist of one, two, three or four arms. The basic principle of operation consists of lowering the sonde to the bottom of an unlined bore where an arm is or arms are extended sideways from the sonde and held against the borehole wall by springs. Movements of the arm(s) are recorded by a displacement transducer(s) and transmitted to the surface as the caliper log sonde is pulled up the bore hole.

The use of multiple arms rather than single arms results in more information relating to the bore geometry. A single arm leads to considerable averaging and may miss sub-vertical fractures completely. Some sondes average the signal from multiple arms, but the best approach is to record the displacement from the arms separately and transmit this information to the surface. The calibration of caliper sondes is achieved by placing them in cylinders of known diameter both before and after logging.

Caliper logs can be used to check borehole construction after the event. The slight change in internal diameter will record the junctions between casing lengths. In Figure 13.4, the 6 m lengths of steel casing are clearly visible in the caliper log. The bore casing appears to have been impacted by compression forces below the base of the alluvial material.

13.3   VERTICALITY LOGS

It is often assumed that because a drill rig is set up to drill a hole down into the earth, that the resulting hole will be vertical. This is frequently not the case, and there is now an industry

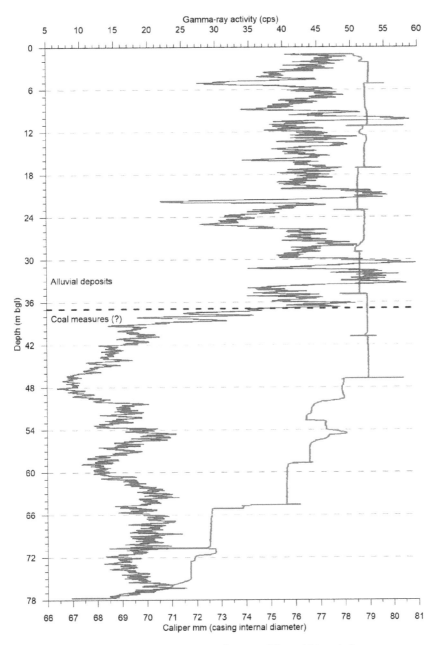

Figure 13.4 Plot of caliper and gamma-ray activity logs in a 30 yr old borehole

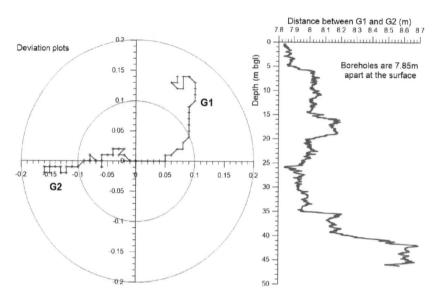

Figure 13.5 Verticality surveys at Cattle Lane

devoted to drilling inclined holes or holes that are directed away from a central location. CSG and oil field drilling is frequently designed to complete a number of bores starting at the same surface location but then spreading out beneath the ground. From a groundwater resources perspective, the absence of verticality can be a major problem when it comes to installing a free-hanging submersible pump in a bore. If the pump lies against the casing, it will frequently not operate to design capacity and may soon develop mechanical problems.

The verticality log is run to establish the exact location of the bores below ground level. An accurate magnetometer is lowered into the bore that records the dip and inclination of the magnetic field. These data can be processed to give accurate x-y-z coordinates down the borehole.

An example of verticality data is shown as measured between two bores at Cattle Lane, Liverpool Plains, NSW, in Figure 13.5.

13.4 FLUID PROPERTY LOGS

13.4.1 Fluid temperature

The commonly available groundwater temperature logging tools use a glass bead thermistor to measure the temperature of the bore water. Two logs are sometimes recorded, a temperature log and a differential temperature log. The differential temperature is more sensitive to small temperature changes and therefore better at locating the movement of water into or out of the borehole column. Temperature logging systems can be calibrated in thermal baths against a mercury thermometer. The accuracy of the log is typically as high as 0.01 °C.

Movement of the temperature logging sonde disturbs the thermal profile in the water column. For this reason, temperature logs are often run first in the borehole. If there is no flow in or adjacent to the borehole, the temperature of the bore water will be a function

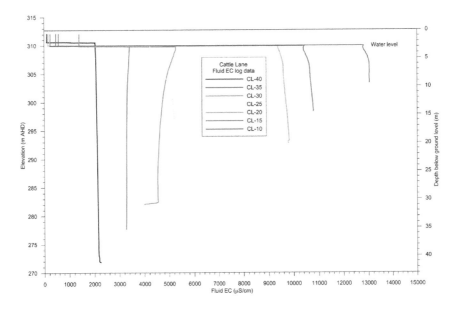

Figure 13.6 Fluid EC logs for Cattle Lane

of the local geothermal gradient. Typical geothermal gradients vary between 2.5 °C per 100 metres to 3 °C per 100 metres.

If the flow of groundwater in a bore is rapid, there will be no change in temperature with depth. This property is used to indicate flow into or out of the bore and to show the length of bore over which the fluid is moving. It is generally the case in a fractured rock or a cemented aquifer that the construction of a screened bore or open hole will provide the highest hydraulic pathways for the movement of water. The entry and exit points will be easily seen on the temperature log, as shown in Figures 13.7 and 13.8.

Temperature logs can be used to locate the position of a cement grout which has been placed behind a casing string. This is due to the exothermic reaction associated with cementing.

13.4.2   Fluid conductivity

Fluid conductivity is the reciprocal of fluid resistivity and can be measured by many commercially available sondes. The units of measurement are typically mSiemens/metre (mS/m). The sonde may be readily calibrated by measurements in specially prepared solutions of known electrical conductivity. Fluid conductivity is directly affected by fluid temperature. Therefore the use of the two logs in combination is necessary. The fluid conductivity is often corrected to a constant 25 °C for comparison with other logs.

The impact of fluid with different electrical conductivity occurring in different fracture zones is seen in Figures 13.7 and 13.8. There is a clear concurrence with changes in fluid temperature at the same depth.

Fluid conductivity logs can be used to map areas of saline contamination emanating either from the ocean or from sites of contamination. Most contaminant plumes from landfill sites exhibit a higher fluid electrical conductivity caused by the increase in total dissolved solids

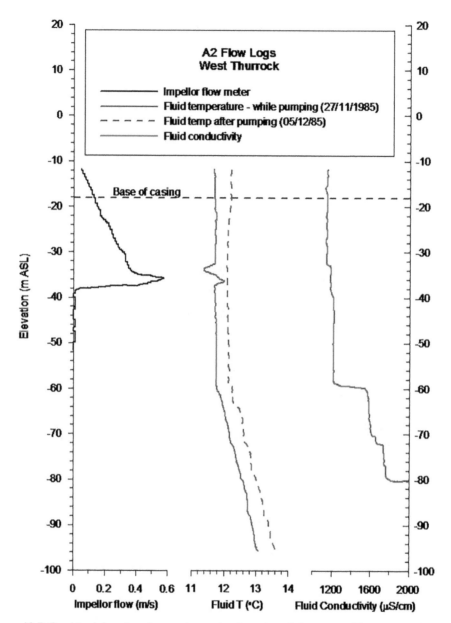

Figure 13.7 Combined flow logs for an abstraction bore in soft limestone. The pump is located at 35 m depth and a major contributing fracture is shown at −60 m by the pronounced change in thermal gradient from a regional increase to a constant temperature. Increases in fluid conductivity confirm the temperature logs

(TDS) in the groundwater. The TDS is linearly related to the fluid conductivity for a given water chemistry. If sufficient chemical analyzes have been carried out in an area to obtain this relationship, then further measurements of fluid conductivity can be used to reasonably accurately map the TDS. Figure 13.6 shows fluid conductivity profiles measured inside 50 mm

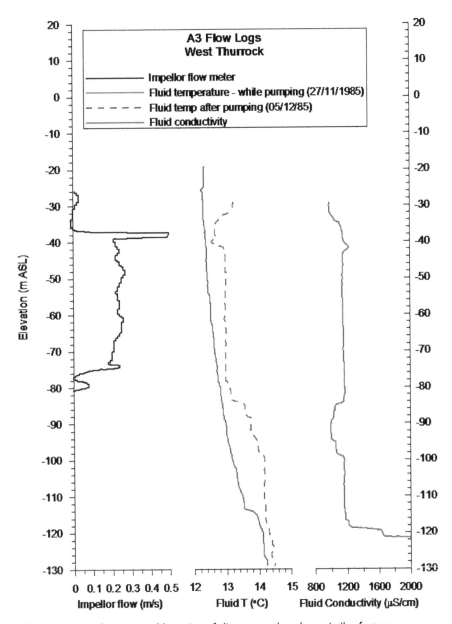

Figure 13.8 Flow logs for a second bore in soft limestone that show similar features

PVC piezometers at Cattle Lane on the Liverpool Plains of NSW. The open area for each piezometer is a 1 m long screen section at the base. These are the same piezometers as shown in Figure 11.2 with the logs taken at the same time. The results show a great variation in the fluid EC of water contained in the formation.

13.4.3   Fluid flow

The rate of fluid movement in a borehole may sometimes be sufficiently high that an impeller flow meter can be used in the bore to map changes in fluid flow rate. Impellers have a certain inertia to overcome before they start to rotate. For this reason, they are not useful for monitoring very small flows. The speed of the groundwater flow in the bore can be increased by the use of a shroud around the meter to act as a constriction to the flow.

Heat pulse flow meters have been developed for detecting low flows. These devices work by emitting a pulse of heat and then detecting its arrival at sensors mounted a fixed distance above and below the heat source.

The fluid flow, temperature and conductivity logs taken in an observation bore at 5 m distance from an abstraction bore during a pumping test are shown in Figures 13.7 and 13.8.

13.4.4   Borehole video cameras

Examination of the walls of a borehole and status of screens can be carried out using a submersible video camera (black-and-white or colour). The focus, orientation, illumination and direction of the camera can all be varied from the surface. The signal is sent back to the surface and viewed on a monitor. The signal is also recorded for later examination.

Figure 13.9 is taken with a colour down-hole camera looking sideways at the borehole screen. Spreading deposits of bacterially derived iron deposits can be clearly seen. In Figure 13.10, a previously unknown cave at Wellington is seen from a borehole camera.

13.5   FUNDAMENTALS OF NUCLEAR GEOPHYSICS

The understanding and interpretation of gamma, gamma spectrometry, gamma-gamma and neutron logs requires an appreciation of the nature of subatomic particles and their decay characteristics.

The nucleus of an atom consists of protons with a mass of 1 unit and an electrically positive charge, and neutrons with a neutral charge and a mass of 1 unit. The mass number (A) of the atom is equal to the sum of the protons and neutrons. The atomic number (Z) is equal to the number of protons. Electrons orbit the nucleus of the atom in orbitals which have

Figure 13.9 The spread of iron deposits onto a screen in a groundwater bore (radial view)

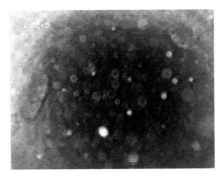

Figure 13.10 Iron deposits on the side of the casing above the screen (axial view)

shapes determined by the energy of the electron. Electrons have a negative charge to balance the protons and a mass of 1/1,840 the mass of a proton. Electrically neutral atoms have the same number of electrons as protons. Under some circumstances, an atom can lose, or gain, one or more electrons and therefore become a charged molecule or ion.

Isotopes are one of two or more different states of the atom, with the same atomic number (Z), and therefore the same chemical properties, but different mass numbers (A) due to the presence of additional neutrons in the nucleus of the atom. For example, uranium consists of three isotopes with different mass numbers of 234, 235 and 238, which can be separated by their difference in weight. These isotopes are naturally occurring in rocks.

Stable isotopes are those that do not change their mass structure with time. Unstable isotopes decay by changing their nuclear mass until they reach a stable state. These decay processes can take a very long time. Almost 1400 isotopes are known, of which 1130 are unstable. However, only 65 unstable isotopes occur naturally (Keys, 1989). There are 104 known elements, of which 83 exist as more than one isotope.

Unstable isotopes change their mass structure spontaneously by the release of radiation. Radiation from the nucleus occurs in the form of alpha particles, positive and negative beta particles and gamma photons. X-rays are released by orbital electrons as they change from an orbital with a higher energy level to an orbital with a lower energy level.

Alpha particles are stopped by a piece of paper; beta particles are stopped by a thin sheet of aluminum. Gamma photons can penetrate 100 mm of lead. Gamma photons are therefore monitored in geophysical logging because they can pass through both rocks and the well casing to be detected by the logging equipment. Neutrons can penetrate dense materials but are slowed and captured by materials that have a high concentration of hydrogen, such as water.

The half-life of an isotope is the time taken for the isotope to lose one half of its radioactivity by decay. Half-lives may vary from a few seconds to millions of years depending upon the nuclear mass structure.

13.5.1   Detection of radiation

The products of the radioactive decay of unstable isotopes (alpha, beta and gamma photon radiation) can be measured by capturing the radiation and converting it to electronic pulses which can be counted and sorted as a function of their energy content. The energy of radiation is measured in electron volts (ev). Energy levels of thousands (kev) or millions (Mev) of electron volts are common. Radiation intensity is measured directly as the number of pulses detected in unit time.

The most common type of radiation detector is a scintillation counter. Scintillation counters are laboratory-grown crystals which produce a flash of light when a gamma photon or neutron passes through them. The scintillations are amplified in a photo-multiplier and the output is a pulse of electrical energy whose amplitude is proportional to the energy of the radiation.

The number of pulses detected by a crystal in a particular radiation field will be proportional to the size of the crystal. Sodium-iodide crystals are used to detect gamma photons and lithium-iodide crystals used to detect neutrons.

Most detectors used in geophysical logging tools are side-collimated. That is, they receive energy from a window open to the side of the borehole. The tool is often designed to be held against the side of the bore as it is pulled to the surface. This is more important with neutron logging than gamma logging due to the greater absorption of neutrons in water compared to gamma photons.

The pulses that are detected in the logging sonde are transmitted to the surface and counted by a rate meter. An analogue rate meter converts the pulse to a DC voltage which is logged onto a chart recorder. A digital rate meter counts the pulses for a pre-selected time period and then transmits a signal to a digital-recording system. Clearly, the efficiency of the rate meter is fundamental to the effectiveness of the radiation count. If the rate meter is too slow, two pulses coming close together will be counted as one pulse, leading to an underestimation of the true radioactivity of the ground.

13.5.2 Counting statistics and logging speed

Although the half-lives of unstable (radioactive) isotopes are known accurately, it is not possible to predict the actual decay event. Gamma photon emission has a Poisson distribution, which means that the standard deviation is equal to the square root of the number of radioactive decay events recorded. Therefore, the greater the number of events recorded, the greater the accuracy of the measurement.

The statistical variations in the radioactive decay cause the rate meter counter to vary even when the sonde is held at the same position. There is therefore a conflict between signal measurement accuracy and logging speed. If the count rate is rapid enough and the measuring period sufficiently long, then the logs will be repeatable. The trick is in determining the system parameters to achieve the log repeatability.

Digital rate meters produce a better representation of the true radioactivity than analogue rate meters.

The effect of the combination of logging speed and the rate meter selection strongly influences the accuracy of the recorded log. Thin beds will be missed if the logging speed is too high. An example of the difference between two logging speeds in the same formation is shown in Figure 13.11.

Keys (1989) gives a detailed discussion of the variation of log quality with the choice of analogue rate meter time constant and logging speed.

The speed of logging also introduces an error in the depth estimation of lithological boundaries, an underestimation of the thickness of thin beds and an overestimation of

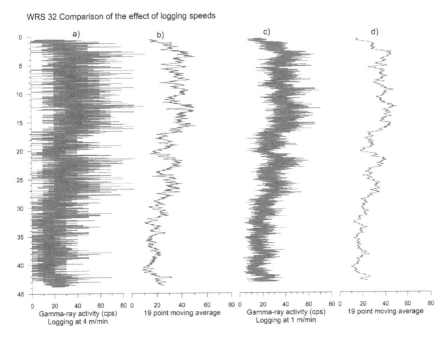

WRS 32 Comparison of the effect of logging speeds

Figure 13.11 Gamma log variations with logging speed

larger bed thicknesses. If accurate resolution of thin beds is required, there is little option but to log at a slow and constant speed. This is less a problem with motor-controlled winches than hand-wound winches.

13.5.3 Use of radioactive sources in well logging

Radioactive sources are required for making gamma-gamma and neutron logs. The movement of these sources is often controlled by state agencies. No source is required for the gamma log as it measures natural gamma photon activity.

13.5.4 Gamma-ray activity logs

The gamma log provides a record of the total natural gamma photon energy within a selected wave band that is recorded within the vicinity of the logging probe. The energy of gamma photon radiation can be used to identify the isotope that emits the radiation.

Isotopes of uranium and thorium produce gamma radiation as they decay to produce end-member stable isotopes. However, the main source of gamma rays monitored during hydro-geological logging is often the isotope potassium-40. Potassium is abundant in feldspar material which weathers down to produce clays. Uranium and thorium can also be concentrated in clay by the processes of adsorption and ion exchange. Gamma logs are therefore fundamentally a means of monitoring lithology which has a high clay content and is usually an aquitard.

Table 13.3 Characteristics of three commonly occurring isotopes which emit gamma radiation

Isotope	Energy of major gamma peaks (Mev)	Number of photons per second per gram	Average content in 200 shale samples	Percentage of total gamma intensity of shale samples
Potassium-40	1.46	3.4	2% Total K	19
Uranium-238 Series in Equilibrium	1.76	2.84×10^4	6 ppm	47
Thorium-232 Series in Equilibrium	2.62	1.0×10^4	2 ppm	34

Some characteristics of the three isotopes are given in Table 13.3 (Keys, 1989). These characteristics were determined by the American Petroleum Institute (API) on 200 shale samples.

Under most conditions, 90% of the gamma activity arises from material within 150 mm to 300 mm of the borehole wall. The location of the gamma probe in the borehole may therefore cause problems in large diameter (> 150 mm) bores. The bores are seldom absolutely vertical and as the sonde is pulled up the bore, it will vary in distance from the side wall, leading to a variability in log activity. This can be overcome by centering the probe in the bore.

The material used to gravel pack a water well can strongly influence a gamma log. For example, if the gravel pack material is a river sand with significant potassium feldspar material, a constant high background activity level will be added to the log. By contrast, a thick gravel pack composed of low activity material will subdue the gamma activity reaching the detector.

The interpretation of the gamma log is qualitative and best by synergistic comparison with other logs. The amplitude of the gamma log is changed by any borehole conditions which alters the density of the material that the photons have to pass through. Gamma logs can be used through steel casing, although there may be a drop in signal strength through the steel. Gamma logs work well through plastic casing and do not require a saturated column of water. This property makes the gamma log a very useful log to run in conjunction with the electromagnetic induction log in contamination studies.

13.5.5   Gamma spectrometry

Gamma spectrometry can be used to detect which radioactive isotope is present in the bore. This is achieved by recording the gamma photon activity at several different frequencies. Gamma spectrometry is widely used in the oil and minerals industry but has not become accepted in the groundwater industry. This is probably due to a combination of lack of training and increased cost of the equipment.

Gamma-ray activity logs are most often dominated by potassium-40 which is found in illite associated with clay. In some areas, thorium is more dominant than potassium. Thorium is often associated with sandstone-derived grains than from clay, and the working hypothesis that high gamma count indicates clay is then no longer valid. Figure 13.12 from the Liverpool Plains (Acworth *et al.*, 2015b) demonstrates this point.

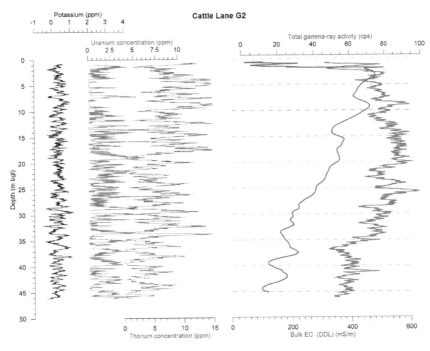

Figure 13.12 Gamma-ray activity spectra from the Southern Liverpool Plains

13.5.6 Gamma-gamma logs

The gamma-gamma log is also known as the density log. The log records the gamma radiation received at a detector from a gamma source in the probe. The requirement for a gamma source makes this log less attractive than the passive gamma log described above. However, the advantages of the tool have long been recognized in the oil industry where it is a standard application.

Gamma-gamma probes contain a stable radiation source which is usually Cesium-137. The detectors in the sonde are shielded from the source by a heavy metal such as lead or tungsten.

Gamma-gamma logging is based upon the principle that the attenuation of gamma radiation as it passes through the rock is directly proportional to the electron density of the material. Gamma radiation is absorbed in the rock material by three processes:

- Compton scattering,
- photoelectric adsorption, and
- pair production.

Compton scattering is the most important of these processes because the probe shell shields the low-frequency energy associated with photoelectric processes and pair production cannot occur at energies characteristically produced by Cesium-137. The energy loss caused by Compton scattering is inversely proportional to the electron density. This is related to the bulk density of the material by the ratio Z/A (atomic number/mass number).

13.5.7   Log interpretation

A properly calibrated log can be used to distinguish both the lithology and the borehole construction. The main use is to measure bulk density and to convert this to porosity if the density of the mineral composition is known. This relationship is shown in Equation 13.1 for clean water:

$$\rho_b = p_m^{(1-\phi)} \qquad\qquad (13.1)$$

where:
ρ_b is the density derived from the log,
ρ_m is the matrix density – known for a range of materials, and
ϕ is the formation porosity.

Density logs can be used to locate voids behind casing or in the formation. They can also be used to locate the presence of cement grout behind casing.

Where information on fluid density is available, it is possible to calculate a log of formation porosity from the Equation 13.2:

$$\phi = \frac{\rho_g - \rho_b}{\rho_g - \rho_f} \qquad\qquad (13.2)$$

where:
ρ_g is the grain density,
ρ_b the bulk density, and
ρ_f the fluid density.

13.6   NEUTRON LOGS

13.6.1   Introduction

Neutron logs require a source of neutrons in the sonde. Neutron detectors are also located in the sonde, and the difference between the source and received signals provides information on the lithology and moisture content.

Two different configurations of sondes are used in the groundwater industry. The first uses a large neutron source and detectors spaced at long spacing for the measurement of porosity. The second uses a small source detector range and is used for measuring soil moisture in the unsaturated zone.

Neutron probes contain a source that can emit neutrons with a high frequency. The source material varies but can contain beryllium and americium.

Two or more detectors are used in some modern probes. The probes are usually collimated and decentralized against the borehole wall.

The flux of neutrons around a probe source can be thought of as a cloud of varying neutron density. Fast neutrons emitted by the source undergo three possible interactions with the surrounding matter:

- inelastic scatter,
- elastic scatter, and
- absorption or capture.

These processes are shown diagrammatically in Figure 13.13.

Inelastic scattering can only take place with energetic (fast) neutrons shortly after their release from the source. When this process occurs, the nucleus of the atom, which has been impacted by the fast neutron, receives energy which it releases as a gamma photon as it decays back to a stable state again. As part of this process, the neutron energy is

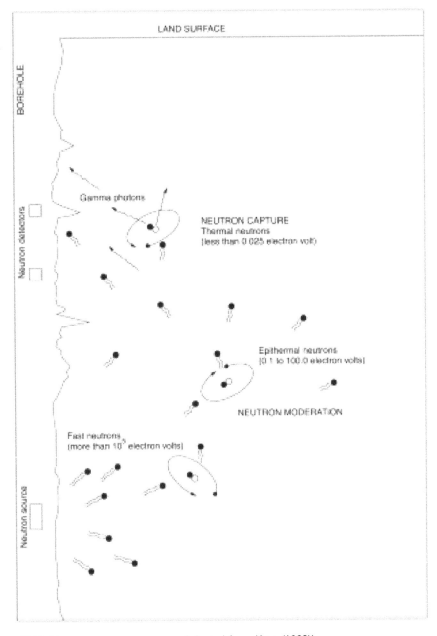

Figure 13.13 Neutron capture processes (adapted from Keys (1989))

decreased. After a number of inelastic encounters, the neutrons have slowed sufficiently for the second process to become significant. The second process is that of elastic scatter.

In elastic scatter, the mass of the atom which the neutron hit determines the loss of energy by the neutron. Light elements are the most effective at slowing (moderating) neutrons. As hydrogen is the lightest element, it is the most effective at neutron moderation. The mass of the hydrogen atoms is the same as that of a neutron.

Most neutron capture takes place after the neutrons have slowed significantly. This occurs in zones which have the highest water content. When a thermal neutron is captured by a nucleus, the nucleus becomes excited and gives off a gamma photon so that it can return to a more stable, lower energy, state. The cross-sections for thermal-neutron capture are highly dependent upon type of nucleus involved.

The process of neutron capture and gamma photon generation produce a gamma photon count which is inversely proportional to the amount of water present in the rock for source detector spacings greater than 300 mm. If the detectors are placed close to the source, then the number of moderated and captured neutrons increases with increasing moisture content because the neutrons are not able to penetrate so far into the surrounding formation. This is the process used in soil moisture probes. A neutron log is not capable of detecting the difference between water bound on a clay surface and water occupying interstitial porosity.

Neutron logs can be made through steel or plastic casing.

13.6.2 Neutron probes for soil moisture measurement

The neutron soil moisture device consists of a compact radiation source and detector designed to be installed in a narrow access tube in the ground. The access tubes are installed in the ground by conventional methods such as augering. Soil samples are taken during the augering for the measurement of soil moisture by conventional methods.

The neutron logging device is mobilized to the field to take a series of measurements as required. In general the equipment is lowered down the hole in increments of depth so as to build up a soil moisture profile at that location. In this manner, the use of soil moisture by crops can be monitored and the value of the soil moisture deficit established as a function of depth. The quantification of the soil moisture deficit allows accurate irrigation scheduling and avoids the over-application of irrigation water.

13.6.3 Equipment design

The radiation source in the neutron logging device is usually radium-beryllium or americium-beryllium (Jury *et al.*, 1991). The source emits high-energy neutrons in the energy range 5 MeV (1600 km/s). These neutrons collide with the nuclei of atoms in the surrounding soil. Since the nuclei of most atoms other than hydrogen are much heavier than the emitted neutrons, most collisions will not slow the neutrons. When the neutrons collide with hydrogen, their similar mass causes them to be slowed substantially and to reach energies characteristic of the thermal motion of hydrogen atoms (0.03 eV or 2.7 km/s) in the soil after only a few such collisions.

The detector, which is mounted alongside the source, is only sensitive to thermal neutrons. The detector cell is filled with BF_3 gas. When the boron nucleus absorbs a slow neutron, it emits an alpha particle which is detected and counted. A count of thermal neutrons is therefore proportional to the moisture content in the soil. A calibration curve can

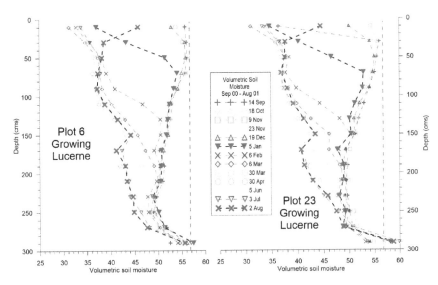

Figure 13.14 Soil moisture profiles changing beneath a crop of lucerne over a year, grown on smectite-derived soils on the southern Liverpool Plains

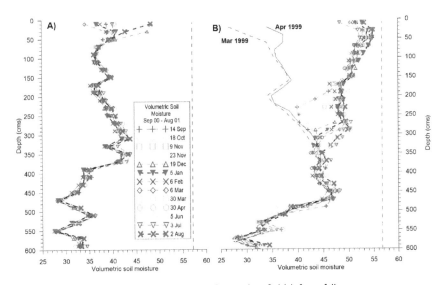

Figure 13.15 Soil moisture profiles over a year beneath a field left to fallow

then be constructed from the soil moisture measurements initially made on the soil samples and the thermal neutron count. Background hydrogen, present in organic matter or in clays, will register as a constant factor on the thermal neutron count. Examples of soil moisture profiles obtained using a neutron log are shown in Figure 13.14 for a growing crop and in Figure 13.15 for a plot left to fallow.

13.7 RESISTIVITY AND ELECTROMAGNETIC LOGGING

13.7.1 Introduction

The short and long normal resistivity logs have been used as standards in bore logging for many years. The principles are based upon the use of four electrodes in a similar manner to surface resistivity work. Five electrode configurations are used:

- spontaneous potential,
- single point,
- short and long normal, and
- lateral.

13.7.2 Spontaneous-potential (SP) and single-point resistance (SPR) logs

This is one of the oldest and simplest logs. The down-hole sonde can comprise a simple lead electrode. Whereas the equipment is simple, interpretation of these logs can be difficult, particularly in freshwater environments. Due to the simplicity of the down-hole sonde, it is often run first down old bores, particularly if conditions are unknown and there exists the possibility for getting the sonde caught in the bore.

The SP log is a record of electrical potentials that develop at the contacts between shale or clay beds and a sand aquifer. The chief sources of spontaneous potential in the borehole are electrochemical and electrokinetic or streaming potentials. Electrochemical effects, which are probably the more significant, can be further subdivided into membrane and liquid-junction potentials. The SP log is often run as part of a wider logging suite including the 16-inch and 64-inch logs described below. However, in its simplest form, it is the output of a millivolt meter connected to a chart recorder with inputs provided by the down-hole electrode and a surface electrode.

This logging technique was developed in the oil industry, where potentials created by the differential invasion of drilling mud into sand and shales provided a ready means of discrimination. The log can provide unpredictable and uninterpretable results when used in some freshwater environments. The log is also particularly prone to problems caused by stray currents and equipment problems. Stray ground currents from distant electrical storms can render the log useless. Electric currents related to corrosion of buried pipelines or well casing can produce anomalous potentials in the ground, as can nearby electric motors.

13.7.3 Normal resistivity logs

Many water well logging systems have the capability to measure short and long normal resistivity logs. The arrangement of electrodes is similar to that for surface electrical resistivity work where the potential difference due to a positive and negative current electrode pair is measured. In the borehole application, one current and one potential electrode are positioned remotely. They may be some distance from the active pair of electrodes on the same sonde, or they can be positioned at the surface. The short normal resistivity log is then acquired by measuring the potential due to a current source at 16 inches distant. The long normal log is similarly acquired using an electrode separation of 64 inches.

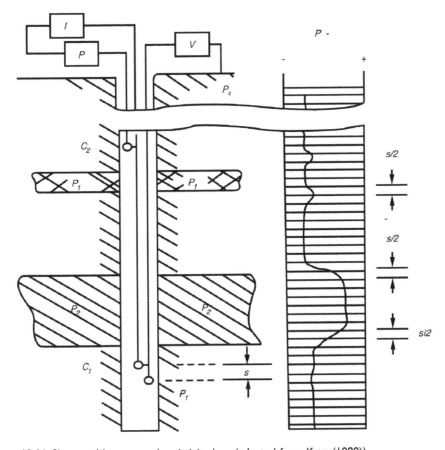

Figure 13.16 Short and long normal resistivity logs (adapted from Keys (1989))

These separations have no theoretical significance, but rather a historical basis, and are maintained purely to allow standardization of logs.

A system schematic is shown in Figure 13.16 (Telford *et al.*, 1976). Each of these electrical logs can be recorded while logging rapidly as they depend only upon potential field effects.

The volume of investigation of the normal resistivity sondes is considered to be a sphere with diameter approximately twice the distance between the down-hole electrodes. Thus, examination of the 16-inch and 64-inch logs together gives a measure of both the deep formation resistivity (from the long normal log) and the formation close to the bore. Where drilling muds have been used to advance the bore, the differential penetration of mud, as deduced from the short normal log, can be used as an indication of porosity.

While the normal resistivity logs can be calibrated to measure accurately, it should be stressed that the measurements made are of apparent resistivity and require correction to obtain true formation resistivities. In practice, the logs are most often used for qualitative analysis. Correction factors include:

- the resistivity of the invaded zone (if mud was used),
- the diameter of the bore,

- the bed thickness,
- the mud resistivity,
- the resistivity of adjacent beds, and
- the electrode separation (AM).

The resistivity of the formation is sometimes calculated automatically by the equipment. The true value can also be calculated from a knowledge of the electrode separation using the relationship shown in Equation 13.3:

$$\Delta\Phi = \frac{I\rho}{2\pi}\left(\frac{1}{r_1} - \frac{1}{r_4} - \frac{1}{r_3} + \frac{1}{r_4}\right) = \frac{I\rho}{2\pi}G \qquad (13.3)$$

where:

I is the current,

ρ is resistivity,

Φ is the potential difference,

r_1 is the distance between the first potential electrode and the first current electrode,

r_2 is the distance between the first potential electrode and the second current electrode,

r_3 is the distance between the second potential electrode and the first current electrode, and

r_4 is the distance between the second potential electrode and the second current electrode.

The resistivity of the formation is then given by:

$$\rho = \frac{\Delta\Phi}{I}\frac{2\pi}{G} = R \times \frac{2\pi}{G} \qquad (13.4)$$

where:

$\frac{2\pi}{G}$ is the geometric factor and varies for each specific array configuration as the distances between the electrodes change and

R is resistance.

Modern field equipment will measure the resistance directly. Temperature corrections for the effect on the fluid electrical conductivity may also be required. Normal resistivity logs can only be acquired in open formation and also require a saturated and conducting fluid in the bore to complete the electrical circuits.

Resistivity logging is used extensively in the oil industry where many variations of the simple sondes described above have been developed.

It is possible to make resistivity logs in the field using a cable with the required electrodes hard-wired to the cable. ABEM use this technique with the Terrameter series of equipment.

13.7.4 Electromagnetic induction logs

The popularity of the surface non-contacting electromagnetic systems produced by GEONICS (EM-31, EM-34.3, EM-38) expanded rapidly and the system was extended for

Figure 13.17 EM-39 logging in progress using GEONICS equipment (Photo: Ian Acworth)

borehole use by the EM-39. This logging tool has been specifically designed for use in small diameter (50 mm) PVC-lined bores that are typical of the groundwater contamination industry. This approach allows a continuous log of bulk formation electrical conductivity to be measured and does not require a water-saturated column.

Other manufacturers (such as GEOVISTA and Robertson) have also released induction logging equipment as the extensive patents taken out by GEONICS expired.

Basically, the EM-39 is a slim diameter (36 mm), short (1300 mm) three-coil borehole induction bulk electrical conduction logger which has been designed specifically for groundwater contamination work. The GEONICS equipment is also available with a separate gamma log capability. This provides the possibility of discrimination between the effects of clay content and fluid electrical conductivity anomalies. This configuration is of particular use in the contamination area.

An EM-39 logging setup is shown in Figure 13.17.

Operating principles

The operating principles of the EM-39 sonde are the same as those for the surface-mounted systems. A small transmitter coil (Tx) in the probe induces eddy currents in the material surrounding the monitoring well as shown in Figure 13.18. A coaxial receiver coil (Rx) measures the magnetic field generated by these currents, which is directly proportional to the electrical conductivity of the ground.

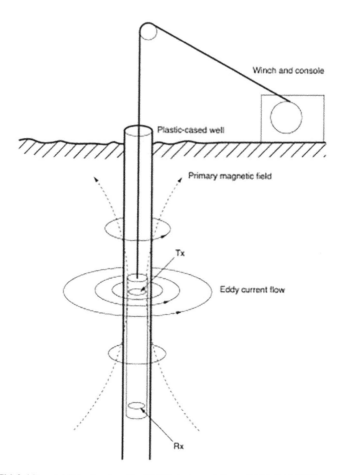

Figure 13.18 EM fields established using the EM39 (adapted from McNeil (1980))

An inter-coil spacing of 50 cm is used to allow good vertical resolution of thin beds. A single centrally located focusing coil is used to minimize the signal contribution from the borehole fluid. This approach produces a log which has a minimal contribution from the fluid and is focussed out in the formation as shown in Figure 13.19 (McNeil, 1986). The sonde can be calibrated by measurements in air or by measurements in a lake of known fluid electrical conductivity.

Examples of field data

Combined conductivity and gamma logs for parts of the Botany Aquifer (Acworth, 1998), where a deep channel extends through the aquifer, are shown in Figure 13.20. Figure 13.21 shows logs from a deep sequence of acid sulphate soils.

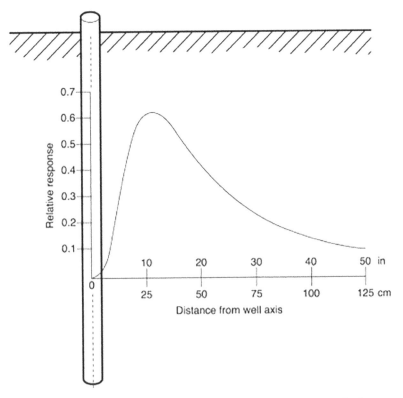

Figure 13.19 Signal contribution using the EM-39 showing that a minimum contribution occurs at the borehole, minimizing the influence from borehole fluids (adapted from McNeil (1980))

13.8   LOGGING VIA PUSHED OR HAMMERED SONDES

In unconsolidated materials that are predominantly composed of clays and fine sands, it is possible to push electrodes into the ground. This technique was an extension of the Dutch cone penetration test developed by FUGRO. It is also possible to hammer a sonde into the ground and this was the method developed later by GEOPROBE.

13.8.1   Conductivity cone penetration logs

Cone penetration testing (CPT) is a widely used geotechnical tool whereby a cone is pushed into the ground through unconsolidated formations. Depths of 30 m can be reached but a shallower depth is more common before too much cone resistance is encountered. A truck carrying additional weight to act as a stable mass is used at the surface as the probe is hydraulically jacked into the ground. CPT rigs use a steady push into the ground to collect pressure and sleeve friction data. However, if the geotechnical measurements are not required, the

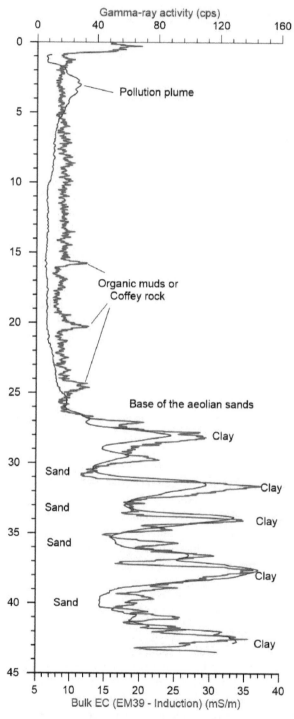

Figure 13.20 Combined gamma and bulk EC logs for Kensington Racecourse Bore, Sydney, NSW. Note the aeolian sands to 26 m overlying the alternating marine clays and sands at the base

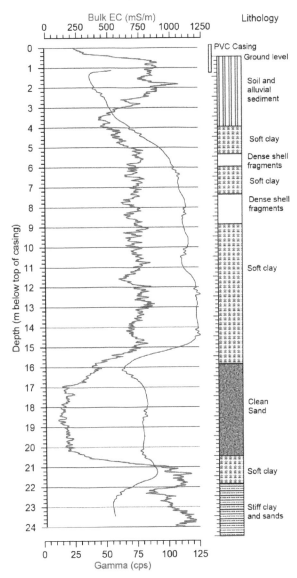

Figure 13.21 Combined gamma and bulk EC logs through acid-sulphate soils at the Oak Factory, Newcastle, NSW

cone can be hammered into the ground. There are a variety of probes that can be used and in some cases a conductivity cone with two or four electrodes is used. A cone probe rig is shown in Figure 13.22 and a two-electrode cone in Figure 13.23.

CPT testing is a standard technique developed by FUGRO in the Netherlands (https://www.fugro.com) and available from many sub-contractors. Commercially available

Figure 13.22 A CPT rig from the Sydney firm GROUNDTEST pushing a CPT (Photo: Ian Acworth)

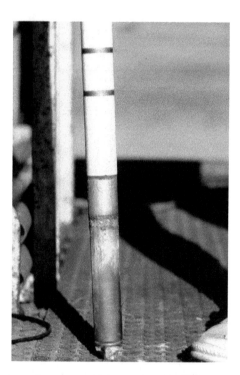

Figure 13.23 Metal ring electrodes separated by a ceramic sleeve are set above the CPT cone elements that are typically 35mm diameter. The cone nose is seen at the base (Photo: Ian Acworth)

equipment for using a hammer to push the probes into the ground is available from GEO-PROBE (http://geoprobe.com/7822dt-direct-push-rig).

The cone can comprise two circular ring electrodes separated by a ceramic or four equi-spaced (Wenner configuration) electrodes. If only two electrodes are used, the current and

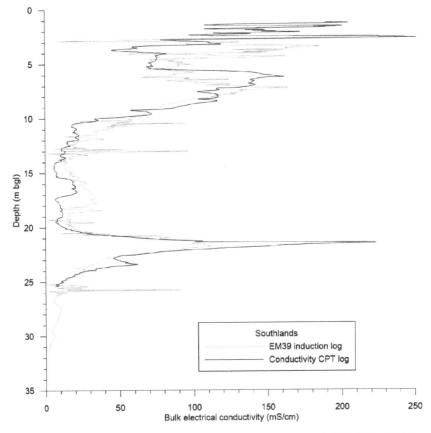

Figure 13.24 Comparison of induction derived bulk conductivity (GEONICS EM-39) with a conductivity CPT at the Southlands DNAPL site. The strong peak in the CCPT log exactly correlates with a zone of elevated (10, 140 mg) DNAPL recovered from a core at the same depth and recovered using a centrifuge (Acworth, 2001b)

potential returns are attached at the surface using metal stakes. A comparison of bulk conductivity from the EM-39 induction sonde and the four-electrode CPT sonde at a contaminated site in sands at Southlands, Sydney, NSW, is given in Figure 13.24. Note that the very high peak in the CCPT at 21.5 m corresponds exactly with analyzes of dense non-aqueous phase liquids recovered from core at the same depth. The induction log has 'smeared' this anomaly because the separation of the coils is much greater than the separation of the electrodes.

13.8.2   GEOPROBE profiles

The US company GEOPROBE manufacture small rigs that are capable of installing piezometers using augers or for taking a core or sampling sediments (the equipment will be described in more detail in Chapter 12). A four-electrode sonde is also available that can

Figure 13.25 Four-electrode cone – with side-mounted electrodes used with the GEOPROBE equipment (Photo: Ian Acworth)

Figure 13.26 GEOPROBE (7822DT) rig used for sampling and bulk EC profiling. The EC probe can be seen on the probe rack at the left, while a sample is inspected (Photo: Ian Acworth)

collect continuous bulk EC profiles. Figure 13.25 shows the sonde. Figure 13.26 shows the deployment of the conductivity sonde.

A GEOPROBE EC profile is shown in Figure 13.27 and compared with a gamma-ray activity log and an induction log using the EM53 sonde from GeoVista. As with the CPT equipment, the GEOPROBE produces a much more clearly defined log through the beach sands at Anna Bay, NSW. A thin anomaly at 6.5 m is completely missed by the induction sonde due to the length of the sonde and the distance between the coils. The high EC anomaly is associated with clayey peat that extends beneath an adjacent pond and under the aeolian sands above.

13.9 LOG INTERPRETATION

Most hydrogeological and contaminant log interpretation remains qualitative. The single most important point is that the logs should be interpreted synergistically. No one log

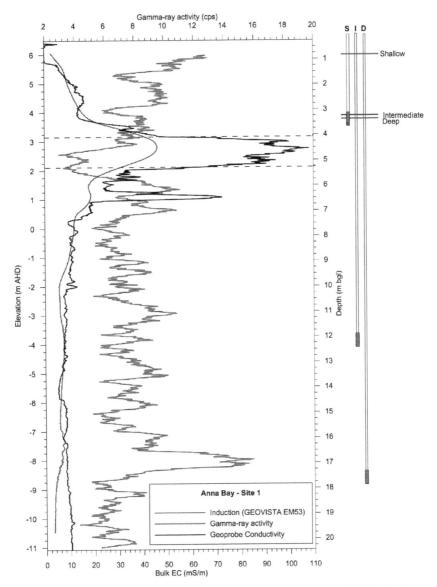

Figure 13.27 Comparison between an induction log (GeoVista EM-53) and GEOPROBE log showing the much improved vertical resolution that the four-electrode GEOPROBE sonde (Figure 13.25) makes. The gamma-ray activity log is shown for further comparison

can provide a complete interpretation. It is only by a consideration of information from all available logs that an interpretation can be made. The combination of electromagnetic induction and gamma logs for contamination studies is particularly useful.

A combined gamma ray and electromagnetic induction log is shown in Figure 13.29.

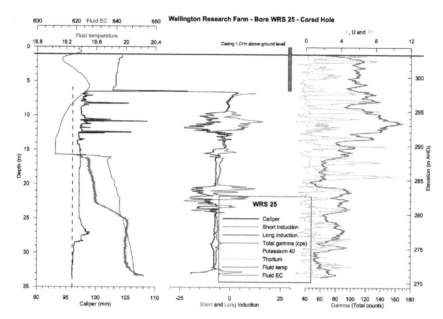

Figure 13.28 Composite logs at WRS-25, Wellington, NSW

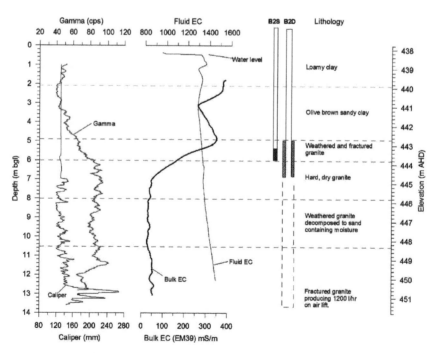

Figure 13.29 Composite logs for a bore drilled in granite at Baldry in western NSW. Clay colluvium covers the top of the weathered granite that passes down into fractured granite at 7 m depth

Chapter 14

Hydrochemistry and groundwater isotopes

14.1 INTRODUCTION

The source of rainwater is either by direct evaporation from the ocean or by transpiration from vegetation. In either case, the chemical quality of the water vapour held in clouds is initially a pure solution of H_2O. Condensation of water vapour then requires a nuclei or cloud seed. This can be dust or clay, soot (carbon) from grassland fires or pollution, sea salt from ocean spray or sulphate from volcanic eruptions. The chemistry of the initial pure H_2O is immediately compromised by the addition of the chemical associated with the nuclei. Further solution of dust or salt can occur as the liquid water falls through the atmosphere. Dust in particular can be washed out of the atmosphere by rainfall and coloured rain or snow are commonly found when the air mass from which the rain falls has passed over or mixed with a source of dust, typically in a desert. A major dust storm in Sydney occurred in 2009 with dust deposits and coloured rain occurring. Figure 14.1 shows the storm approaching the east coast of Australia on 23 September 2009. Figure 14.2 shows the dust plume blowing away from the east coast of Australia. Tozer and Leys (2013) provided an economic analysis of the cost of this storm.

Rain falling to the surface of the earth also reacts with the atmospheric CO_2 and So_2 from pollution to become mildly acidic. In the soil, the partial pressure of CO_2 increases the pH. The rainfall also mixes with dry deposition (dust) accumulated at the surface. In some cases, rainwater can contain as much as 50% salt where large dust influxes have occurred (Kiefert and McTainsh, 1995). In general, rain falling close to the ocean contains a significant component of sea salt. The chemistry of the rainfall that enters the soil is therefore not pure water but has already been significantly modified.

The rainwater moving through the unsaturated zone and into an aquifer undergoes significant further chemical change. Contact with different rock types, weathering reactions, cation exchange reactions, solution and precipitation all occur as the result of movement of groundwater through the subsurface.

The analysis of an accurately taken groundwater sample from a location along a groundwater flow path can therefore indicate the chemical processes that the water sample has undergone and be used as an important indication of process. The chemistry of the groundwater can therefore be used as a further important tool for investigating groundwater process.

The major characteristics of groundwater are mainly established in the unsaturated zone. In the saturated zone the geochemical evolution, though less intense than in the soil and unsaturated zones, follows progressive changes in water quality towards areas of discharge. These processes are time-dependent, and the chemical changes as well as isotopic variations

Figure 14.1 The dust storm approaching the east Australian coast on 23 September 2009. Image from the NASA MODIS experiment on board the Tera satellite

Figure 14.2 The dust plume that originated over the Lake Eyre Basin in Central Australia is seen passing over the coast and moving towards New Zealand. It left a light covering of dust over Sydney. Image from the NASA MODIS experiment on board the Tera satellite

may be used to identify this evolution and provide input into the overall hydrogeological investigations.

The chemistry of groundwater is important if a source is to be used as a public supply. The quality of spring discharge is normally very good – leading to numerous spring-water bottling enterprises. Groundwater quality is frequently compromised by pollution, and much attention is given to understanding the various processes involved in dispersion and diffusion of a source plume into moving groundwater (Appello and Postma, 2005). The field of hydrochemistry (hydrogeochemistry) is evolving rapidly, and it is now possible to model the chemical reactions that may take place. Public domain software (PHREEQC) is available and used extensively in some texts (Appello and Postma, 2005). Additional information is available in other texts (Eriksson, 1985; Stumm and Morgan, 1996; Drever, 1996; Langmuir, 1997; Mazor, 1997).

It is only intended that this chapter give an introduction to hydrochemistry and guidance for sample selection preparation. A brief discussion of the occurrence and use of water isotopes is also provided.

14.2 BASIC AQUATIC CHEMISTRY

The groundwater geochemical system is a dynamic system comprising

1   solid phase (minerals and organic matter),
2   a soil gas phase, and
3   an aqueous solution phase.

14.2.1   Groundwater solution and concentration units

Groundwater is an aqueous solution comprising a solvent (water) and dissolved inorganic or organic constituents (solutes). For most groundwater, 95% of the ions are represented by eight major ionic species: the positively charged cations: sodium (Na^+), potassium (K^+), calcium (Ca^{2+}) and magnesium (Mg^{2+}), and the four negatively charged anions: chloride (Cl^-), sulphate (SO_4^{2-}), bicarbonate plus carbonate ($HCO_3^- + CO_3^{2-}$) and nitrate (NO_3^-). Si is generally occurring as the uncharged dissolved species $H_4SiO_4^0$. These species, when added together, make up most of the salinity that is commonly referred to as the total mineralization or total dissolved solids (TDS).

The concentration of ions is now commonly reported from the laboratory in the units of milligrams per liter (mg/L), virtually equivalent to parts per million (ppm). However, for the purpose of understanding geochemical reactions, it is necessary to convert the units of mg/L to the molarity (mmol/L) (or simply mM when the density of the solution is 1000 kg m^{-3} or ~ 1g/cm^3) to relate concentrations back to the chemical stoichiometry of minerals. A mole refers to 6.022×10^{23} (Avogadro's number) of atoms or molecules of a constituent and has a mass equal to the atomic or molecular weight (g/mol). The molarity is then obtained by dividing by the molecular weight:

$$mmol/L = \frac{(mg/L)}{(g/mol)}$$

Table 14.1 Conversion factors for chemical units

Major Ion	Valency	mg/L to meq/L	mg/L to mmoles/L
Cations			
K^+	1	0.02557	0.02557
Na^+	1	0.04350	0.04350
Ca^{2+}	2	0.04990	0.02495
Mg^{2+}	2	0.08226	0.04113
Fe^{2+}	2	0.03581	0.01791
Anions			
Cl^-	1	0.02821	0.02821
SO_4^{2-}	2	0.02082	0.01041
HCO_3^-	1	0.01639	0.01639
CO_3^{2-}	2	0.03333	0.01666
NO_3^-	1	0.01613	0.01613

The other unit frequently used is milli-equivalents per liter (meq/L), where for this conversion the valency Z, (positive or negative) charge, is also taken into account. The conversion is then simply:

$$meq/L = (mmol/L) \cdot Z$$

See Table 14.1 for conversion factors.

Given that the groundwater solution is electrically neutral, the dissolved anions and cations must balance on an equivalent charge basis. The electrical balance of the solution can be calculated from the chemical analysis to check the accuracy of analysis of the major ions. The sum of the equivalents of all the cations must equal the sum of the equivalents of all the anions for the condition of electrical neutrality to be preserved. This can be formalized in the charge imbalance (CI) error on the chemical analysis:

$$CI = \frac{\Sigma\, cations - \Sigma\, anions}{\Sigma\, cations + \Sigma\, anions} \times 100 \tag{14.1}$$

where:

Σ *cations* is the sum of the equivalents (meq/L) of sodium, potassium, calcium and magnesium ions. This may be extended to include ammonia, iron and other cations in cases where they are present in appreciable amounts and

Σ *anions* is the sum of the equivalents (meq/L) of chloride, sulphate, bicarbonate and carbonate ions.

CIs of less than 2% are desirable, whereas CIs greater than 5% are unacceptable.

Table 14.2 Atomic weights and valences

Element	Symbol	Atomic weight	Valency
Hydrogen	H	1.0	+1
Helium	He	4.0	0
Carbon	C	12.0	+4
Nitrogen	N	14.0	−3 to +5
Oxygen	O	16.0	−2
Sodium	Na	23.0	+1
Magnesium	Mg	24.3	+2
Silicon	Si	28.0	+4
Sulfur	S	32.0	−2 to +6
Chlorine	Cl	35.5	−1
Potassium	K	39.1	+1
Calcium	Ca	40.1	+2
Bromine	Br	80.0	−1

Table 14.2 summarizes some useful atomic weights and valences (Mazor, 1997). Note that fractional values for atomic weights in Table 14.2 indicate varying proportions of the different isotopes for each element.

14.2.2   Field parameters

While the laboratory analyzes generally include the common ions, it is usual practice in groundwater studies to measure those parameters which can alter between sampling and laboratory analysis, or in the case of electrical conductivity a further check that there are no sampling identification errors. The sampling methods for these parameters include:

- temperature,
- Hydrogen ion activity (pH): $pH = -\log H^+$, where H^+ is the activity of the hydrogen ion (protons),
- alkalinity is a measure of the total acid neutralizing capacity of the water,
- redox potential (Eh): the Eh or redox potential is needed to calculate the distribution between redox sensitive species between their possible redox states, and
- fluid electrical conductivity (EC): the fluid electrical conductivity is usually measured in $\mu S/cm$ and relates to the concentration of total dissolved solids.

It should be remembered that if water is being recovered from some depth, the pressure of gasses dissolved in the water will be significantly reduced when the water is delivered to the surface. There may be outgassing of Co_2 that can impact upon the alkalinity. Or there may be reactions with o_2 that cause the precipitation of insoluble components. It is therefore necessary to make measurements of these unstable parameters as soon as possible and to use methods that prevent as much change as possible. It is accepted practice that the sample measurements should be made in flow-through cells that allow the measuring instruments to be inserted but without contact with the atmosphere. Sundaram *et al.* (2009) provide a detailed review of sampling techniques, and typical sampling is shown in Figure 14.3.

14.2.3   Sampling variability of hydrochemistry with depth

An important indication of the hydrochemical variability beneath the ground is demonstrated by Table 14.3. Note that a single borehole with a 3.0 m screen set between 10.0 and 13.0 m would have produced a very mixed sample that would have served to confuse an interpretation of hydrochemical processes at the site! The use of bundled piezometers (see Chapter 12) is strongly advised in this type of environment.

The selection of an appropriate depth to sample in otherwise uniform lithology, as shown by the aeolian sands in Figure 14.4, is difficult without some guidance. The top piezometer in Figure 14.4 would yield a sample contaminated by a landfill plume emanating from an adjacent site. The lower piezometer (Piezo 2) would give a clean groundwater sample from groundwater that is beneath the contaminant plume. The 3D variability in hydrochemistry is often difficult to represent on 2D maps and commonly gets overlooked.

If a geophysical log was first run, the presence of the contaminant plume would have been indicated by the elevated bulk electrical conductivity and the low gamma-ray activity that is characteristic of sands.

14.2.4   Graphical presentation of chemical analyzes

Major ion data may be presented in a graphical format to help with analysis. Hem (1985) gives examples of the many types of graphical analysis that have been used. Possibly the most useful and most popular summary plot is the combination of trilinear fields which show the total major anion and cation composition in a Piper diagram. The evolution of the hydrochemical composition of groundwater is illustrated by the relationship between major dominant ion compositions (e.g., Ca^{2+}, HCO_3^- to Na^+, Cl^-), indicating trends

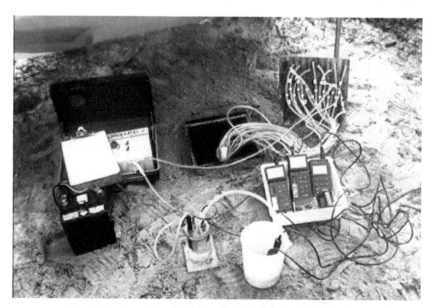

Figure 14.3 Sampling from a multi-channel piezometer installed into sand dunes at Hat Head, northern NSW, using a peristaltic pump (Photo: Ian Acworth)

Table 14.3 Field chemistry for piezometers at the top of a sandy beach. Shown in pink is a section that would contribute to a standard 3 m long section of screen – illustrating the major mixing that would occur. Seawater composition from the same beach is shown in green.

Depth (m)	Temp (°C)	Fluid EC (µS/cm)	Eh (mV)	pH
4.5	18.8	640	−121	7.54
5.0	19.4	599	−199	7.56
5.5	19.5	678	−230	7.65
6.0	19.6	794	−226	7.74
6.5	19.4	684	−236	7.81
7.0	19.0	767	−235	7.97
7.5	19.3	1013	−253	7.84
8.0	19.3	1063	−278	7.80
8.5	17.7	1365	−	7.80
9.0	18.7	9220	−256	7.70
9.5	19.8	9580	−285	7.69
10.0	20.8	12370	−325	7.66
10.5	21.0	9360	−340	7.75
11.0	21.0	6270	−330	7.58
11.5	20.4	4790	−320	7.13
12.0	21.0	724	−256	7.03
13.0	20.6	777	−242	6.67
13.5	20.2	7450	−255	6.92
14.0	19.9	20400	−215	7.05
14.5	19.7	26900	−230	7.10
15.0	20.4	31400	−230	7.13
15.5	20.7	34800	−239	7.08
16.0	21.1	37300	−239	7.08
16.5	20.5	41500	−281	7.07
17.0	20.5	45000	−280	7.07
17.5	20.3	45700	−273	7.12
18.0	20.1	46200	−284	7.10
Seawater-1	11.1	55680	80	8.13
Seawater-2	11.7	55400	74	8.14

along flow paths or mixing between water bodies. Similarities at the regional scale may suggest a degree of homogeneity in groundwater evolution or hydraulic connection. Other types of diagram include the mixing plot as well as the Schoeller-type diagram and both Durov and expanded Durov diagrams.

If the hydrogeology is well constrained, and appropriate sampling points are available, the presentation of data along flow paths is strongly recommended for interpreting or verifying flow systems and locating points of inter-aquifer mixing. Plotting of chemical data for one or more aquifers is possible (using different symbols) as well as trends of ion ratios such as Cl⁻ (where Cl⁻ is used as a conservative reference ion). The shape of such plots, of which several are used in this chapter, may be used to infer mixing processes via faults or fractures (e.g., by sudden changes in concentrations along a flow system) or the gradual addition of local recharge.

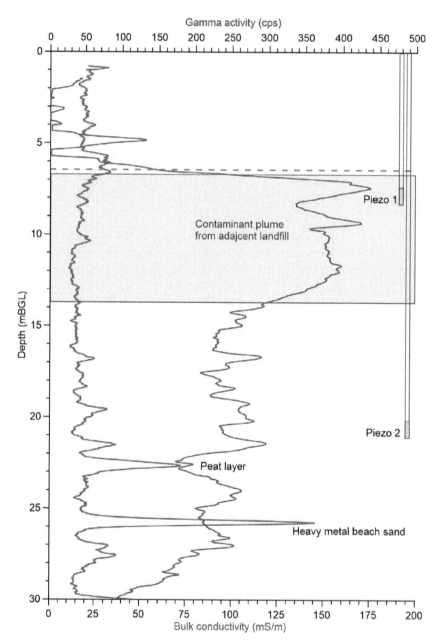

Figure 14.4 Two possible piezometer locations in uniform sands. Piezo 1 will sample the contamination plume emanating from an adjacent landfill, whereas Piezo 2 will sample clean groundwater from beneath the landfill plume

Examples of two commonly used diagrams are given in Figures 14.5 and 14.6. These waters all have a cation-anion balance of < 2% with green triangular symbols representing samples from a bundled piezometer on one side of a tidal creek in sands at Hat Head, northern NSW, while the red diamonds represent samples from a bundled piezometer on the other side of the same creek. The purple diamond is from a seawater analysis.

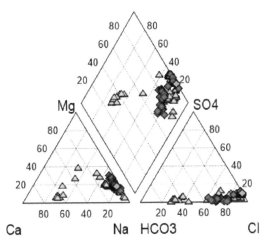

Figure 14.5 The *Piper Diagram*: Cations are plotted on the lower left trilinear plot with the three axes representing Na + K, Ca and Mg expressed as percentages of the total cations. Similarly, the lower right plot shows anions with the three axes representing HCO_3, SO_4 and Cl. The upper parallelogram is formed by extrapolating the data points from each of the lower trilinear plots to form an intersection. Similar water types will plot in the same areas

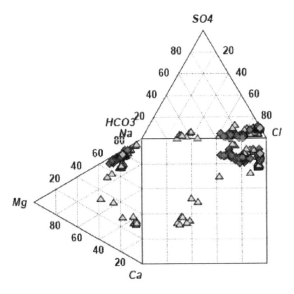

Figure 14.6 The *Durov Diagram* follows a similar approach to the Piper plot with the trilinear plots to the left (cations) and above (anions). Projection from these two trilinear plots to an intersection in the lower square field forms the Durov diagram

14.3   ORIGIN OF SOLUTES

Solutes that are present in groundwater are derived from three main sources:

1   inputs from atmospheric precipitation. The condensation of water vapour into rain drops frequently requires an initial solid particle which is usually surface-derived dust or marine salts mixed with air from wind blowing over the sea. Surface-derived dust may contain as much as 50% salts (NaCl) if the wind has blown over salt lakes (Aryal *et al.*, 2012),
2   acquisition during weathering and water-rock interaction, and
3   the mixing with formation waters, usually saline, seawater, or via diffusion of ions from adjacent aquitards.

14.3.1   Rainfall

In 2007, the CSIRO commenced a program of sampling at 20 stations throughout Australia. Samples were accumulated monthly and the major ion chemistry and isotopes determined (Crosbie *et al.*, 2012). Unfortunately, this program was only funded for a few years, with the results reported in 2012.

The evaporation of rainfall from the ocean is equivalent to distillation. Only pure water is evaporated. However, because the process occurs over the ocean, winds blowing over the ocean pick up sea spray which mixes with the evaporated water. The relative composition of rainfall close to the ocean is therefore very similar to that of sea water. This can be seen in Figure 14.7 where the composition of seawater plots at the beginning of the arrow shown in the diamond of the Piper diagram. There is then a transition through to the rainwater chemical composition of Alice Springs. The rainfall composition for Sydney is close to seawater composition with other coastal stations such as Esperance, Adelaide and Perth plotting close by.

Evaporation from inland water bodies and evapotranspiration from plants also provides a significant source of water away from the ocean. Again, the evaporation, or evapotranspiration, is equivalent to distillation, so that additional salts in rainfall have to be sourced from surface processes, such as the entrainment of dust. This dust can contain significant quantities of salts from playa lakes or simply desiccated soils when drought conditions are prevalent. Inland stations show the increasing input of aeolian dust in the bicarbonates (sodium and calcium) and the decrease of chlorine. It is a little surprising that Cape Grim, on the northwest coast of Tasmania, is not closer to Sydney in chemical composition. The Cape Grim samples show the highest concentration of salts (Crosbie *et al.*, 2012) with direct input of significant sea spray indicated by the high concentrations of strontium (Sr), but there must also be a significant input of aeolian material blowing off the mainland. Mt Isa is an outlier that indicates the very considerable sources of dust in this mining community.

The declining significance of the ocean-derived chloride ion is clearly shown in the anions trilinear plot by a trend arrow (lower right in Figure 14.7) and has also been noted in other studies (Gambell and Fisher, 1966; Nagamoto *et al.*, 1990; Robson *et al.*, 1993). It should also be noted that the accuracy of the laboratory determinations for these readings was not high. An acceptable accuracy for the cation-anion balance (Equation 14.1) is normally ±5% (Hem, 1985). The average cation-anion balance for these 20 groups of analyzes is −21.7 (SD = 14.4). It is clear that we could significantly improve understanding of process by extensive and accurate measurement of rainfall chemistry.

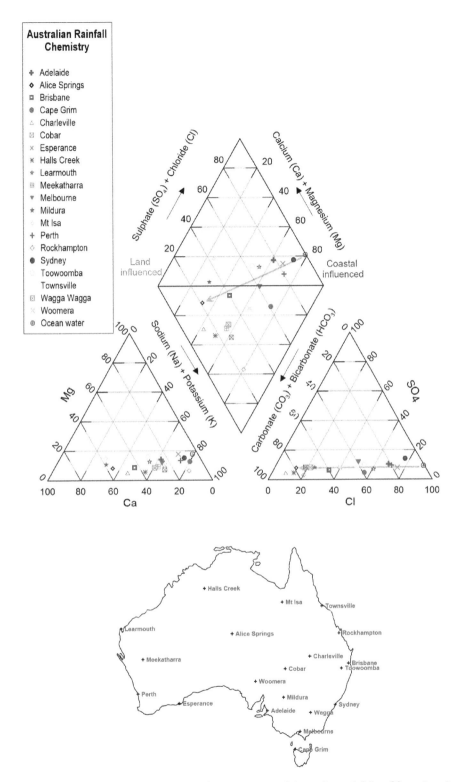

Figure 14.7 Piper plot showing the chemical composition of Australian rainfall at 20 stations in the period 2008 to 2010 (Crosbie *et al.*, 2012). Each data point represents the average of approximately 40 monthly rainfall totals. Note that the millennium drought ended during this period. The significance of the trend lines is described in the text

Once rainfall enters the terrestrial hydrological cycle, it undergoes evaporation from the soil or surface water or is transpired through vegetation. Because only pure water can be returned to the atmosphere by evaporation or transpiration, the residual solutions are concentrated according to the amount of water lost by evapotranspiration. All ions are concentrated equally until saturation for certain minerals (e.g., $CaCO_3$) is reached. In areas where potential evaporation exceeds precipitation, concentrations of dissolved ions in surface water and soil water can be orders of magnitude higher than that in rainfall, due to the effects of evapo-concentration. Modification of the rainfall composition also occurs due to the addition of carbon dioxide (CO_2) via the respiration of plants and microbiological degradation of soil organic matter. CO_2 dissolves to form a weak acid – carbonic acid (H_2CO_3), causing the dissolution of minerals within the soil zone and deeper in the aquifer.

Considerable care is required in the collection of rainfall data for analysis. Contamination of the sample by dust that may carry salts can significantly alter the composition over and above the concentration that occurs from direct evaporation. It is necessary to collect water only during a storm event if the contribution from dust is to be avoided.

14.3.2 Rock weathering

Release of cations (Na^+, Ca^{2+}, Mg^{2+}, K^+ and $Fe^{2+/3+}$) and bicarbonate (HCO_3^-) ions via dissolution of carbonate and silicate minerals are largely driven by the action of CO_2 and to a lesser extent by organic acids produced within the soil zone. It is important to recognize that typically only a few different carbonate and silicate rocks are necessary to explain the release and uptake of solutes. These include:

congruent dissolution of calcite:

$$CO_2 + CaCO_3 + H_2O \rightleftharpoons 2\,HCO_3^- + Ca^{2+} \tag{14.2}$$

incongruent dissolution of impure calcite (where Y could be Mg^{2+}, Mn^{2+}, Fe^{2+} or Sr^{2+}):

$$xCa^{2+} + Ca_{(1-x)}Y_xCO_3 \rightleftharpoons CaCO_3 + xY^{2+}$$

Weathering of primary silicate minerals to form clays, e.g., the transformation of albite to kaolinite:

$$2\,CO_2 + 2\,NaAlSi_3O_8 + 11\,H_2O \rightleftharpoons Al_2Si_2O_5(OH)_4 + 2\,Na^+ + 2\,HCO_3^- + 4\,H_4SiO_4 \tag{14.3}$$

Most of these reactions involve the consumption of CO_2 (or carbonic acid H_2CO_3) and an increase in pH. During the early evolution of the chemical composition (in the soil zone or upper aquifer), open-system or closed-system behaviour with respect to CO_2 is important. In the unsaturated zone, the infiltrating water remains in contact with gaseous CO_2 (open system), but below the water table contact to the gas phase is lost and the subsequent evolution takes place with a depleting pool of dissolved CO_2 (closed system). However, further *in situ* production of CO_2 from breakdown of organic matter may still occur from geologically young aquifers containing reactive organic matter (Appello and Postma, 2005). In some cases additional CO_2 may come from volcanic sources.

The reactions of carbonate rocks are usually quite rapid, whereas those involving silicate minerals tend to be rather slow and irreversible. In other words, carbonate minerals can dissolve and re-precipitate (in slightly modified form) according to the prevailing temperature, CO_2 partial pressure and ionic strength, on a relatively short time scale. This is demonstrated by the deposition of calcrete close to springs (seen in Figure 1.45).

Silicate minerals are involved in slower, incongruent reactions that involve precipitation of clay minerals. An unstable mineral such as albite (formed under high temperature conditions) reacts with carbonic acid (represented above only by H^+) and forms a more stable mineral (here kaolinite). In the process, cations are released into solution (both major and trace elements, since natural minerals are generally impure) along with bicarbonate and silica. It is important to note that this is the primary source of silica in most groundwater systems and not quartz, which is an almost inert mineral.

14.4 GEOCHEMICAL MODELLING

Geochemical modelling codes now available allow once laborious calculations of chemical activities, speciation and saturation indices with respect to specified minerals, to be rapidly determined. They also allow testing and facilitate the quantification of processes involved in water-rock interaction. Details of models available and their limitations are given, for example, by Parkhurst and Plummer (1993) and Appello and Postma (2005). Geochemical modelling first needs a clear hydrogeological and geochemical conceptual model, together with a sound database.

The widely available model NETPATH (Parkhurst and Plummer, 1993) uses groundwater chemical data along a hydraulic transect, and user-specified mineral phases from the aquifer material, to infer mass transfer between the solid and aqueous phases. For example, two sets of water analyzes along a flow path can be compared. The difference in chemical composition can be quantitatively modelled as addition or removal of solutes according to a variety of possible combinations (depending on the types of minerals present). It is referred to as an inverse model because it back-calculates the reactions that must have occurred to the water during transit through the aquifer to produce the final observed chemical composition.

PHREEQC is the most advanced and up-to-date code for geochemical modelling that has been developed by the USGS. Windows version is freely available at https://www.usgs. gov/software/phreeqc. PHREEQC calculates the distribution of different aqueous species; mineral complexes; mineral saturation indices, incorporates some inverse mass balance capabilities. It also incorporates a transport component that is a powerful tool for many groundwater problems. This type of model is useful to test the end result of a set of imposed conditions such as the mixing of different water types, or exposure to oxygen, or to certain minerals or contaminants. The choice of PHREEQC or NETPATH depends on the nature of the problem and the availability of data. Similarly, isotope data can be incorporated into NETPATH and PHREEQC to assess the relative importance of possible water sources sampled at a given location.

Direct incorporation of chemical data into coupled flow-solute transport models has been viewed as the holy grail for groundwater geochemistry, though some skeptics have suggested that they may be too complex, both numerically and conceptually, to be useful. That is, the large degree of chemical heterogeneity and the large number of constraints required for solute transport modelling may preclude their general use through large uncertainties in the resultant output. In that case, use of separate or sequential flow, particle path-tracking and chemical modelling routines may provide the balance between utility and reliability.

14.5   EXAMPLE OF HYDROCHEMICAL ANALYSIS

14.5.1   Introduction

While the study of groundwater contamination is not a part of *Investigating Groundwater*, there are some interesting examples of the use of hydrochemical observations to interpret field data.

This example comes from a part of the Botany Sands Aquifer in Sydney that has had very significant contamination from chlorinated hydrocarbons (Acworth, 2001b). The groundwater analyzes, upon which Figure 14.8 are based, were obtained from core collected with a core barrel pushed into the sands just behind a rotating auger head (see Chapter 12). The core was recovered to the surface using a wire line. The core was cut into 100 mm increments in the laboratory and subject to various geophysical tests (Acworth, 2001b). One reason for listing the data from this investigation is to further emphasize the variation in major ion chemistry and physical properties that can occur within only 100 mm in an otherwise uniform sand aquifer. It is obvious that there is no point in simply installing a piezometer and sending a sample for analysis to a laboratory if the chemical variation at a site is not well understood.

The main method to obtain a basic understanding of hydrochemical variation is to drill and install a piezometer with no screen that can first be used for geophysical logging (see Chapter 13). A simple logging run with an EM sonde can indicate changes in fluid chemistry. The addition of a gamma sonde will allow detection of clay bands. With this information to hand, a program of targeted piezometer installation can be carried out.

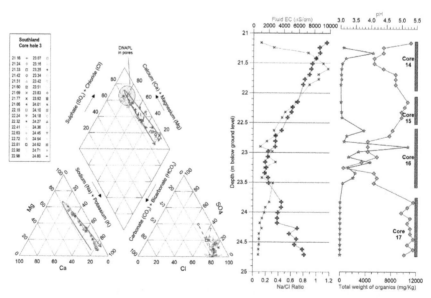

Figure 14.8 Piper diagram and selected chemical parameters from a sequence of core sub-sections at a DNAPL contaminated site on Southland in Sydney (Acworth, 2001b). The greatly reduced sodium to chloride ratio and the acidic pH occur in the same core (Core 16) as extensive DNAPL was recovered. The location of the cores (14 to 17) shown is seen at the right of the figure

14.5.2   Process description

On completion of this testing, the core subsamples were split in half and placed in a centrifuge to recover the fluid content. In some cores, dense non-aqueous phase liquids (DNAPL) were recovered that consisted primarily of trichloroethylene (TCE C_2HCl_3), tetrachloroethylene (PCE C_2Cl_4) and tetrachloroethane (PCA $C_2H_2Cl_4$). The results of the inorganic chemistry and the locations of the recovered DNAPL are shown in Figure 14.8.

The chemical analyzes of the pore fluid recovered from each core sub-section is shown in Table 14.4.

When investigating contamination, even in a uniform hydraulic conductivity matrix, there can still be major changes in chemistry over distances of 0.1 m. The data from a single core (shown in Figure 12.28), where the core was first cut into 100 mm lengths and then the pore fluids spun from each segment, are shown in Table 14.4. It is clear from the chemical makeup of TCE, PCE and PCA that each molecule has either three or four chloride ions attached. Although some texts (Cohen and Mercer, 1993) suggest that the DNAPLS are stable in the environment, it would seem clear that under favorable conditions they can begin to break down and release the attached chloride ions. Butler and Barker (1996) suggest that under strongly reducing conditions and in the presence of high iron content, anaerobic or abiotic breakdown of DNAPL can occur. The Eh conditions in the core and the iron content are not known. However in an adjacent bundled piezometer, the Eh was -168 mV with a total iron content of 154 mg/L (Acworth, 2001b).

The very high levels of chloride would indicate that this is occurring. The free chloride radical will cause water molecules to disassociate and release protons (H^+) to electrically balance the solution. This process leads to the formation of HCl that reduces the pH, as seen in the data for Core 16.

The ratio of sodium to chloride in the core subsamples is shown in Figure 14.9. The normal ratio of sodium to chloride is approximately 0.88 in sea water. Ratio values above

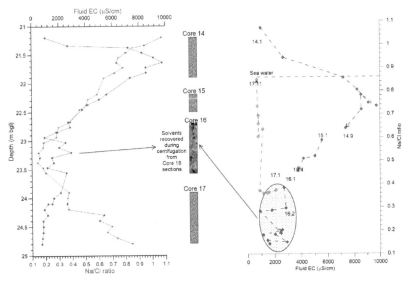

Figure 14.9 Plot showing that the ratio of sodium to chloride in the cores from a borehole (Acworth, 2001a) is reduced to a minima where DNAPL was recovered

this indicate cation exchange impacts where sodium is released from clay into the solution. Reverse cation exchange is indicated with lower values of the ratio. In these cores, the ratio is reduced to <0.2. This indicates either a major loss of sodium to cation exchange or a major influx of chloride. There is no evidence of increased magnesium or calcium (Table 14.4) and therefore the evidence points to a significant source of chloride – presumably released from breakdown of the DNAPL!

Table 14.4 Inorganic chemistry for pore fluids extracted from Borehole SG3 sub-cores

Core	Depth (m)	pH	EC $\mu S/cm$	Na (mg/L)	K (mg/L)	Ca (mg/L)	Mg (mg/L)	TotFe (mg/L)	SO_4 (mg/L)	Cl (mg/L)
14.1	21.16	5.3	1014	163	25.4	17.5	9.1	0.3	93.0	239
14.2	21.24	4.4	2680	460	13.8	39.2	21.6	17.1	250	695
14.3	21.33	4.4	7210	1450	50.6	135	77.8	66.4	654	2181
14.4	21.42	4.1	8240	1672	27.8	154	116	90.0	774	2635
14.5	21.51	4.1	9120	1760	32.2	168	157	93.0	816	2949
14.6	21.60	4.4	9760	1732	35.4	151	194	132	762	2961
14.7	21.69	4.8	8540	1470	31.0	136	203	120	572	2590
14.8	21.77	5.8	8560	1340	26.8	118	204	82.5	473	2569
14.9	21.86	4.8	7360	1166	41.8	107	224	136	403	2336
15.1	22.15	5.2	5580	782	26.6	80.2	204	101	202	1748
15.2	22.24	5.1	5040	624	29.0	72.6	196	112	132	1594
15.3	22.32	5.0	4180	516	31.2	63.6	171	100	76.2	1362
15.4	22.41	5.0	3760	410	27.2	78.8	171	85.0	51.6	1217
16.2	22.63	4.7	2710	240	22.4	52.4	119	78.0	27.0	871
16.3	22.72	4.6	2860	204	25.8	57.6	147	112	15.0	956
16.4	22.81	3.9	1803	220	30.6	64.6	168	142	16.8	1053
16.5	22.90	3.9	928	203	31.0	68.0	159	123	15.0	995
16.6	22.98	3.9	2490	128	24.6	56.4	142	122	16.6	944
16.7	23.07	4.2	2030	127	27.0	63.5	155	162	15.0	960
16.8	23.16	3.7	2920	104	26.2	63.4	169	154	9.0	960
16.9	23.25	3.1	1473	110	31.2	63.2	162	158	6.0	963
16.10	23.34	4.1	1615	97.6	25.2	73.0	182	164	12.6	948
16.11	23.42	4.2	1145	121	33.5	75.0	173	165	34.5	929
16.12	23.51	4.1	2570	111	34.0	67.5	150	117	45.0	836
17.1	23.83	5.4	2120	109	28.8	64.8	132	98.0	504	435
17.2	23.92	5.2	1606	93.8	29.4	39.0	89	38.4	161	387
17.3	24.01	5.0	1478	93.6	20.0	29.2	69	46.0	61.2	419
17.4	24.10	5.3	1175	80.8	17.0	30.2	67	24.6	38.8	368
17.5	24.18	5.2	896	66.0	12.0	26.4	60	27.8	62.4	296
17.6	24.27	5.3	1137	117	19.7	26.3	51	22.1	143	283
17.7	24.36	5.3	807	86.6	16.2	24.2	43	24.2	124	259
17.8	24.45	5.4	811	92.5	16.5	14.6	28	13.4	31.5	229
17.9	24.54	5.2	804	90.0	16.0	13.0	27	15.8	52.5	233
17.10	24.62	5.4	747	87.0	16.0	13.4	26	13.8	78.0	190
17.11	24.71	5.1	702	1437	23.0	28.3	48	19.5	225	300

14.6   ISOTOPES

14.6.1   Atomic sub-structure

The atom is composed of a small central nucleus containing protons and neutrons surrounded by energy levels (or shells) in which electrons move. The number of protons in the nucleus is referred to as the atomic number (Z). The combined number of protons and neutrons in the atom is the atomic weight (A). While the number of protons defines the chemical and physical characteristics of the element, the same element can have different numbers of neutrons. These are referred to as isotopes of the element. Elements, their mass and their isotopes can be shown symbolically as $^A_Z E$ where E is the symbol of the element. Common isotopes of hydrogen, carbon and oxygen are shown in Table 14.5. In most cases the atomic number (Z) is not shown as the symbol also defines this quantity.

The isotopes of hydrogen have been given special names. The most common isotope is hydrogen which, as can be seen from Table 14.5, is just a proton. The second isotope is called deuterium and the third is tritium. The isotopes of hydrogen, carbon and oxygen are all used extensively in hydrogeology.

Many isotopes are stable. This means that the number of neutrons remains fixed. However, as the atomic number increases, there is a tendency for the nucleus to be unstable and a neutron is lost from the nucleus combined with the production of some form of energy. These isotopes are referred to as unstable or radioactive isotopes. The instability can be predicted statistically, and the isotope has what is referred to as a half-life, the time taken for half the atoms to release their excess neutron.

The instability of the nucleus is resolved by either a β^- or a β^+ process. In the β^- process, a neutron changes to a proton as shown by Equation 14.4. The neutron emits a high-energy electron, an electron anti-neutrino and gamma radiation to leave a new proton. Note that the atomic weight (sum of protons and neutrons) remains the same:

$$n \rightarrow p + e^- + \bar{v}_e + \gamma \tag{14.4}$$

where:
n is a neutron and
p is a proton.

Table 14.5 Isotopes

Isotopes	Symbol	Atomic composition			Average abundance	Half-Life (years)
		Protons	Neutrons	Electrons		
Hydrogen-1	$^1_1 H$	1	0	1	99.985	Stable
Hydrogen-2	$^2_1 H$	1	1	1	0.0115	Stable
Hydrogen-3	$^3_1 H$	1	2	1	10^{-15}	12.43
Carbon-12	$^{12}_6 C$	CD	6	6	98.93	Stable
Carbon-13	$^{13}_6 C$	CD	7	6	1.078	Stable
Carbon-14	$^{14}_6 C$	CD	8	6	1.2×10^{-10}	5730
Oxygen-16	$^{16}_8 O$	CO	8	co	99.759	Stable
Oxygen-17	$^{17}_8 O$	co	9	co	0.037	
Oxygen-18	$^{18}_8 O$	co	10	co	0.204	Stable

where:

e^- is an electron,

$\bar{v}_e$ is an electron anti-neutrino, and

γ is gamma radiation.

An example of electron emission (β^- decay) is the decay of carbon-14 to nitrogen-14 with a half-life of about 5730 years:

$$^{14}_{6}C \rightarrow ^{14}_{7}N + e^- + \bar{v}_e \qquad (14.5)$$

In this form of decay, the original element becomes a new chemical element in a process known as nuclear transmutation. This new element has an unchanged mass number A, but an atomic number Z that is increased by one.

An example of positron emission (β^+ decay) is the decay of magnesium-23 into sodium-23 with a half-life of about 11.3 seconds:

$$^{23}_{12}Mg \rightarrow ^{23}_{11}Na + e^+ + v_e \qquad (14.6)$$

β^+ decay also results in nuclear transmutation, with the resulting element having an atomic number that is decreased by one.

Many hydrological studies use the ratios of stable isotopes of hydrogen ($^2H/^1H$) and oxygen ($^{18}O/^{16}O$) to determine the origin, recharge mechanisms and hydraulic inter-connection of the hydrological cycle. These isotopes provide us with information on the actual flow of water, rather than inferences derived from water-level data. Similarly, isotopes of carbon, nitrogen and sulphur ($^{13}C/^{12}C$, $^{15}N/^{14}N$ and $^{34}S/^{32}S$) can give valuable information about reactions involving these elements in biogeochemical reactions, or they can be used as tracers for pollutants if they display a distinctive signature deviating from the natural background.

Radioactive isotopes of some elements can also be used as tracers, but more importantly due to their constant decay rate they can add the dimension of time. Most commonly, radiocarbon (^{14}C) is used to estimate groundwater residence time, often referred to as carbon dating. Isotopes of the uranium series (e.g., ^{234}U, ^{238}U, ^{226}Ra and ^{222}Rn) also show promise as tracers, but are not precise enough to establish water ages due to mineral-water interactions. Development of accelerator mass spectrometry allowed for measurement of the radioactive chlorine-36 isotope which, because of its long half-life (>300,000 years), is applicable to dating groundwater in large confined systems such as the Great Artesian Basin. Local groundwater tracer studies using the short half-life ^{82}Br have also proved useful (see below).

Atmospheric nuclear weapons testing during the 1950s and 1960s released a wide array of nuclides into the atmosphere that overwhelmed the abundance levels of some natural isotopes, adding anthropogenic tracers to the hydrosphere. In the early 1960s, a large network of monitoring stations were set up to measure tritium (^3H) in precipitation which gave hydrologists a benchmark for dating recent groundwater. Other bomb-test nuclides which have applications for groundwater hydrology include the radioactive isotopes ^{14}C and ^{36}Cl.

14.6.2   Stable isotopes of water

Different numbers of neutrons within the nucleus of an element result in there being a whole range of isotopes having different mass numbers. The advantage of stable isotopes is that the two isotopes of the same element should take part in the same chemical reactions in the environment, but at slightly different rates. Slight variations in isotopic abundance are caused by small differences in reactivity of the different isotopes because of their mass differences.

Notation

Absolute stable isotope concentrations are very difficult to measure sufficiently accurately for general applications in nature. However, their isotopic abundance can be measured very precisely relative to a given standard. The measurements are conveniently expressed as a ratio relative to an internationally accepted standard so that all laboratories can compare their isotopic values. The general formulation, where R refers to the ratio of heavy (or less abundant) to the more abundant light isotope (e.g., $^2H/^1H$, $^{13}C/^{12}C$ etc.) is $\frac{R_{sample}}{R_{standard}}$. Because the differences in ratios from one sample to another are quite small, using the above ratios would be awkward because we would be comparing values in the third or fourth decimal place. Instead, the delta (δ) formulation is now universally adopted so that comparison can be made more easily. δ is expressed in per mill (parts per thousand expressed as ‰) notation according to the general formula:

$$(\delta, \permil) = \left(\frac{R_{sample}}{R_{standard}} - 1 \right) \times 1000 \tag{14.7}$$

For hydrogen and oxygen isotopes of waters, the standard adopted is the Vienna Standard Mean Ocean Water (VSMOW). For carbon the standard is a marine carbonate fossil, the Vienna Pee Dee Belemnite (VPDB).

14.6.3 Fractionation

Fractionation of stable isotopes occurs as a result of different physical processes and because of the mass differences between isotopes of the same element. Two fundamental processes control the fractionation. The first of these, known as equilibrium fractionation, results from chemical bonds of molecules containing the lighter isotope being slightly weaker than those containing the heavier isotope. Molecules containing the lighter isotope therefore react more quickly during phase changes or reversible chemical reactions. For example, the exchange between isotopes of hydrogen and oxygen between liquid water and vapour (evaporation):

$$H_2O_{liq} \Leftrightarrow H_2O_{vap} \tag{14.8}$$

with the associated fractionation defined by:

$$\alpha_{liq-vap} = \frac{(^2H/^1H)_{liq}}{(^2H/^1H)_{vap}} \ or \ \frac{(^{18}O/^{16}O)_{liq}}{(^{18}O/^{16}O)_{vap}} \tag{14.9}$$

with values of 1.084 and 1.0098 (at 20 °C) respectively and with the liquid phase heavier than the vapour phase. In per mill terms, these values translate to vapour being 84‰ and 9.8‰ isotopically lighter than liquid for the respective isotopes.

A fractionation factor (α) can be used to describe the naturally occurring abundance of different isotopes at equilibrium with certain reactions. The general convention in formulation of fractionation factors is to have values of $\delta > 1$, where the isotope ratio of the lighter phase is multiplied by the value of α to give the equilibrium isotopic ratio of the heavier phase and one defines the specific equilibrium reaction as above.

The second type of fractionation, known as kinetic fractionation, also relates to the mass differences between the nuclei of respective isotopes. It is generally applicable to certain irreversible reactions in nature, such as evaporation into air that has relative humidity < 100%. Such a process is related to differences in the rate of diffusion (the molecule containing the lighter isotope moving faster than that containing the heavier isotope) through a boundary layer. Although there is bi-directional movement of water molecules via the exchange reaction described above, there is also a net transfer of water from liquid to vapour. In the case of evaporation, the lower the relative humidity (or vapour pressure deficit) the faster the rate of evaporation and the greater the kinetic fractionation. The kinetic fractionation is defined in the same way as the equilibrium fractionation, except sometimes it is denoted with the subscript k (i.e. α_k). Diffusion rates are approximately proportional to the square root of respective masses, and the kinetic fractionation effect therefore acts unequally on water molecules bearing the oxygen and hydrogen isotopes because of the different relative mass difference.

Since the $H_2{}^{18}O$ is about 10% heavier than $H_2{}^{16}O$, and $^2H^{16}O$ is only 5% heavier than $H_2{}^{16}O$, it becomes clear why the effect of kinetic fractionation during evaporation is greater for oxygen isotopes than for hydrogen isotopes. This explains the slope of the trend of δ^2H versus $\delta^{18}O$ for evaporated waters below and to the right of the equilibrium fractionation-controlled world meteoric water line (Figure 14.17).

14.6.4 Rayleigh distillation

The process known as Rayleigh distillation is common to many natural phenomena observed in stable isotope studies. In general terms, it refers to the changing isotopic composition of the products or different phases during reaction progress, as the respective reservoirs change in size without re-equilibrating. The classic example is the progressive depletion in the heavy isotopes of water in a cloud mass as water vapour progressively rains out. The isotopically heavier water molecules in the vapour will tend to condense and form droplets first.

During evaporation of water from the sea surface to form clouds, there will be a fractionation of oxygen isotopes (about $-9‰$ at 20 °C). This cloud mass will have an isotopic composition of 9‰ given that the ocean has a $\delta^{18}O = 0‰$. As the vapour condenses, the very first rainfall will have a composition of about 0‰ because of the fractionation of $+9‰$ during condensation. However, the loss of water from the cloud results in the residual water vapour becoming lighter because of preferential rain-out of the heavier isotope, and subsequent rain will have an isotopic composition more negative than 0‰ (Clarke and Fritz, 1997) as shown in Figure 14.10. This occurs because the cloud mass is not being replenished and continues to preferentially lose the heavy isotope of water as the condensation process continues. This process can be approximated by the expression:

$$\frac{R}{R_o} = f^{\alpha-1} \tag{14.10}$$

where:
R is the $^{18}O^{16}O$ ratio of residual vapour,
R_o is the $^{18}O^{16}O$ ratio of vapour before condensation,

f the fraction of vapour remaining, and
α is the fractionation factor (1.009).

In this way, the first 25% or so of rain from a single rain cloud may be more than 25‰ heavier than rain from the last 25%. The further a rain event occurs from its oceanic water source, the lighter its isotopic composition.

Note that Rayleigh distillation can be applied to any type of fractionation mechanism, provided that the fractionation factor remains constant during the reaction progress. Most fresh water in the world has deuterium and oxygen-18 compositions that plot on a straight-line relationship characterized by:

$$\delta^2 H = 8\delta^{18}O + 10 \tag{14.11}$$

and is known as the world meteoric water line (WMWL). The slope and intercept of this line might vary from place to place because of re-evaporation of regional waters, but it serves as a very close approximation in most instances and is the best fit of the global data set.

For detailed stable isotopic studies within a given geographic region, the local meteoric water line (LMWL) should be established from analysis of rainfall samples collected locally to properly interpret isotopic data from surface and groundwater. Hollins *et al.* (2018) present detailed isotope data for rainfall for 15 stations around Australia from which local LMWL relationships are derived.

All of these processes described above give rise to the following generalizations that can be used for isotopic investigations of groundwater:

- higher altitude rainfall is increasingly depleted in heavier isotopes,
- greater rainfall amount tends to be more depleted in heavier isotopes than light showers,
- rainfall inland is more depleted in heavier isotopes than that at the coast,
- rainfall at cooler temperatures is more depleted in heavier isotopes (e.g., winter rain is lighter than summer rain; groundwater recharged during a colder climate is more depleted), and
- waters that have undergone evaporation are more enriched in the heavier isotopes and tend to lie to the right of the meteoric water line.

The conventional way to interpret stable isotope data is on a plot of $\delta^2 H$ versus $\delta^{18}O$ as shown in Figure 14.17. Groundwater samples often plot on, or around, the so-called local meteoric water line and their composition will depend on a number of factors. These include the following:

- location of the recharge area,
- the storm track and rainfall history of the cloud mass that contributes to the recharge,
- processes affecting the isotopic composition of recharge water during its passage through the soil zone, and
- water-rock interactions along the groundwater flow path.

In general, the majority of samples on a plot of $\delta^2 H$ versus $\delta^{18}O$ have values of less than unity. They are negative. The various processes outlined above have reduced the concentration of ^{18}O over the standard SMOW because ^{16}O has been increased due to fractionation

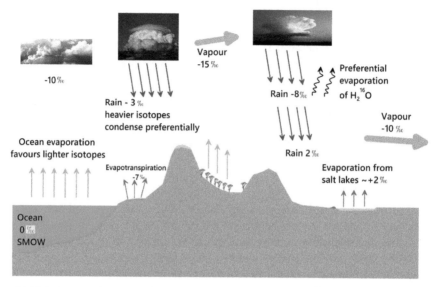

Figure 14.10 Schematic showing the possible changes in fractionation of water isotopes in the hydrological cycle

processes. In general, the most negative values that lie on the meteoric water line can indicate localized recharge by large rainfall events or passage of a cloud mass across continents.
These samples are said to be *depleted* in that there is less ^{18}O in the sample, the lighter ^{16}O
having been preferentially increased. Where samples have ratios of greater than unity, the
isotopic composition is said to be enriched.

For samples that have a positive isotopic ratio, ^{18}O is increased over the SMOW standard.
These samples are said to be *enriched*. Fewer processes lead to enrichment. The most
common is evaporation of rainfall in a low-humidity atmosphere where ^{16}O is preferentially
removed leaving the heavier isotope in the rain drops.

Isotopic compositions of groundwater that lie to the right of the LMWL are indicative of
evaporation during recharge or lateral flow, either at the surface or within the soil zone. For
a given rainfall regime, the further away a sample lies to the right of the LMWL, the lower
the recharge rate and the greater the degree of evaporative concentration.

The possible changes to the isotopic composition of water through the hydrological cycle
are shown in Figure 14.10. Panga *et al.* (2017) summarized the possible changes that can
occur.

14.6.5 General applications of stable isotopes of the water molecule

There have been several books and reviews concerning the use of stable isotopes in understanding aspects of the hydrological cycle. Mazor (1997) describes an applied approach to
chemical and isotopic hydrology. Galewsky *et al.* (2016) review the occurrence of stable
isotopes and their use in hydrology.

Mechanism and sources of recharge

In a given groundwater basin, the source(s) of groundwater may be fingerprinted using isotopes that recharge at varying altitudes. Higher altitudes will have lighter isotopic composition because they tend to be colder (greater equilibrium fractionation). In large groundwater basins, water sources may have different isotopic compositions resulting from different climatic conditions or recharge mechanisms and may also be used to fingerprint water sources.

Similarly, for large groundwater basins where rainfall gradients may be significant, there may be differences in isotopic composition of respective recharge areas due to the continental effect. This is because as rainfall traverses across the continent and rains out, the cloud mass gets progressively depleted in ^{18}O and ^{2}H, which in turn causes groundwater recharge isotopic compositions to become more and more negative.

Water that recharges in arid or semi-arid regions may have a long residence time within the top few meters of soil and therefore be subject to partial evaporation. The soil moisture will be more enriched in heavy isotopes (Allison et al., 1984) and lie off the meteoric water line relative to water that recharges rapidly during flood events. In the latter case, water may or may not be subjected to much evaporative enrichment and would tend to be more depleted in ^{18}O and ^{2}H than the water that enters the soil by direct infiltration or lie on a steeper slope than the slope caused by evaporation in the soil zone.

Higher rainfall amounts also tend to have isotopic compositions that are quite different to the more frequent lighter event because of the Rayleigh effect. Because the stable isotopic composition of groundwater is an integrator of all recharge events, it will reflect a mixed isotopic signature roughly in proportion to the amount of water derived from flood recharge and normal recharge respectively.

Appello and Postma (2005) identify the deuterium excess as a parameter that could explain enrichment above the meteoric water line. The deuterium excess is calculated from the global meteoric water line (GMWL) in Equation 14.11 as:

$$d = \delta^2 H_{rain} - 8.0\,\delta^{18}O_{rain} \tag{14.12}$$

Values of the deuterium excess > 10 would indicate additions of the heavier isotope ^{18}O. Appello and Postma (2005) suggest that the deuterium excess can be an indication of a contribution from evaporated water in an arid climate with lower humidity, as would occur from rainfall beginning to evaporate before it gets to the ground, or a contribution from previously evaporated inland water to the rain. The latter process is indicated on the right-hand side of Figure 14.10.

Discharge rate estimation

The rate of evaporation (or discharge) can be estimated from bare soil surfaces by the shape of the isotope profile. Evaporation tends to force the isotopic composition of water away from the meteoric water line. The upward flux of water (the evaporation rate) is balanced by a downward diffusive flux. Knowing the water content and depth distribution of isotopic composition of the water, the evaporation rate can be calculated.

For example, Allison and Barnes (1985) used such a technique to estimate evaporation from a dry salt lake, Lake Frome, in South Australia. ^{18}O and ^{2}H concentrations in interstitial water in the lake bed were found to rise approximately exponentially from about 0.5 m below the surface to a peak value at or near the surface. The isotope profiles were explained as being due to a balance between the upwards advective flux of isotopes caused by evaporation, and the downwards diffusive flux of isotopes as a result of the evaporative enrichment at the surface, and consequent concentration gradients.

Hydrograph separation

The source of water to streams and lakes is sometimes important when investigating the yield of catchments, potential for nutrient input to surface waters and other applications. Distinguishing the sources of surface water in a flood spate can be problematic. Is it the recent rainfall or displaced groundwater that enters the stream? The stable isotopic composition of discrete rain events is usually quite different from that of groundwater or soil water, because the latter have integrated the isotopic composition of several previous storm events. When using this data, account needs to be taken of the intra-storm variations of rain isotopic concentrations, which can be as large as inter-storm variations, and reflect cloud dynamics.

Differences between the proportions of stream flow along particular pathways estimated from event hydrographic and isotopic responses can be explained in terms of differences between the two response functions. The hydrograph is essentially a hydraulic response to precipitation, where precipitation input to the groundwater at one point rapidly results in exfiltration at another; whereas the isotopic response is determined by the residence time of the actual water molecules in the system and measures true water velocities.

Paleoclimate and paleorecharge

In confined aquifers, where we can assume piston or tube flow, the isotopic composition of discrete water samples along a flow line may reflect changes in recharge conditions over time. There is strong evidence from around the world that recharge temperatures during the last glacial maximum (18,000 years BP) were several degrees colder than the present, as evidenced by the lower oxygen-18 concentrations of paleowaters. The interpretation of the existence of paleowaters that have recharged in, say, a cooler climate is not necessarily straightforward. One might intuitively think that cooler recharge temperatures inferred from lower ^{2}H and ^{18}O concentrations indicate lower evaporation rates and therefore higher recharge rates. Climatic conditions in many parts of the world were actually drier during glacial periods, indicating lower rainfall, but changes in vegetation type and amount during climate changes may be the main controlling factor that determines recharge, rather than precipitation amount and potential evaporation alone.

Stable isotopes are a key part of paleoclimate reconstruction based upon the composition of water recovered from ice cores.

14.7 EXAMPLE: USE OF RADIOACTIVE AND STABLE ISOTOPES IN A SAND AQUIFER AT HAT HEAD, NSW

14.7.1 Radioactive potassium bromide as a tracer

Acworth *et al.* (2007) described the use of a short half-life (35.28 hrs) KBr tracer to determine groundwater flow directions in a tidally influenced sand aquifer at Hat Head, NSW. The site for the investigation was on the banks of Korrogorro Creek in Hat Head (Acworth and Dasey, 2003). The aquifer material was aeolian-transported beach sand containing groundwater

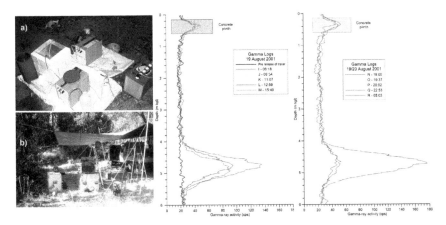

Figure 14.11 (a) The isotope insertion equipment; (b) the gamma-ray logging (GEONICS) equipment used to detect the moving source; time-lapse gamma-ray logs showing the waxing and waning gamma-ray signal detected at the observation borehole in response to the moving source

recharged into dunes between the tidal creek and the ocean. The investigation was undertaken to determine the direction of groundwater movement during a king tide period where the creek was flooded with sea water in response to a 2 m tidal event. Under lower range tides, groundwater moves from the aquifer to discharge into the creek at low tide.

The bromine anion had been radiated in the reactor at Lucas Heights (ANSTO) to produce a quantity of the unstable isotope ^{82}Br. The ^{82}Br decays to ^{82}Kr with the release of high energy β^- radiation (0.44 MeV) and a γ photon (0.7788 MeV). The half-life of this process is 1.4701 d or 35.28 hr. The atomic structural changes can be represented by $^{82}_{35}$Br $\rightarrow ^{82}_{36}$Kr.

The gamma photon was detected using standard gamma-ray logging equipment as described in Chapter 13. Equipment to store and introduce the KBr is shown in Figure 14.11a. The isotope was stored in a lead shield enclosure made up of interlocking solid lead blocks to absorb the very high energy gamma radiation and prevent gamma rays from escaping. It was mixed with a small volume of bore water recovered from the piezometer using a peristaltic pump. The same peristaltic pump was used to slowly introduce the water into the screened section of the bore. A slow speed was required so as not to disturb the natural flow field in the aquifer. Gamma-ray logging equipment (GEONICS) is shown set up at an observation bore 0.5 m from the source bore (Acworth *et al.*, 2007).

Eighty mBq of this tracer was introduced to the sand through a specially constructed piezometer. Time-lapse measurements of the gamma-ray activity in observation bores were then conducted regularly over the following 48 hours. Examples of the logs recorded are shown in Figure 14.11 in the two panels on the right-hand side of the figure. The log before the isotope was inserted is shown as a black line. The shallow peak relates to the gamma activity of the cement used to stabilize the 50 mm piezometer in the loose sands. The gamma-ray peak at ≈5 m waxes and wanes with time as the source moves firstly closer to the observation borehole and then away again as the head responds to the high tide. Directional sensors were also installed to show the distance to and the direction of the source from the observation borehole (Acworth *et al.*, 2007). Knowing the decay rate of the radioactive isotope, it was possible to correct the observed signal.

14.7.2   Stable isotopes of water

Two piezometer bundles were also installed on either side of the creek as a part of the investigation at Hat Head (Acworth and Dasey, 2003). The location is approximately 50 m upstream of the location for the KBr study described above. The naturally occurring isotopes ^{18}O and ^2H were sampled at each depth. A limited set of tritium (^3H) samples was also recovered and analyzed.

The plot of oxygen-18 vs deuterium in Figure 14.12 shows that two slightly different populations exist on either side of the creek suggesting a different source for the waters. The local meteorological water lines for the two have very similar gradients but different offsets. Both LMWL lines have a different gradient to the GMWL and probably indicate mixing with sea water, possibly as spray (Panga *et al.*, 2017). There is a greater range of values from the

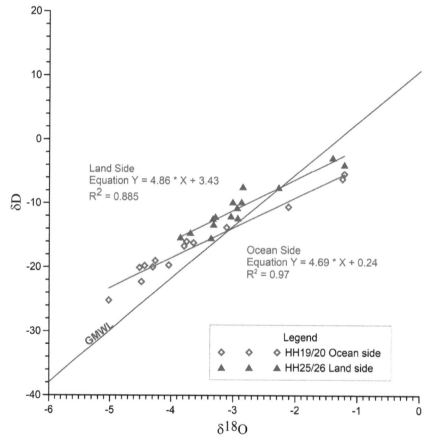

Figure 14.12 Oxygen-18 vs deuterium plots for the two banks of bundled piezometers on either side of the creek. Separate straight-line equations are shown with the GMWL for comparison

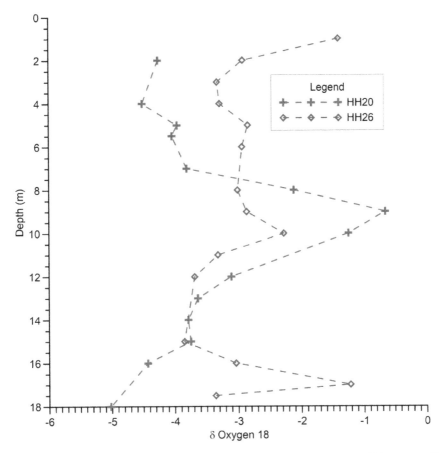

Figure 14.13 Plot of oxygen 18 vs depth for the ocean side and the land side of Korrogoro Creek

samples taken from the bundled piezometer on the ocean side of the creek, suggesting a similarity with the range of rainfall samples at Maules Creek (Figure 14.17) described below.

The plot of oxygen 18 vs depth in Figure 14.13 shows marked variation with depth. Four regions can be recognized in the ocean-side data, indicative of four different water types.

1   From the surface to 7 m $- \delta^{18}O$ lies between -4 and -5 ‰.
2   Between 8 and 12 m δ, ^{18}O approaches 0 ‰ indicating the influence of sea water mixing with a maximum seawater content at 9 m.
3   Between 12 and 15 m, a return to similar values of the surface $- \delta^{18}O$ lies between -3 and -4 ‰.
4   Beneath 15 m, a sharp reduction with a value of -5 $\delta^{18}O$ at 18 m.

The data for the landward side of the creek show both similarities and distinct differences to the ocean side. The shallowest sample from 1 m shows the pronounced influence of sea water, which is not surprising as the surface is flooded with sea water at king tides. There is also a suggestion of seawater influence at 10 m – matching that on the ocean side but much

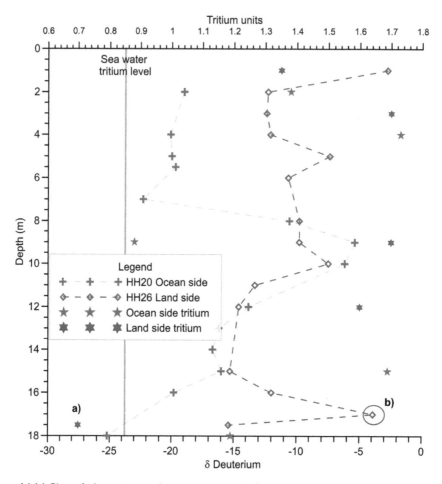

Figure 14.14 Plot of deuterium and tritium vs depth for the ocean side and the land side of Korrogorro Creek

less marked. The oxygen isotope values below 13 m are very similar apart from the last three samples (16, 17 and 18 m depth). It is unfortunate that sea water overtopped the piezometer on the landward side during the king tide event with the result that sea water ran down the 25 mm core pipe of the bundled piezometer and into the formation at the base.

The ^2H and ^3H data show broadly similar features to the ^{18}O data. Four depth intervals can again be seen in the landward data:

1 saline influence at the top due to overbank flooding during high tides,
2 between 2 and 9 m – slowly increasing ^2H between 11 ‰ and 8 ‰,
3 below 10 m, down to 15 m – a zone of reduced ^2H of approximately −15 ‰, and
4 the same zone of seawater contamination noted in the ^{18}O profile.

An inspection of the fluid electrical conductivity data in Table 14.6 confirms the general interpretation of the presence of sea water.

Table 14.6 Field chemistry, oxygen 18, deuterium and tritium values for piezometers 19/20 at Korrogorro Creek and adjacent to the ocean at Hat Head, NSW

Depth (m)	Oxygen 18 ‰	Deuterium ‰	Tritium (TU)	Temperature (°C)	Fluid EC mS/m	pH	Redox (mV)
2.0	−4.26	−19.0	1.38	16.9	86.5	6.90	−165
3.0				17.1	46.8	6.64	−26
4.0	−4.51	−20.1	1.73	17.8	156.0	6.30	−88
5.0	−4.30	−20.0					
5.5	−4.04	−19.7		18.1	768.0	7.01	−130
6.0				17.8	914.0	7.00	−145
7.0	−4.48	−22.3		18.1	1112.0	6.42	−85
8.0	−2.12	−10.6		18.3	2740.0	6.35	−94
9.0	−1.22	−5.4	0.88	18.7	4020.0	6.51	−165
10.0	−1.25	−6.2		19.3	3300.0	5.45	−87
12.0	−3.11	−13.8		19.2	897.0	5.42	−80
13.0	−3.64	−16.2		18.8	92.1	5.70	−72
14.0	−3.79	−16.7		18.6	116.0	5.67	−68
15.0	−3.75	−16.0	1.69	18.5	107.0	5.41	−31
16.0	−4.43	−19.8		18.4	20.9	5.70	−67
17.0				18.5	23.3	5.65	−42
18.0	−5.02	−25.2	1.19	18.2	29.7	5.67	−5

^{3}H values were of less interest, showing that the water in the aquifer is recent and confirming the contamination by sea water at the base of the landward piezometer bundle (b) in Figure 14.14. The value at the base of the bundle (a) in Figure 14.14 is very similar to sea water marked by the grey vertical line.

The isotope data confirm again the importance of sampling from a small area of the aquifer if there is any suspicion that significant vertical variation could be present. The isotope data is shown with the available field chemistry for the two piezometer bundles in Table 14.6 and Table 14.7.

14.8   EXAMPLE: THE USE OF STABLE ISOTOPES OF WATER FROM THE MAULES CREEK CATCHMENT IN NORTHERN NSW

Andersen and Acworth (2009) describe surface water and groundwater interaction in Maules Creek, a tributary to the Namoi River in northern NSW. The location of this area is shown in Figure 14.15. Andersen *et al.* (2008) presented detailed results of a study of ^{18}O and ^{2}H that shows how the isotopes can be used to detect exchange between surface water of the Namoi River and groundwater. Results for rainfall, surface water and groundwater are included.

14.8.1   Rainfall

The local meteoric water line ($\delta^2 H = 8.43\ \delta^{18}O + 16.26$) for the area was based on 16 rain samples collected between March 1998 and March 2001 at the Gunnedah Research Station

Table 14.7 Field chemistry, oxygen 18, deuterium and tritium values for piezometers 26/27 at Korrogorro Creek and adjacent to the land at Hat Head, NSW

Depth (m)	Oxygen 18 (‰)	Deuterium (‰)	Tritium (TU)	Temperature (°C)	Fluid EC mS/m	pH	Redox (mV)
1.0	−1.41	−2.8	1.35	17.1	1351.0	6.02	−127
2.0	−2.93	−12.3		18.4	569.0	5.14	−24
3.0	−3.33	−12.4	1.70	18.3	88.0	5.56	−56
4.0	−3.29	−12.1		19.0	321.0	5.00	−65
5.0	−2.85	−7.4					
5.5				19.0	16.1	5.40	7
6.0	−2.94	−10.7		19.7	17.0	5.42	−18
7.0				19.4	17.5	5.22	−29
8.0	−3.01	−9.8	0	19.5	16.7	5.20	12
9.0	−2.87	−9.8	1.70	19.8	16.9	5.19	−28
10.0	−2.28	−7.5		19.9	560.0	5.44	−16
11.0	−3.32	−13.3		18.9	25.0	5.58	−36
13.0				19.5	16.1	5.45	35
14.0				19.4	16.6	5.45	17
15.0	−3.85	−15.3		1.2	15.6	5.51	15
16.0	−3.04	−12		18.6	16.3	5.50	−163
17.0	−1.22	−3.9		19.1	16.2	5.50	25
17.5	−3.36	−15.4	0.70				
18.0				19.2	22.0	5.64	−49

in the central part of the Namoi River catchment (BoM Station 055024). These analyzes can be seen in Figure 14.17 and are listed in Table 14.8. The collection equipment used to gather the data is shown in Figure 14.16. The collection device was activated at the start of rainfall and closed immediately after cessation of rainfall. The water sample was collected and placed in a refrigerator pending analysis. This approach avoided contamination from dust and other sources. The pH, fluid EC and sodium concentration were recorded. The data for these 16 rainfall events are given in Table 14.8.

The LMWL was calculated using individual rainfall events ranging from 3.6 to 57.6 mm (Table 14.8). The stable isotope plot for the rainfall data (Figure 14.17) shows that there is a large amount of variability in results with $\delta^{18}O$ ranging from −11.4‰ to 1.5‰ and δ^2H ranging from −76.1‰ to 27.8‰. The slope of the data is well correlated with a correlation coefficient of 0.98. The LMWL is close to, but slightly above, the WMWL shown in Figure 14.17. There are two groups of outliers in the data plot marked (a) and (c) on Figure 14.17.

The four samples in November (Group [b] – orange) in Table 14.8 are for significant rainfall events on almost consecutive days and plot in a similar field indicating a fairly uniform composition of the air. The synoptic analysis shows that a steady north-easterly on-shore flow from the Coral Sea and SE Queensland was occurring at the time. The fluid electrical conductivity of these samples was all low 1.7 to 3.9 $\mu S/cm$ and similarly the sodium concentration was low (0.06−1.12 mg/L). Both parameters indicate clean air without dust contamination. There is a slight but not significant deuterium excess on 3 days of this rainfall event. The pH values are low, with the lowest pH recorded in the 16 samples of 3.6. The significance of this is not clear.

Figure 14.15 General location for the Maules Creek area, south of Narrabri in the north-west of New South Wales. The white line shows the catchment boundary. Mount Kaputar is a Tertiary basalt volcanic complex that is 1500 m high and forms the northern and eastern boundary to the catchment. The Namoi River flows from south to north and forms the western boundary of the catchment. Open-cast coal is mined in the hills along the southern boundary

Rainfall events in January 2001 (Group [c] – samples 15 and 16) show a strongly depleted isotopic composition (δ^{18}O of -10.4 and -11.04; δ^2H of -69.2 and -76.1) and plot towards the base of the LMWL. These rainfall events were associated on the synoptic analyzes with cold fronts moving up from the south-west with their origin in sub-polar latitudes.

Rainfall samples in December 2000 (Group [a] – samples 13 and 14) plot towards the right of the LMWL and are significantly enriched (δ^{18}O of 0.77 and 1.52; δ^2H of 22.1 and 27.8). This can be explained by the synoptic analysis that shows the remnants of tropical cyclone Sam that had entered Australia associated with the equatorial trough on 10 December and traveled a long way across Australia before rain fell at Gunnedah. The enrichment can be explained by evaporation of rainfall from clouds in a low-humidity environment and/or evaporation from more saline water than sea water from the many saline lakes in the Great Artesian Basin. Enrichment past the SMOW appears to be a common feature in Australian inland rainfall (Hollins *et al.*, 2018).

There would seem to be little seasonality to the data with samples from winter (Sample 6) plotting in the middle of the series with the synoptic analysis indicating a source of air from well to the south of Australia. In general, it is assumed that summer rainfall is associated with troughs breaking south from Queensland and winter rainfall with cold outbreaks from the sub-polar regions. However, as usual, it is a little more complicated. It would appear, however, that the position along the LMWL can be explained by the source of the air from which the condensation and rainfall occurred, an observation supported by Hollins *et al.* (2018). There would seem to be little relationship between the quantity of rainfall and the location along the LMWL.

Figure 14.16 Components of a rainfall collector for isotope analysis. The lid to the dustbin was removed at the beginning of the rainfall event and the water sample collected at the end of the event and transferred to a refrigerator (Photo: Wendy Timms)

Table 14.8 Stable isotope values for rainfall measured at the Gunnedah Research Station (BoM Station No 055024). Colours relate to Figure 14.17.

Sample	Date	Rainfall (mm)	$\delta^{18}O$ (‰)	$\delta^{2}H$ (‰)	Deuterium (Excess)	pH	EC (µS/cm)	Na (mg/L)
1	23 Oct 1999	3.6	−4.16	−11.9	21.4			
2	9 Nov 1999	16.0	−2.97	−3.8	20.0			
3	28 Jan 2000	3.6	−4.22	−14.1	19.7		20.3	0.27
4	10 Mar 2000	38.4	−8.28	−60.9	5.3	5.0	18.8	0.04
5	5 May 2000	8.0	−2.66	−11.8	9.5	5.4	14.5	0.70
6	9 Aug 2000	9.4	−1.66	3.3	16.6	5.0	11.0	0.50
7	14 Oct 2000	57.6	−2.47	−0.5	19.3	5.4	12.9	0.35
8	25 Oct 2000	28.6	−5.14	−30.1	11.0	4.0	3.9	0.14
9	15 Nov 2000	21.8	−4.53	−21.4	14.8	4.4	3.4	0.06
10	16 Nov 2000	31.6	−4.76	−24.7	13.4	4.2	1.7	0.05
11	18 Nov 2000	12.6	−3.00	−14.2	9.8	4.3	3.9	0.07
12	19 Nov 2000	29.2	−4.06	−19.9	12.6	3.6	3.2	1.12
13	1 Dec 2000	9.4	1.52	22.1	9.9	3.9	11.2	0.16
14	14 Dec 2000	31.2	0.77	27.8	21.6	3.8	4.2	0.08
15	31 Jan 2001	33.4	−10.40	−69.2	14.0	4.9	2.2	0.07
16	13 Mar 2001	30.8	−11.04	−76.10	12.2	3.7	1.8	0.06

14.8.2 Surface water

Surface water samples were collected from the main Namoi River channel at Narrabri and from the unregulated tributary Maules Creek draining the Great Dividing Range. The location of surface and groundwater sampling points is shown in Figure 14.18.

The majority of the surface-water samples along of the Namoi River plot below the LMWL, on a line which is referred to as the local evaporation line (LEL: $\delta^{2}H = 5.7$

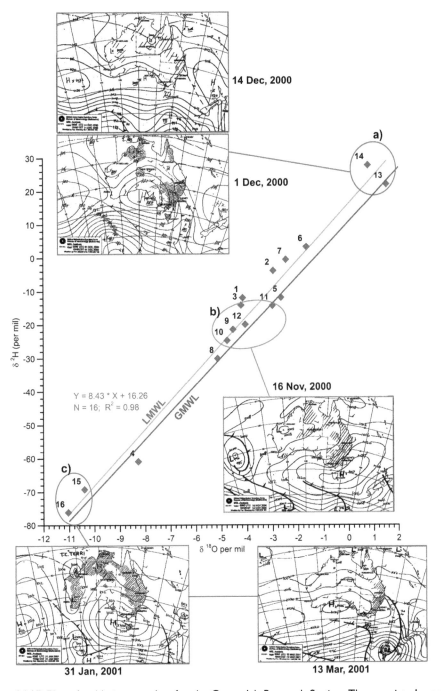

Figure 14.17 Plot of stable isotope data for the Gunnedah Research Station. The associated synoptic maps are included to show possible source areas for the rainfall. The station location is shown as a red dot. The three data sets: (a) red, (b) orange and (c) magenta refer to the groupings of isotope data shown in Table 14.8

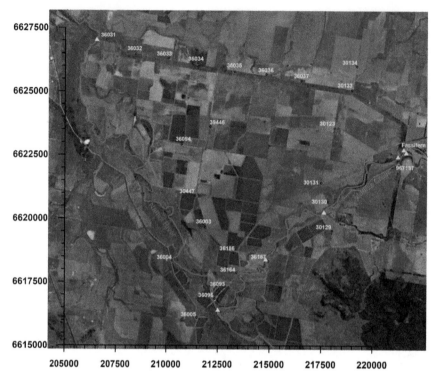

Figure 14.18 Map showing the location of surface- and groundwater samples in the Maules Creek subcatchment. Red circles with an associated number are groundwater sample locations. Cyan coloured solid triangles represent surface water locations. Fassifern is a shallow well. The area of this map is shown on Figure 14.15

$\delta^{18}O - 3.91$) for the area (Figure 14.19). Samples that plot on this line are indicative of waters that have undergone isotopic enrichment due to evaporation. A similar evaporative pattern is also seen for surface-water samples in other parts of Australia. In contrast to the Namoi River samples, the surface-water samples from the tributary of Maules Creek plot in a narrow cluster ($\delta^{18}O$: -5.3‰ to -4.5‰ and δ^2H: -28.3‰ to -24.8‰) above the LEL, but on the LMWL in Figure 14.19. The plot position indicates that these Maules Creek surface samples represent discharging groundwater as described by Andersen *et al.* (2008) and Andersen and Acworth (2009).

The seeps from Mount Kaputar plot slightly above the LMWL line, indicating the impact of elevation change. The seeps occur above 1400 m in comparison to the majority of the groundwater samples on the plain at approximately 300 m (Mazor, 1997).

14.8.3 Groundwater

The location of the groundwater and surface-water samples within the Maules Creek subcatchment are shown in Figure 14.18.

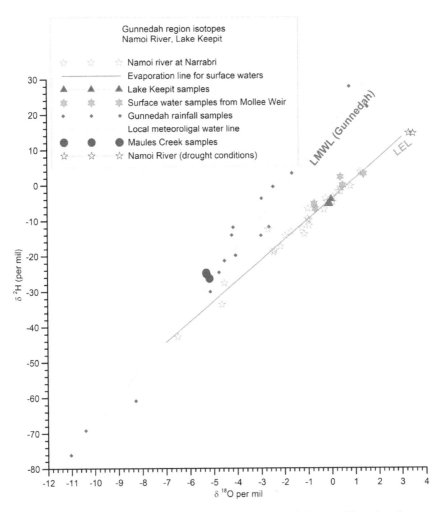

Figure 14.19 Comparison of deuterium and oxygen 18 (SMOW) for rainfall and surface water at Maules Creek, northern NSW and the WMWL

Groundwater samples from the Maules Creek catchment generally range from $-7.1‰$ to $-5.2‰$ for $\delta^{18}O$ and $-41.5‰$ to $-26.6‰$ for $\delta^2 H$ with average values of $\delta^{18}O = -6.22‰$ and $\delta^2 H = -35.31‰$.

Most samples plot on or close to the LMWL and form a coherent group on the $\delta^{18}O$ vs $\delta^2 H$ plot. However, three groundwater samples show considerable enrichment compared to the bulk of samples. These are encircled in Figure 14.20 and marked (a) and (b). These enriched groundwater samples were from shallow wells (11.5 to 22.8 m deep) close to the Namoi River (< 1 km). Sample (a) plots close to the LWL line and is an indication that Namoi River water has infiltrated directly as far as the well. The other two samples (b) on Figure 14.20 indicate mixed waters.

In general, the grouping of the groundwater water isotope results is very close, indicating a similar mode of origin. This is significant given that the samples come from different

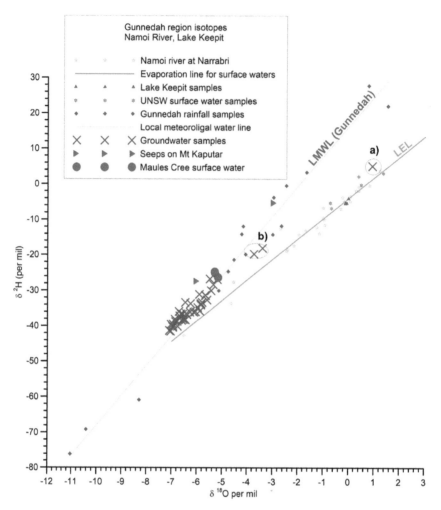

Figure 14.20 Comparison of deuterium and oxygen 18 (SMOW) for surface water and groundwater at Maules Creek

depths in an alluvial sequence. The most depleted groundwater samples were collected from bores located near the Nandewar Range in the north. However, even those were less depleted than the strongly depleted rainfall data from Group (c) (Figure 14.17). The majority of the groundwater samples plot close to the central group of rainfall samples (Group [b]) indicating that the major recharge to the aquifer is from storms associated with on-shore north-east winds.

In the southern part and below Maules Creek, shallow groundwater samples have isotopic signatures that are higher and range from −5.7‰ to −5.2‰ for $\delta^{18}O$ and −31.4‰ to −26.9‰ for δ^2H. These groundwater samples are slightly lower but comparable to the surface-water samples from Maules Creek ($\delta^{18}O$: −5.3‰ to −4.5‰ and δ^2H: −28.3‰ to −24.8‰). Groundwater discharge occurs into Maules Creek except under very long drought conditions. The Maules Creek surface samples (Figure 14.20) show that there is already some isotopic enrichment due to evaporation, as these samples plot at the top of the groundwater group.

14.8.4   Summary

It has been clearly demonstrated that there is much that can be understood about both hydrology and hydrogeology from collecting representative water samples for stable isotopes. In a similar manner to the degree of vertical variability noted in the previous discussions of hydrogeochemical techniques, it is clear that field campaigns must be well planned and executed. Simple bulk samples are more likely to confuse than to provide evidence of important processes. Importantly, bulk rainfall samples, where rain is collected for a month before analysis, will almost certainly obscure significant and important variability.

14.9   CARBON ISOTOPES

Carbon has two stable, non-radioactive isotopes: carbon-12 (^{12}C) and carbon-13 (^{13}C). In addition there are trace quantities of the unstable isotope carbon-14 (^{14}C) on the earth's surface. Carbon-14 has a half-life of only 5730 years if it did not get constantly replenished in the upper atmosphere. The means of renewal is the impact of cosmic rays from the sun reacting on nitrogen in the upper atmosphere of the earth via the reaction:

$$n + {}^{14}_{7}N \rightarrow {}^{14}_{6}C + p \tag{14.13}$$

In effect, a proton in the atomic nucleus is exchanged for a high-energy neutron. This reaction takes place between 9 and 14 km towards the poles of the earth. The carbon forms carbon dioxide and disperses evenly throughout the atmosphere. It also permeates to the oceans, dissolving in water. Plants take up carbon dioxide by photosynthesis which then propagate into higher tropic levels of the food chain. This process keep plants and animals in equilibrium with atmospheric ^{14}C. However, when the plant or animal dies, the uptake of ^{14}C stops and the ^{14}C content decays by beta decay:

$$^{14}_{6}C \rightarrow {}^{14}_{7}N + e^{-} + v_{e} \tag{14.14}$$

The radioactive decay of carbon-14 follows an exponential decay:

$$\frac{dN}{dt} = -\lambda N \tag{14.15}$$

where:
N_0 is the number of radiocarbon atoms at $t = 0$,
N is the number of radiocarbon atoms remaining after radioactive decay during time t, and
λ is the radiocarbon decay or disintegration constant.

The solution to this equation is:

$$N = N_0 e^{-\lambda t} \tag{14.16}$$

It can be shown that $t_{average} = \frac{1}{\lambda} = 8033 \ years$ or that $t_{\frac{1}{2}} = t_{average} \times \ln 2 = 5568 \ years$.

The radiocarbon age can thus be described as: $t(BP) = \frac{1}{\lambda} \ln \frac{N}{N_o}$.

14.9.1   Principles of ^{14}C dating of groundwater

Measurements of ^{14}C are made by counting beta decay in either a gas or liquid phase, or more recently by accelerator mass spectrometry of a graphite target. Sampling for radiocarbon determination is carried out in one of two ways. The first involves precipitation of about 60 mmoles of total dissolved carbon as barium or strontium carbonate at pH > 9. Another method involves acidification of the water, gas stripping the CO_2 with pure nitrogen, and trapping the evolved CO_2 in a solution of sodium hydroxide. The latter technique is used when samples have low concentration of dissolved inorganic carbon.

Radiocarbon activities are expressed as percent modern carbon (pmc) which represents the activity of carbon prior to the dilution by post-industrial dead fossil-fuel CO_2 and enrichment from thermonuclear weapons testing. A carbon sample having 0% modern carbon is labeled ^{14}C-dead or has an age beyond that resolvable by the radiocarbon technique (>40,000 years). A sample which contains >100% modern carbon must have some component of bomb-radiocarbon and thus probably is less than 50 years old. The age of a water sample can be calculated from the rearrangement of the decay equation to give:

$$t = \frac{1}{\lambda} \ln \left(\frac{C}{C_o} \right)$$
(14.17)

where:
t is the age in years,
λ is the decay constant of ^{14}C,
C is measured ^{14}C activity, and
C_o is the initial ^{14}C activity.

By measuring the carbon-14 content of a water sample alone, we cannot assign an absolute age to the water. This is because dissolved inorganic carbon (DIC) interacts with several other carbon reservoirs such as soil CO_2, calcium carbonate in the soil zone and $CaCO_3$ within the aquifer.

Consider a situation where water percolates through the soil zone. The assimilation of CO_2 by land plants is accompanied by production of CO_2 in the soil zone as plant litter decays and roots respire. The bicarbonate in groundwater arises from the action of dissolved CO_2 (the partial pressure of which can be orders of magnitude higher than that in surface waters) or organic matter on calcium carbonate:

$$CO_2 + CaCO_3 + H_2O = 2\ HCO_3^- + Ca^{2+}$$
(14.18)

where for every two atoms of bicarbonate produced, one atom will have been derived from the solid carbonate and the other from soil CO_2.

The soil CO_2 or dissolved organic carbon represents carbon essentially of modern ^{14}C activity (at the time of reaction). If $CaCO_3$ was deposited within a soil matrix more than 40,000 years ago, then all of its ^{14}C will have decayed away and it will have no radiocarbon activity. Following from the above stoichiometry, a modern water sample will, in this case, have a radiocarbon content of about (100+0) / 2 = 50% modern carbon and thus if taken at

Table 14.9 Range of $\delta^{13}C$ of some sources of dissolved inorganic carbon

Source zones	$\delta^{13}C$ (‰, PDB)
Atmospheric CO_2	−8‰
Soil-CO_2 or organic matter	−15 to −26‰
$CaCO_3$	−6 to +1‰
Groundwater $\delta^{13}C$	−25 to +2‰

face value would have an apparent age of:

$$t = -8270 \ln\left(\frac{50}{100}\right) = 5730 \text{ years} \tag{14.19}$$

This would be a gross overestimation of its age. The way in which we can correct these apparent ages for true radiocarbon ages is by measuring the isotopic composition of stable carbon isotopes dissolved in the water and in the soil matrix.

A useful way of correcting ^{14}C apparent ages is by measuring the ^{13}C content. The $\delta^{13}C$ composition of the various carbon reservoirs in the figure above are usually quite different, hence we can apply isotope mass-balance calculations to determine the relative amount of carbon derived from biological and inorganic sources. Table 14.9 gives some $\delta^{13}C$ values for different carbon reservoirs involved in the geochemistry of carbon in soil water and groundwater.

An example of such a calculation would be if we had a sample of soil $CO_2 = -18‰$ (say in an arid region) and a $CaCO_3 = 0‰$, the resulting HCO_3^- would be about −9‰. Alternatively, weathering of silicate minerals, such as feldspars, also gives rise to HCO_3^-:

$$NaAlSi_3O_8 + 7H_2O + H_2CO_3 \rightleftharpoons Na^+ + Al(OH)_3 + 3H_4SiO_4 + HCO_3^- \tag{14.20}$$

where the above process contributes carbon solely from biological sources, therefore no ^{13}C correction would be needed. The contribution of this reaction to the dissolved HCO_3^- in groundwater is subordinate to the one involving dissolution of $CaCO_3$.

14.9.2 Correction procedures for estimating carbon-14 ages

We have listed below a summary of various correction schemes that need to be applied to measured ^{14}C activities according to the prevailing soil conditions and the presence or absence of a gas phase.

Open system models (systems open to a gas phase)

An open system for the purposes of radiocarbon dating of groundwater refers to a situation where the dissolved CO_2 and HCO_3^- isotopic composition is driven by the high CO_2 concentrations within the soil gas. If it can be demonstrated that such conditions occurred during recharge, this represents the ideal conditions for ^{14}C dating. The following three situations represent some of the physical conditions that would be amenable to application of an open system model and requires no formal correction to estimate a ^{14}C age.

Non-carbonate aquifer – high air filled porosity, high pCO_2. The dissolved bicarbonate within the groundwater equilibrates completely with the gaseous CO_2. Measured

$\delta^{13}C = -17‰$. This represents the usual +8‰ fractionation between gaseous CO_2 and dissolved bicarbonate. $A_i = 100$ pmc (percent modern carbon). No correction is required.

Same as above but low air filled porosity. Total DIC in the form of uncharged CO_2. The $\delta^{13}C = -25‰$. $A_i = 100$ pmc and no correction is required.

Carbonate aquifer, high air filled porosity with high pCO_2. The $\delta^{13}C = -17‰$. $A_i = 100$ pmc and no correction is required.

There may be some situations, however, that display ^{13}C and DIC characteristics that give the appearance of open system behaviour, but have A_o values less than 100 pmc. Such circumstances may occur through the dissolution of secondary calcite that had precipitated under open system conditions within the soil zone. The resultant $\delta^{13}C$ and chemical parameters would indicate apparent open system situation, but the ^{14}C activity would reflect that of the time of carbonate precipitation. This secondary calcite need not be dead and could even enrich ^{14}C activity of dissolved HCO_3^- if precipitated over the past 50 years (i.e., during the post-bomb era).

Closed-system models (system closed to gas phase)

These usually apply to low-porosity media, in deep unsaturated zones with low organic matter or in deep confined aquifers. Correction scenarios include:

Dissolution of modern $CaCO_3$ in the soil matrix. No carbonate within the aquifer matrix, therefore no correction is required.

Reaction between biogenic CO_2 and ancient $CaCO_3$. Same as for case described above. System is closed to CO_2 in the unsaturated zone. The initial ^{14}C activity is written in the form:

$$A_0 = \frac{\delta_d - \delta_c}{\delta_g - \delta_c}(A_g - A_c) + A_c \qquad (14.21)$$

where δ and A represent the ^{13}C and ^{14}C contents respectively, and d, c and g refer to ΣDIC, solid carbonate and soil CO_2 respectively. The value for δ_g is between $-25‰$ in temperate regions to $-12‰$ for vegetated arid zone vegetation. δ_c is usually around $0‰$ but can be as light as $-6‰$ for freshwater carbonates.

In an open system with CO_2 in the unsaturated zone, HCO_3^- resulting from reaction between CO_2 and $CaCO_3$ is allowed to equilibrate with the soil gas. The equation to calculate A_o is then:

$$A_0 = \frac{\delta_d - \delta_c}{\delta_g - \epsilon_g - \delta_c}(A_g - A_c) + A_c \qquad (14.22)$$

where ϵ_g = fractionation between gaseous CO_2 and dissolved $HCO_3^- \approx -8‰$.

The most successful application of ^{14}C to estimate water residence time (as opposed to explicit ages) is where samples can be collected along flow paths. Calculations of ^{14}C ages along the flow path may then be made and the mean groundwater velocity calculated by dividing the ages difference between two sampling points by the distance between them. This approach minimizes errors induced by the assumptions and approximations that propagate through the various correction schemes.

Particular care is needed when there is potential for mixing between two aquifers. Consider two groundwaters. Water from aquifer A is 40,000 years old ($A = 0$ pmc); water from aquifer B is 100 years old ($A = 100$ pmc). If the two were mixed in equal molar proportions, the average age would be $\frac{40,000+100}{2} = 20,000$. However by taking the radiocarbon data at face value, we would obtain: $\frac{100+0}{2} = 50$ pmc, which gives a ^{14}C age of 5,568 years! Much younger than the average age.

14.9.3 Summary of ^{14}C methods

The use of radiocarbon to estimate water residence time should be done with great care and requires high-quality hydrochemical data, stable isotope data and a good knowledge of the regional hydrogeology. The following procedure should be adopted to make maximum use of radiocarbon data.

1 Establish flow paths and hydraulic connectivity and take several samples along that flow path. The groundwater residence time between two sampling points is far more useful than a single radiocarbon age.
2 Evaluate the importance of mixing of different groundwater sources and their effect on the ^{14}C activity of the end mixture.
3 Samples should be collected from the recharge area to calibrate the initial radiocarbon activity (i.e., soil CO_2, δ^{13}C of carbonates, alkalinity and total DIC).
4 Use the hydrochemical data to accurately determine the carbonate speciation, the partial pressure of CO_2 and the saturation state (sub- or supersaturation) with respect to carbonate minerals.

14.10 CHLORINE-36

Chlorine-36 application to hydrology has attracted considerable interest over the past decade. This is due to the ability of tandem accelerator mass spectrometry to measure very low levels of ^{36}Cl with reasonable accuracy. Some of the advantages of chlorine-36 are (1) it behaves conservatively in most hydrological environments and (2) has a half-life of 300,000 years, making it suitable for a wide range of applications. These include groundwater dating and determination of recharge rates. Collection and sample preparation is fairly straightforward, but the cost of analysis remains quite high (about $1000 per sample).

^{36}Cl is produced naturally by cosmic ray spallation of ^{40}Ar. During the bomb-testing of the 1960s, a significant amount of ^{36}Cl was released to overwhelm the natural abundance. The bomb ^{36}Cl can thus be used to delineate recharge rates and compare with tritium data. It has an advantage over tritium in that it is not affected by vapour transport processes in the unsaturated zone.

One of the classic early studies involving ^{36}Cl was carried out in the Great Artesian Basin (GAB) in Australia, where it was used to date very old groundwater. By making an estimate of the initial ^{36}Cl/^{35}Cl ratio in the recharge area of the GAB, and correcting for *in situ* processes affecting ^{36}Cl abundance, the age at any sampling point can be obtained from the expression:

$$t = -\frac{1}{\lambda} \ln\left(\frac{C_t(R_t - R_{se})}{C_0(R_o - R_{se})}\right) \tag{14.23}$$

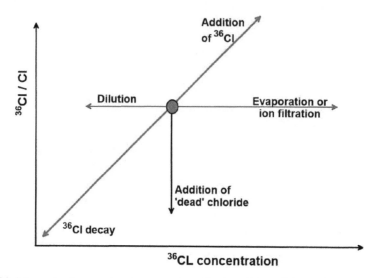

Figure 14.21 Schematic diagram showing processes affecting chlorine-36 and Cl in groundwater

where:

λ is the decay constant of ^{36}Cl (yr^{-1}),
C_t is the measured chloride concentration (atoms/L),
C_o is the initial chloride concentration (atoms/L),
R_t is the measured ^{36}Cl/^{35}Cl activity ratio,
R_{se} is the ratio at secular equilibrium with the aquifer U and Th minerals, and
R_o is the initial ^{36}Cl/^{35}Cl ratio.

The ages calculated for the GAB indicate that water which recharges just east of Charleville in Queensland reaches Innaminka on the Queensland–SA border after about 1.5 million years.

The distribution of ^{36}Cl/^{35}Cl ratios as a function of ^{36}Cl concentration can yield considerable information on the processes that affect chloride in groundwater. Processes such as dissolution of old halite, evaporation, membrane filtration and mixing with new, dilute meteoric waters can be recognized (Figure 14.21).

Bomb-produced ^{36}Cl may be particularly useful for estimating recharge rates in the unsaturated zone. The fallout pattern is reasonably well known at some locations, therefore representative cumulative water content and ^{36}Cl/^{35}Cl ratios in soil profiles can be obtained. These profiles are best obtained at the driest time of year to ensure a steady-state soil moisture profile.

Chapter 15

Well hydraulics, radial flow modelling and single-well tests

15.1 INTRODUCTION

In this chapter we will present the derivation of groundwater flow equations that allow prediction of the change in hydraulic head that occurs as a result of pumping from an abstraction well. This area is normally referred to as groundwater hydraulics. Some terminology has been developed that is specific to groundwater hydraulics, so a review of these terms and the conceptual models that are associated with them will be provided first.

The groundwater flow equation can be expressed in radial coordinates and solved analytically in an approach first described by Theis (1935). This solution method was developed using the analogy to heat flow in a conducting plate. The analysis is briefly reviewed and then compared to numerical solutions to the same equation. The numerical solutions, using either finite differences or finite elements, allow for consideration of far more complex flow models.

It has been recognized (Kruseman and de Ridder, 1990) that flow to a well may not be linear in regions close to the well. This leads to significant complications in the use of Darcy-type solutions and is investigated in more detail later using experimental and numerical analysis approaches. The concept of non-linear flow close to an abstraction well leads to a consideration of single-well tests that allow a better indication of the efficiency of an abstraction well to be determined. All pumping tests should incorporate analysis of data from observation wells if the possible complications of non-linear flow around the abstraction well are to be avoided. This is not always possible, and techniques for the analysis of single-well tests are described.

There is a rich literature concerned with the inversion of pumping-test data (Kruseman and de Ridder, 1990) and sophisticated computer packages available such as AQTESOLV, WTAQ and Aquifer Test Pro, for example, to analyze the data. This area will not be described in any further detail.

15.2 REVIEW OF DEFINITIONS

Before proceeding with the analysis, a number of terms are defined. These terms can be found in the Glossary of Terms but are included here for convenience:

Standing Water Level (SWL): The distance from ground level to the water level in a non-pumping bore outside the area of influence of any pumping bore.

Static Head: This is the flowing bore equivalent of SWL It is the height above ground level that water at a particular temperature would stand if the casing were extended upwards. It is expressed as metres or kilopascals.

Drawdown: The depth (distance) that the water level in a borehole or well has been lowered from the SWL during pumping. It is measured at specified time after pumping commenced.

Residual Drawdown: The depth (distance) that the water level in a borehole or well remains lowered from the SWL after pumping ceases. It is measured at specified times after pumping stopped.

Recovery: The amount by which the water level in a borehole has risen at a given time after pumping ceased. It is the difference between the residual drawdown after the given time and the hypothetical drawdown if pumping had not ceased. When the water level returns to SWL, recovery is said to be complete.

Available Drawdown: For a particular pump installation, this is the distance between the SWL and the pump suction or the depth of water over the pump suction.

15.2.1   Flowing and non-flowing boreholes

The basic difference between a flowing borehole and a non-flowing borehole in a confined aquifer is that the ground level is lower than the standing water level (or static head) in the case of a flowing borehole and higher than in a non-flowing borehole as shown in Figure 15.1. If, in the case of a flowing artesian borehole, the casing were extended high enough above the ground, the water would stand in the casing and a pump would have to be installed to discharge the water. The groundwater hydraulics are the same for both types of boreholes, and no differentiation will be made between the hydraulics of flowing and non-flowing boreholes.

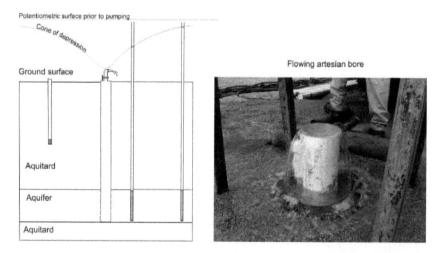

Figure 15.1 Flowing artesian bore (Photo: Colin Hazell)

15.2.2   Unconfined aquifer

An unconfined aquifer occurs where the top of the saturated zone is the water table. The hydraulic pressure at the water table is equal to atmospheric pressure (Figure 15.2a). Changes in atmospheric pressure are immediately transmitted to the water table through the unsaturated porosity above the water table. Therefore, there is no differential movement when the atmospheric pressure changes, in contrast to a confined aquifer system.

15.2.3   Confined aquifer

The water pressure in a confined aquifer (Figure 15.2b) is indicated by a potentiometric surface that is above the top of the aquifer.

15.2.4   Leaky aquifer

The concept of a confined aquifer is good mathematically, but they rarely exist in practice. Figure 15.3 shows two possibilities where leakage occurs. In the first, clay in an aquitard

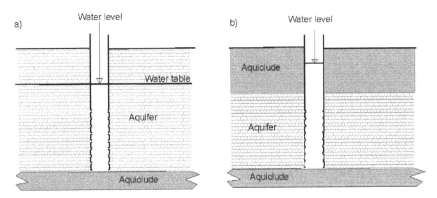

Figure 15.2 Unconfined and confined aquifers

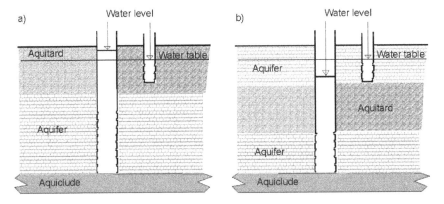

Figure 15.3 Leaky aquifers

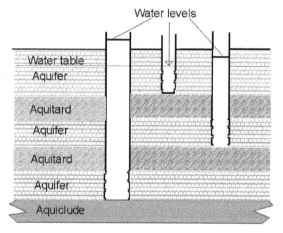

Figure 15.4 Sequence of aquifers and aquitards

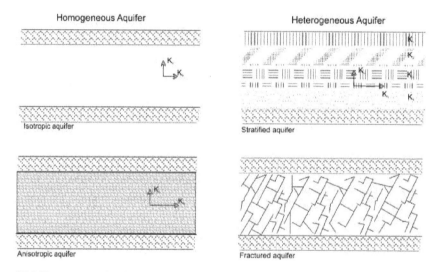

Figure 15.5 Variations in the distribution of hydraulic conductivity

caps the aquifer and water drains slowly down from storage in the clay. In Australia, this is a particularly important concept because many of the aquifers are overlain by silts rather than clays. The silts can leak large quantities of poor-quality water over time. The second concept in Figure 15.3 shows an aquifer overlying the aquitard that overlies the aquifer. In this case we can have water draining down through the aquitard from the aquifer as well as drainage from the aquitard. If there are fractures in the clay/silt, flow will be considerable.

The final figure (Figure 15.4) shows a sequence of aquifers and aquitards – similar to that which would exist in a fluvial or estuarine environment.

15.2.5   Anisotropy and heterogeneity

Aquifers may be isotropic and homogeneous but are more usually anisotropic and inhomogeneous. The various classes are shown in Figure 15.5.

15.2.6   Bounded aquifers

A common assumption is that the aquifer is horizontal, is of constant thickness and extends to infinity. In fact, this assumption is necessary to allow modification to the flow equation such that it can be solved analytically. Clearly, in practice that is not possible, and three types of bounded aquifers are recognized:

- barrier boundary (Figure 15.6a),
- recharge boundary (Figure 15.6b), and
- aquifer of non-uniform thickness (Figure 15.6c).

Figure 15.6 illustrates these three concepts.

15.2.7   Steady state and non-steady state

There are two types of equations derived in groundwater hydraulics. The first describes conditions in the aquifer when the cone of depression has stopped expanding and conditions are

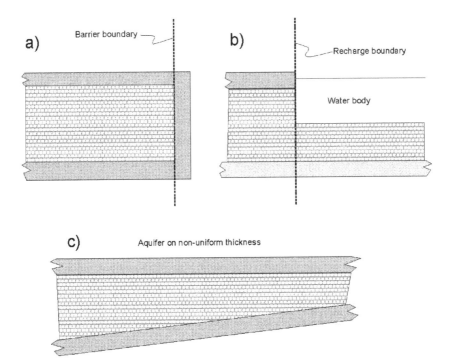

Figure 15.6 Types of boundary condition

said to be in steady state. The water being abstracted by the water well is no longer coming from the aquifer by the development of storage but is flowing across the outer boundary of the cone of depression, or down from above in the leaky case. Every pumping test will tend towards a steady-state condition after a long enough period of time.

Non-steady state analysis has to take account of the release of water from storage. By definition, the cone of depression is still expanding in size.

15.2.8   Bulk compressibility β_b and water compressibility β_w

Bulk compressibility is important in the analysis on non-steady state flow. It describes the change in volume or the strain induced in the aquifer or aquitard under a given stress:

$$\beta_b = \frac{dV_T/V_T}{d\sigma_e} \qquad (15.1)$$

Bulk compressibility is expressed in m^2/m or Pa^{-1}. Its value for clay varies from 10^{-6} to 10^{-8} Pa^{-1}, for sand from 10^{-7} to 10^{-9} Pa^{-1} and for gravel and fractured rock from 10^{-8} to 10^{-10} Pa^{-1}. In the original definition (Jacob, 1940) for which much of the interpretation of pumping-test data is based, bulk compressibility was defined on the assumption that the individual grains were not compressible. A more complete analysis is given in Chapter 7.

The compressibility of water is:

$$\beta_w = \frac{dV_T/V_T}{dP} \qquad (15.2)$$

The compressibility of water at normal groundwater temperatures is approximately 4.4×10^{-10} Pa^{-1}.

15.2.9   Specific storage (S_s)

The volume of water that is yielded from a unit volume of a confined aquifer for a unit decline in hydraulic head:

$$S_s = \rho g(\beta_b + \phi\beta_w) \qquad (15.3)$$

The dimensions of specific storage are L^{-1}. Note that the bulk compressibility term β_b was also defined using the term by Jacob (1940).

15.2.10   Storativity or storage coefficient

This term relates to specific storage in the same manner that transmissivity relates to hydraulic conductivity. It is the integration of the specific storage over the thickness of the aquifer b:

$$S = \rho gb(\beta_b + \phi\beta_w) = S_s b \qquad (15.4)$$

15.2.11   Diffusivity ($\frac{Kb}{S}$)

The hydraulic diffusivity is the ratio of the transmissivity and the storativity of a saturated aquifer. It governs the propagation of changes in head in the aquifer. Diffusivity has dimensions of $L^2/(T)ime$.

15.2.12   Leakage factor (L)

The leakage factor, or characteristic length, is a measure for the spatial distribution of the leakage through an aquitard into a leaky aquifer and vice versa (Kruseman and de Ridder, 1990). It is defined as:

$$L = \sqrt{\frac{Kbb'}{K'}} = \sqrt{\frac{Tb'}{K'}} \tag{15.5}$$

The primes denote the property of the aquitard where K' is the hydraulic conductivity and b' is the thickness. Large values of L indicate a low leakage rate, whereas small values of L indicate a high leakage rate.

15.3   GROUNDWATER FLOW EQUATIONS

The groundwater flow equations represent a mathematical model of the groundwater flow process. For certain simple situations, the mathematical model, as represented by the equations, can be solved by analytical methods. The Theis method is an example of an analytical solution to the transient flow equation (the diffusion equation) in two dimensions.

The equation for non-steady state flow of groundwater through an anisotropic medium was developed in Equation 7.8 and is repeated here as:

$$\frac{\partial}{\partial x}\left(K_x \frac{\partial(h)}{\partial x}\right) + \frac{\partial}{\partial y}\left(K_y \frac{\partial(h)}{\partial y}\right) + \frac{\partial}{\partial z}\left(K_z \frac{\partial(h)}{\partial z}\right) = S_s \frac{\partial h}{\partial t} \tag{15.6}$$

For an isotropic medium where $K_x = K_y = K_z$ and for a homogeneous medium where $K_{(x,y,z)}$ is constant, Equation 15.6 reduces to:

$$K\left(\frac{\partial}{\partial x}\left(\frac{\partial h}{\partial x}\right) + \frac{\partial}{\partial y}\left(\frac{\partial h}{\partial y}\right) + \frac{\partial}{\partial z}\left(\frac{\partial h}{\partial z}\right)\right) = K\left(\frac{\partial^2 h}{\partial x^2} + \frac{\partial^2 h}{\partial y^2} + \frac{\partial^2 h}{\partial z^2}\right) = S_s \frac{\partial h}{\partial t} \tag{15.7}$$

Using the ∇ representation for the partial derivatives, and dividing through by the hydraulic conductivity K, gives:

$$\nabla^2 h = \frac{S_s}{K}\frac{\partial h}{\partial t} \tag{15.8}$$

This equation is known as the *diffusion equation*. The solution $h(x, y, z, t)$ describes the value of the hydraulic head at any point in a flow field at any time. This equation cannot be solved using analytical methods. A full numerical analysis is required using either finite differences or finite elements.

The problem can be reduced from three dimensions to two dimensions if the following assumptions are made:

- the thickness of the confined aquifer is constant,
- the aquifer is of infinite extent, and
- there are no vertical flow components in the aquifer which therefore makes the derivative of the head with respect to vertical distance (z) equal to zero.

Then, multiplying the top and bottom of the right-hand side of Equation 15.8 by the aquifer thickness (b) gives the transient flow equation in two dimensions as:

$$\frac{\partial^2 h}{\partial x^2} + \frac{\partial^2 h}{\partial y^2} = \frac{S}{T}\frac{\partial h}{\partial t} \tag{15.9}$$

where:

T is the aquifer transmissivity – the product of the hydraulic conductivity and the aquifer thickness and

S is the aquifer storage coefficient – the product of the specific storage and the aquifer thickness.

Note that the introduction of the transmissivity term is simply a mathematical necessity required to simplify the equation for groundwater flow by moving from 3D to 2D. It does not have a geological or hydrogeological reality in the same way that hydraulic conductivity does. A suitable value of transmissivity must be established from consideration of the available information on hydraulic conductivity.

The transmissivity is defined above as the product of the aquifer hydraulic conductivity and the aquifer depth (thickness). This could be interpreted as meaning that the transmissivity can only be defined for homogeneous aquifer material. This is not the case. The transmissivity can be defined Rushton (2003) as:

$$T = \sum_{sat-depth} K_x \Delta z \tag{15.10}$$

or:

$$T = \int_{z=base}^{z=top} K_x dz \tag{15.11}$$

Consider an aquifer comprised of three horizontal layers that acts as a single unit. The bottom layer of 15 m thickness has a hydraulic conductivity of 10 m/day; a 30 m middle zone has a hydraulic conductivity of 1 m/day and the 5 m thick upper zone a K value of 20 m/day – as shown in Figure 15.7. Applying Equation 15.10, the transmissivity function with respect to depth can be generated and is shown on the right-hand side of Figure 15.7, in which case the transmissivity of the single unit would be 280 m²/day.

15.4 ANALYTICAL SOLUTION METHODS TO THE RADIAL FLOW EQUATION

It is clearly shown in Figure 15.8 that the flow to a well in a homogeneous aquifer will be radially symmetric.

It is therefore convenient to convert Equation 15.9 to radial coordinates, *viz.*:

$$\frac{\partial^2 h}{\partial r^2} + \frac{1}{r}\frac{\partial h}{\partial r} = \frac{S}{T}\frac{\partial h}{\partial t} \tag{15.12}$$

where:

r is the radial distance from the well $r = \sqrt{x^2 + y^2}$.

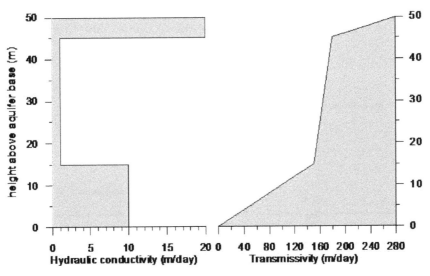

Figure 15.7 Transmissivity calculation for aquifers with layering (adapted from Rushton (2003))

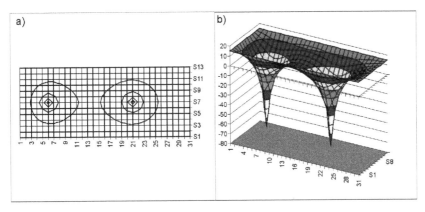

Figure 15.8 2D results from a finite difference model developed using an EXCEL spreadsheet

The mathematical region of flow is a one-dimensional line through the aquifer, from $r = 0$ at the well to $r = \infty$ at the boundary. A schematic of the flow regime is shown in Figure 15.9.

15.4.1   Forward and inverse solutions to the radial groundwater flow equation

Theis (1935) utilized the heat flow analogy to arrive at an analytical solution to the radial flow equation (Equation 15.12) subject to initial and boundary conditions. His solution, written in terms of the drawdown, is:

$$s = h_0 - h_{r,t} = \frac{Q}{4\pi T} \int_u^\infty \frac{e^{-u}}{u} du \qquad (15.13)$$

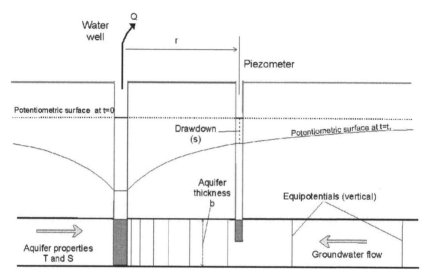

Figure 15.9 Radial flow to a well

where:

$$u = \frac{r^2 S}{4Tt} \tag{15.14}$$

The solution is valid for the initial condition that:

$$h_{r,0} = h_0 \tag{15.15}$$

for all radial distances, and where h_0 is the initial water level, or standing water level (SWL).

The solution requires that on the boundary at infinity, there is no drawdown in hydraulic head:

$$h_{\infty,t} = h_0 \tag{15.16}$$

for all times (t). A constant abstraction (pumping) rate Q occurring at $r = 0$ (the well) is also assumed – however, note that the well has no finite diameter and therefore contains no water. The solution $h_{r,t}$ describes the hydraulic head at any radial distance r at any time t after the start of pumping. The integral in the equation is the exponential integral and is well known in mathematics. Tables of its values are available or they can be readily calculated using sufficient terms in the approximation:

$$W(u) = \int_u^\infty \frac{e^{-u}}{u} du = -0.577216 - \log_e u + u - \frac{u^2}{2 \times 2!} + \frac{u^3}{3 \times 3!} - \frac{u^4}{4 \times 4!} + \dots \tag{15.17}$$

Substituting this function into Equation 15.13 yields:

$$s = \frac{Q}{4\pi T} W(u) \tag{15.18}$$

If the aquifer transmissivity ($T = Kb$), storativity (S) and the pumping rate (Q) are known or estimated, it is then possible to predict the drawdown at any distance (r) from a well at any time (t) after the start of pumping. The appropriate value of the well function (u) from Equation 15.14 is calculated and a value of $W(u)$ derived from Equation 15.18 (or by the use of tables). The drawdown is calculated from Equation 15.18.

Inverse solutions to the radial flow equation can be determined if data from a pumping test is available where measurements of drawdown at varying times and distances are made for different flow rates. This is the area of pumping-test interpretation described in detail by Kruseman and de Ridder (1990). A simple analysis of data is given here as modifications to this are used in the interpretation of single-well tests described below.

Taking logs to the base 10 of Equation 15.18 gives:

$$\log_{10} s = \log_{10} \frac{Q}{4\pi T} + \log_{10} W(u) \tag{15.19}$$

and taking logs to the base 10 for Equation 15.14 gives:

$$\log_{10} \frac{r^2}{t} = \log_{10} \frac{4T}{S} + \log_{10} u \tag{15.20}$$

Hence, for a constant pumping rate Q, s is related to $\frac{r^2}{t}$ in the same manner as $W(u)$ is related to u. Theis (1935) proposed a graphical procedure for the analysis as follows:

1 Plot the function $W(u)$ against $\frac{1}{u}$ on log-log paper. This curve is known as a type curve.
2 Plot s vs t or $\frac{t}{r^2}$ on a second sheet of log-log paper of the same size and scale as the $W(u)$ vs 1/u plot.
3 Superimpose the field curve on the type curve, keeping the principal axes parallel. Adjust the curves until most of the observed data points fall on the type curve.
4 Select an arbitrary match point and read off paired values $W(u)$, 1/u, s and t or $\frac{t}{r^2}$ at the match point.
5 Calculate values of transmissivity and storativity from the equations above using the values of u, $W(u)$, t and s derived from the match point.

Example application of the Theis method

Table 15.1 lists data supplied by J. G. Ferris and presented by S. W. Lohman at the Australian Water Resources Council Adelaide Groundwater School in 1967. The same data set appears in Kruseman and de Ridder (1990). Table 15.1 gives drawdowns in three bores in an aquifer at varying distances from a bore being pumped at a rate of 2720 m³/day. Individual plots of log-drawdown vs log-time are shown in Figure 15.10. Drawdown data for all 3 bores is shown in Figure 15.11 where the drawdown data is plotted vs t/r^2.

The data for either of the curves in Figure 15.10 can be used to obtain a solution for transmissivity and storage coefficient. However, where more than one observation point is available, it is sensible to combine the data to one curve so that a better match point to the type curve can be obtained. The plot in Figure 15.11 will produce a better solution for the transmissivity and storage coefficient.

Straight line solution – the Jacob approximation: Under certain circumstances, the non-steady state flow equations can be modified to give straight-line solutions. If u is very small, then all terms beyond and including u in Equation 15.14 can be neglected. Jacob assumed that this would be valid if $u = 0.01$. However, Driscoll (1986) and Misstear

Table 15.1 Drawdown at observation bores N-1, N-2, N-3

Elapsed Time (minutes)	Bore N-1 r = 61 m Drawdown (m)	Bore N-2 r = 122 m Drawdown (m)	Bore N-3 r = 244 m Drawdown (m)
1.0	0.20	0.05	0
1.5	0.27	0.08	0.01
2.0	0.30	0.12	0.01
2.5	0.34	0.14	0.02
3.0	0.37	0.16	0.03
4.0	0.41	0.20	0.05
5.0	0.45	0.23	0.07
6.0	0.48	0.27	0.08
8.0	0.53	0.30	0.11
10	0.57	0.34	0.14
12	0.60	0.37	0.16
14	0.63	0.40	0.18
18	0.67	0.44	0.22
24	0.72	0.48	0.27
30	0.76	0.52	0.29
40	0.81	0.57	0.34
50	0.85	0.61	0.37
60	0.88	0.64	0.40
80	0.93	0.68	0.45
100	0.96	0.73	0.49
120	1.00	0.76	0.52
150	1.04	0.80	0.56
180	1.07	0.83	0.59
210	1.10	0.86	0.62
240	1.12	0.88	0.64

et al. (2006) have found that this is still valid for values of $u = 0.05$. The smaller the value of u the smaller is the error involved.

For values of u less than 0.05, all but the first two terms in the series expansion for the well function $W(u)$ (Equation 15.17) can be neglected. The Theis equation can then be approximated without recourse to the well function and tables as:

$$s = \frac{Q}{4\pi T}\left(-0.577216 - \log_e \frac{r^2 S}{4Tt}\right) \tag{15.21}$$

Noting that $\log_e u = 2.30 \log_{10} u$
and $-\log_e u = \log_e \frac{1}{u}$
and that $\log_e 1.78 = 0.5772$
the equation can be rewritten as:

$$s = \frac{2.30\,Q}{4\pi T}\log_{10}\frac{2.25\,T\,t}{S\,r^2} \tag{15.22}$$

Since Q, r, T and S are constants, it is clear that s vs $\log_{10} t$ should plot as a straight line as long as u is less than 0.05.

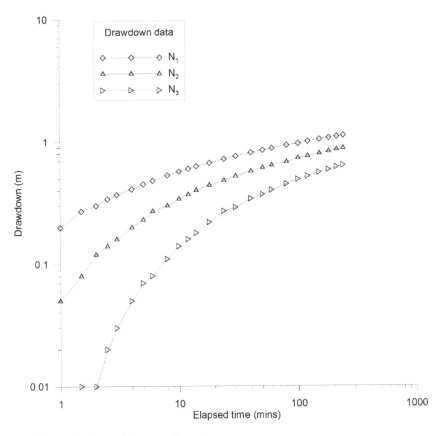

Figure 15.10 Log-log plots of the drawdown data

Two main methods of solution may be derived from this approximation, *viz.*:

- constant radius (r), varying time (t) approach and
- constant time (t), varying radius (r) approach.

Constant radius (r), varying time (t): This method can be used to interpret aquifer test observations made at a single observation point for the duration of a test. A plot of drawdown versus the logarithm of time will yield a straight line. At time t_1, the drawdown s_1 is expressed from Equation 15.22:

$$s_1 = \frac{2.30\, Q}{4\pi T} \log_{10} \frac{2.25\, T\, t_1}{S\, r^2} \tag{15.23}$$

At time t_2, the drawdown s_2 will be:

$$s_2 = \frac{2.30\, Q}{4\pi T} \log_{10} \frac{2.25\, T\, t_2}{S\, r^2} \tag{15.24}$$

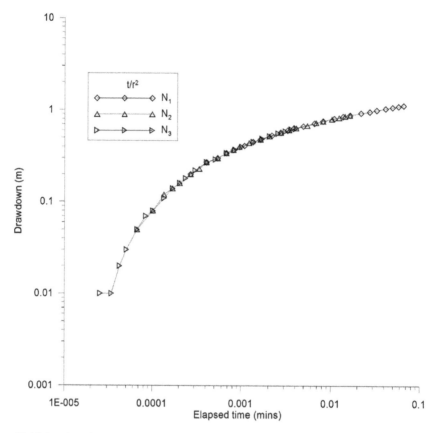

Figure 15.11 Log-log plots normalized by t/(r*r)

It follows that:

$$s_2 - s_1 = \frac{2.30Q}{4\pi T}\log_{10}\frac{t_2}{t_1}$$

(15.25)

If t_1 and t_2 are selected one log cycle apart, then $\log_{10}(t_2/t_1) = 1$. Then Equation 15.25 becomes:

$$\Delta s = \frac{2.30Q}{4\pi T}$$

(15.26)

where Δs is the drawdown for one log cycle of time. Equation 15.26 can be used to solve for T.

The storativity can be obtained by setting the drawdown to zero in Equation 15.22, such that:

$$s = 0 = \frac{2.30Q}{4\pi T}\log_{10}\frac{2.25Tt}{Sr^2}$$

(15.27)

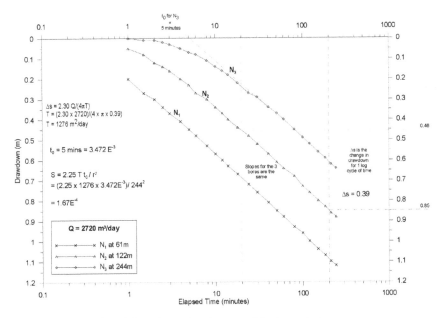

Figure 15.12 Semi-log plot of drawdown showing a straight-line solution

This can only occur if $\frac{2.25T\,t_0}{Sr^2} = 1$. Therefore:

$$S = \frac{2.25Tt_0}{r^2} \tag{15.28}$$

where:

t_0 is the intercept of the straight line at zero drawdown, i.e., the time immediately prior to drawdown in the observation well.

The solution to this problem is shown in Figure 15.12.

Note that the drawdown data for N3 only becomes a straight line after about 60 minutes. Values of the well function (Equation 15.14) are greater than 0.5 for times less than 60 minutes with the result that the straight-line approximation inherent in the Jacob analysis has not been met.

Constant time (t), varying radial distance (r): The time drawdown analysis presented above requires one pumping well and one observation well, where a plot of drawdown vs time on semi-logarithmic paper yields a straight line. It is also possible to obtain the hydraulic properties by examining drawdown at two or, preferably, more points.

At time t, the drawdown s_1 at a distance r_1 is:

$$s_1 = \frac{2.30\,Q}{4\pi T} \log_{10} \frac{2.25Tt}{S\,r_1^2} \tag{15.29}$$

and the drawdown s_2 at a distance r_2 is:

$$s_2 = \frac{2.30Q}{4\pi T}\log_{10}\frac{2.25Tt}{Sr_2^2} \qquad (15.30)$$

It follows that:

$$s_1 - s_2 = \frac{2.30Q}{4\pi T}\log_{10}\frac{r_2^2}{r_1^2} \qquad (15.31)$$

Recognising that $\log_{10}\left(\frac{r_2^2}{r_1^2}\right) = \log_{10}\left(\frac{1}{r_1^2}\right) - \log_{10}\left(\frac{1}{r_2^2}\right)$ and that $\log_{10}\frac{1}{r^2} = 2 \times \log_{10}\frac{1}{r}$, Equation 15.31 becomes:

$$s_1 - s_2 = \frac{2.30Q}{2\pi T}\log_{10}\frac{r_2}{r_1} \qquad (15.32)$$

The graphical procedure calls for plotting the drawdowns against log distance at two or more wells for the same time t. By considering again the drawdown per log cycle:

$$\Delta s = \frac{2.30\,Q}{2\pi T} \qquad (15.33)$$

By extrapolating the distance-drawdown curve to its intersection with the zero drawdown axis, the storativity can be determined in a similar manner to the time drawdown method:

$$S = \frac{2.25\,T\,t}{r_0^2} \qquad (15.34)$$

where:
r_0 is the intersection of the straight-line slope with the zero drawdown axis.

Figure 15.13 shows the distance drawdown data for 240 minutes from Table 15.1.

15.4.2 Summary of assumptions implicit in the Theis solution to the radial flow equation

It is convenient to summarize the assumptions implicit in the Theis solution to the radial flow equation (Equation15.12), *viz.*:

1 The aquifer has an infinite areal extent.
2 The aquifer is homogeneous, isotropic and of uniform thickness over the area influenced by the pumping test.
3 Prior to pumping, the potentiometric surface and/or water table surface are horizontal over the area influenced by the pumping test.
4 The aquifer is pumped at a constant discharge rate.
5 The pumped well penetrates the entire aquifer and thus receives water from the entire thickness of the aquifer by horizontal flow.

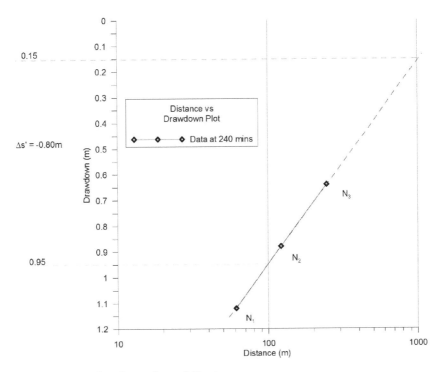

Figure 15.13 Distance drawdown plot at 240 minutes

6   The aquifer is confined.
7   The flow to the well is in unsteady state, i.e., the drawdown differences with time are not negligible nor is the hydraulic gradient constant with time.
8   The water removed from storage is discharged instantaneously with decline of head.
9   The diameter of the pumped well is very small, i.e., the storage in the well can be neglected.
10   Water passes from the aquifer through the well with no loss of head.

The large number of assumptions in the Theis analytical method demonstrates the difficulty of developing methods for more complicated conditions using analytical methods. However, this has not stopped the publication of numerous papers in the literature dealing with partial penetration, leakage, unconfined flow, large diameter wells, etc. and the use of many different sets of type curves. These are extensively reviewed by Kruseman and de Ridder (1990) and implemented in many software packages (e.g., AQTESOLV).

15.5   NUMERICAL SOLUTIONS TO THE RADIAL FLOW EQUATION

Rushton and Redshaw (1979) published a simple finite difference radial flow model which was capable of overcoming many of the restrictions imposed by the analytical solution to the flow equation established by Theis. In its simplest form, it is a single layer model. Due to the assumption of radial symmetry, the finite difference discretization can be represented by a tridiagonal matrix equation which can be readily solved using the Thomas algorithm. The radial flow model provides an alternative approach to the analysis of pumping tests and to the prediction of drawdowns for various scenarios (Rushton and Redshaw, 1979).

The method is reviewed here as it provides a good example of the use of numerical methods and a method for the interpretation of a set of data from a large physical model of flow near to an abstraction bore described later. Modifications to the treatment of the inner and outer boundary conditions are also presented that significantly improve the numerical accuracy of the original code presented by Rushton and Redshaw (1979).

15.5.1   Finite difference discretization

The nature of radial flow to a well is that conditions close to the well have a far greater influence than conditions in the aquifer at distance. The hydraulic head change close to the well is also much greater than change at distance. For these reasons, rather than use the radial (linear) distance r as the spatial variable in the radial flow equation (Equation 15.12), the variable $a = \ln r$ is substituted for r.

If Δa is used for a mesh interval representing radial distance in a finite difference model, it is closely spaced near the well and increases logarithmically away from the well towards the boundary. This variable has the characteristics required to specify conditions in detail close to the well and with less detail as distance from the well increases.

The analysis is further simplified by using the drawdown as a calculated variable rather than the hydraulic head:

$$s = \Delta(h_0 - h_t) \tag{15.35}$$

With the substitution that $a = \ln r$, we can write:

$$\frac{da}{dr} = \frac{d}{dr}\ln r = \frac{1}{r} \tag{15.36}$$

and therefore by cross multiplying, that also:

$$dr = r \times da \tag{15.37}$$

These results will be used later.

Referring back to Equation 15.12 and substituting s for h, we can write:

$$\frac{\partial^2 s}{\partial r^2} + \frac{1}{r}\frac{\partial s}{\partial r} = \frac{S}{T}\frac{\partial s}{\partial t} \tag{15.38}$$

A change of variables in Equation 15.38 allows incorporation of the logarithmic spacing (a) rather than the linear spacing (r). Recognising that $\frac{\partial s}{\partial r}$ can be expanded by multiplying the numerator and denominator by ∂a so that:

$$\frac{\partial s}{\partial r} = \frac{\partial a}{\partial r} \times \frac{\partial s}{\partial a} = \frac{1}{r}\frac{\partial s}{\partial a} \tag{15.39}$$

The second term in Equation 15.38 ($\frac{1}{r}\frac{\partial s}{\partial r}$) is then $\frac{1}{r} \times \frac{1}{r}\frac{\partial s}{\partial a} = \frac{1}{r^2}\frac{\partial s}{\partial a}$

Returning to the first term in Equation 15.38 ($\frac{\partial^2 s}{\partial r^2}$) and using the result described by Equation 15.39, we can write:

$$\frac{\partial^2 s}{\partial r^2} = \frac{\partial}{\partial r}\left(\frac{1}{r}\frac{\partial s}{\partial a}\right) \tag{15.40}$$

Using the chain rule of expansion on the RHS of Equation 15.40, we can write:

$$\frac{\partial^2 s}{\partial r^2} = \frac{1}{r} \times \frac{\partial}{\partial r}\left(\frac{\partial s}{\partial a}\right) + \frac{\partial}{\partial r}\frac{1}{r} \times \frac{\partial s}{\partial a} \tag{15.41}$$

The first term on the RHS of Equation 15.41 may be expressed in terms of the log distance a by making the change of variables based on $\partial r = r\,\partial a$, then:

$$\frac{1}{r} \times \frac{\partial}{\partial r}\left(\frac{\partial s}{\partial a}\right) = \frac{1}{r} \times \frac{1}{r}\frac{\partial}{\partial a}\left(\frac{\partial s}{\partial a}\right) = \frac{1}{r^2}\frac{\partial^2 s}{\partial a^2} \tag{15.42}$$

The second term on the right-hand side of Equation 15.41 can be modified as:

$$\frac{\partial}{\partial r}\left(\frac{1}{r}\right) \times \frac{\partial s}{\partial a} = -\frac{1}{r^2}\frac{\partial s}{\partial a} \tag{15.43}$$

Collecting and substituting these terms back in Equation 15.38 gives:

$$\frac{1}{r^2}\frac{\partial^2 s}{\partial a^2} - \frac{1}{r^2}\frac{\partial s}{\partial a} + \frac{1}{r} \times \frac{1}{r}\frac{\partial s}{\partial a} = \frac{S}{T_r}\frac{\partial s}{\partial t} + R_{r,t} \tag{15.44}$$

which can be simplified to:

$$T_r\frac{\partial^2 s}{\partial a^2} = r^2 S\frac{\partial s}{\partial t} + R_{r,t}r^2 \tag{15.45}$$

The LHS of Equation 15.45 is amenable to a finite difference discretization as shown in the linear representation in Figure 15.14.

Incorporating a backward difference representation of the time derivative, we can write an equation for the *ith* node as:

$$\frac{T_r}{\Delta a^2}(s_{i-1} - 2s_i + s_{i+1})_{t+\Delta t} = \frac{Sr_i^2}{\Delta t}(s_{i,t+\Delta t} - s_{i,t}) + R_{t+\Delta t}r_i^2 \tag{15.46}$$

The first terms on both the left and right sides of Equation 15.46 can be visualised as equivalent to hydraulic resistances such that:

$$H_i = \frac{\Delta a^2}{T_r} \tag{15.47}$$

and:

$$T_i = \frac{\Delta t}{Sr_i^2} \tag{15.48}$$

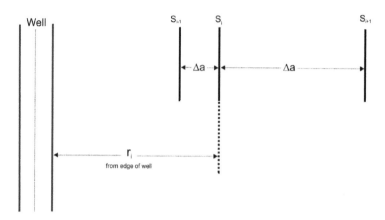

Figure 15.14 Discretization of log radial distance. r = 0 is at the outer edge of the well

where:
H_i is the hydraulic resistance and
T_i is the time resistance.

Using these relationships, Equation 15.46 becomes:

$$\frac{(s_{i-1} - 2s_i + s_{i+1})_{t+\Delta t}}{H_i} = \frac{s_{i,t+\Delta t} - s_{i,t}}{T_i} + R_{t+\Delta t}r_i^2 \qquad (15.49)$$

This is the basic equation which is written for each node to form a set of n equations in the n unknown values of s, the drawdown.

Inclusion of the abstraction rate in the code

The abstraction rate Q is represented as a negative recharge at the first node. In the flow equations above, the flux term R has been positive, representing rainfall recharge or leakage inward to the aquifer. Abstraction is from the aquifer. This is denoted by the term W, where:

$$W = -R \qquad (15.50)$$

Abstraction only occurs from the first node, so we can write:

$$R_1 = -\frac{Q}{A_1} \qquad (15.51)$$

where A_1 is the area represented by the first node.
The discharge term for the first node is therefore:

$$R_1 = -\frac{Q}{A_1} \times r_i^2 \qquad (15.52)$$

Note that the units of R are L/T. The values of R for the other nodes are positive.

Table 15.2 Radial distances

$r(0)$	=	$r_{well} \times e^{-2\Delta a}$	=	0.4642
$r(1)$	=	$r_{well} \times e^{-1\Delta a}$	=	0.6813
$r(2)$	=	$r_{well} \times e^{0}$	=	1.0000
$r(3)$	=	$r_{well} \times e^{+1\Delta a}$	=	1.4678
$r(4)$	=	$r_{well} \times e^{+2\Delta a}$	=	2.1544
$r(5)$	=	$r_{well} \times e^{+3\Delta a}$	=	3.1623
$r(6)$	=	$r_{well} \times e^{+4\Delta a}$	=	4.6416
$r(7)$	=	$r_{well} \times e^{+5\Delta a}$	=	6.8129
$r(8)$	=	$r_{well} \times e^{+6\Delta a}$	=	10.000
$r(9)$	=	$r_{well} \times e^{+7\Delta a}$	=	14.678

15.5.2 Example discretization and assembly of a set of equations

$$[A]|x| = |B|$$

A natural log transformation of the linear radial distance has been used in the above discretization, where:

$$a = \ln r \tag{15.53}$$

Therefore if we know a we can derive the linear radial distances from the simple log/power relationship:

$$r = e^{a} \tag{15.54}$$

If $n \times \Delta a$ is an increment of log distance, then $r = e^{n \times \Delta a}$.

Consider a simple problem of a well of 1000 mm radius and an outer boundary at 10 m. If we choose to discretize each log cycle of radial distance by six, we can write:

$$\Delta a = \frac{1}{6} \times \ln(10) = 0.3837641 \tag{15.55}$$

The node distances are then calculated as shown in Table 15.2.

Note that the first ($r[0]$) and last ($r[9]$) nodes are required to calculate the areas represented by nodes 1 and 8. Note also that the nodes on the boundary will not generally fall at a distance described by the series progression $r = e^{n \times \Delta a}$. This series has the well diameter as a base which determines the values in the radial distance series. In the case that the user specifies an outer boundary not equal to one of the series values, the drawdown is calculated for the last node within the boundary and the flux to the last node modified accordingly. Each node represents an area which may be defined as shown in Figure 15.15. The area represented by the node extends from an inner radial distance halfway between *node [i-1]* and *node [i]* to an outer radial distance halfway between *node [i]* and *node [i+1]*. Special modifications are required to fluxes entering the areas at the outer boundary and within the well.

The area A_i represented by node i may be written as:

$$A_i = \pi \left(\frac{r_{i+1} + r_i}{2} \right)^2 - \pi \left(\frac{r_i + r_{i-1}}{2} \right)^2 \tag{15.56}$$

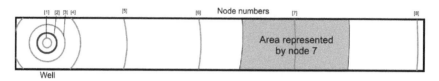

Figure 15.15 Area represented by a node is shown by the shading. The area is a disc starting halfway between two nodal radii and ending halfway between the next two nodal radii

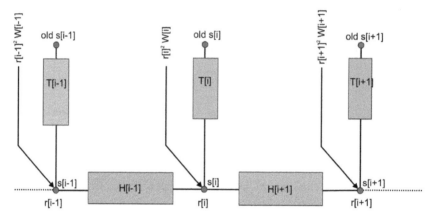

Figure 15.16 Finite difference representation for internal nodes in the mesh

Finite difference equation for the central part of the mesh

We can write finite difference equations for the central part of the mesh as shown in Figure 15.16.

In Figure 15.16, *Old s* represents the head at the previous time step, and other variables are as previously defined.

The finite difference equation may be written as:

$$\left(\frac{s[i+1] - s[i]}{H[i]}\right) - \left(\frac{s[i] - s[i-1]}{H[i-1]}\right) = \frac{s[i] - Olds[i]}{T[i]} + R[i] \times r[i]^2 \qquad (15.57)$$

collecting terms, this gives:

$$-\frac{s[i-1]}{H[i-1]} + s[i]\left(\frac{1}{H[i]} + \frac{1}{H[i-1]} + \frac{1}{T[i]}\right) - \frac{s[i+1]}{H[i]} = \frac{Olds[i]}{T[i]} + R[i] \times r[i]^2 \quad (15.58)$$

The same procedure can be written for each node and the eight equations written as a single matrix Equation 15.59 where all the unknown values of drawdown (*s*) are contained in a single column matrix. Expansion of each line of the matrix equation recovers all the terms of Equation 15.58. Inversion of Equation 15.59 allows all the unknown values of drawdown for a single time step to be solved. The inversion process of the square matrix in Equation 15.59 has only to be carried out once, as this matrix contains all the information about the geometry and hydraulic properties of the problem. Back-substitution using the values in the right-hand-side column matrix provides new values of the unknown draw-downs as the increment of time in the RHS column matrix is increased.

$$
\begin{bmatrix}
\frac{1}{H[1]}+\frac{1}{T[1]} & \frac{-1}{H[1])} & 0 & 0 & 0 & 0 & 0 & 0 \\
\frac{-1}{H[1]} & A & \frac{-1}{H[2]} & 0 & 0 & 0 & 0 & 0 \\
0 & \frac{-1}{H[2]} & B & \frac{-1}{H[3]} & 0 & 0 & 0 & 0 \\
0 & 0 & \frac{-1}{H[3]} & C & \frac{-1}{H[4]} & 0 & 0 & 0 \\
0 & 0 & 0 & \frac{-1}{H[4]} & D & \frac{-1}{H[5]} & 0 & 0 \\
0 & 0 & 0 & 0 & \frac{-1}{H[5]} & E & \frac{-1}{H[6]} & 0 \\
0 & 0 & 0 & 0 & 0 & \frac{-1}{H[6]} & F & \frac{-1}{H[7]} \\
0 & 0 & 0 & 0 & 0 & 0 & \frac{-1}{H[7]} & \frac{1}{H[7]}+\frac{1}{T[8]}
\end{bmatrix}
\begin{vmatrix} s[1] \\ s[2] \\ s[3] \\ s[4] \\ s[5] \\ s[6] \\ s[7] \\ s[8] \end{vmatrix}
=
\begin{vmatrix}
\frac{Olds[1]}{T[1]}+Qr[1]^2 \\
\frac{Olds[2]}{T[2]}+W[2]r[2]^2 \\
\frac{Olds[3]}{T[3]}+W[3]r[3]^2 \\
\frac{Olds[4]}{T[4]}+W[4]r[4]^2 \\
\frac{Olds[5]}{T[5]}+W[5]r[5]^2 \\
\frac{Olds[6]}{T[6]}+W[6]r[6]^2 \\
\frac{Olds[7]}{T[7]}+W[7]r[7]^2 \\
\frac{Olds[8]}{T[8]}+W[8]r[8]^2
\end{vmatrix}
$$

$$(15.59)$$

The central coefficients for [2,2], [3,3], [4,4], [5,5], [6,6] and [7,7] have been shown in Equation 15.59 by the letters A, B, C, D, E and F for the sake of readability. These coefficients can be calculated from the following:

$$
A = \frac{1}{H[2]}+\frac{1}{H[1]}+\frac{-1}{T[2]} ; B = \frac{1}{H[3]}+\frac{1}{H[2]}+\frac{1}{T[3]} ; C = \frac{1}{H[4]}+\frac{1}{H[3]}+\frac{1}{T[4]} ;
$$

$$
D = \frac{1}{H[5]}+\frac{1}{H[4]}+\frac{1}{T[5]} ; E = \frac{1}{H[6]}+\frac{1}{H[5]}+\frac{1}{T[6]} ; F = \frac{1}{H[7]}+\frac{1}{H[6]}+\frac{1}{T[7]}.
$$

Note the structure of the matrix is symmetrical, sparse and banded with only two off-diagonal bands containing non-zero terms. The principal diagonal contains the terms $\frac{1}{H[1]}+\frac{1}{T[1]}$, A, B, C, D, E, F and $\frac{1}{H[7]}+\frac{1}{T[8]}$. This type of matrix is readily inverted using a simple numerical method called the Thomas algorithm (Rushton and Redshaw, 1979). Note that the first and last equations incorporated into the matrix equation (Equation 15.59) include fewer terms than the internal nodes. This is because these represent boundary conditions to the numerical problem that require specific attention.

15.5.3   Solution of equation sets

The eight equations that represent the change of head at each of the eight nodes can be assembled into a matrix equation:

$$[A]|x| = |B|$$

where:

A is the square matrix that contains all the known coefficients describing the aquifer properties.

The Inverse of a Matrix

The matrix inverse $[A]^{-1}$ is defined by:

$$[A] \times [A]^{-1} = [A]^{-1} \times [A] = [I] \tag{15.60}$$

where:
I is the identity matrix.

The identity matrix can be written as:

$$\begin{bmatrix} 1 & 0 & 0 \\ 0 & 1 & 0 \\ 0 & 0 & 1 \end{bmatrix} \tag{15.61}$$

The multiplication of a matrix by the identity matrix is similar to multiplying a number by 1. Referring back to the representation of a set of equations:

$$[A]|x| = |B|$$

If we can form the inverse of the matrix ($[A]^{-1}$) and then multiply both sides of the equation by the inverse, we get:

$$[A][A]^{-1}|x| = [A]^{-1}|B|$$

or:

$$[I]|x| = |x| = [A]^{-1}|B|$$

Consequently, a set of simultaneous linear algebraic equations is solved through the use of the inverse matrix.

There are many possible ways of inverting the matrix, and matrix inversion methods form a major branch of numerical analysis. To illustrate the method we will consider in detail a computationally inefficient but conceptually simple method. The inverse of a matrix can be found as follows:

$$[A]^{-1} = \frac{\text{adj } |a|}{|A|}$$

This is equivalent to division of each member of the adjoint of matrix A by the determinant of the matrix.

The method can be illustrated by considering the manipulations on a simple set of three equations in three unknowns, *viz.*:

$$2x_1 + x_3 = 7$$
$$x_2 + 5x_3 = 4$$
$$4x_1 + 6x_2 - 5x_3 = 2$$

This can be written in a matrix form as:

$$\begin{bmatrix} 2 & 0 & 1 \\ 0 & 1 & 5 \\ 4 & 6 & -5 \end{bmatrix} \begin{vmatrix} x_1 \\ x_2 \\ x_3 \end{vmatrix} = \begin{vmatrix} 7 \\ 4 \\ 2 \end{vmatrix}$$

(15.62)

The first step is to find the determinant of the matrix:

$$2(-5 - 30) - 0(0 - 20) + 1(0 - 4) = -74$$

The second step is to calculate the cofactor matrix of $[A]$:

$$cof[A] = \begin{bmatrix} +(-5 - 30) & -(-0 - 20) & +(-0 - 4) \\ -(0 - 6) & +(-10 - 4) & -(12 - 0) \\ +(0 - 1) & -(10 - 0) & +(2 - 0) \end{bmatrix} = \begin{bmatrix} -35 & 20 & -4 \\ 6 & -14 & -12 \\ -1 & -10 & 2 \end{bmatrix}$$

(15.63)

The third step is to calculate the adjoint of $[A]$:

$$adj[A] = [cof\ A]^T = \begin{bmatrix} -35 & 6 & -1 \\ 20 & -14 & -10 \\ -4 & -12 & 2 \end{bmatrix}$$

(15.64)

The fourth step is to calculate the inverse matrix $[A]^{-1}$:

$$[A]^{-1} = \frac{adj[A]}{|A|} = \frac{1}{-74} \begin{bmatrix} -35 & 6 & -1 \\ 20 & -14 & -10 \\ -4 & -12 & 2 \end{bmatrix}$$

(15.65)

The fifth step is to compute the vector of unknowns using the inverse matrix:

$$|x| = [A]^{-1}|B| = \frac{1}{-74} \begin{bmatrix} -35 & 6 & -1 \\ 20 & -14 & -10 \\ -4 & -12 & 2 \end{bmatrix} \begin{bmatrix} 7 \\ 4 \\ 2 \end{bmatrix} = \begin{bmatrix} 3.01 \\ -0.865 \\ 0.973 \end{bmatrix} = \begin{bmatrix} x_1 \\ x_2 \\ x_3 \end{bmatrix}$$

(15.66)

The column vector $|x|$ now contains the values of the unknowns x_1, x_2, and x_3. This is a very rigorous method for calculating the inverse and is used here to demonstrate the numerical analysis technique. In reality, there are many different techniques used. The choice of technique depends on the structure of the matrix, the type of banding etc.

15.5.4 Special types of matrix

Introduction

Many special types of matrix occur, and particular problems give rise to particular types of matrix. The banded matrix occurs commonly in the numerical analysis of groundwater problems. The finite difference solutions for two- and three-dimensional groundwater flow problems give rise to banded matrices. Matrix solution inversion routines have been especially developed for inverting these banded matrices. The inversion time is very much a function of the size and bandwidth of the matrix.

Tridiagonal matrices

A tridiagonal matrix is a banded matrix with non-zero elements along the principal diagonal and the adjacent diagonal on either side of the principal diagonal. The *Thomas algorithm* is a particularly efficient method for inverting matrices of this type.

Tridiagonal matrices are generated in many one-dimensional problems. The solution to the radial flow equation provides a good example.

Consider the set of equations:

$$
\begin{array}{llll}
+b_1 x_1 & +c_1 x_2 & & & = & f_1 \\
+a_2 x_1 & +b_2 x_2 & +c_2 x_3 & & = & f_2 \\
& +a_3 x_2 & +b_3 x_3 & +c_3 x_4 & = & f_3 \\
& & +a_4 x_3 & +b_4 x_4 & = & f_4
\end{array}
\tag{15.67}
$$

This may be expressed in matrix notation as:

$$
\begin{bmatrix}
b_1 & c_1 & 0 & 0 \\
a_2 & b_2 & c_2 & 0 \\
0 & a_3 & b_3 & c_3 \\
0 & 0 & a_4 & b_4
\end{bmatrix}
\times
\begin{vmatrix}
x_1 \\ x_2 \\ x_3 \\ x_4
\end{vmatrix}
=
\begin{vmatrix}
f_1 \\ f_2 \\ f_3 \\ f_4
\end{vmatrix}
\tag{15.68}
$$

Now if $[A]$ is decomposed into $[L] \times [U]$, where $[L]$ is of lower triangular form and equals:

$$
[L] =
\begin{bmatrix}
\alpha_1 & 0 & 0 & 0 \\
\alpha_2 & \alpha_2 & 0 & 0 \\
0 & \alpha_3 & \alpha_3 & 0 \\
0 & 0 & \alpha_4 & \alpha_4
\end{bmatrix}
\tag{15.69}
$$

and $[U]$ is of upper triangular form, where:

$$
[U] =
\begin{bmatrix}
1 & \beta_1 & 0 & 0 \\
0 & 1 & \beta_2 & 0 \\
0 & 0 & 1 & \beta_3 \\
0 & 0 & 0 & 1
\end{bmatrix}
\tag{15.70}
$$

If we form the product of $[L] \times [U]$ we get:

$$\begin{bmatrix} \alpha_1 & 0 & 0 & 0 \\ a_2 & \alpha_2 & 0 & 0 \\ 0 & a_3 & \alpha_3 & 0 \\ 0 & 0 & a_4 & \alpha_4 \end{bmatrix} \times \begin{bmatrix} 1 & \beta_1 & 0 & 0 \\ 0 & 1 & \beta_2 & 0 \\ 0 & 0 & 1 & \beta_3 \\ 0 & 0 & 0 & 1 \end{bmatrix} = \begin{bmatrix} a_1 & \alpha_1\beta_1 & & \\ a_2 & a_2\beta_1 + \alpha_2 & \alpha_2\beta_2 & \\ & a_3 & a_3\beta_3 + \alpha_3 & \alpha_3\beta_3 \\ & & a_4 & a_4\beta_3 + \alpha_4 \end{bmatrix}$$

(15.71)

Equating coefficients in the matrix $[A]$ with the coefficients above then gives us:

$$\alpha_1 = b_1$$
$$a_2\beta_1 + \alpha_2 = b_2$$
$$a_3\beta_2 + \alpha_3 = b_3$$

and:

$$a_4\beta_3 + \alpha_4 = b_4$$

for the coefficients on the principal diagonals. These equations can be rearranged to give:

$$\alpha_2 = b_2 - a_2\beta_1$$
$$\alpha_3 = b_3 - a_3\beta_2$$

and:

$$\alpha_4 = b_4 - a_4\beta_3$$

A general relationship then becomes $\alpha_i = b_i - a_i\beta_{i-1}$ where $i = 2, 3, \cdots, n$. The values of β can be found from:

$$\alpha_1\beta_1 = c_1$$
$$\alpha_2\beta_2 = c_2$$
$$\alpha_3\beta_3 = c_3$$

and rearranging for β in the same manner.

So, knowing α_1 we can find β_1. Then use this to find α_2, and then β_2 etc., until all the coefficients of the upper and lower diagonal matrices are known.

Then, if we set $[L] \times [U] \times |x| = |f|$ and form an intermediate solution $|y| = [U] \times |x|$ then $[L] \times |y| = |f|$.

The elements of $|y|$ can be easily determined by performing a forward solution of this equation:

$$\begin{bmatrix} \alpha_1 & 0 & 0 & 0 \\ a_2 & \alpha_2 & 0 & 0 \\ 0 & a_3 & \alpha_3 & 0 \\ 0 & 0 & a_4 & \alpha_4 \end{bmatrix} \times \begin{vmatrix} y_1 \\ y_2 \\ y_3 \\ y_4 \end{vmatrix} = \begin{vmatrix} f_1 \\ f_2 \\ f_3 \\ f_4 \end{vmatrix} = \begin{bmatrix} \alpha_1 y_1 & & & \\ y_1 a_2 & y_2\alpha_2 & & \\ & y_3 a_3 & y_3\alpha_3 & \\ & & y_3 a_4 & y_4\alpha_4 \end{bmatrix}$$

(15.72)

Now, noting that $f_1 = a_1 y_1$, and $f_2 = y_1 a_2 + y_2 a_2$, etc., solving these for values of y from the general relationship that:

$$y_i = \frac{f_i - y_{i-1} a_i}{\alpha_1} \tag{15.73}$$

leads to values for all of y. Then, knowing $|y|$ we can get values of the unknowns $|x|$ from backward substitution in the equation:

$$[U] \times |x| = |y|$$

$$\begin{bmatrix} 1 & \beta_1 & 0 & 0 \\ 0 & 1 & \beta_2 & 0 \\ 0 & 0 & 1 & \beta_3 \\ 0 & 0 & 0 & 1 \end{bmatrix} \times \begin{vmatrix} x_1 \\ x_2 \\ x_3 \\ x_4 \end{vmatrix} = \begin{vmatrix} y_1 \\ y_2 \\ y_3 \\ y_4 \end{vmatrix} = \begin{bmatrix} x_1 & x_2\beta_1 & 0 & 0 \\ 0 & x_2 & x_3\beta_2 & 0 \\ 0 & 0 & x_3 & x_4\beta_3 \\ 0 & 0 & 0 & x_4 \end{bmatrix} \tag{15.74}$$

Then $x_4 = y_4$. Knowing this, $y_3 = x_3 + x_4\beta_3$, therefore $x_3 = y_3 - x_4\beta_3$. Then from $y_2 = x_2 + x_3\beta_2$ we can solve $x_2 = y_2 - x_3\beta_2$. Lastly, from $y_1 = x_1 + x_2\beta_1$ we can solve $x_1 = y_1 - x_2\beta_1$.

All values of the unknowns $|x|$ are then solved. The major advantage of this technique is that all the zero elements of the original matrix can be ignored. It is only necessary to work with the non-zero diagonal elements of $a, b, c, x,$ and f of the original data and intermediate values of y, α and β.

The algorithm briefly described above can be simply coded. See if you can follow the program logic!

15.5.5   FORTRAN code for the Thomas algorithm

This is a FORTRAN routine for a tridiagonal equation solver using the Thomas algorithm.

```
C N is the order of the matrix
SUBROUTINE TRIDIA (N)
COMMON A, B, C, X, F
DIMENSION A(50), B(50), C(50), X(50), F(50)
DIMENSION ALPHA(50), BETA(50), Y(50)

ALPHA(1) = B(1)
BETA(1) = C(1)/ALPHA(1)
Y(1) = F(1)/ALPHA(1)

DO 1O I = 2, N
ALPHA(I) = B(I) – A(I)* BETA(I-1)
BETA(I) = C(I) / ALPHA (I)
Y(I) = (F(I) – A(I)* Y(I-1)) / ALPHA (I)
10 CONTINUE

C BEGIN BACKWARD SUBSTITUTION FROM LAST ROW

X(N) = Y(N)
NM = N-1
DO 20 I = 1, NM
```

J = N-I
X(J) = Y(J) – BETA(J) * X(J+1)
20 CONTINUE
RETURN
END

Modification to account for conditions near the boundary

The eighth node at a radial distance of 10 m lies on the boundary. We can draw the finite difference representation at the boundary as shown in Figure 15.17. As the ith hydraulic resistance connects the ith node with the $i^{th} + 1$ node, it can be seen that there will be no hydraulic resistance out past the boundary. In our example, H[8] will not be required as examination of the matrix equation (Equation 15.59) shows there is no available position for the term!

The finite difference equation can then be written leaving out H[8] as follows:

$$-\frac{s[7]}{H[7]} + s[8]\left(\frac{1}{H[7]} + \frac{1}{T[8]}\right) = \frac{Olds[8]}{T[8]} + R[8] \times r[8]^2 \tag{15.75}$$

The time resistance T[8] incorporates the storage contributing to the eighth node. The recharge term R[8] represents recharge to that part of the area represented by the eighth node. Both terms require modification to represent the fact that the area around node 8 is truncated by the boundary. If they are not modified, far too much water will be allowed to enter the model at the eighth node. This is particularly serious as we are dealing with a logarithmically increasing mesh radius.

The correction factor will be the ratio of the true area (incorporating the boundary condition) represented by the final node:

$$\pi r_{boundary}^2 - \pi\left(\frac{r[8] - r[7]}{2}\right)^2 \tag{15.76}$$

and the unmodified area for the node which is calculated using Equation 15.56. Note that in this example, the outer boundary coincides with the radial distance to the eighth node although this will not generally be the case.

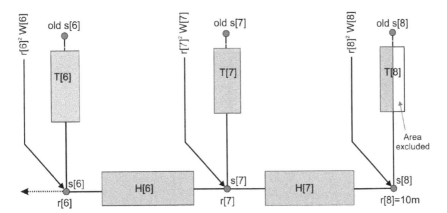

Figure 15.17 Modifications to the general mesh to account for the outer boundary condition

If the outer boundary represents a fixed head condition, this can be readily incorporated in the time resistance T[8] by making it very small. This is equivalent to saying that the storativity at the boundary approaches infinity and maintains a limitless source of water resulting in no change of head, whatever quantity of water is caused to flow from the boundary. It is clear that implementation of this type of boundary condition should be treated cautiously. No modification is required to implement a no-flow boundary, as this is automatically carried out by the algorithm.

Modification to account for the inner boundary

A similar treatment is required for the inner boundary. Consider the model cells at the inner boundary as shown in Figure 15.18.

The finite difference equation may be set down as:

$$s[1]\left(\frac{1}{H[1]} + \frac{1}{T[1]}\right) - \frac{s[2]}{H[1]} = \frac{Olds[1]}{T[1]} - \frac{Q}{A_1} \times r[1]^2 \tag{15.77}$$

where the well flux term is the abstraction rate divided by the area represented by the first node as described earlier. The first node represents water in the well, with the boundary between the well and the aquifer represented by the boundary between the two nodes (Figure 15.14).

The first hydraulic resistance H[1] is set to a very small value, as this represents very high transmissivity of water in the well. The first value of the time resistance requires modification as the storativity of the well will be unity. The area represented by the well also requires modification, as A_1 does not represent the area inside the well as shown by Equation 15.56.

The second value of the time resistance requires modification as it now represents only that part of the formation from the well edge. The factor is calculated as a ratio of areas in a similar fashion to that for the outer boundary.

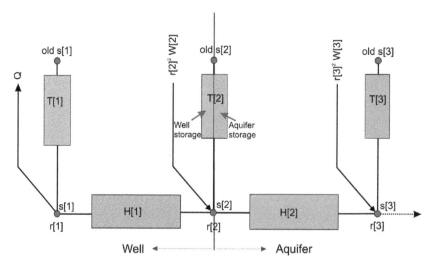

Figure 15.18 Modifications to the general mesh to account for the inner boundary condition at the well

Representation of the time derivative

Discretization of either time or space must be very fine wherever the function is changing rapidly. An appropriate discretization for time during a pumping test is to use a logarithmic time step with 10 time steps per decade.

The initial time value is given by reference to the analytical solution (Theis, 1935) using the well radius (r_w in Equation 15.14). Kruseman and de Ridder (1990) show that for a value of u = 10.0, the drawdown everywhere will be very small. Substitution of typical values into Equation 15.14 gives an initial value of t of approximately 10^{-10} days. This is a sufficiently small increment of time for the analysis. After this initial value, the time increases by a factor $10^{0.1}$ for each time step.

It is necessary to reset the model time step to the initial small value whenever a change in flow rate occurs. In this way, multiple flow periods can be accommodated and a recharge period is simply a further pumping period with the abstraction rate set to zero.

Excessive drawdown

It is necessary to halt the model run if the drawdown in the well exceeds the saturated depth of the aquifer. It would be possible to continue the mathematical approximation producing completely unrealistic drawdowns, but to constrain the model, a check is run at the end of each time step to make sure that there is at least 10% of the original saturated depth left.

Confining conditions

The model will work for confined and unconfined conditions. The type of condition is determined by the program by comparing the input data for top and base of the aquifer with the water level. If the water level is above the top of the aquifer at a node, the confined storage condition is used. If the water level is below the top of the aquifer, the unconfined storage value is used for that node. In each case the datum is taken at ground surface.

Variable saturated depth

The horizontal hydraulic resistance term H_n is a function of the saturated depth of the aquifer, *viz.*:

$$H_n = \frac{\Delta a^2}{bK_r} \tag{15.78}$$

While the aquifer is confined, specification of the depth is simple.

If an unconfined aquifer is being modelled, or a confined aquifer in which the drawdown around the well begins to dewater the aquifer, the saturated depth becomes a function of the unknown drawdown $S(N)$ for each node and each timestep. This requires that the drawdown be resolved iteratively for each time step; to account for this, the matrix equation is solved four times for each time step to achieve an accurate solution. However, it may be that if the well becomes unconfined at late (long time step) times, there will remain some inaccuracy close to the well and recalculation using a smaller time step is required.

Accuracy test for the model code

It is always necessary to carry out an analysis of the accuracy of a model prediction. Results will be presented in the following section; however, an initial test can be posed as well.

Consider an aquifer with a radius of 1000 mm and an outer boundary of 10 m – as used in the code derivation. To simplify matters, consider the aquifer to have a storage of 0.1 and a very high (10,000 m/day) hydraulic conductivity so that the water level is flat. An abstraction of 100 m³ for 1 day will be sourced from storage in the well (where $S = 1$) and the remainder from storage in the aquifer. It is a simple matter to calculate that the water level in this system will decline by 2.9202 m after the abstraction. The radial flow code described above predicts a drawdown of exactly the same (2.9202 m). This is quite a severe test of the model as a significant quantity of the water ($\approx$30%) comes from storage in the well – that is not possible to represent using the assumptions inherent in the Theis analysis.

15.6   RADIAL FLOW MODEL RESULTS

The RADFLOW program can be used to investigate the impact of changing the abstraction well radius or the distance to and type of boundary condition. Examples of data output are shown below.

15.6.1   Results for an infinite confined aquifer

In the model results which follow, the aquifer has been modelled as:

*   transmissivity of 500 m²/day,
*   storage Coefficients of 0.001 and 0.0001,
*   well diameter of 1 mm (insignificantly small),
*   a no-flow outer boundary at 10,000 m,
*   abstraction rate of 500 m³/day, and
*   abstraction for 100 days.

The results for the approximation to a Theis analysis are shown in Figure 15.19.

The results in Figure 15.19 can be used in a Jacob approximation analysis to recover the transmissivity data. Note the curvature of the observation well data at relatively early times. This is the area where the Jacob approximation to the Theis method is invalidated as a result of truncating the series expansion of $W(u)$. Note also the down turn at approximately 10 days in the results for a value of $S = 0.0001$. This is the result of dewatering of this aquifer due to the low storage, even though the outer boundary is at a radius of 10 km. The data presented as a log-log plot is shown in Figure 15.20.

15.6.2   Results for a finite well diameter

The impact of a finite well diameter of 300 mm is shown in Figure 15.21, with data presented as a log-log plot in recovery shown in Figure 15.22. The outer boundary is still set as a no-flow boundary at 10 km.

Note that the drawdown in all wells is less than predicted by the first set of data. Conditions at the abstraction well could also be interpreted as a recharge boundary close to the well. This aspect of well storage makes monitoring of the abstraction well data important if the well hydraulics are to be sufficiently understood. The change in aquifer storage also has significant impacts.

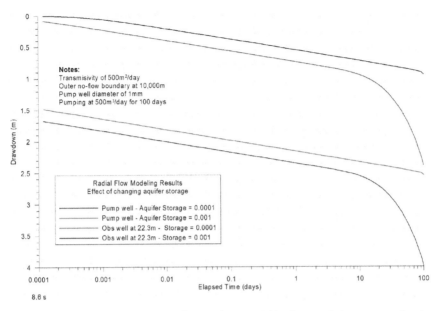

Figure 15.19 Semi-log plot of drawdown for conditions within the restrictions set out by the Theis analysis

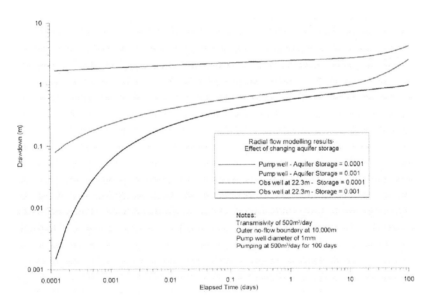

Figure 15.20 Log-log plot of drawdown: Theis conditions

15.6.3   Results for a bounded system

The impact of a recharge boundary at 2000 m is shown in Figure 15.23 for a semi-log plot and in Figure 15.24 for a log-log plot. A real abstraction well diameter of 300 mm has been kept.

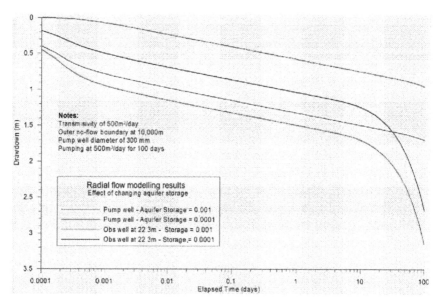

Figure 15.21 Semi-log plot: well storage

The recharge boundary at a distance of 2000 m is shown at each observation point at the same time. The decline in water levels throughout the aquifer stops and a steady state is reached where abstraction from the well is matched by flow across the boundary. Note that this model would be the equivalent to pumping from a 2 km radius island in a lake.

The impact of a no-flow boundary at 2000 m is shown in Figure 15.25 (300 mm abstraction well). The impact of changing the outer boundary at 2000 m from a recharge condition to a no-flow condition is clearly seen. Note that the change in gradient occurs at all points at the same time again.

15.6.4   2D radial flow models

Johnson *et al.* (2001) have published a 2D radial flow model that solves the 2D radial flow equation. Users can enter their own pumping test data and see if they can match the field data by changing the layer hydraulic conductivities, thicknesses and storages. The ability to vary hydraulic conductivity with depth means that the hydraulic head distribution can be calculated as a function of depth.

Two-dimensional finite element code was also developed (Cox, 1977) and can be used to vary the hydraulic properties throughout the problem domain.

15.7   NON-DARCY FLOW CLOSE TO AN ABSTRACTION WELL

15.7.1   Introduction

The common assumption made for groundwater flow is that the flow domain is linear. An increase in hydraulic gradient will lead to a proportional increase in groundwater flow

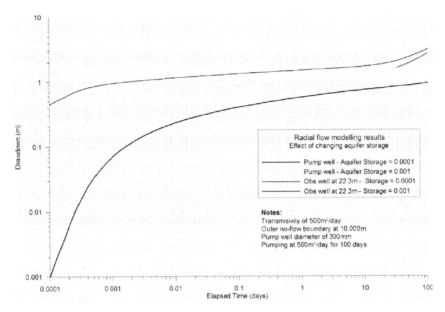

Figure 15.22 Log-log plot: well storage

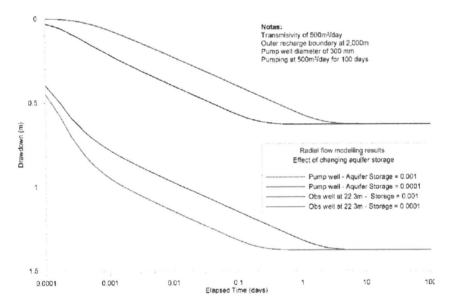

Figure 15.23 Semi-log plot: recharge boundary

where the constant of proportionality is the hydraulic conductivity. This is Darcy's Law and is the case for regions where flow is laminar, but is not the case for regions where turbulent flow occurs. Unfortunately, turbulent flow is probably common close to an abstraction bore. The problem has received very little attention in the literature, probably because there is very little information available. Unexpectedly low discharge rates are frequently blamed

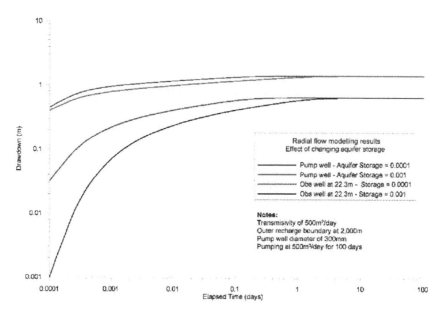

Figure 15.24 Log-log plot: recharge boundary

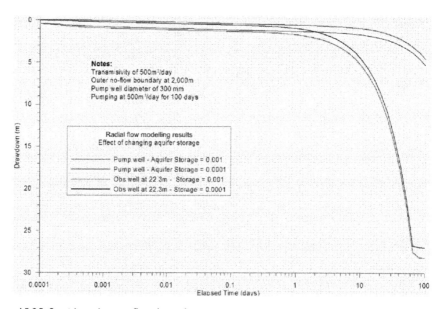

Figure 15.25 Semi-log plot: no-flow boundary

on inefficiencies or malfunctions in the pump rather than problems with the flow regime. The problem is confounded by a lack of data. It is frequently the case that there is no dip-tube access in a pumping well in which drawdown in the well can be measured. This despite design recommendations to the contrary! There is also very little data available

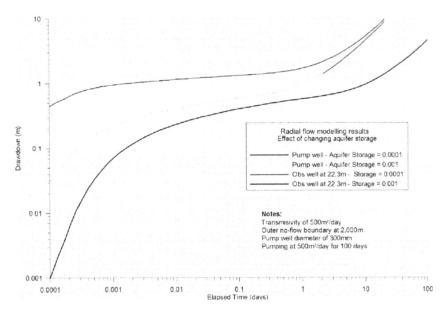

Figure 15.26 Log-log plot: no-flow boundary

in which the hydraulic head in the gravel pack or the formation close to the abstraction bore is measured. The increased investigation and construction cost is always blamed for the lack of monitoring points. It is therefore particularly unfortunate that the increased head loss caused by turbulent flow is directly paid for – on a continuing basis – by the cost of pumping. Data in Figure 15.27 and Figure 15.28 illustrate the problem. This data is from an unconfined fine sand (aeolian) aquifer at Centennial Park in Sydney. It is clear that the majority of drawdown occurs within 6 m of the abstraction bore with a massive 56% through the gravel pack and screen.

The inefficiency of the bore in Centennial Park is only obvious from the additional data available. Clearly, the design of the bore could be improved to reduce the head loss close to the bore by providing a better designed gravel pack and a larger screen opening. The lack of performance could be associated with turbulent flow close to the bore or to reduced hydraulic conductivity close to the bore. The abstraction rate at this site was only 17 L/s and the impacts on pumping costs not that significant. However, on larger irrigation bores in NSW, the abstraction rates are sometimes higher than 300 L/s, and in this case the impact on running costs is very great.

In the late 1960s, the group at the Water Research Laboratory of The University of New South Wales, led by Colin Dudgeon and including PhD students Ron Cox and Peter Huyakorn, designed and constructed a large well tank facility that was capable of testing the response of a confined aquifer close to an abstraction bore. They were particularly concerned about the validity of Darcy's Law as groundwater speeds up to enter the bore (Chapter 4). This major research project was initiated in 1968 as a part of the Australian Water Resources Council Research Project 68/8 and was completed and reported upon in 1972. The reports were published as UNSW Water Research Laboratory Research reports split into two volumes, with Volume 1 comprising four parts each written by different team members

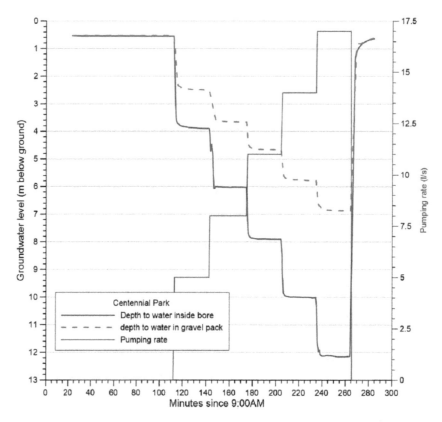

Figure 15.27 Data for a step test at Centennial Park showing the abstraction rate and the ground-water level response at the outside of the gravel pack and inside the abstraction bore

(Dudgeon *et al.*, 1972; Huyakorn, 1972a, b; Dudgeon and Swan, 1972). Huyakorn (1972a) reported on extensive developments of finite element work to model the flow through two zones close to the bore where non-Darcy flow occurred. The team took their experimental analysis and interpretation to the field to conduct a major pumping test at Gummly Gummly near Wagga in NSW (Dudgeon *et al.*, 1972). This project was at the forefront of groundwater research at the time, but has been largely overlooked due to a lack of reporting and the staff moving onto other fields of investigation, as no further funding was approved. Among other developments, Dudgeon and Swan (1972) include detail of the first dip meter developed for recording depth to water in narrow piezometer tubes. The success of this innovation is noted and the suggestion made that they be commercially developed!

The well tank was configured as a quadrant of a confined aquifer with water from the dam at Manly Vale used to keep a constant head above the top of the aquifer (Dudgeon and Swan, 1972). Discharge was through a well screen set into the corner of the tank and drainage to a stream bed 3 m lower. Figure 15.29 shows the overall setup of the tank and Figure 15.30 shows the configuration for a quadrant test. The tank was filled with coarse river sand that had been collected from an active river channel and then sorted and graded. A grain size analysis for the sand used is given in Table 15.3. Using Hazen's relationship (Equation 5.41) gives a hydraulic conductivity of approximately 2200 m/day.

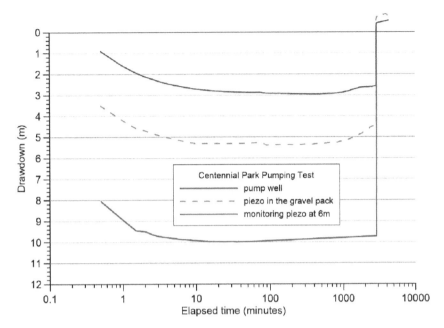

Figure 15.28 Test data for a day of pumping showing that the majority of drawdown occurs close to the abstraction bore. A lake close by provides a source of recharge that dominates after approximately 10 minutes of pumping. Recovery before the end of pumping was due to a heavy rainfall event!

Table 15.3 Grain size analysis for the well tank sediments

Sieve size (mm)	% passing
0.1	0.0
1.5	8.0
3.5	27.5
7.5	86.0
10.0	100.0

The well tank had the following dimensions (converted to SI units):

1 the radius of the well was 128 mm,
2 the outer radius of the tank was 4.93 m,
3 the top of the aquifer was at 0.914 m,
4 the base was at 2.438 m,
5 the depth to water was 0.0 m,
6 there was a recharge boundary – a constant source of water from the dam, and
7 the experiment was conducted on a quarter segment of the aquifer.

The top of the tank was sealed by a plate and water inside the tank held at direct pressure above the top of the tank allowing confined conditions to be simulated. Piezometer tubes

Figure 15.29 The well tank building at The Water Research Laboratory, UNSW (Dudgeon and Swan, 1972)

were let into the tank at nine radial distances as shown in Table 15.5. The change of head with time could then be established. In reality, steadystate was rapidly reached and the experiment provided the steady-state heads at eight points in the aquifer and in a well, for 21 different flow rates. The effective porosity of sands in the tank was calculated by measurements of the volume of water taken to raise the head during filling of the tank. Measurements were made after the tank had been filled and allowed to drain. Effective porosity measurements are listed in Table 15.4.

A set of data observed in the tank is given in Table 15.5 (Dudgeon and Swan, 1972).

The data for the steady state at the abstraction rates shown in Table 15.5 are shown in Figure 15.31. A value of hydraulic conductivity can be established from the steady-state data.

15.7.2   Hydraulic conductivity determination from tank data

An expression for calculating the hydraulic conductivity from drawdowns measured at different distances was developed from the Theis solution to the radial flow equation and using the Jacob approximation (Equation 15.33). A simpler and perhaps more elegant mathematical solution can be derived directly from Darcy's Law.

If Darcy's Law is expressed in cylindrical coordinates, then:

$$Q = -KA\frac{dh}{dr} \tag{15.79}$$

where:

$A = 2\pi rb$ – the area of the cylinder around the abstraction well with an aquifer height of 'b'.

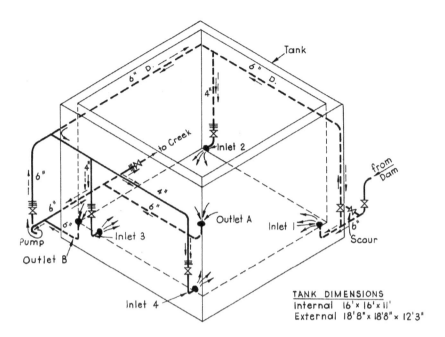

== Orifice Plates shown thus

1. With ¼ well under investigation Inlet I only to be used and Outlet A,
   to be blank flanged; valves on Inlets 2,3,4 to be enclosed.
2. With full well under investigation use all Inlets and blank flange
   outlet B
3. With pump in operation flow shown——►; gravitate———►

Figure 15.30 Schematic to show general plumbing in the WRL well tank (Dudgeon and Swan, 1972)

Table 15.4 Effective porosity data for the tank sand fill

Depth (m)		Effective porosity (%)
From	*To*	
0.000	0.152	33.75
0.152	0.305	31.36
0.305	0.457	32.40
0.457	0.610	30.80
0.610	0.762	30.50
0.762	0.914	30.50
0.914	1.067	36.75
1.067	1.219	38.68

In the tank experiment, only a quadrant of the aquifer was replicated (Dudgeon and Swan, 1972). The area of the cylinder in Equation 15.79 was therefore reduced by a factor of four, as noted by Dudgeon and Swan (1972) in their Equation 5.1. However, the quantity of flow from the quadrant was also reduced by the same factor and therefore there was no net change to Equation 15.79.

Substituting in Equation 15.79 gives:

$$Q = -2\pi b r K \frac{dh}{dr} \tag{15.80}$$

separating variables gives:

$$\frac{dr}{r} = -\frac{2\pi bK}{Q} dh \tag{15.81}$$

If the head at different radial locations (r_1 and r_2) in the cone of depression formed by pumping (Figure 15.9) is measured to give h_1 and h_2 then integrating between r_1 and r_2 gives:

$$\int_{r_1}^{r_2} \frac{dr}{r} = -\frac{2\pi bK}{Q} \int_{h_1}^{h_2} dh \tag{15.82}$$

which gives:

$$[log_e r]_{r_1}^{r_2} = -\frac{2\pi bK}{Q} [h]_{h_1}^{h_2} \tag{15.83}$$

and:

$$log_e \frac{r_2}{r_1} = -\frac{2\pi bK}{Q}(h_2 - h_1) \tag{15.84}$$

and solving for K gives:

$$K = -\frac{Q \, log_e(r_2/r_1)}{2\pi b \, (h_2 - h_1)} \tag{15.85}$$

Equation 15.85 is the same (after conversion of the r_2/r_1 term to logs to base 10 from logs to base e) to that derived for a straight-line solution to the Theis equation (Equation 15.33) but using a different analytical approach.

Applying Equation 15.85 to the data given in Table 15.5 provides an estimate of the hydraulic conductivity of the tank material. The value of the hydraulic conductivity is a function of the aquifer material and not a function of the pumping rate. It will therefore be constant unless the aquifer material settles during testing.

Examination of Figure 15.31 indicates that it is only for low pumping rates that linearity in the plot of drawdown vs distance is maintained. Using data for the first five pumping rates and a variety of distance drawdown combinations in Equation 15.85 gives a value of hydraulic conductivity of 1375 m/day^{-1}. This is lower than the disturbed material figure predicted using Hazen but to be expected as great care was taken to settle the sands during placement (Dudgeon and Swan, 1972).

Table 15.5 Tank: test data

Test	Q m³day⁻¹	Radial Distance (m)								
		well	0.152	0.229	0.305	0.457	0.914	1.524	2.134	3.048
		Drawdown								
1	127.01	0.024	0.021	0.019	0.017	0.014	0.008	0.003	0.001	0.000
2	203.04	0.055	0.050	0.044	0.041	0.034	0.025	0.017	0.012	0.009
3	315.36	0.094	0.077	0.065	0.058	0.049	0.033	0.020	0.012	0.006
4	447.55	0.143	0.111	0.094	0.083	0.069	0.049	0.033	0.023	0.012
5	499.39	0.180	0.142	0.123	0.109	0.090	0.062	0.044	0.031	0.016
6	520.99	0.223	0.162	0.148	0.128	0.105	0.075	0.053	0.035	0.021
7	543.46	0.232	0.165	0.133	0.116	0.096	0.067	0.043	0.028	0.013
8	665.28	0.244	0.185	0.158	0.137	0.112	0.078	0.055	0.039	0.021
9	690.34	0.302	0.213	0.181	0.157	0.127	0.089	0.059	0.040	0.020
10	765.50	0.375	0.271	0.216	0.186	0.149	0.100	0.067	0.043	0.021
11	834.62	0.418	0.306	0.243	0.206	0.166	0.112	0.074	0.049	0.025
12	897.70	0.479	0.349	0.273	0.233	0.187	0.129	0.086	0.055	0.029
13	976.32	0.533	0.381	0.302	0.254	0.203	0.138	0.094	0.060	0.030
14	1059.26	0.619	0.443	0.344	0.294	0.218	0.157	0.106	0.067	0.035
15	1184.54	0.725	0.517	0.407	0.340	0.267	0.180	0.119	0.076	0.038
16	1345.25	0.899	0.629	0.483	0.407	0.318	0.211	0.141	0.089	0.045
17	1450.66	1.021	0.717	0.546	0.461	0.359	0.236	0.158	0.100	0.051
18	1487.81	1.055	0.716	0.556	0.465	0.362	0.237	0.159	0.100	0.052
19	1526.69	1.082	0.748	0.570	0.476	0.370	0.240	0.159	0.105	0.051
20	1548.29	1.097	0.770	0.592	0.500	0.392	0.270	0.173	0.102	0.057
21	1732.32	1.341	0.920	0.711	0.589	0.458	0.297	0.199	0.127	0.065

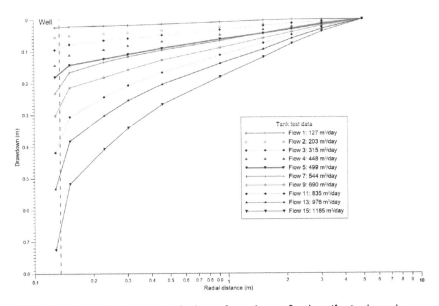

Figure 15.31 Drawdowns for increasing discharge from the confined aquifer in the tank

15.7.3 Radial flow modelling (RADFLOW) of the tank test results

The data was analyzed at the time using finite-element code developed for the purpose by Huyakorn (1972a, b) and Cox (1977). It is interesting to repeat the data analysis using the RADFLOW model code developed by Rushton and Redshaw (1979). In the RADFLOW code, the flow is always assumed to be linear. Departures from linearity will therefore indicate the onset of non-linear flow. This is demonstrated by the modelling of two data sets from the tank test using RADFLOW. Figure 15.32 shows the tank test data for an abstraction of 447.5 m^3/day^{-1} while Figure 15.33 shows the response for an abstraction of 765.5 m^3/day^{-1}. The lower abstraction rate appears to be well matched by RADFLOW results for a hydraulic conductivity of 1375 m/day. It can be assumed that the flow regime is linear – i.e., that Darcy flow occurs almost all the way to the well face. By contrast, the results for the abstraction of 765.5 m^3/day^{-1} show a pronounced break away that commences close to the outer edge of the test cell and increases towards the well. This indicates that non-Darcy flow commences at approximately a radial distance of 4 m from the well at this extraction rate. The deviation from linear flow increases steadily with drawdown at a pumping rate of 1732 m^3/day^{-1} of 1.34 m in the well.

15.7.4 Non-linear conditions

At the boundary between solid and liquid, a layer of water molecules is held close to the surface by tension. Moving out into the fluid, the next layer of molecules is free to slide over the first layer by shear. Subsequent layers can also shear and slide smoothly. This is laminar flow, and the dissipation of energy with hydraulic gradient follows Darcy's Law and is linear.

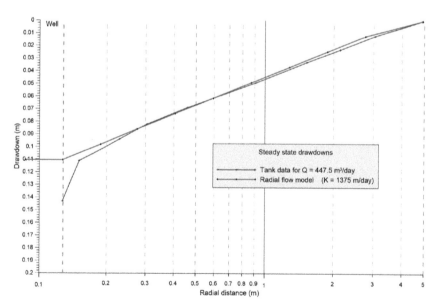

Figure 15.32 Tank test data and RADFLOW model data for an abstraction of 447.5 m^3/day (Flow 4). The hydraulic conductivity value used was 1375 m/day

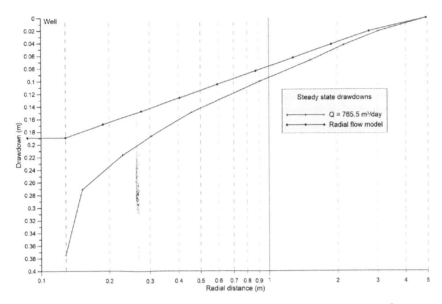

Figure 15.33 Tank test data and RADFLOW model data for an abstraction of 765.5 m³/day (Flow 10). The hydraulic conductivity value used was 1375 m/day

Turbulence occurs when laminar flow breaks down as a result of increasing fluid velocity. Eddies form and as a result, the resistance to friction is greatly increased (Smith and Sayre, 1964). The dissipation of energy with hydraulic gradient becomes non-linear and Darcy's Law breaks down. An example of the onset of turbulence is shown in Figure 15.34.

Smith and Sayre (1964) demonstrate that flow through medium sands and smaller grain-size materials is expected to be laminar, with turbulence only developing under large hydraulic gradients in very coarse sands and gravels. Table 15.4 shows very coarse sand to have a grain size between 1 mm and 2 mm. The d_{10} for the tank test sands is approximately 1.6 mm, indicating that all the sands in the tank were coarse sand size to larger. Under these conditions, turbulent flow will occur when there is a significant hydraulic gradient, as is likely close to the screen.

The onset of turbulent flow in groundwater systems can be related to a critical Reynold's number (Equation 15.86). The critical velocity above which turbulence commences is usually related to a critical value of the Reynolds number (Chapter 4). For porous media, the R_e is considered to lie between 1 and 10 (Bear, 1972).

$$R_e = \frac{\rho u L}{\mu} \tag{15.86}$$

where:
R_e is the Reynold's number (dimensionless),
ρ is fluid density kg/m³,
u is the fluid velocity m/s,
L a characteristic length of the system. In pipe flow it is the internal diameter of the pipe. In a porous medium it is the d_{10} 10% passing or the 90% retained, and
μ is the dynamic viscosity [Pa·s or Ns/m² or kg/m·s].

Figure 15.34 Transition from laminar to turbulent flow in the hot air stream above a candle (Creative Commons Attribution-Share Alike 3.0)

The Reynold's number can be calculated for the tank-test data (Table 15.5) using known values of the parameters in Equation 15.86 and the calculated velocity derived from Q/A, where the area is the area of the cylinder through which water is passing. Reynold's numbers can then be compared to the start of non-linearity in the tank test data. Reynold's numbers for the range of flows and radial distances are given in Table 15.6.

An inspection of the deviation from linearity, as indicated by the increase in gradient when compared to the RADFLOW results suggests that the critical value of the Reynold's number for the material tested is approximately 3.0. This is at the lower end but within the range (1 to 10) suggested by Bear (1972). Turbulent flow extends out to the piezometer at 0.914 m from the abstraction well. The Reynold's number indicating the onset of turbulent flow are shown in red in Table 15.6 and demonstrate that turbulent conditions are restricted to the area close to the abstraction bore with the distance out into the formation increasing as abstraction rate increases.

Close inspection of the drawdown data shown in Figure 15.31 also shows a significant head loss as water passes from the formation to the well. Dudgeon and Swan (1972) recognized this and recommended further work be carried out.

The Forchheimer equation

The use of the Forchheimer equation has been recommended (Cox, 1977; Dudgeon, 1985a, b) to more accurately reflect non-Darcy (non-linear) flow at higher hydraulic gradients. The

Table 15.6 Tank: Reynold's numbers – characteristic length of the $d_{10} = 1.6$ mm

Test	Q m³day⁻¹	Radial Distance (m)							
		0.152	0.229	0.305	0.457	0.914	1.524	2.134	3.048
		Reynold's Number							
1	127.01	1.81	1.21	0.91	0.60	0.30	0.18	0.13	0.09
2	203.04	2.90	1.93	1.45	0.97	0.48	0.29	0.21	0.14
3	315.36	4.50	3.00	2.25	1.50	0.75	0.45	0.32	0.23
4	447.55	6.38	4.26	3.19	2.13	1.06	0.64	0.46	0.32
5	499.39	7.12	4.74	3.56	2.37	1.19	0.71	0.51	0.36
6	520.99	7.43	4.95	3.72	2.48	1.24	0.74	0.53	0.37
7	543.46	7.74	5.16	3.87	2.58	1.29	0.77	0.55	0.39
8	665.28	9.49	6.33	4.74	3.16	1.58	0.95	0.68	0.47
9	690.34	9.48	6.56	4.92	3.28	1.64	0.98	0.70	0.49
10	765.50	10.92	7.28	5.46	3.64	1.82	1.09	0.78	0.55
11	834.62	11.90	7.93	5.95	3.97	1.98	1.19	0.85	0.59
12	897.70	12.80	8.53	6.40	4.27	2.13	1.28	0.91	0.64
13	976.32	13.92	9.28	6.96	4.64	2.32	1.39	0.99	0.70
14	1059.26	15.10	10.07	7.55	5.03	2.52	1.51	1.08	0.76
15	1184.54	16.88	11.26	8.44	5.3	2.81	1.69	1.21	0.84
16	1345.25	19.19	12.79	9.59	6.40	3.20	1.92	1.37	1.03
17	1450.66	20.69	13.79	10.34	6.90	3.45	2.07	1.48	1.03
18	1487.81	21.21	14.14	10.60	7.07	3.53	2.12	1.51	1.06
19	1526.69	21.77	14.51	11.04	7.36	3.68	2.21	1.58	1.10
20	1548.29	22.08	14.72	11.04	7.36	3.68	2.21	1.58	1.10
21	1732.32	24.70	16.47	12.35	8.23	4.12	2.47	1.76	1.23

onset of non-linearity can also occur close to a large pit. In this case, dewatering projections based upon Darcy flow can greatly overestimate the amount of groundwater pumping required to dewater a proposed pit for open-cast mining. This can sometimes make the economic basis for mining unacceptable. The problem is shown diagrammatically in Figure 15.35.

The Forchheimer equation can be expressed as:

$$i = \frac{dh}{dL} = av + bv^2 \tag{15.87}$$

where:

a, b are numerical coefficients, and
v is the specific discharge.

In practice, single pairs of coefficients can be fitted to experimental data to cover both the linear and non-linear portions of the curve. The Darcy equation can be rearranged to give:

$$\frac{dh}{dL} = \frac{1}{K}v \tag{15.88}$$

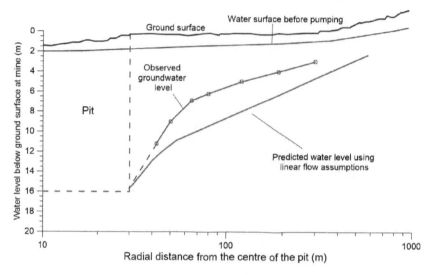

Figure 15.35 Section showing reduced actual drawdown around a pit as the result of turbulent flow (Dudgeon, 1985b)

Table 15.7 Forchheimer coefficients

Type	Material median diameter (mm)	median diameter (mm)	porosity %	Forcheimer Coefficients a (s/m)	b (s² m)⁻²
Unsorted river sand	0.27	0.53	38.7	700	9700
0 mm to 5 mm gravel	0.95	2.3	41.8	70	2400
2 mm to 5 mm dolerite	1.9	3.2	41.7	13	1100
5 mm to 10 mm dolerite	4.7	6.4	45.8	6.5	700

For all practical purposes, the coefficient a in the Forchheimer equation can be equated to $\frac{1}{K}$ over the linear flow range since the term bv^2 will be very small.

Dudgeon (1985a, b) gives a number of sets of Forchheimer coefficients, some of which are reproduced in Table 15.7. Appropriate values of the Forchheimer coefficients should be used to predict the drawdown wherever the possibility of additional head loss due to the onset of turbulent flow is a possibility.

15.8 SPECIFIC CAPACITY

Jacob (1944) suggested that the drawdown in an abstraction bore is composed of two components:

$$s_w = B(r_{ew,t})Q + CQ^2 \qquad (15.89)$$

and:

$$B(r_{ew,t}) = B_{1(r_w,t)} + B_2 \qquad (15.90)$$

where:

s_w is the drawdown in the well (m),
Q is the pumping rate,
$B_1(r_w,t)$ is the linear aquifer loss (m),
B_2 is the linear well loss,
C is the non-linear well loss (m),
r_{ew} is the effective radius of the well,
r_w is the actual radius of the well, and
t is the pumping time.

The proposed equation suggested by Jacob (1944) includes parameters that are not easy to specify, particularly the effective radius (r_{ew}). Rorabaugh (1953) suggested that Jacob's original equation could simplified and expressed as:

$$s_w = BQ + CQ^P \qquad (15.91)$$

where:

P has values of between 1.5 and 3.5 depending upon the value of Q. However, several
authors (Hazel, 2009; Kruseman and de Ridder, 1990) argue strongly that the value of
$P = 2$. Software packages such as AQTESOLV allow this parameter to vary.

The analysis was further refined by Bierschenk (1963) and Hantush (1964). It has become common practice to vary the discharge in an abstraction well to form a step test (Clark, 1977; Kruseman and de Ridder, 1990), as shown in Figure 15.27.

The ratio of drawdown to pumping rate is called the specific capacity. In Equation 15.91 dividing by Q gives:

$$\frac{s_w}{Q} = B + CQ \qquad (15.92)$$

where:
B is the aquifer loss and
C is the well loss.

The specific capacity values for the tank-test data and the RADFLOW model data for s_w/Q are shown in Figure 15.36. The straight-line plot for the model data illustrates the occurrence of linear flow at all abstraction rates. The parameter C in Equation 15.92 is zero, indicating no flow-dependent well loss and only a constant aquifer loss (B). It is pertinent to remember that all applications of Theis theory assume this condition and that all pumping test interpretations that do not take account of well losses will be in error. This is why it is necessary to install observation wells! By contrast, the tank-test data shows a good linear relationship between specific capacity and abstraction rate for the 21 steps in the tank-test step test. The value of B (the aquifer losses) and C (the well losses) in Equation 15.92 are 3.457E-7 and 0.0002 (Figure 15.36).

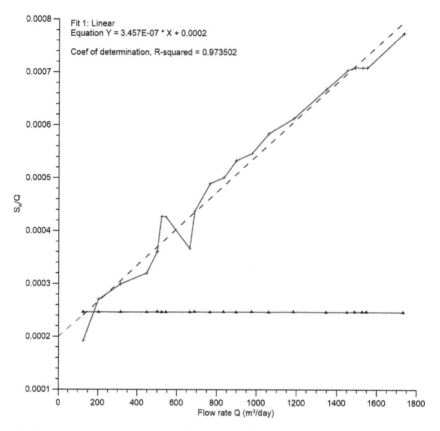

Figure 15.36 Plot of specific capacity data (Equation 15.92) for the tank test data (black line) and the output from the radial flow model (red line)

The efficiency for a specific abstraction rate can be calculated as:

$$E_w = \left[\frac{BQ}{BQ + CQ^2}\right] \times 100\% \qquad (15.93)$$

If a well has no well losses, it has an efficiency of 100%.

Skin effects

As used in well hydraulics, the concepts of linear and non-linear head loss components $(BQ + CQ^2)$ relate to the concepts of skin effect and non-Darcy flow. The total drawdown inside a well due to well losses (also indicated as the apparent skin effects) can be expressed as:

$$s = BQ + CQ^2 = \frac{1}{2\pi T}(\text{skin} + C'Q)Q \qquad (15.94)$$

where:

$C' = C \times 2\pi T$ is the non-linear well loss coefficient or high-velocity coefficient and **skin** $= B \times 2\pi T$ is the skin factor.

The skin factor appears in recent computer packages such as AQTESOLV. The effective well radius (r_{ew}) is related to the well radius (r_w) as:

$$r_{ew} = r_w \exp^{-\text{skin}} \qquad (15.95)$$

If the effective radius of the well r_{ew} is larger than the real radius of the well r_w it is called a positive skin effect. If it is smaller, the well is usually poorly developed or the screen is clogged and the effect is referred to as a negative skin effect.

15.9 INTERPRETATION OF STEP-TEST DATA

Various methods are available to analyze step-drawdown tests. The Hantush-Bierschenk method and the Eden-Hazel method are based upon developments of Jacob's approach. The Hantush-Bierschenk method can determine values of B and C and can be applied to confined, leaky or unconfined aquifers. The Eden-Hazel method can be applied in confined aquifers and gives values of the well-loss parameters as well as estimates of transmissivity. It is also a more generally applicable method available for use where pumping has been intermittent.

Clark (1977) presented a detailed data set based upon work on a deep confined sandstone aquifer in Saudi Arabia. The data set has also been reported by Kruseman and de Ridder (1990). The data set is presented in Table 15.8 and will be used to illustrate both approaches.

15.9.1 Hantush – Bierschenk analysis

Hantush (1964) applied the principle of superposition to Jacob's equations to express the drawdown $s_{w(n)}$ in a well during the nth step of a step-drawdown test as:

$$s_{w(n)} = \sum_{i=1}^{n} \Delta Q_i B(r_{ew}, t - t_i) + C Q_n^2 \qquad (15.96)$$

Where:

$s_{w(n)}$ is the total drawdown in the well during the nth step at time t,

r_{ew} is the effective radius of the well,

t_i time at which the ith step begins ($t_1 = 0$),

Q_n is the constant discharge during the nth step,

Q_i is the constant discharge at the preceding step, and

$\Delta Q_i = Q_i - Q_{i-1}$ is the discharge increment beginning at time t_i.

The sum of increments of drawdown taken at a fixed interval of time from the beginning of each step can be obtained from Equation 15.96:

$$\sum_{i=1}^{n} \Delta s_{w(i)} = s_{w(n)} = B(r_{ew}, \Delta t) Q_n + C Q_n^2 \qquad (15.97)$$

Table 15.8 Step drawdown test data for a deep sandstone aquifer in Saudi Arabia (Clark, 1977)

Step	1	2	3	4	5	6
Discharge	1306	1693	2423	3261	4094	5019
Time	Level	Level	Level	Level	Level	Level
1	0	5.458	8.17	10.881	15.318	20.036
2	0	5.529	8.24	11.797	15.494	20.248
3	0	5.564	8.346	11.902	15.598	20.389
4	0	5.599	8.451	12.008	15.740	20.529
5	1.303	5.634	8.486	12.078	15.846	20.600
6	2.289	5.669	8.557	12.149	15.881	20.660
7	3.117	5.669	8.557	12.149	15.952	20.741
8	3.345	5.705	8.592	12.184	16.022	20.811
9	3.486	5.74	8.672	12.219	16.022	20.882
10	3.521	5.74	8.672	12.325	16.093	20.917
12	3.592	5.81	8.663	12.36	16.198	20.952
14	3.627	5.81	8.698	12.395	16.268	21.022
16	3.733	5.824	8.733	12.43	16.304	21.128
18	3.768	5.845	8.839	12.43	16.374	21.163
20	3.836	5.81	8.874	12.501	16.409	21.198
25	3.873	5.824	8.874	12.508	16.586	21.304
30	4.014	5.824	8.979	12.606	16.621	21.375
35	3.803	5.881	8.979	12.712	16.691	21.480
40	4.043	5.591	8.994	12.747	16.726	21.551
45	4.261	5.591	9.05	12.783	16.776	21.619
50	4.261	6.092	9.05	12.818	16.797	21.656
55	4.19	6.092	9.12	12.853	16.902	21.660
60	4.12	6.176	9.12	12.853	16.938	21.663
70	4.12	6.162	9.155	12.888	16.973	21.691
80	4.226	6.176	9.191	12.923	17.079	21.762
90	4.226	6.169	9.191	12.994	17.079	21.832
100	4.226	6.169	9.226	12.994	17.114	21.903
120	4.402	6.176	9.261	13.099	17.219	22.008
150	4.402	6.374	9.367	13.205	17.325	22.184
180	4.683	6.514	9.578	13.24	17.395	22.325

Note discharge is in cubic metres per day.

where:

$\Delta s_{w(i)}$ is the drawdown increment between the ith step and that preceding it, taken at time $t_i + \Delta t$ from the beginning of the ith step.

Equation 15.97 can also be written as:

$$\frac{S_{w(n)}}{Q_n} = B(r_{ew}, \Delta t) + CQ_n \qquad (15.98)$$

This representation allows a graphical solution to B and C. The increments of drawdown have to be extrapolated and measured at a fixed time after the beginning of each step.

A plot of $S_{w(n)}/Q_n$ versus Q_n on an arithmetic scale paper will yield a straight line whose slope is equal to C and the intercept at $Q = 0$ will give B.

Note that if the test has reached equilibrium (constant drawdown) at the end of each step, then no extrapolation is required.

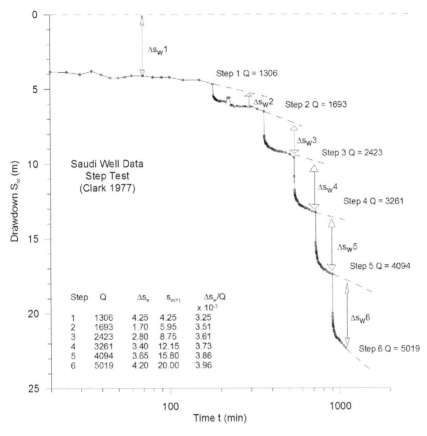

Figure 15.37 Analysis for Saudi step-test data using the Hantush-Bierschenk approach

The steps for this analysis have been reproduced here from Kruseman and de Ridder (1990).

1 On semi-log paper, plot the observed drawdown in the well (s_w) against the corresponding time t. Use the log axis for time as shown in Figure 15.37.
2 Extrapolate the curve through the plotted data of each step to the end of the next step.
3 Determine the increments of drawdown $\Delta s_{w(i)}$ for each step by taking the difference between the observed drawdown at a fixed time interval (Δt), taken from the beginning of each step, and the corresponding drawdown on the extrapolated curve from the preceding step.
4 Determine the values of $s_{w(n)}$ corresponding to the discharge Q_n from step $s_{w(n)} = \Delta s_{w(1)} + \Delta s_{w(2)} + ... + \Delta s_{w(n)}$.
5 Calculate the ratios of $s_{w(n)}/Q_n$ for each step.
6 On arithmetic paper, plot the values of $s_{w(n)}/Q_n$ versus the corresponding values of Q_n and fit a straight line through the points as shown in Figure 15.38. If the data does not fit a straight line, it is possibly due to the exponent in Equation 15.97 not equal to 2. Eden and Hazel (1973) and Clark (1977), however, caution against this approach.
7 Determine the slope C and intercept B of the straight line.

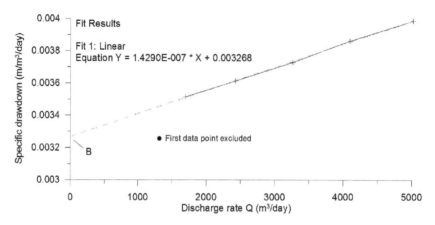

Figure 15.38 Plot of specific drawdown vs discharge rate for Saudi data

The data for the deep confined sandstone test conducted by Clark (1977) in Saudi is shown in Figure 15.37. The drawdown data has been calculated for $t = 100$ minutes into each step. The first data point is poor as it plots way off the line for the remaining five points. Clark argued that this point was impacted by the water initially being cooled as it stood in the bore. The location of this point is shown in Figure 15.38.

The equation of best fit shown in Figure 15.38 defines the drawdown equation for this bore for $t = 100$ minutes:

$$s_w = 0.003268Q + 1.429 \times 10^{-7}Q^2 \tag{15.99}$$

The AQTESOLV package can be used to derive a solution as well (somewhat more quickly) and can also be used to test the sensitivity of the results. A solution based upon a Theis step test is shown in Figure 15.39. Note that the meaning of some of the symbols changes. In the AQTESOLV program, S_w refers to the skin thickness.

15.9.2   Eden-Hazel method

The basic assumption in all step test analysis recognizes that the drawdown is a scalar quantity and that drawdown components arising from different well functions can be added or subtracted. The discharge rate at the beginning of each step is equivalent to a new pump in the well with a discharge rate equal to the increase in abstraction. The total drawdown in a step drawdown test is equal to the sum of the drawdowns caused by the theoretical pumps responsible for each discharge step in the test.

The multiple stages of a step-drawdown test are equivalent to:

- pump 1 discharging at rate Q_1 from time t_1 to t_5,
- pump 2 switching in from time t_2 to t_5 with a discharge rate of $Q_2 - Q_1$,
- pump 3 switching on from time t_3 to t_5 with a discharge rate of $Q_3 - Q_2$, and
- pump 4 switching on from time t_4 to t_5 with a discharge rate of $Q_4 - Q_3$.

Hazel (2009) devised a rigorous graphical analysis of step drawdown tests in which the true drawdown-discharge rate curve for each step is reconstructed (Clark, 1977). The

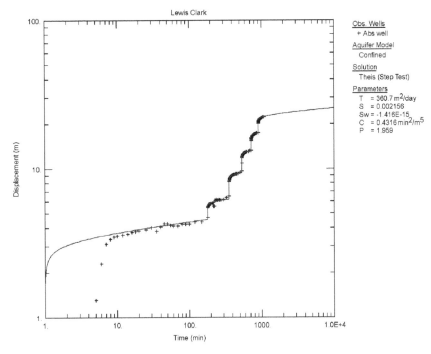

Figure 15.39 AQTESOLV solution for the Saudi confined sandstone step test

analysis is similar to much of the above but the complete analysis is repeated here for clarity and is based upon Kruseman and de Ridder (1990). The method is based upon Jacob's approximation to the Theis equation and can be used to determine well losses and also the transmissivity of the aquifer.

The drawdown is given (again) as:

$$s_w = \frac{2.3Q}{4\pi T} \log \frac{2.25Tt}{r_{ew}^2 S}$$

(15.100)

This can be written as:

$$s_w = (a + b \log t)Q$$

(15.101)

where:

$$a = \frac{2.30}{4\pi T} \log \frac{2.25T}{r_{ew}^2 S}$$

(15.102)

and:

$$b = \frac{2.30}{4\pi T}$$

(15.103)

The drawdown in a well at the end of the step test is:

$$
\begin{aligned}
s_{wt_5} \ = \ & (a + b\log(t_5 - t_1))Q_1 \\
+ \ & (a + b\log(t_5 - t_2))(Q_2 - Q_1) \\
+ \ & (a + b\log(t_5 - t_3))(Q_3 - Q_2) \\
+ \ & (a + b\log(t_5 - t_4))(Q_4 - Q_3) \\
+ \ & CQ_4^2
\end{aligned}
\tag{15.104}
$$

Using the principle of superposition, the drawdown at the time t during the nth step is:

$$
s_{w(n)} = \sum_{i=1}^{n} (\Delta Q_i)(a + b\log(t - t_i)) = aQ_n + b\sum_{i=1}^{n} \Delta Q_i \log(t - t_i)
\tag{15.105}
$$

where:

Q_n is the constant discharge during the n-th step,
Q_i is the discharge in the preceding step,
$\Delta Q_i = Q_i - Q_{i-1}$ is the discharge increment beginning at time t_i,
t_i time at which the i-th step begins, and
t time since the step-drawdown test started.

The drawdown at any time t is thus a summation of the individual drawdowns of each of the hypothetical pumps.

Equation 15.105 does not account for non-linear well losses (the CQ^2 term); however, these can be simply introduced. A simplification can also be included by defining:

$$
H_n = \sum_{i=1}^{n} \Delta Q_i \log(t - t_i)
\tag{15.106}
$$

The final result is:

$$
s_{w(n)} = aQ_n + bH_n + CQ_n^2
\tag{15.107}
$$

Step test analysis can now be directed to the evaluation of a, b and C in Equation 15.107. These factors can then be used to evaluate the well loss component of the drawdown (C) and the aquifer transmissivity (T) and to estimate the aquifer storage coefficient.

Since a, b and C are constants, a plot of $s_w(n)$ versus H_n on natural-scale graph paper will give a series of parallel straight lines (one for each value of Q_i) with slope 'b' and intercepts of $aQ_n + CQ_n^2$ on the $s_w(n)$ axis.

If these intercepts are divided by Q_i, then a and C can be determined.

This method of analysis gives a very ready means of allowing for antecedent pumping conditions and for pump stoppages. It also is in a form which is readily amenable to spreadsheet analysis. This analysis is not dependent on the antecedent drawdowns, as is the previous method, but rather on antecedent discharges.

If the pump stops during a step for which the discharge rate had been Q_n, then a recharge bore delivering Q_n to the bore is superimposed in the analysis. A similar operation is carried out for any change in rate, so all rate changes are able to be accounted for in the analysis.

By means of a regression analysis, the constants a, b and C are able to be calculated and the drawdown equation can be determined for any required time of discharge. This technique is used extensively in Queensland to analyze pumping tests but has not received general acceptance due to a lack of published examples.

Equation 15.107 is laborious to calculate. It basically means the evaluation of a new transformed time for each measurement point. A different value of H_n will be calculated for each value of head as they each occur at different times. This is laborious by hand but fairly straightforward in EXCEL.

Table 15.9 Pumping rate quantities for the Eden–Hazel method – applied to the Saudi example

Step	Pumping Rates m^3/day	Pumping Increments	Start Time	Time Steps	Step Duration
1	$Q_1 = 1306$	$\Delta Q_1 = 1306 - 0 = 1306$	$t_1 = 0$	$1 - 180$	$t - t_1$
2	$Q_2 = 1693$	$\Delta Q_2 = 1693 - 1306 = 387$	$t_2 = 180$	$181 - 360$	$t - t_2$
3	$Q_3 = 2423$	$\Delta Q_3 = 2423 - 1693 = 730$	$t_3 = 360$	$361 - 540$	$t - t_3$
4	$Q_4 = 3261$	$\Delta Q_4 = 3261 - 2423 = 838$	$t_4 = 540$	$541 - 720$	$t - t_4$
5	$Q_5 = 4094$	$\Delta Q_5 = 4094 - 3261 = 833$	$t_5 = 720$	$721 - 900$	$t - t_5$
6	$Q_6 = 5019$	$\Delta Q_6 = 5019 - 4094 = 925$	$t_6 = 900$	$901 - 1080$	$t - t_6$

Table 15.10 Values of H_i for the confined sandstone example of the Eden–Hazel analysis

Step 1		Step 2		Step 3		Step 4		Step 5		Step 6	
Time	H_1i	time	H_2i	time	H_3i	time	H_4i	time	H_5i	time	H_6i
1	0.000	181	2.048	361	2.926	541	4.311	721	5.937	901	7.628
2	0.273	182	2.131	362	3.081	542	4.488	722	6.114	902	7.824
3	0.433	183	2.180	363	3.172	543	4.593	723	6.218	903	7.940
4	0.546	184	2.216	364	3.237	544	4.668	724	6.293	904	8.024
5	0.634	185	2.244	365	3.287	545	4.726	725	6.352	905	8.089
6	0.706	186	2.267	366	3.329	546	4.775	726	6.442	907	8.189
8	0.819	188	2.305	368	3.396	548	4.852	728	6.478	908	8.229
9	0.865	189	2.321	369	3.424	549	4.884	729	6.511	909	8.298
12	0.979	192	2.361	372	3.492	552	4.963	734	6.635	914	8.403
16	1.092	196	2.403	376	3.562	556	5.044	736	6.674	916	8.446
18	1.138	198	2.420	378	3.591	558	5.079	738	6.709	918	8.485
20	1.180	200	2.437	380	3.618	560	5.109	740	6.740	920	8.597
30	1.340	210	2.503	390	3.723	570	5.233	750	6.868	930	8.662
35	1.400	215	2.530	395	3.765	575	5.282	755	6.919	935	8.719
40	1.453	220	2.555	400	3.802	585	5.365	765	7.006	945	8.816
50	1.541	230	2.599	410	3.866	590	5.401	770	7.044	950	8.859
55	1.578	235	2.618	415	3.894	595	5.435	775	7.080	955	8.936
70	1.673	250	2.671	430	3.968	610	5.523	790	7.174	970	9.004
80	1.726	260	2.702	440	4.011	620	5.575	800	7.230	980	9.066
90	1.772	270	2.730	450	4.050	630	5.665	820	7.328	1000	9.176
120	1.886	300	2.805	480	4.151	660	5.743	840	7.413	1020	9.273
150	1.974	330	2.869	610	4.236	690	5.845	870	7.526	1050	9.400
180	2.045	360	2.924	540	4.308	720	5.934	900	7.625	1080	9.511

The data set for the Saudi example is presented in Table 15.8. The pumping rates and derived values for time are summarized in Table 15.9. Note that the H factor will be calculated in minutes for this analysis.

For the first step, Equation 15.106 becomes:

$$H_1 = \frac{1306}{1440} \log(t) \tag{15.108}$$

and is calculated for each time for 1 to 180 minutes as shown by way of an example in Table 15.10. For $t = 1, 2, 3, 4, ..., H = 0, 0.273, 0.433, 0.546, ...$

The second step is calculated using:

$$H_2 = \frac{1306}{1440} \log(t) + \frac{387}{1440} \log(t - 180) \tag{15.109}$$

and is again calculated for all values of t in the second step. In this step, the time (t) varies between 181 and 360 minutes. The first few values of H for the second step of this test are 2.048, 2.131, 2.180, etc.

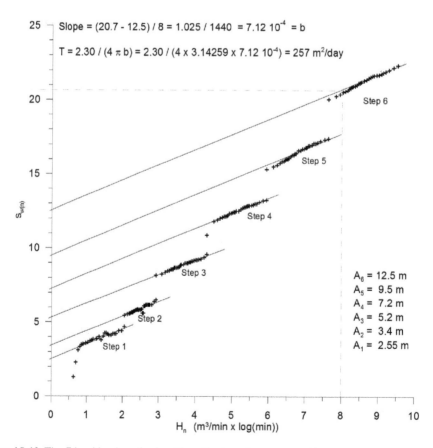

Figure 15.40 The Eden-Hazel method: arithmetic plot of $s_{w(n)}$ versus H_n

The third step is calculated using:

$$H_3 = \frac{1306}{1440}\log(t) + \frac{387}{1440}\log(t-180) + \frac{730}{1440}\log(t-360) \qquad (15.110)$$

Here, t varies between 361 and 540 minutes.

Similarly for the fourth and fifth steps.

The last step (sixth) in this test is calculated by:

$$\begin{aligned}
H_6 &= 1306/1440\log(t) + 387/1440\log(t-180) \\
&\quad + 730/1440\log(t-360) + 838/1440\log(t-540) \\
&\quad + 833/1440\log(t-720) + 925/1440\log(t-900)
\end{aligned} \qquad (15.111)$$

Values of H generated by this application for each of the six steps in the test are given for this example in Table 15.10.

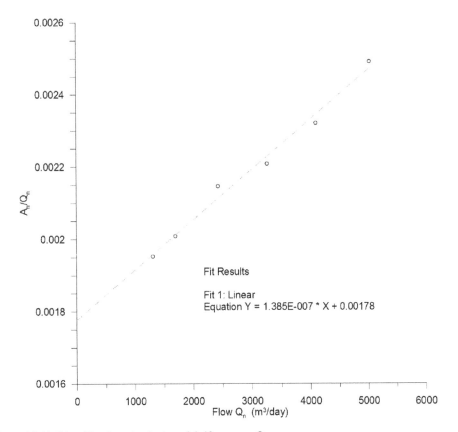

Figure 15.41 Eden-Hazel method: plot of A_n/Q_n versus Q_n

The analysis procedure can be listed as:

1 calculate the values of H_n using Equation 15.106 and the measured discharges and times,
2 on arithmetic paper, plot the observed drawdowns $(s_w(n))$ versus the corresponding values of H_n as shown in Figure 15.40,
3 draw parallel straight lines of best fit through the plotted points, one straight line through each set of points (Figure 15.40),
4 determine the slopes of the lines $\Delta s_{w(n)}/\Delta H_n$, which gives values of b ($b = \frac{2.30}{4\pi T}$ from Equation 15.103),
5 extend the lines so that they intercept the $H_n = 0$ axis. The interception point (A_n) of each line is given by $A_n = aQ_n + CQ_n^2$,
6 read the values of A_n,
7 calculate the ratio A_n/Q_n for each step (i.e., for each value of Q_n),
8 on arithmetic paper, plot the values of A_n/Q_n versus the corresponding values of Q_n. Fit a straight line through the plotted points (Figure 15.41),
9 determine the slope of the straight line – which is the value of C,
10 extend the straight line until it intersects the axis at $Q_n = 0$ to give values of a, and
11 knowing b, calculate T.

The slope of the parallel straight lines in Figure 15.40 is 6.9×10^{-4}. Using this slope in Equation 15.103 gives a value of transmissivity of 257 m²/day. Substituting values from the slope and intercept of Figure 15.41 into the drawdown equation (Equation 15.107) gives:

$$s_w = ((1.78 \times 10^{-3}) + (6.9 \times 10^{-4})\log t)Q + (1.4 \times 10^{-7})Q^2 \qquad (15.112)$$

The international system of units – SI

SI is the accepted symbol for Système International d'Unités (International System of Units), the modern form of the metric system finally agreed in 1960 by the supreme international authority on units of measurement, the General Conference on Weights and Measures (CGPM).

DEFINITIONS OF FUNDAMENTAL UNITS

metre (m) The metre is the length equal to 1,650,763.73 wavelengths in vacuum of the radiation corresponding to the transition between levels $2_1 0^p$ and 5^d_5 of the krypton-86 atom.

kilogram (kg) The kilogram is the unit of mass; it is equal to the mass of the international prototype of the kilogram.

second (s) The second is the duration of 9,192,631,770 periods of the radiation corresponding to the transition between the two hyperfine levels of the ground state of the caesium-133 atom.

ampere (A) The ampere is that constant current which, if maintained in two straight parallel conductors of infinite length of negligible circular cross-section and placed 1 metre apart in vacuum, would produce between these conductors a force equal to 2×10^{-7} newtons per metre of length.

kelvin (K) The kelvin, unit of thermodynamic temperature, is the fraction 1/273.16 of the thermodynamic temperature of the triple point of water.

candela (cd) The candela is the luminous intensity, in the perpendicular direction, of a surface of 1/600,000 square metre of a black body at the temperature of freezing platinum under a pressure of 101,325 newtons per square metre.

mole (mol) The mole is the amount of substance of a system which contains as many elementary entities as there are atoms in 0.012 kilograms of carbon-12.

SUPPLEMENTARY AND DERIVED UNITS

Some supplementary and derived units employed in the system are given below; the degree Celsius is not strictly an SI unit but is used with the system.

DERIVED SI UNITS WITH SPECIAL NAMES

hertz (Hz)   The hertz is the number of repetitions of a regular occurrence in 1 second.
newton (N)   The newton is that force which, applied to a mass of 1 kilogram, gives it an acceleration of 1 metre per second squared.
pascal (Pa)   The pascal is the pressure produced by a force of 1 newton applied, uniformly distributed, over an area of 1 square metre.
litre (L)   1000 litres = 1 cubic metre.
joule (J)   The joule is the work done when the point of application of a force of 1 newton is displaced through a distance of 1 metre in the direction of the force.
watt (W)   The watt is the power which in 1 second gives rise to energy of 1 joule.
coulomb (C)   The coulomb is the quantity of electricity carried in 1 second by a current of 1 ampere.

Table A.1 SI quantities

Quantity	Unit Name	Unit Symbol
Area	square metre	m^2
Volume	cubic metre	m^3
Velocity	metre per second	ms^{-1}
Acceleration	metre per second per second	ms^{-2}
Frequency	hertz	Hz
Density (mass density)	kilogram per cubic metre	$kg\,m^{-3}$
Momentum	kilogram metre per second	$kg\,m\,s^{-1}$
Moment of inertia	kilogram metre squared	$kg\,m^2$
Force	newton	$N\,(kg\,m\,s^{-2})$
Moment of force (torque)	newton metre	Nm
Pressure, stress	pascal	$Pa\,(Nm^{-2})$
Viscosity: kinematic	metre squared per second	$m^2 s^{-2}$
Viscosity: dynamic	pascal second	$Pas\,(N\,sm^{-2})$
Surface tension	newton per metre	Nm^{-1}
Energy, work, quantity of heat	joule	J(Nm)
Power, radiant flux	watt	$W(J\,s^{-1})$
Temperature	degree Celsius	°C
Thermal conductivity	watt per metre kelvin	$W\,m^{-1}K$
Heat capacity	joule per degree kelvin	JK^{-1}
Specific heat capacity	joule per kilogram per degree Kelvin	$J(kgK)^{-1})$
Quantity of electricity	coulomb	CAs
Electric potential	volt	$V(WA^{-1})$
Electric field strength	volt per metre	$V\,m^{-1}$
Electric resistance	ohm	$\Omega(V\,A^{-1})$
Electric conductance	siemens	$S(AV^{-1})$
Electric capacitance	farad	$F\,(A\,s\,V^{-1})$
Magnetic flux	weber	Wb(V s)
Inductance	henry	$H(V\,s\,A^{-1})$
Magnetic flux density	tesla	$T(W\,m^{-2})$
Magnetic field strength	ampere per metre	Am^{-1}

volt (V)	The volt is the difference of electric potential between two points of a conducting wire carrying a constant current of 1 ampere, when the power dissipated between these points is equal to 1 watt.
ohm Ω	The ohm is the electrical resistance between two points of a conductor when a constant potential difference of 1 volt, applied to these points, produces in the conductor a current of 1 ampere, the conductor not being the seat of any electromotive force.
farad (F)	The farad is the capacitance of a capacitor between the plates of which there appears a difference of electric potential of 1 volt when it is charged by a quantity of electricity of 1 coulomb.
weber (Wb)	The weber is the magnetic flux which, linking a circuit of 1 turn, would produce in it an electromotive force of 1 volt if it were reduced to zero at a uniform rate in 1 second.
henry (H)	The henry is the inductance of a closed circuit in which an electromotive force of 1 volt is produced when the electric current in the circuit varies uniformly at the rate of 1 ampere per second.
tesla (T)	The tesla is equal to 1 weber per square metre of circuit area.

THE MULTIPLYING PREFIXES

The SI prefixes for multiplication are given in Table A.2. Examples: gigahertz (GHz), megawatt (MW), kilometre (km), centimetre (cm), Gigalitre (Gl) milligram (mg), microsecond (μs), nanometre (nm).

THE GREEK ALPHABET

Many quantities are shown with Greek alphabet symbols. Table A.3 gives a summary of the Greek alphabet and English language equivalents.

Table A.2 The multiplying prefixes

Prefix Name	Prefix Symbol	Multiplication Factor
tera	T	10^{12}
giga	G	10^{9}
mega	M	10^{6}
kilo	k	10^{3}
hecto	h	10^{2}
deca	da	10^{1}
deci	d	10^{-1}
centi	c	10^{-2}
milli	m	10^{-3}
micro	μ	10^{-6}
nano	n	10^{-9}
pico	P	10^{-12}

Table A.3 The Greek alphabet

Greek Character		Greek Name	English Equivalent	
Upper Case	*Lower Case*		*Upper Case*	*Lower Case*
A	α	alpha	A	a
B	β	beta	B	b
Γ	γ	gamma	G	g
Δ	δ	delta	D	d
E	ε	epsilon	Ĕ	ĕ
Z	ζ	zeta	Z	z
H	η	eta	Ē	ē
Θ	θ	theta	Th	th
I	ι	iota	I	i
K	κ	kappa	K	k
Λ	λ	lambda	L	l
M	μ	mu	M	m
N	ν	nu	N	n
Ξ	ξ	xi	X	X
O	ο	omnicron	Ŏ	ŏ
Π	π	pi	P	P
P	ρ	rho	R	r
Σ	σ	sigma	S	s
T	τ	tau	T	t
Υ	υ	upsilon	Y	y
Φ	φ	phi	Ph	ph
X	χ	chi	Ch	ch
Ψ	ψ	psi	Ps	ps
Ω	ω	omega	Ō	ō

TIME

The fundamental SI unit of time is a second (s).

The SI unit for a minute is (min) = 60 s

The SI unit for an hour is (h) = 60 min = 1440 s

The SI unit for a day is (d) = 24 h = 86,400 s

The SI unit for a year is not an exact quantity but is usually defined as 1 year (a) = 365.25 days.

Geological time is often written in terms of millions or billions of years. Neither the term million or the term billion are included in the SI system. The age of the earth is better written as 4.53 Ga than 4.53 billion years as there is a difference of a factor of 1 million between the American billion and the European billion!

Note that 'BP' means before present.

Bibliography

Abu-Hamdeh, N.H. & Reeder, R.C. (2000) Soil thermal conductivity: effects of density, moisture, salt concentration, and organic matter. *Soil Science Society of America Journal Abstract Division S-1-Soil Physics*, 64(4), 1285–1290. doi:10.2136/sssaj2000.6441285x.

Acworth, R.I. (1981) *The Evaluation of Groundwater Resources in the Crystalline Basement of Northern Nigeria*. PhD Thesis, Geology. Available from: http://etheses.bham.ac.uk/3576/.etheses.bham.ac.uk/357611/Acworth81PhD.pdf

Acworth, R.I. (1987) The development of crystalline basement aquifers in a tropical environment. *Quarterly Journal of Engineering Geology and Hydrogeology*, 20(4), 265–272. ISSN 1470-9236. doi:10.1144/GSL.QJEG.1987.020.04.02. Available from: http:l/qjegh.lyellcollection.org/cgi/doi/10.1144/GSL.QJEG.1987.020.04.02.

Acworth, R.I. (1998) Electromagnetic induction logs from selected bores in the Botany Sands aquifer. In: McNally, G.H. & Jankowski, J. (eds) *Environmental Geology of the Botany Basin*. Environmental, Engineering and Hydrogeology Specialist Group (EEHSG) of the Geological Society of Australia. pp. 143–159.

Acworth, R.I. (1999) Investigation of dryland salinity using the electrical image method. *Australian Journal of Soil Research*, 37(4), 623–636. ISSN 00049573. doi:10.1071/SR98084.

Acworth, R.I. (2001a) The electrical image method compared with resistivity sounding and electromagnetic profiling for investigation in areas of complex geology: a case study from groundwater investigation in a weathered crystalline rock environment. *Exploration Geophysics*, 32(2), 119–128. doi:10.1071/EG01119.

Acworth, R.I. (February 2001b) Physical and chemical properties of a DNAPL contaminated zone in a sand aquifer. *Quarterly Journal of Engineering Geology and Hydrogeology*, 34(1), 85–98. ISSN 14709236. doi:10.1144/qjegh.34.1.85. Available from: http://qjegh.lyellcollection.org/content/34/1/85.abstract.

Acworth, R.I. (November 2007) Measurement of vertical environmental-head profiles in unconfined sand aquifers using a multi-channel manometer board. *Hydrogeology Journal*, 15(7), 1279–1289. doi:10.1007/s10040-007-0178-9. First Online: 13 April 2007.

Acworth, R.I. (April 2009) Surface water and groundwater: understanding the importance of their connections. *Australian Journal of Earth Sciences*, 56, 1–2. ISSN 0812-0099. doi:10.1080/08120090802541853.

Acworth, R.I. & Beasley, R. (1998) Investigation of em-31 anomalies at yarramanbah/pump station creek, on the Liverpool plains of New South Wales. Research Report 195, Water Research Laboratory. Available from: http://lhandle.unsw.edu.au/1959.4/36224. ISBN 0 85824 0289.

Acworth, R.I. & Brain, T. (2008) Calculation of barometric efficiency in shallow piezometers using water levels, atmospheric and earth tide data. *Hydrogeology Journal*, 14(8), 1469–1481. doi:10.1007/s10040-008-0333-y.

Acworth, R.I. & Dasey, G.R. (2003) Mapping of the hyporheic zone around a tidal creek using a combination of borehole logging, borehole electrical tomography and cross-creek electrical imaging,

New South Wales, Australia. *Hydrogeology Journal*, 11(3), 368–377. ISSN 1431-2174. doi:10. 1007/s10040-003-0258-4.

Acworth, R.I. & Griffiths, D.H. (1985) Simple data processing of tripotential apparent resistivity measurements as an aid to the interpretation of subsurface structure. *Geophysical Prospecting*, 33, 861–867.

Acworth, R.I., Halloran, L.J.S., Rau, G.C., Cuthbert, M.O. & Bernadi, T.L. (November 2016a) An objective method to quantify groundwater compressible storage using earth and atmospheric tides. *Geophysical Research Letters*, 43(22), 11, 671–11, 678. doi:10.1002/2016gl071328.

Acworth, R.I., Hughes, C.E. & Turner, I.L. (2007) A radioisotope tracer investigation to determine the direction of groundwater movement adjacent to a tidal creek during spring and neap tides. *Hydrogeology Journal*, 15(2), 281–296. ISSN 1431-2174. doi:10.1007/s10040-006-0085-5.

Acworth, R.I. & Jankowski, J. (1997) The relationship between bulk electrical conductivity and dryland salinity in the narrabri formation at breeza, Liverpool plains, New South Wales, Australia. *Hydrogeology Journal*, 5(3), 109–122. ISSN 1431-2174. doi:10.1007/s100400050259.

Acworth, R.I. & Jankowski, J. (2001) Salt source for dryland salinity: evidence from an upland catchment on the southern tablelands of New South Wales. *Australian Journal of Soil Research*, 39(1), 39–59. ISSN 00049573. doi:10.1071/SR99120.

Acworth, R.I. & Jorstad, L.B. (2006) Integration of multi-channel piezometry and electrical tomography to better define chemical heterogeneity in a landfill leachate plume within a sand aquifer. *Journal of Contaminant Hydrology*, 83(3–4), 200–220. ISSN 01697722. doi:10.1016/ jjconhyd2005.11.007.

Acworth, R.I., Rau, G.C., Cuthbert, M.O., Jensen, E. & Leggett, K. (2016b) Long-term spatio-temporal precipitation variability in arid-zone Australia and implications for groundwater recharge. *Hydrogeology Journal*, 24(4), 905–921. doi:10.1007/s10040-015-1358-7.

Acworth, R.I., Rau, G.C., Halloran, L.J.S. & Timms, W.A. (April 2017) Vertical groundwater storage properties and changes in confinement determined using hydraulic head response to atmospheric tides. *Water Resources Research*, 53(4), 2983–2997. doi:10.1002/2016WR020311.

Acworth, R.I., Rau, G.C., McCallum, A.M., Andersen, M.S. & Cuthbert, M. (2015a) Understanding connected surface-water/groundwater systems using Fourier analysis of daily and sub-daily head fluctuations. *Hydrogeology Journal*, 23(1), 143–159. doi:10.1007/s10040-014-1182-5.

Acworth, R.I., Timms, W.A., Kelly, B.F.J., McGeeney, D.E., Ralph, T.J., Larkin, Z.T. & Rau, G.C. (2015b) Late Cenozoic paleovalley fill sequence from the southern Liverpool plains, New South Wales: implications for groundwater resource evaluation. *Australian Journal of Earth Sciences*, 62, 657–680. doi:10.1080/08120099.2015.1086815.

Acworth, R.I., Young, R.R. & Bernardi, T. (2005) Monitoring soil moisture status in a black vertosol on the Liverpool plains, NSW, using a combination of neutron scattering and electrical imaging methods. *Australian Journal Soil Research*, 43, 1–13.

Agriculture, Resource Management Council of Australia, and New Zealand. (1997) Minimum construction requirements for water bores in Australia. Technical Report DNRQ97080, Agriculture and Resource Management Council of Australia and New Zealand. ISBN 0 7242 7401 4.

Allen, A.D., Laws, A.T. & Commander, D.P. (1992) A review of major water resources in western Australia. Technical Report, Kimberly Water Resources Development Office.

Allen, R.G., Jensen, M.E., Wright, J.L. & Burman, R.D. (1989) Operational estimates of reference evapotranspiration. *Agronomy Journal*, 81(4), 650. ISSN 0002-1962. doi:10.2134/agronj1989. 00021962008100040019x.

Allen, R.G., Pereira, L.S., Raes, D. & Smith, M. (1998) Crop evaporation: guidelines for computing crop water requirements. Irrigation and Drainage Paper 56, Food and Agriculture Organisation of the United Nations (FAO), FAO, Rome. Available from: http://www.fao.org/docrep/x0490e/ x0490e00.htm.

Alley, W.M. & Alley, R. (2017) *High and Dry: New Book Examines World's Biggest Groundwater Challenges*. Yale University Printers, New Haven, CT, USA.

Alley, W.M. & Konikow, L.F. (2015) Bringing grace down to earth. *Groundwater*, 35(6), 826–829. doi:10.1111/gwat.l2379.

Allison, G.B. & Barnes, C.J. (1985) Estimation of evaporation from the normally dry Lake Frome in South Australia. *Journal of Hydrology*, 78, 229–242.

Allison, G.B., Barnes, C.J., Hughes, M.W. & Leaney, F.W.J. (1984) Effect of climate and vegetation on Oxygen-18 and Deuterium profiles in soils. Technical Report, International Atomic Energy Agency, Vienna, Austria.

Amin, S. & Salimi, M.A. (1996) Manshadi. Controlling outflow rate of ghanats: case study (Yazd province). *Iranian Journal of Science and Technology*, 20(3), 285–297.

Andersen, M.S. & Acworth, R.I. (2009) Stream-aquifer interactions in the Maules Creek catchment, Namoi Valley, New South Wales, Australia. *Hydrogeology Journal*, 17(8), 2005–2021. doi:10. 1007/sl 0040-009-0500-9.

Andersen, M.S., Meredith, K., Timms, W.A. & Acworth, R.I. (2008) Investigation of 18O and 2H in the Namoi River catchment: elucidating recharge sources and the extent of surface water/groundwater interaction. Unpublished paper presented at the IAH 2008 Congress in Toyama.

Appello, C.A.J. & Postma, D. (2005) *Geochemistry, Groundwater and Pollution*. A.A. Balkema, Netherlands, 2nd edition, Leiden.

Archie, G.E. (1942) The electrical resistivity log as an aid in determining some reservoir characteristics. *Transactions ASCE*, 146, 54–62.

Archie, G.E. (1950) Introduction to petrophysics of reservoir rocks. *B.A.A.P.G.*, 34(5), 943–961.

Aryal, R., Kandel, D., Acharya, D., Chong, M.N. & Beecham, S. (2012) Unusual Sydney dust storm and its mineralogical and organic characteristics. *Environmental Chemistry*, 9, 537–546. doi:10. 107l/EN12131.

Australian Drilling Industry Training Committee Limited. (2015) *The Drilling Manual*. CRC Press, 5th edition. Wangara, Western Australia ISBN 9781439814208.

Barker, R.D. (1981) The offset sounding system of electrical resistivity sounding and its use with a multi-core cable. *Geophysical Prospecting*, 29(1), 128–143.

Barker, R.D., White, C.C. & Houston, J.T.F. (1992) Borehole siting in an African accelerated drought relief project. In: Wright, E.P. & Burgess, W.G. (eds) *Hydrogeology of Crystalline Basement Aquifers in Africa*. Geological Society Special Publication number 66. The Geological Society, London, UK. pp. 183–201.

Barlow, P.M. (2003) Ground water in freshwater-saltwater environments of the Atlantic coast. Circular 1262, U.S. Geological Survey.

Bear, J. (1972) *Dynamics of Fluids in Porous Media*. American Elsevier, New York, NY, USA.

Bear, J. & Verruijt, A. (1987) *Modelling Groundwater Flow and Pollution*. D. Reidel, Dordrecht.

Beringer, J., Hutley, L.B., McHugh, I., Arndt, S.K., Campbell, D. *et al.* (2016) An introduction to the Australian and New Zealand flux tower: network-OzFlux. *Biogeosciences*, 13, 5895–5916 doi:10.5194/bg-13-5895-2016.

Bierschenk, W.H. (1963) Determining well efficiency by multiple step-drawdown tests. *International Association of Scientific Hydrology*, 63, 493–507.

Biot, M. (1941) General theory of three-dimensional consolidation. *Journal of Applied Geophysics*, 12, 155–164.

Blatt, H., Middleton, G. & Murray, R. (1980) *Origin of Sedimentary Rocks*. Prentice Hall, 3rd edition, Second edition. Englewood Cliffs, New Jersey, USA.

Boiten, W. (2000) *Hydrometry*. IHE Delft Lecture Note Series. A. A. Balkema, Delft.

Brandes, I.M. (2005) *The Negative Chargeability of Clays*. PhD Thesis, School of Civil and Environmental Engineering, University of New South Wales. Available from: http://hand.le.unsw.edu.au/1959.4/21847.

Brown, C.M. (1989) Structural and statigraphic framework of groundwater occurrence and surface discharge in the Murray Basin, southeastern Australia. *BMR Journal of Australian Geology and Geophysics*, II, 127–146.

Burba, G. (2013) Eddy covariance methods for scientific, industrial, agricultural and regulatory applications. LI-COR Biosciences, LI-COR Biosciences, 4647 Superior Street, P.O. Box 4425, Lincoln, Nebraska 68504 USA. ISBN 978-0-615-76827-4.

Burger, R.H. (1992) *Exploration Geophysics of the Shallow Subsuiface*. Prentice Hall, Englewood Cliffs, N.J., USA.

Burke, J.J. (1995) Hydrogeological provinces in central Sudan: morphostructural and hydrogeological controls. In: Brown, A.G. (ed) *Geomorphology and Groundwater*. John Wiley & Sons Ltd. pp. 177–208.

Bussian, A.E. (September 1983) Electrical conductance in a porous-medium. *Geophysics*, Chichester, New York, 48(9), 1258–1268. doi:10.1190/1.1441549.

Butler, A.P., Hughes, A.G., Jackson, C.R., Ireson, A.M., Parker, S.J., Wheater, H.S. & Peach, D.W. (2012) Advances in modelling groundwater behaviour in chalk catchment. In: *Groundwater Resources Modelling: A Case Study From the UK*, volume 364, chapter 9, pp. 113–127. Geological Society of London, Special Publications, Lyell Collection. doi:10.1144/SP364.9.

Butler, B.J. & Barker, J.F. (1996) Chemical and microbiological transformations and degradation of chlorinated solvent compounds. In: Pankow, J.F. & Cherry, J.A. (eds) *Dense Chlorinated Solvents and Other DNAPL s in Groundwater*. Waterloo Press. Chapter 9. pp. 267–304, P.O. Box 91399, Portland, Oregon 97291-1399.

Campbell, G.S. (1985) *Soil Physics with Basic: Transport Models for Soil-Plant Systems*. Elsevier, Amsterdam.

Carruthers, R.M. & Smith, I.F. (1992) The use of ground electrical survey methods for siting water-supply boreholes in shallow crystalline basement terrains. In: Wright, E.P. & Burgess, W.G. (eds) *Hydro-geology of Crystalline Basement Aquifers in Africa*. Geological Society Special Publication number 66. The Geological Society, London, UK.pp. 203–220.

Cartwright, D.E. (1999) *Tides: A Scientific History*. Cambridge University Press, Cambridge, UK.

Cerveny, V., Langer, J. & Psencik, I. (1974) Computation of geometric spreading of seismic body waves in laterally inhomogeneous media with curved interfaces. *Geophysical Journal of the Royal Society*, 38, 9–19.

Charnley, H. (1989) *Clay Sedimentology*. Springer-Verlag, Berlin.

Clark, L. (1977) The analysis and planning of step drawdown tests. *Quarterly Journal of Engineering Geology*, 10, 125–143.

Clarke, I. & Fritz, P. (1997) *Environmental Isotopes in Hydrogeology*. Lewis Publishers, Boca Raton.

Clayton, C.R.I. (2011) Stifness at small strain: research and practice. *Geotechnique*, 61(1), 5–37. ISSN 0016-8505. doi:10.1680/geot.2011.61.1.5.

Cleverly, J., Eamus, D., Gorse, E. Van, Chena, C., Rummana, R., Luo, Q. & Restrep, N. (2016) Productivity and evapotranspiration of two contrasting semiarid ecosystems following the 2011 global carbon land sink anomaly. *Agricultural and Forest Meteorology*, 220, 151–159. doi:10.1016/j.agrformet.2016.01.086.

Cohen, R.M. & Mercer, J.W. (1993) *DNAPL Site Evaluation*. CRC Press, 1st edition Corporate Blvd. N.W., Boca Raton, F. ISBN 0873719778.

Cook, P.G. (2003) A guide to regional groundwater flow in fractured rock aquifers. Technical Report, CSIRO.

Cook, P.G., Walker, G.R., Buselli, G., Potts, I. & Dodds, A.R. (January 1992) The application of electromagnetic techniques to groundwater recharge investigations. *Journal of Hydrology*, 130(1–4), 201–229. doi:10.1016/0022-1694(92)90111-8.

Cooper, H.H. (1966) The equation of groundwater flow in fixed and deforming coordinates. *Journal of Geophysical*, 71(20), 4785–4790. doi:10.1029/JZ071i020p04785.

Corwin, R.F. (1987) The selp-potential method for environmental and engineering applications. In: Nabighian, M.N. & Corbett, J.D. (eds) *Electromagnetic Methods in Applied Geophysics – Theory*. Society of Exploration Geophysicists, Houston.

Coster, H.G.L. & Chilcott, T.C. (February 1999) *Surface Chemistry and Electrochemistry of Membranes-Impedance of Multilayer Membrane Systems*, *The Characterization of Membranes and Membrane*

Surfaces Using Impedance Spectroscopy. Number 79 in Surfactant Science Series. M. Dekker, New York, NY, USA. Chapter 19. pp. 749–792. ISBN 9780824719227. Ed. Sorensen, Torben Smith.

Cox, R.I. (March 1977) A study of near well groundwater flow and the implications for well design. Research Report 148, UNSW Water Research Laboratory. Available from: http://handle.unsw.edu. au/1959.4/36131.I.

Crosbie, R., Morrow, D., Cresswell, R., Leaney, F., Lamontagne, S. & Lefoumour, M.M. (2012) An analysis of the chemical and isotopic composition of rainfall across Australia. Csiro Water for a Healthy Country Flagship Report, CSIRO. ISSN: 1835-095X.

Crowder, R.B. (1995) *The Wonders of the Weather.* Australian Government Publishing Service, Canberra, Australia.

Cull, J.P. & Conley, D. (1983) Geothermal gradients and heat flow in Australian sedimentary basins. *BMR Journal of Australian Geology & Geophysics*, 8, 329–337.

Cuthbert, M.O., Gleeson, T., Reynolds, S.C., Bennett, M.R., Newton, C.J., McCormack, C.J. & Ashley, G.M. (2017) Modelling the role of groundwater hydro-refugia in East African hominin evolution and dispersal. *Nature Communications*, 8, 05. doi:10.1038/ncomms15696.

Dahlin, T. & Zhou, B. (2006) Gradient array measurements for multi-channel 2D resistivity imaging. *Near Surface Geophysics*, 4, 113–123.

Danis, C., O'Neill, C.O. & Lackie, M.A. (2010) Gunnedah Basin 3D architecture and upper crustal temperatures. *Australian Journal of Earth Sciences*, 57. doi:10.1080/08120099.2010.481353.

Danis, C., O'Neill, C.O., Lackie, M., Twigg, L. & Danis, A. (2011) Deep 3D structure of the Sydney Basin using gravity modelling. *Australian Journal of Earth Sciences*, 58(5), 517–542. doi:10.1080/08120099.2011.565802.

Darcy, H. (1856) *Les fontaines publiques de la ville de Dijon.* Victor Dalmont, Ghent.

Dasey, G.R. (2010) *Geophysical and Hydrogeological Assessment of the Interaction of Saline and Fresh Groundwater Neas a Tidal Creek.* PhD Thesis, School of Civil and Environmental Engineering, UNSW, Sydney, Australia. Available from: http://lhandle.unsw.edu.au/1959.4/44732.

Davis, A. & Taylor-Smith, D. (1980) Dynamic elastic moduli loging of foundation materials. In: Ardus, D.A. (ed) *Offshore Site Investigation.* Graham & Trotman. pp. 21–132.

de Wiest, R.J.M. (1969) *Flow Through Porous Media.* Academic Press, New York, NY, USA and London, UK.

Dey, A. & Morrison, H.F. (1979) Resistivity modeling for arbitrarily shaped three-dimensional structures. *Geophysics*, 44(4), 753–780. doi:10.119011.1440975.

Domenico, P.A. & Schwartz, F.W. (1997) *Physical and Chemical Hydrogeology.* John Wiley and Sons, 2nd edition, New York.

Drever, J.I. (1996) *The Geochemistry of Natural Waters.* Prentice-Hall, New York, NY, USA, 3rd edition.

Driscoll, F.G. (1986) *Groundwater and Wells.* Johnson Division, St. Paul, MN, USA, 55112, 2nd edition.

Dudgeon, B. (1993) *The Hydrogeology, Hydrology and Hydrochemistry of the Botany Sands Recharge Zone in Centennial Park, Sydney.* PhD Thesis, Faculty of Applied Science.

Dudgeon, C.R. (1985a.) Non-Darcy flow of groundwater – part 2 – inflows and water levels for dewatered circular pits in unconfined aquifers. Technical Report, Water Research Laboratory, Faculty of Engineering, UNSW. Available from: http://lhandle.unsw.edu.au/1959.4/56253.

Dudgeon, C.R. (1985b) Non-Darcy flow of groundwater – part 1 – theoretical, experimental and numerical studies. Research Report 162, Water Research Laboratory, UNSW. Available from: http://handle. unsw.edu.au/1959.4/56254.

Dudgeon, C.R., Huyakorn, P.S. & Swan, W.H.C. (1972) Hydraulics of flow near wells in unconsolidated sediments – volume 1 – part a vol 1: theoretical and experimental studies vol 2: field studies. Research Report 126, The University of New South Wales Water Research Laboratory. Australian Water Resources Council Research Project 68/8 Extraction of water from unconsolidated sediments.

Duncan, A.C., Roberts, G.P., Buselli, G., Pik, J.P., Williamson, D.R., Roocke, P.A., Thorn, R.G. & Anderson, A. (June 1992) Saltrnap? Airborne em for the environment. *Exploration Geophysics*, 23(2), 123–126. doi:10.1071/EG992123.

Dudgeon, C.R. & Swan, W.H.C. (1972) Hydraulics of flow near wells in unconsolidated sediments – volume I part d: design, construction and operation of the experimental facility. Research Report 126, The University of New South Wales Water Research Laboratory. Australian Water Resources Research Project 68/8. Available from: http://handle.unsw.edu.au/1959.4/56291.

Eamus, D., Cleverly, J., Boulaina, N., Grant, N., Faux, R. & Villalobos, R. (2013) Vega. Carbon and water fluxes in an arid-zone Acacia savanna woodland: an analyses of seasonal patterns and responses to rainfall events. *Agricultural and Forest Meteorology*, 182–183, 225–238. doi:10.1016/j.agrformet.2013.04.020.

Eamus, D., Hatton, T., Cook, P. & Colvin, C. (2006) *ECOHYDOLOGY-Vegetation Function, Water and Resource Management*. CSIRO Publishing, 150 Oxford Street, Collingwood, Victoria, Australia.

Eberhard, S. & Spate, A. (1995) Cave invertebrate survey: toward an atlas of NSW cave fauna. Technical Report, NSW Heritage Assistance Program.

Eden, R.N. & Hazel, C.P. (1973) Computer and graphical analysis of variable discharge pumping tests of wells. Technical Report, Civil Engineering. Trans. Inst. Eng. Austr..

Edmunds, W.M., Darling, W.G., Kinniburgh, D.G., Dever, L. & Vachier, P. (1992) Chalk Groundwater in England and France: hydrogeochemistry and water quality. Research Report SD/92/2, British Geological Survey.

English, P., Richardson, P., Glover, M., Creswell, H. & Gallant, J. (2004) Interpreting airborne geophysics as an adjunct to hydrogeological investigations for salinity management: honeysuckle creek catchment, Victoria. CSIRO Land and Water Technical Report 18/04, CSIRO Land and Water.

Eriksson, E. (1985) *Principles and Applications of Hydrochemistry*. Chapman and Hall, London, UK.

Ese Encrenaz. (2008) Water in the solar system. *Annual Review of Astronomy and Astrophysics*, 46(3), 57–87. Australian Academy of Science. ISBN 1545-4282. doi:10.1146/annurev.astro.46.060407.145229.

Fetter, C.W. (2001) *Applied Hydrogeology*. Prentice Hall, Upper Saddle River, NJ 07458, USA, 4th edition.

Feynman, R.P., Leighton, R.B. & Sands, M. (1964) *The Feynman Lectures on Physics*. Addison-Wesley Publishing Company, New York. Volume 2.

Freeze, R.A. & Cherry, J.A. (1979) *Groundwater*. Prentice-Hall, Inc., Englewood Cliffs, NJ, USA.

Galewsky, J., Steen-Larsen, H.C., Field, R.D., Worden, J., Risi, C. & Schneider, M. (2016) Stable isotopes in atmospheric water vapor and applications to the hydrologic cycle, 54(4), 809–865. *Reviews a/Geophysics*. doi:10.1002/2015RG000512.

Galperin, A.M., Zaytsev, V.S., Yu, A. & Norvatov, Y.A. (1993) *Hydrogeology and Engineering Geology*. A. A. Balkema, Rotterdam. Translated from Russian and edited by R.B. Zeidler of H*T*S*, Gdansk, Poland.

Gambell, A.W. & Fisher, D.W. (1966) Chemical composition of rainfall eastern North Carolina and Southeastern Virginia. Geological Survey Water-Supply Paper 1535-K, USGS.

George, R.J., Beasley, R.B., Gordon, I., Heislers, D., Speed, R., Brodie, R., Mc-Connell, C. & Woodgate, P. (1998) National airborne geophysics project – evaluation for catchment management national report. Technical Report, National Dryland Salinity Program for Agriculture Fisheries and Forestry Australia, December. Available from: https://www.researchgate.net/publication/281545129.

Ghosh, D.P. (1971a) The application of linear filter theory to the direct interpretation of geoelectrical resistivity sounding measurements. *Geophysical Prospecting*, 19, 192–217.

Ghosh, D.P. (1971b) Inverse filter coefficients for the computation of apparent resistivity curves over horizontally stratified earth. *Geophysical Prospecting*, 19, 769–775.

Gibert, J., Danielopol, D.L. & Stanford, J.A. (1994) *Groundwater Ecology*. Academic Press, London, UK.

Gilbert, J. (1996) Do groundwater ecosystems really matter ? In: Barber, C. & Davis, G. (eds) *Groundwater and Land-Use Planning Conference Proceedings*. CSIRO Division of Water Resources. Centre for Groundwater Studies, Perth, Australia.

Gonthier, G.J. (2007) A graphical method for estimation of barometric efficiency from continuous data-concepts and application to a site in the Piedmont, Air Force Plant 6, Marietta, Georgia. Science Investigation Report 2007-5111, US Geol, Survey.

Grant, F.S. & West, G.F. (1965) *Interpretation Theory in Applied Geophysics*. McGraw-Hill, New York, NY, USA.

Green, D.H. & Wang, H.F. (July 1990) Storage as a poroelastic coefficient. *Water Resources Research*, 26(7), 1631–1637. doi:10.1029/WR026i007p01631.

Greve, A.K. (2009) *Detection of Subsurface Cracking Depth Through Electrical Resistivity Anisotropy*. PhD Thesis, Civil and Environmental Engineering, University of New South Wales, Australia.

Greve, A.K., Acworth, R.I. & Kelly, B.F.K. (2011) 3D cross-hole resistivity tomography to monitor water percolation during irrigation on cracking soil. *Soil Research*, 49, 661–669.

Greve, A.K., Andersen, M.S. & Acworth, R.I. (2010) Investigations of soil cracking and preferential flow in a weighing lysimeter filled with cracking clay soil. *Journal of Hydrology*, 393(1–2), 105–113. ISSN 00221694. doi:10.1016/j.jhydrol.201O.Q3.007.

Greve, A.K., Rohan, H., Kelly, B.F.K. & Acworth, R.I. (2013) Electrical conductivity of partially saturated porous media with surface conduction: an improved formulation. *Journal of Geophysical Research: Solid Earth*, 118, 3297–3303. doi:10.1002/jgrb.50270.

Guo, B., Lackie, M.A. & Flood, R.H. (2007) Upper crustal structure of the Tamworth Belt, New South Wales: constraints from new gravity data. *Australian Journal of Earth Sciences*, 54. doi:10.1080/08120090701615725. Published Online: 02 May 2008.

Habermehl, M.A. (1982) Springs in the Great Artesian Basin – their origin and nature. Report 234, Bureau of Mineral Resources, Geology and Geophysics, Australian Government Publishing Service.

Hagedoorn, J.G. (1959) The plus-minus method of interpreting seismic refraction sections. *Geophysical Prospecting*, 7, 158–181. doi:10.1111/j.1365-2478.1959.tb01460.x.

Halloran, L.J.S., Rau, G.C. & Andersen, M.S. (2016a) Heat as a tracer to quantify processes and properties in the vadose zone: a review. *Earth-Science Reviews*, 159, 358–373. doi:10.1016/j.earscirev.2016.06.009.

Halloran, L.J.S., Roshan, H., Rau, G.C., Andersen, M.S. & Acworth, R.I. (March 2016b) Improved spatial delineation of streambed properties and water fluxes using distributed temperature sensing. *Hydrological Processes*, 30. doi:10.1002/hyp.l0806.

Hammer, S. (1939) Terrain corrections for gravimeter stations. *Geophysics*, 4, 184–194.

Hancock, P.J., Boulton, A.J. & Humphreys, W.F. (2005) Aquifers and hyporheic zones: towards an ecological understanding of groundwater. *Hydrogeology Journal*, 13, 98–111. Special Edition on the Future of Hydrogeology.

Hanks, R.I. & Ashcroft, G.L. (1980) *Applied Soil Physics – Soil Water and Temperature Applications*. Advanced Series in Agricultural Sciences – 8. Springer-Verlag, Berlin.

Hantush, M.S. (1964) Hydraulics of wells. In: *Hydraulics of Wells*. Academic Press, New York, NY, USA and London, Volume 1. pp. 281–432.

Hays, R.I. & Ullman, W.L. (2007) Dissolved nutrient fluxes through a sandy estuarine beach face: contributions from fresh groundwater discharge, seawater recycling, and diagenesis. *Estuaries and Coasts*, 10(4), 710–724.

Hazel, C.P. (2009) *Groundwater Hydraulics*. Groundwater School in Adelaide, Australia.

Hazell, J.R.T., Cratchley, C.R. & Preston, A.M. (1988) The location of aquifers in crystalline rocks and alluvium in Northern Nigeria using combined electromagnetic and resistivity techniques. *Quarterly Journal of Engineering Geology*, 21, 159–176.

Hazen, A. (1911) Discussion: dams on sand foundations. *Transactions, American Society of Civil Engineers*, 73, 199.

Helander, D.P. (1983) *Fundamentals of Formation Evaluation*. OGI Publications, Oil and Gas Consultants International, Tulsa, OK, USA.

Hem, J.D. (1985) Study and interpretation of the chemical characteristics of natural water. Water Supply Paper 2254, US Geological Survey.

Hill, J.A. (1992) Field techniques and instrumentation in shallow seismic reflection. *Quarterly Journal of Engineering Geology and Hydrogeology*, 25(3), 183–190.

Hollins, S.E., Hughes, C.E., Crawford, J., Cendon, D.I. & Mere, K.M. (2018) Dith. Rainfall isotope variations over the Australian continent Implications for hydrology and isoscape applications. *Science of the Total Environment*, 645(15), 630–645. doi:10.1016/j.scitotenv.2018.07.082.

Holzbecher, E. (1998) *Modeling Density-driven Flow in Porous Media*. Springer Publishing, New York, NY, USA.

Howard, K.W.F. & Lloyd, J.W. (1979) The sensitivity of parameters in the penman evaporation equation and direct recharge balance. *Journal of Hydrology*, 41.

Hubbert, M.K. (1940) Theory of ground-water motion. *Journal of Geology*, 48(8), 795–944.

Hughes, B.J.D. & Sanford, W.E. (2004) Sutra-ms: a version of sutra modified to simulate heat and multiple-solute transport. Technical Report, US Geological Survey, USGS, Reston, VA, USA.

Huizar-Alvarez, R., Carrillo-Rivera, J.J., Angeles-Serrano, G., Hergt, T. & Cardona, A. (2004) Chemical response to groundwater extraction southeast of Mexico City. *Hydrogeology Journal*, 12, 436–450.

Humphreys, W.F. (1999a) Relict stygofaunas living in sea salt, karst and calcrete habitats in arid northwestern Australia contain many ancient lineages. In: Ponder, W. & Lunney, D. (eds) *The Conservation and Biodiversity of Invertebrates*. Royal Zoological Society of New South Wales, Mossman, NSW 2088. pp. 219–227.

Humphreys, W.F. (1999b) Physico-chemical profile and energy fixation in Bundera Sinkhole, an anchialine remiped habitat in north-western Australia. *Journal of the Royal Society of Western Australia*, 82, 89–98.

Humphreys, W.F. (2000) The hypogean fauna of the Cape Range peninsula and Barrow Island, northwest Australia. In: Wilkens, H., Culver, D.C., & Humphreys, W.F. (eds) *Ecosystems of the World-Sub-terranean Ecosystems*. Elsevier, Amsterdam, Netherlands, Volume 30.

Huyakorn, P.S. (1972a) Hydraulics of flow near wells in unconsolidated sediments – Volume 1 Part B Theoretical and Numerical analysis of two regime flow toward wells. Research Report 126, The University of New South Wales Water Research Laboratory. Australian Water Resources Council Research Report 68/8 Extraction of Water from Unconsolidated Sediments.

Huyakorn, P.S. (1972b) Hydraulics of flow near wells in unconsolidated sediments – Volume 1 – Part C: behaviour of wells in confined aquifer and well design procedures. Technical Report, The University of New South Wales Water Research Laboratory. Australian Water Resources Council Research Project 68/8.

Hvorslev, M.J. (1951) Time lag and soil permeability in groundwater observations. Bulletin 36, U.S. Army Corps Engineers. Waterways Experimental Station, Vicksburg, MS, USA.

Israelsen, O.W. & Hasen, V.E. (1962) *Irrigation Principles and Practices*. Wiley, New York, NY, USA.

Issar, A.S. (1990) *Water Shall Flow from the Rock – Hydrogeology and Climate in the Lands of the Bible*. Springer-Verlag, Berlin.

Jacob, C.E. (July 1939) Fluctuations in artesian pressure produced by passing railroad-trains as shown in a well on Long Island, New York. *EoS Archives*, 20(4), 666–674. doi:10.1029/TR020i004p00666.

Jacob, C.E. (1940) On the flow of water in an elastic artesian aquifer. *Transactions American Geophysics Union*, 21(2), 574–586. doi:10.1029/TR021i002p00574.

Jacob, C.E. (1944) Notes on determining permeability by pumping tests under water table conditions. Open File Report, U.S. Geological Survey.

Jankowski, J. & Beck, P. (2000) Aquifer heterogeneity: hydrogeological and hydrochemical properties of the botany sands aquifer and their impact on contaminant transport. *Australian Journal of Earth Sciences*, 47, 45–64.

Jiao, J.J. (2007) A 5,600-year-old wooden well in Zhejiang Province, China. *Hydrogeology Journal Geological Science (Czech)*, 15, 1021–1029.

Johannes, R.E. (1980) The ecological significance of the submarine discharge of groundwater. *Marine Ecology-Progress Series*, 3, 363–373.

Johnson, G.S., Cosgrove, D.M. & Frederick, D.B. (July 2001) A numerical model and spreadsheet interface for pumping test analysis. *Groundwater*, 39(4). doi:10.1111/j.1745-6584.2001.tb02346.x.

Jones, M.J. (1985) The weathered zone aquifers of the basement complex areas of Africa. *Quarterly Journal of Engineering Geology*, 18, 35–46.

Jury, W.A., Gardner, W.R. & Gardner, W.H. (1991) *Soil Physics*. John Wiley & Sons, Inc., 5th edition, New York.

Kearey, P. & Brooks, M. (1991) *An Introduction to Geophysical Exploration*. Blackwell Science, Cambridge, UK, 2nd edition.

Keller, C.K., Van Der Kamp, G. & Cherry, J.A. (1989) A Multiscale study of the permeability of a thick clayey till. *Water Resources Research*, 25(11), 2299–2317. ISSN 19447973. doi:10.1029/WR025i011p02299.

Kelly, B.F.J. (1994) *Electrical Properties of Sediments and the Geophysical Detection of Ground Water Contamination*. PhD Thesis, The University of New South Wales.

Keys, S.W. (1989) Borehole geophysics applied to ground-water investigations. U.S. Geological Survey Open-File Report; 87-539. National Water Well Association, 6375 Riverside Drive, Dublin, OH 43017, USA. ISBN 0565-596X.

Kiefert, L. & McTainsh, G. (1995) Clay minerals in east Australian dust. In: Churchman, G.J., Fitzpatrick, R.W., & Eggleton, R.A. (eds) *In Clays Controlling the Environment. 10th International Clay Conference*. CSIRO Publishing, Melbourne, Australia.

King, R.F. (1992) High-resolution shallow seismology: history, principles and problems. *Quarterly Journal of Engineering Geology and Hydrogeology*, 25(3), 177–182. doi:10.1144/GSL.QJEG.1992.025.03.01.0.

Koefoed, O. (1979) *Geosounding Principles 1 – Resistivity Sounding Measurements*. Methods in Geochemistry and Geophysics – Number 14A. Elsevier Scientific Publishing Company, Amsterdam.

Kowal, J.M. & Kassam, A.H. (1978) *Agricultural Ecology of Savanna, A Study of West Africa*. Clarendon Press, Oxford University Press, Walton Street, Oxford, OX2 6DP, UK.

Kowal, J.M. & Omolokum, A.O. (1970) The hydrology of a small catchment basin at Samaru, Nigeria: seasonal fluctuations in the height of the groundwater table. *Nigerian Agriculture Journal*, 7.

Kruseman, G.P. & de Ridder, N.A. (1994) Analysis and evaluation of pumping test data. Publication 47, International Institute for Land Reclamation and Improvement, P.O. Box 45, 6700 AA Wageningen, The Netherlands, 1990. Second Edition (Completely Revised).

Langmuir, D. (1997) *Aqueous Environmental Geochemistry*. Prentice-Hall, Englewood Cliffs, NJ, USA.

Lau, J.E., Commander, D.P. & Jacobson, G. (1987) Hydrogeology of Australia. Bulletin 227, Bureau of Mineral Resources, Geology and Geophysics, Australian Government Publishing Office, Canberra, Australia.

Lerner, D.N., Issar, A.S. & Simmers, I. (1990) *Groundwater Recharge – A Guide to Understanding and Estimating Natural Recharge*. Verlag Heinz Heise, Volume 8 of the International Contributions to Hydrogeology, Hanover.

Levitskaya, T.M. & Sternberg, B.K. (1994) Electrical properties of rocks in the frequency range 0.01 hz to 100 mhz. Technical Report LASI-94-3, Laboratory for Advanced Subsurface Imaging (LASI), Department of Mining and Geological Engineering, University of Arizona, Tucson, Arizona 85721.

Lisiecki, L.E. & Raymo, M.E. (2005) A pliocene-pleistocene stack of 57 globally distributed benthic d18o records. *Paleoceanography*, 20(PA 1003). doi:10.1029/2004PA001071.

List, R.J. (1947) *Smithsonian Meterological Tables*, Volume 114 – Publication 4014. Smithsonian Institution Press, City of Washington, 6th Revised edition. 4th reprint issued in 1968.

Liu, Y. & Yin, C. (November–December 2016) 3D inversion for multipulse airborne transient electromagnetic data. *Geophysics*, 81(6), E401–E408. doi:10.1190/geo2015-0481.1.

Lliffe, T.M. (2000) Anchialine cave ecology. In: Wilkens, H., Culver, D.C., & Humphreys, W.F. (eds) *Ecosystems of the World – Subterranean Ecosystems*. Elsevier, Amsterdam, Netherlands, Volume 30.

Loke, D.M. (1999) Electrical imaging surveys for environmental and engineering studies. A Practical Guide to 2-D and 3-D Surveys. RES2DINV Manual. IRIS Instruments. Available from: www.iris-intruments.com.

Loke, M.H. (1995) Least-squares deconvolution of apparent resistivity pseudosections. *Geophysics*, 60(6), 1682. ISSN 1070485X. doi:10.1190/1.1443900.

Loke, M.H. (2016) Tutorial: 2-d and 3-d electrical imaging surveys. Available from: www.geotomosoft.com (accessed 26 June 2016).

Loke, M.H. & Barker, R.D. (1996) Rapid least-squares inversion of apparent resistivity pseudosections by a quasi-newton method. *Geophysical Prospecting*, 44, 131–152.

Long, D., Chen, X., Scanlon, B.R., Wada, Y., Hong, Y., Singh, V.P., Chen, Y., Wang, C., Han, Z. & Yang, W. (2016) Have grace satellites overestimated groundwater depletion in the Northwest India aquifer? *Scientific Reports*, 6, 24398.

Lusczynski, N.J. (1961) Head and flow of groundwater of variable density. *Journal of Geophysical Research*, 66, 4247–4255.

Macumber, P.G. (1991) *Interaction Between Ground Water and Surface Systems in Northern Victoria*. Department of Conservation and Environment, Victoria, Canada.

Mather, J.D. (2004) 200 years of British hydrogeology – an introduction and overview. In: Mather, J.D. (ed) *200 Years of British Hydrogeology*, number 225 in Geological Society Special Publication, p. 114. The Geological Society of London, The Geological Society Publishing House, Unit 7, Brassmill Enterprise Centre, Brassmill Lane, Bath BA1 3Jn, UK.

Mazor, E. (1997) *Chemical and Isotopic Groundwater Hydrology – The Applied Approach*. Marcel Deckker, Inc., New York, NY, USA, 2nd edition.

McFarlane, M.J. (1992) Groundwater movement and water chemistry associated with weathering profiles of the African surface in parts of Malawi. In: Wright, E.P. & Burgess, W.G. (eds), *Hydrogeology of Crystalline Basement Aquifers in Africa*. Geological Society Special Publication number 66. The Geological Society, London, UK. pp. 101–129.

McFarlane, M.J., Chilton, P.J. & Lewis, M.A. (1992) Geomorphological controls on borehole yields: a statistical study in an area of basement rocks in central Malawi. In: Wright, E.P. & Burgess, W.G. (eds) *Hydrogeology of Crystalline Basement Aquifers in Africa*. Geological Society Special Publication number 66. The Geological Society, London, UK. pp. 131–154.

McNeil, J.D. (1980) Electromagnetic terrain conductivity measurement at low induction numbers – technical note tn-6. Technical Report, Geonics Limited.

McNeil, J.D. (1986) Geonics EM39 borehole conductivity meter-theory of operation. Technical Report, Geonics Limited.

McNeil, J.D. (1990) Use of electromagnetic methods for groundwater studies. In: Ward, S.H. (ed) *Geotechnical and Environmental Geophysics*. Society of Exploration Geophysicists. Volume 1: Review and Tutorial. pp. 191–218.

Meinzer, O.E. (1923) The occurrence of groundwater in the united states, with a discussion of principles. Water-Supply Paper 489, U.S. Geological Survey.

Meinzer, O.E. (1928) Compressibility and elasticity of artesian aquifers. *Economic Geology*, 23(3), 263–291. ISSN 03610128. doi:10.2113/gsecongeo.23.3.263.

Milburn, J.A. (1979) *Water Flow in Plants – (Integrated Themes in Biology)*. Longman, London, UK.

Misstear, B., Banks, D. & Clark, L. (2006) *Water Wells and Boreholes*. John Wiley & Sons, Ltd, West Sussex, P019 8SQ, UK.

Monteith, J.L. (1965) Evaporation and the environment. *Symposium Society Exploration Biologists*, 19, 205–234.

Moore, W.S. (1999) The subterranean estuary: a reaction zone of groundwater and seawater. *Marine Chemistry*, 65, 111–125.

Morgan, K.H. (1993) Development, sedimentation and economic potential of paleoriver systems of the Yilgam Craton of Western Australia. *Sedimentary Geology*, 85, 637–656.

Morton, F.I. (1983) Operational estimates of areal evapotranspiration and their significance to the science and practice of hydrology. *Journal of Hydrology*, 66, 1–76.

Nagamoto, C., Parungo, F., Kopcewicz, B. & Zhou, M.Y. (December 1990) Chemical analysis of rain samples collected over the Pacific Ocean. *Journal of Geophysical Research – Atmospheres*, 95(D13), 22343–22354. doi:10.1029/JD095iD13p22343.

Oilier, C.D. & Pain, C.F. (1995) *Regolith, Soils and Landforms*. John Wiley & Sons, Brisbane, Australia.

Olayinka, A. & Barker, R.D. (1990) Borehole siting in crystalline basement areas of Nigeria with a micro-processor-controlled resistivity traversing system. *Ground Water*, 28, 178–193.

Olhoeft, G.R. (1985) Low frequency electrical properties. *Geophysics*, 50(12), 2492–2503.

Palacky, G.J. (1987) Clay mapping using electromagnetic methods. *First Break*, 5(8), 295–306.

Palmer, D. (1980) *The Generalised Reciprocal Method of Seismic Refraction Interpretation*. Society of Exploration Geophysicists, Tulsa, OK, USA.

Palumbo, A. (1998) Atmospheric tides. *Journal of Atmospheric and Solar-Terrestrial Physics*, 60(3), 279–287.

Panga, Z., Kong, Y., Lia, J. & Tiana, J. (2017) An isotopic Geoindicator in the hydrological cycle. *Procedia Earth and Planetary Science*, 17, 534–537. doi:10.1016/j.proeps.2016.12. 135. Available from: http://creativecommons.org/licenseslby-nc-nd/4.0/.

Parkhurst, D.L. & Plummer, L.N. (1993) Geochemical models. In: Alley, W.M. (ed) *Regional Ground Water Quality*. Van Nostrand Reinhold, New York, NY, USA. pp. 199–226.

Patterson, A.J. (2010) Song of the artesian water. Wikisource. (accessed 11 June 2018). http://www.connectedwaters.unsw.edu.au/articles/2008/03/song-artesian-water-banjo-patterson.

Penman, H.L. (1949) The dependence of transpiration on weather and soil conditions. *Journal of Soil Science*, 1, 74–89.

Polac, E.J. & Horsfall, C.L. (1979) Geothermal gradients in the Great Artesian Basin, Australia. *Bulletin of the Australian Society of Exploration Geophysicists*, 10, 144–148.

Pool, D.R. & Eychaner, J.H. (1995) Measurements of aquifer-storage change and specific yield using gravity surveys. *Ground Water*, 33(3), 425–432. doi:10.1111/j.l745-6584.1995.tb00299.x.

Price, M. (2004) Dr John Snow and an early investigation of groundwater contamination. In: Mather, J.D. (ed) *200 Years of British Hydrogeology*. Geological Society Special Publication number 225. The Geological Society, London, UK.

Pruitt, W.O. & Lourence, F.J. (1985) *Experience in Lysimetry for ET and Surface Drag Measurements*. American Society of Agricultural Engineers, St. Joseph, MI, USA, Chapter 1. pp. 51–69.

Rancic, A., Salas, G., Kathuria, A., Acworth, I., Johnston, W., Smithson, A. & Beal, G. (2009) Climatic influences on shallow fractured-rock groundwater system in the Murray-Darling Basin. Technical Report, Department of Environment and Climate Change, NSW, 59–61 Goulburn Street, PO Box A290, Sydney South 1232, Australia.

Ransic, A., Salas, G., Kathuria, A., Johnston, W., Smithson, A. & Beale, G. (2007) MDB salinity audit: climatic influence on shallow fractured rock groundwater systems in the Murray-Darling Basin, NSW. Technical Report 458, Murra-Darling Basin Authority.

Rau, G.C., Acworth, R.I., Halloran, L.J.S., Timms, W.A. & Cuthbert, M. (2018) Quantifying compressible groundwater storage by combining cross-hole seismic surveys and head response to atmospheric tides. *JGR Earth Surface*. doi:10.1029/2018JF004660. First Published: 03 July 2018.

Rau, G.C., Andersen, M.S. & Acworth, R.I. (March 2012) Experimental investigation of the thermal dispersivity term and its significance in the heat transport equation for flow in sediments. *Water Resources Research*, 48(3). doi:10.1029/2011WR011038.

Rau, G.C., Andersen, M.S., McCallum, A.M. & Acworth, R.I. (2010) Analytical methods that use natural heat as a tracer to quantify surface water-groundwater exchange, evaluated using field temperature records. *Hydrogeology Journal*, 18(5), 1093–1110. doi:10.1007/sl0040-010-0586-0.

Rau, G.C., Andersen, M.S., McCallum, A.M., Roshan, H. & Acworth, R.I. (February 2014) Heat as a tracer to quantify water flow in near-surface sediments. *Earth-Science Reviews*, 129, 40–58. doi:10.1016/j.earscirev.2013.10.015.

Rau, G.C., Halloran, L.J.S., Cuthbert, M., Andersen, M.S., Acworth, R.I. & Tellam, J.H. (2017) Characterising the dynamics of surface water-groundwater interactions in intermittent and ephemeral streams using stream bed thermal signatures. *Advances in Water Resources*, 107, 354–369. doi:10.1016/j.advwatres.2017.07.005.

Revil, A. (2013) Effective conductivity and permittivity of unsaturated porous materials in the frequency range 1 mhz–1 ghz. *Water Resources Research*, 49(1), 306–327. doi:10.1029/2012WR012700.

Rhoades, J.D., Manteghi, N.A., Shouse, P.J. & Alves, W.J. (1989a) Estimating soil salinity from saturated soil-past electrical conductivity. *Soil Science Society America Journal*, 53, 428–433.

Rhoades, J.D., Manteghi, N.A., Shouse, P.J. & Alves, W.J. (1989b) Soil electrical conductivity and soil salinity: new formulations and calibrations. *Soil Science Society America Journal*, 53, 433–439.

Rhoades, J.D., Waggoner, B.L., Shouse, P.J. & Alves, W.J. (1989c) Determining soil salinity from soil and soil-paste electrical conductivities: sensitivity analysis of models. *Soil Science Society America Journal*, 53, 1368–1374.

Robson, A.J., Neal, C., Hill, S. & Smith, C.J. (1993) Linking variations in short- and medium-term stream chemistry to rainfall inputs – some observations at Plynlimon, Mid-Wales. *Journal of Hydrology*, 144, 291–310. doi:10.1016/0022-1694(93)90177-B.

Rorabaugh, M.J. (1953) Graphical and theoretical analysis of step drawdown test of artesian well. *Proceedings of the American Society of Civil Engineers*, 79(362), 1–23.

Rose, E.P.F. (2004) The contribution of geologists to the development of emergency groundwater supplies by the British army. In: Mather, J.D. (ed) *200 Years of British Hydrogeology*. Geological Society Special Publication number 225. The Geological Society, London, UK.

Rose, J.I., Usik, V.I., Marks, A.E., Hilbert, Y.H., Galleti, C.S., Parton, A., Marie, J., Geiling, Cerny, V., Morley, M.W. & Roberts, R.G. (2011) The Nubian complex of Dhofar, Oman: an African middle stone age industry in Southern Arabia. *PLOS One*. doi:10.1371/journal.pone.0028239.

Rushton, K.R. (1988) *Numerical and Conceptual Models for Recharge Estimation in Arid and Semi-arid Zones*. D. Reidel Publishing Company. in Estimation of Natural Groundwater Recharge edited by Ian Simmers. Hamburg, pp. 223–238.

Rushton, K.R. (2003) *Groundwater Hydrology-Conceptual and Computational Models*. Wiley, John Wiley and Sons Ltd, Chichester, West Sussex, P019 8SQ, UK.

Rushton, K.R. & Redshaw, S.C. (1979) *Seepage and Groundwater Flow: Numerical Analysis by Analog and Digital Methods*. John Wiley and Sons, New York.

Rybach, L. (2007) Geothermal sustainability. Geo-Heat Centre Quarterly Bulletin 28, Oregon Institute of Technology.

Scanlon, B.R. & Cook, P.G. (2002) Theme issue: groundwater recharge. *Hydrogeology Journal*, 10.

Schurch, M. & Buckley, D. (2002) Integrating geophysical and hydrochemical borehole-log measurements to characterize the Chalk aquifer, Berkshire, United Kingdom. *Hydrogeology Journal*, 10, 610–627.

Selker, J.S., van de Giesen N., Westhoff, M., Luxemburg, W. & Parlange, M.B. (2006a) Fiber optics opens window on stream dynamics. *Geophysical Research Letters*, 33(L24401). doi:10.1029/2006GL027979.

Selker, J.S., Thévenaz, L., Huwald, H., Mallet, A., Luxemburg, W., van de Giesen, N., Stejskal, M., Zeman, J., Westhoff, M. & Parlange, M.B. (December 2006b) Distributed fibre-optic temperature sensing for hydrologic systems. *Water Resources Research*, 42(12), doi:10.1029/2006WR005326.

Shuttleworth, W.J. (1993) Evaporation. In: *Handbook of Hydrology*. McGraw-Hill, Inc, New York, Chapter 4. pp. 4.1–4.53.

Simmers, I. (1988) *Estimation of Natural Groundwater Recharge, Volume 222 of NATO ASI Series C: Mathematical and Physical Sciences*. D. Reidel Publishing Co, Hamburg.

Simmers, I. (1997) Recharge of phreatic aquifer in (semi-) arid areas. In: Simmers, I. (ed) *Recharge of Phreatic Aquifer in (Semi-) Arid Areas, Volume 19 of International Contributions to Hydrogeology*. A. A. Balkema, Rotterdam, Netherlands.

Sket, B. (1996) The ecology of anchihaline caves. *Trends in Ecology and Evolution*, 11, 221–225.

Smiles, D.E. (2000) Hydrology of swelling soils: a review. *Australian Journal of Soil Research*, 38, 501–521. ISSN 0004-9573. doi:10.1071/SR99098.

Smith, M. (October 1991) Report on the expert consultation on procedures for revision of FAO guidelines for prediction of crop water requirements. Technical Report, Land and Water Development Division, Food and Agriculture Organisation of the United Nations.

Smith, W.O. & Nelson Sayre, A. (1964) Turbulence in ground-water flow. Geological Survey Professional Paper 402-E, USGS, United States Government Printing Office, Washington, DC 20402.

Stauffer, F., Bayer, P., Blum, P., Giraldo, M.N. & Kinzelbach, W. (2017) *Thermal Use of Shallow Groundwater*. CRC Press. ISBN 9, New York.

Steeples, D.W. & Miller, R.D. (1990) Seismic reflection methods applied to engineering, environmental, and groundwater problems. In: Ward, S.H. (ed), *Geotechnical and Environmental Geophysics*. Society of Exploration Geophysicists, Houston, Volume 1. pp. 1–30.

Stewart, M. & Morgenstern, U. (2001) Age and source of groundwater from isotope tracers. In: Rosen, M.R. & White, P.A. (eds) *Groundwaters of New Zealand*. New Zealand Hydrological Society, Wellington, New Zealand. pp. 161–183.

Stonestrom, D.A. & Constanz, J. (2003) Heat as a tool for studying the movement of ground water near streams. Circular, US Geological Survey, U.S. Geological Survey, Reston, VA, USA.

Stumm, W. & Morgan, J.J. (1996) *Aquatic Chemistry: An Introduction Emphasising Chemical Equilibria in Natural Waters*. John Wiley and Sons, New York, NY, USA, 3rd edition.

Stuyfzand, P. (January 1993) *Hydrochemistry and Hydrology of the Coastal Dune Area of the Western Netherlands*, Volume 53. KIWA, Research and Consultancy Division, Groningenhaven 7, P.O. Box 1072, Nieuwegein, The Netherlands. ISBN 9788578110796. doi:10.1017/CB09781107415324. 004.

Sun, G., Noormets, A.M., Chen, J. & McNulty, S.G. (2008) Evapotranspiration estimates from eddy covariance towers and hydrologic modeling in managed forests in Northern Wisconsin, USA. *Agriculture and Forest Meteorology*, 148, 257–267. doi:10.1016/j.agrformet.2007.08.010.

Sundaram, B., Peitz, A.J., de Caritat, P., Plazinska, A., Brodie, R.S., Coram, J. & Ransley, T. (2009) Groundwater sampling and analysis: a field guide. Technical Report RECORD 2009/27, Geoscience Australia.

Sutherland, L. (1995) *The Volcanic Earth*. The University of New South Wales Press, Sydney, Australia.

Talsma, T. (1977) Measurement of the overburden component of total potential in swelling field soils. *Australian Journal of Soil Research*, 15, 95–102.

Talsma, T. & van der Lelij, A. (1976) Infiltration and water movement in an in situ swelling soil during prolonged ponding. *Australian Journal of Soil Research*, 14, 337–349.

Taniguchi, M. (2002) Tidal effects on submarine groundwater discharge into the ocean. *Geophysical Research Letters*, 29(12), 1561–1563.

Taniguchi, M., Burnett, W.C., Cable, I.E. & Turner, J.V. (2002) Investigation of submarine groundwater discharge. *Hydrological Processes*, 16(11), 2115–2129. ISSN 0885-6087. doi:10.1002/Hyp.1145.

Taniguchi, M., Wang, K. & Garno, T. (2003) *Land and Marine Hydrogeology*. Elsevier, Amsterdam.

Taylor, R. & Howard, K. (2000) A tectono-geomorphic model of the hydrogeology of deeply weathered crystalline rock: evidence from Uganda. *Hydrogeology Journal*, 8, 279–294.

Telford, W.M., Geldart, L.P., Sheriff, R.E. & Keys, D.A. (1976) *Applied Geophysics*. Cambridge University Press, Cambridge, UK, 2nd edition.

Terzaghi, K. (1943) *Theoretical Soil Mechanics*. Wiley, New York, NY, USA.

Terzaghi, K. & Peck, R.B. (1948) *Soil Mechanics in Engineering Practice*. John Wiley, New York, NY, USA.

Theis, C.V. (1935) The relationship between the lowering of the piezometric surface and the rate and duration of discharge of a well using groundwater storage. *Transactions American Geophysical Union*, 2, 519–524.

Timms, W.A. & Acworth, R.I. (2005) Propagation of porewater pressure change through thick clay sequences: an example from the Yarramanbah site, Liverpool Plains. *Hydrogeology*, 13, 858–870.

Timms, W.A., Acworth, R.I. & Berhane, D. (2001) Shallow groundwater dynamics in smectite dominated clay on the Liverpool Plains of New South Wales. *Australian Journal of Soil Science*, 39(2), 203–218.

Timms, W.A., Acworth, R.I., Crane, R.A., Arns, C.H., Ams, Y., Ji, McGeeney, D.E., Rau, G.C. & Cuthbert, M. (April 2018) The influence of syndepositional macropores on the hydraulicintegrity of thick alluvial clay aquitards. *Water Resources Research*, 54(4), 3122–3138. doi:10.1029/2017WR021681.

Timms, W.A., Acworth, R.I. & Young, R.R. (2002) Natural leakage pathways through smectite clay: a hydrogeological synthesis of data from the Hudson Agricultural Trial Site on the Liverpool Plains. Research Report 209, Water Research Laboratory, 110 King Street, Manly Vale 2093, NSW, Australia. Available from: http:/lhandle.unsw.edu.au/1959.4/36244. Available from: www.w.rl.unsw.edu.au/research.

Townley, L.R. & Trefry, M.G. (2000) Surface water-groundwater interaction near shallow circular lakes: flow geometry in three dimensions. *Water Resources Research*, 36, 935–948.

Tozer, P. & Leys, J. (2013) Dust storms – what do they really cost? *The Rangeland Journal*, 35, 131–142. doi:10.1071/RJ12805.

Tribble, G. (2008) Ground water on tropical Pacific Islands-understanding a vital resource. Circular 1312, US Geological Survey.

Turner, I.L., Coates, P. & Acworth, R.I. (1996) The effects of tides and waves on water-table elevations in coastal zones. *Hydrogeology Journal*, 4(2), 51–69. doi:10.1007/s100400050090.

USGS. (2013) Water on earth. In: *Water: Planets, Plants and People*. Australian Academy of Science, Chapter 4. ISBN 032813936X. Available from: http://ga.water.usgs.gov/edu/earthhow.much.htrnl.

van Camp, M. & Vauterin, P. (June 2005). Tsoft: graphical and interactive software for the analysis of time series and Earth tides. *Computers and Geosciences*, 31(5), 631–640. ISSN 009 83004. doi:10.1016/j.cageo.2004.11.015. Available from: http://linkinghub.elsevier.com/retrieve/pii/S0098300404002456.

van der Kamp, G. & Gale, J.E. (April 1983) Theory of earth tide and barometric effects in porous formations with compressible grains. *Water Resources Research*, 19(2), 538–544. ISSN 19447973. doi:10.1029/WR019i002p00538.

van der Kamp, G. & Maathuis, H. (1991) Annual fluctuations of groundwater levels as a result of loading by surface moisture. *Journal of Hydrology*, 127, 137–152.

van Olphen, H. (1963) *An Introduction to Clay Colloid Chemistry*. Interscience, New York, NY, USA.

van Overmeeren, R.A. (2001) Hagedoom's plus-minus method: the beauty of simplicity. *Geophysical Prospecting*, 49, 687–696.

Verruijt, A. (2016) Theory and problems of poroelasticity. LaTeX. Available from: http://geo.verruijt.net.

Volland, H. (January 1996) Atmosphere and earth's rotation. *Surveys in Geophysics*, 17(1), 101–144. doi:10.1007/BF01904476.

Vukovic, M. & Soro, A. (1992) *Determination of Hydraulic Conductivity of Porous Media from Grain-Size Composition*. Water Resources Publications, Littleton, Colorado.

Wang, D. & Anderson, D.W. (1998) Direct measurement of organic carbon content in soils by the {Leco CR-12} carbon analyser. *Communications in Soil Science and Plant Analysis*, 29.

Wang, H.F. (2000) *Theory of Linear Poroelasticity, with Applications to Geomechanis and Hydrogeology*. Princeton University Press, Princeton, NJ, USA.

Wang, H.F. & Anderson, M.P. (1982) *Introduction to Groundwater Modelling-Finite Difference and Finite Element Methods*. Academic Press, Inc, New York.

Wang, Q.J., McConachy, F.L.N., Chiew, F.H.S., James, R., de Hoedt, G.C. & Wright, W.J. (2000) Climatic atlas of Australia-Maps of evapotranspiration. Technical Report, Bureau of Meteorology-Australia.

Ward, S.H. (1990) Resistivity and induced potential measurements. In: Ward, S.H. (ed) *Geotechnical and Environmental Geophysics*. Society of Exploration Geophysicists, Houston.

Waxman, M.H. & Smits, L.J.M. (1968) Electrical conductivities in oil bearing shaly sands. *Transactions American Institute Mining Petroleum Engineers*, 243(2), 107–122.

Weaver, T.R., Cartwright, I., Tweed, S.O., Ahearne, D., Cooper, M., Czapnik, K. & Tranter, J. (2006) Controls on chemistry during fracture-hosted flow of cold C02-bearing mineral waters, Daylesford, Victoria, Australia: implications for resource protection. *Applied Geochemistry*, 21(2), 289–304.

Wellings, S.R. & Bell, J.P. (1980) Movement of water and nitrate in the unsaturated zone of the upper chalk near Winchester, Hants, England. *Journal of Hydrology*, 58, 279–306. doi:10.1016/0022-1694 (80)90070-0.

Wenner, F. (1912) The four electrode array and the Thompson Bridge. *U.S Bureau of Standards*, 8, 559–660.

White, C.C., Houston, J.F.T. & Barker, R.D. (1988) The Victoria Province drought relief project, 1, Geophysical siting of boreholes. *Ground Water*, 26, 309–316.

Wiesner, C.J. (1970) *Hydrometeorology*. Chapman and Hall, London, UK.

Wilson, G.D.F. & Keable, S.J. (1999) A new genus of phreatoicidean isopod (Crustacea) from the north Kimberley Region, Western Australia. *Zoological Journal of the Linnean Society*, 126, 51–79.

Wilson, K.B., Mulholland, P.J., Baldocchi, D.D. & Wullschleger, P.J.D. (2001) A comparison of methods for determining forest evapotranspiration and its components: sap-flow, soil water budget, eddy covariance and catchment water balance. *Agricultural and Forest Meteorology*, 106, 153–168. doi:10.1016/S0168-1923(00)00199-4.

Winter, T.C., Harvey, J.W., Franke, O.L. & Alley, W.M. (1998) Ground water and surface water – A single resource. Geological Survey Circular 1139, U.S. Geological Survey.

Wolfgram, P. & Karlik, G. (1995) Conductivity-depth transform of GEOTEM data. *Exploration Geophysics*, 26, 179–185.

Woods, M.A. (2006) UK Chalk Group stratigraphy (Cenomanian – Santonian) determined from borehole geophysical logs. *Quarterly Journal of Engineering Geology and Hydrogeology*, 39, 83–96.

Wright, E.P. (1992) The hydrogeology of crystalline basement aquifers in Africa. In: Wright, E.P. & Burgess, W.G. (eds) *The Hydrogeology of Crystalline Basement Aquifers in Africa*, Geological Society Special Publication number 66. The Geological Society, London, UK.

Zhang, C. & Lu, N. (2018) What is the range of soil water density? critical reviews with a unified model. *Reviews of Geophysics*. Accepted for publication July 2018, 56(3), 532–562.

Zitouna-Chebbi, R., Prevot, L., Chakhar, A., Marniche-Ben Abdallah, M. & Jacob, F. (2018) Observing actual evapotranspiration from flux tower eddy covariance measurements within a hilly watershed: case study of the Kamech site, Cap Bon peninsula, Tunisia. *Atmosphere*, 9(68). doi:10.3390/atrnos9020068.

Index

Note: Page numbers in italics indicate figures on corresponding pages.

SERIES IAH International Contributions to Hydrogeology (ICH)

Milton Keynes UK
Ingram Content Group UK Ltd.
UKHW050457071024
449327UK00015B/421